Models of Random Processes

A Handbook for Mathematicians and Engineers

I.N. Kovalenko
N.Yu. Kuznetsov
V.M. Shurenkov

CRC Press
Boca Raton New York London Tokyo

Svetlana Landau:	Acquiring Editor
Denise Craig:	Cover Designer
Susie Carlisle:	Marketing Manager
Arline Massey:	Associate Marketing Manager
Paul Gottehrer:	Assistant Managing Editor, EDP
Gary Bennett, Kevin Luong, Carlos Esser, Greg Cuciak:	Pre-Press
Sheri Schwartz:	Manufacturing Assistant

Library of Congress Cataloging-in-Publication Data

Kovalenko, Igor′ Nikolaevich.
[Sluchainye protsessy. English]
Models of random processes : a handbook for mathematicians and engineers / I.N. Kovalenko, N. Yu. Kuznetsov, V.M. Shurenkov.
p. cm.
Includes bibliographical references.
ISBN 0-8493-2870-5 (alk. paper)
1. Stochastic processes. I. Kuznetsov, N. IU. (Nikolai IUr′evich) II. Shurenkov, V. M. (Valentin Mikhailovich) III. Title.
QA274.K6813 1996
519.2--dc20 96-14022
CIP

International Standard Book Number 0-8493-2870-5
Library of Congress Card Number 96-14022
Printed in the United States of America 1 2 3 4 5 6 7 8 9 0
Printed on acid-free paper

Contents

Preface

This book is an updated version of the Russian edition: Kovalenko, I. N., Kuznetsov, N. Yu., Shurenkov, V. M., *Random Processes: A Handbook*, Naukova Dumka Publ., Kiev, 1983 (ed. A. V. Skorokhod). Our dear co-author, a well known probabilist, Valentyn Shurenkov unfortunately died in 1994, so that the translation into English and the processing of afkasl;fj the text have been done by the two of us.

I. Kovalenko and N. Kuznetsov
December 1995

Preface to the Russian Edition

A major portion of probability theory applications to technology and science is related to devising and investigating a random process which describes a mathematical model of a phenomenon in question. The latter encouraged the derivation of numerous classes of random processes adapted to specific applications. However the general theory of random processes reduces the investigation to a quite restricted family of random processes (Markov, stationary, processes with independent increments, martingales). What connection is there between processes arising from a logical advance of the theory on the one hand and the ones arising from applications on the other hand? As a rule such a connection is ignored by applied specialists (they can not see the general for the particular). But also specialists in the theory of random processes often manage only to remark that a specific process arising from an application reduces to a class of random processes from the general theory (they can not see the particular in the general).

In this current handbook the authors attempted to summarize a rather wide class of processes arising from applications within the scope of the general theory of random processes. All necessary background of the theory of random processes is also provided. The contents of the handbook are split schematically into three sections: (i) general theory of random processes; (ii) random processes originating from applications; (iii) very specific random processes adapted to special situations and occurring rarely. These sections are not separated spatially since the exposition follows, by and large,the conventional way of presenting random processes.

The manner of the text differs in the three sections. In the first, background random processes theory is presented in detail; basic theorems are stated (without proofs).

In the second section a constructive definition of a process is often given, implied by the essence of an application; definitions and results are illustrated by examples helping to modify the process if the problem is modified. The text in the third section is restricted to the description of the situation and a constructive definition of a specific process in question. Because of the large amount of such processes stating any results would increase unreasonably the volume of the handbook.

The references provide the sources where the proofs of all assertions contained in the handbook are available. The index will help the reader to find a topic of interest.

Chapters 1-5, 17 and Section 7.5 are by I.N. Kovalenko, Chapters 6-11, 16 are by N.Yu. Kuznetsov, Chapters 12-15 are by V.M. Shurenkov.

The authors and the editor are sincerely gr ateful to reviewers N.V. Karta- shov, N.N. Leonenko, N.I. Portenko and M.I.Yadrenko for their valuable remarks and G.A. Marchuk for his help in preparing the manuscript.

A. V. Skorokhod
Member of the Ukrainian
National Academy of Sciences

Introduction

The theory of random processes is the richest in results and eventual applications of a scientific discipline stemming from the probability theory. As the traditional probability theory deals with the properties of finite sets of random events and random variables so the theory of random processes does with infinite sets of random variables depending on time. Thus the traditional probability theory is engaged when investigating instantaneous random events (hitting a target by a shell, destruction of a structure under a random load, resonance of a mechanical system with random parameters etc.) whereas the theory of random processes studies the time behavior of diverse phenomena influenced by random factors. Hence the theory of random processes relates to the probability theory as dynamics relates to statics.

A classical example of a random process is the Brownian motion process (cf. Ch. 5). It describes the motion of an inertialess particle subjected to a chaotic interaction with other particles. The theory of Brownian motion processes gave rise to the theory of diffusion processes describing the diffusion in gases, course of chemical reactions, chaotic drift of electronic components' parameters, noise processes in electronic devices, the behavior of random errors in serial processing of information on a moving object and many other phenomena. Currently a theory of automatic control based on the representation of a process in question as a solution of a stochastic differential equation (with applications to electrical engineering, radio engineering, automation, economics and other areas) which in turn generalizes diffusion processes is in development (cf. Ch. 15). Modern technology applies results of this theory to devise optimum algorithms for the recognition of noisy objects such as targets in radar-location, optimum pursuit and escape in stochastic games, and optimum control of moving objects under uncertain conditions. This theory permits selection a predicted tolerance for time varying parameters of mechanical and electronic components falling outside which makes a device to turn off and adjust parameters. It makes it possibile to compute the reliability of a wide class of devices defined as the probability of the sojourn of basis parameters within a given tolerance during the task execution time. It is worthwhile to note applications of diffusion processes to mathematical statistics (to prove nonparametric criteria for testing statistical hypotheses), investigation of the growth dynamics of large biological populations to ecology. Similar processes are used in mathematical economics to derive economical models with a proper account of numerous real deviations from a steady state process. Thus, diffusion processes describe the behavior of queues to meet the demand for a need under the approximate equilibrium between the demand and system performance (examples: yield computing the size of switchboards, speed and size of cash memory of a computer controlling several manufacturing routines, throughput capacity of traffic networks).

Stationary in the broad sense processes have wide applications (cf. Ch. 11). In practice it is often impossible not only to collect statistical data but even to measure an instantaneous value of a process, e.g., this is the case when investigating processes generated by electromagnetic oscillations of high frequency. However measurements of the time averages of the form $\frac{1}{T}\int\limits_0^T \xi(t)\,dt$ or even $\frac{1}{T}\int\limits_0^T \xi(t)\,\xi(t+\tau)\,dt$ can be obtained by means of physical devices. But these are the averages (more precisely, their stochastic limits as $T \to \infty$) which lie at the base of the theory of stationary in the broad sense processes. The discovery of this class of random processes enhanced first development of statistical radio engineering and later the linear theory of automatic control.The comparative simplicity of the computation of correlation and spectrum characteristics of the stationary in the broad sense process when it passes through linear elements has led to the derivation of a convenient formalism of the computation and then to the optimization of linear systems. The most remarkable results in this domain are the prediction and filtration theories of random processes (the latter is the basis of statistical radio engineering); other applications to mention are the synthesis of radio receiving and transmitting devices, optimum control of reactors, manufacturing routines, speech recognition and synthesis, underwater acoustics, detecting periodicities in natural and economical phenomena, investigation of models of vital activity of systems of the human body and ecological models. An outstanding result of the theory is the Shannon-Kotelnikov Theorem on the exact restoration of a random function with a bounded spectrum through observations at discrete times. This theorem is the basis for the pulse modulation widely used in the radio engineering.

The limpidity of the theory of the stationary in the broad sense processes and the simplicity of the interpretation of the results obtained in applications by its means encouraged a large-scale reduction of real processes to stationary ones through transformations or modification. Thus in statistical radio engineering nonstationary processes are of wide use which can be handled as stationary in small time intervals. Human speech transformed to an electrical signal is of this kind. In the most widespread models of signal detection through noise it is common to consider the noise stationary. Many processes of mathematical economics are of the same kind, removing trend and periodic components.

However the area of the applications of stationary in the wide sense processes is not boundless. The stationarity may be lost when such a process passes through a nonlinear element. To meet this lack, attempts have been made to develop a theory of stationary in a stricter sense processes, namely, processes stationary up to some degree.

Many applications have stationary in the strict sense processes (cf. Ch. 14). They are applied to a wide class of physical phenomena where the stationarity is implied by the essence of the object in question. The qualitative inferences of this theory in the form of ergodic theorems are very important for they guarantee a very important property of the stability in some sense of real systems (provided the model of the process is adequate for the real object). Their most essential application is to the stability of dynamic (in particular, mechanical) systems as well as to the information theory. The creator of the latter C. Shannon has derived the notion of stationary sources and channels as well as important principal assertions as to conditions of errorfree information transmission. This theory is applied to devising communication lines, information transmission devices, as well as coding and decoding algorithms.

Many applications are based on the theory of Gaussian processes (cf. Ch. 2). The Gaussian (normal) property assumption for a joint distribution of process values at different times is wellsuited to the common situation when the process values are the sum of a large number of independent factors influencing the process. Thus is the case in the statistical radio engineering (cf. Ch. 11). Gaussian process forms the base of the theory of nonlinear transformation of a process when it passes through typical nonlinear elements (e.g., detectors) which are used, in particular, in radar-location. An important trend in the investigation of Gaussian processes is studying their local properties: spikes, number of crossings of a level, et al. These topics are applied in radio engineering, reliability theory and other areas. Thus the theory of excesses over a level permits computation of the reliability of a physical device if its parameters behave like a Gaussian process. A similar problem arises when justifying algorithms of the generation of random numbers for use in Monte Carlo methods (cf. Ch. 17) if their source is a record of a physical noise process.

An important milestone in the development of the theory of random processes is the discovery by A.A. Markov of chains (Markov chains, cf. Ch. 3). Their proper definition makes them surprisingly close to many processes of the real world. Consider, e.g., the operation of a digital automaton subjected to random failures. Since the failures at different clock times are independent events, the future behavior of the automaton is fully defined in the probabilistic sense by automaton state at a given clock time. Hence it follows that the behavior of such an automaton is described by a Markov chain. In a similar way Markov chains are good models of the dynamics of birth and death processes in biological populations, failure and renewal of technical and biological systems, formation of queues in information processing systems and so on. The laws of Markov chains govern the probability of traits in formal genetics. Main applications of Markov chains are statistical models of language, computer-aided translation, economy growth models, dynamics of conflict situations, oscillations of the water level in water storage, aging models in engineering, pulse systems of automatic control, ecological models, learning and pattern recognition (automatic character recognition). Markov chains are used in physics to describe cascade cosmic rays, operation of various devices: elementary particle counters, radiovalves, memory units of computers. In chemistry they help to construct models of large molecules; in strength theory they describe models of crack propagation in materials; in communication theory they describe models of burst errors in a transmitted text.

The idea of Markov chains lies in the formulation of the theory of general Markov processes (cf. Ch. 15) and various classes of these processes interesting for the theory and applications. Maintaining the basic property of Markov chains, i.e., the independence of the future from the past provided the present is fixed, Markov processes run in continuous time as well as in an arbitrary space of states. This extends the range of their applications to systems of continuous operation, e.g., queueing theory which is a part of operations research and is based mostly on the theory of Markov processes. The queueing theory is used to compute the throughput capacity of telephone networks, traffic networks and the efficiency of various queueing systems (repair utilities, health entities, supply facilities etc.); it permits estimattion of the probability of undesirable events (catastrophes) in engineering systems of different kinds and losses because of a delay in service. A control point in the queueing theory is the stability of a system under given values of its parameters, i.e., whether a system can perform the task it is designed for. Within the framework of the theory of Markov processes ergodic theorems (cf. Ch. 3) have been established concerning the behavior of queueing systems which imply important practical conclusions. Thus the theory predicts an unbounded growth of waiting time when the device load tends to its maximum admissible value. Queueing theory methods are widely used to devise optimum algorithms to attend a group of tools by a repairman, to distribute vessels to port facilities, to organize road traffic; it permits derivation of optimum preventive maintenance policies, optimum redundancy rules and inspection policies. The theory of Markov processes finds a very important application in the field of computer science ranging from the "race problem" in microchips to the development of huge computer networks. Modern results from the theory allow solving problems of the optimum use of computer memory, optimum information exchange between computers, time sharing, batch information processing etc. Recently problems of deriving an optimum strategy to process a stream of calls having different priorities have come into play and sometimes they find a successful solution.

It is worth noting that parallel to the general theory of Markov processes its various ramifications are intensively developed aimed at specific applications. (However a mutual penetration and interaction of different results occur usually without any connection to applications.) So, the renewal theory (cf. Ch. 12) is oriented first to engineering systems maintenance; branching processes (cf. Ch. 13) deal with the dynamics of biological populations. Meanwhile the same mathematical model of a random process often describes phenomena having quite different natures (birth and death processes in biology and reliability theory).

In practice an application of random processes goes as follows: 1) devising a mathematical model; 2) statistical testing of the model's adequacy and parameters estimations; 3) exploitation of the model to infer opinions on the real object. The first step involves the investigation of physical processes in the object as well as analysis of a wide class of eventual mathematical models known to a researcher. The second step falls into the domain of random processes statistics (cf. Ch 16). The third step requires skills to develop criteria to estimate system characteristics in the language of the mathematical model as well as to apply analytical and numerical methods. In the last two decades the Monte Carlo method (cf. Ch. 17) has become the main numerical method of the applied analysis of random processes.

A more deliberate presentation of the applications of random processes to various fields of science and technology may be found in References 1 to 21.

References

1. Aoki, M., *Optimization of Stochastic Systems: Topics in Discrete-Time Systems,* Acad. Press, New York, 1967.
2. Bolshakov, I. A., *Extraction of the Signals Flow Out of Noise,* Sov. Radio, Moscow, 1969 (in Russian).
3. Borovkov, A. A., *Probability Processes in the Queueing Theory,* Nauka, Moscow, 1972 (in Russian).
4. Buslenko, N. P., Kalashnikov, V. V., Kovalenko, I. N., *Lectures on the Theory of Complex Systems,* Sov. Radio, 1973 (in Russian).
5. Chinčin, A. J., *Mathematical Methods in the Theory of Queueing,* Griffin, London, 1969.
6. Dahl, O. J., Myhrhaug, B., Nygaard, K., *SIMULA 67. Common Base Language,* Norwegian Computing Center, Oslo, 1968.
7. Gnedenko, B. V., Belyayev, Yu. K., Solovyev, A. D., *Mathematical Methods of Reliability Theory,* Acad. Press, New York, 1969.
8. Gnedenko, B. V., Kovalenko, I. N., *Introduction to Queueing Theory,* Birkhäuser, Boston, 1989.
9. Haight, F. A., *Applied Probability,* Plenum Press, New York, 1981.
10. Karlin, S., *A First Course in Stochastic Processes,* Acad. Press, New York, 1968.
11. Karlin, S., Taylor H. M., *A Second Course in Stochastic Processes,* Acad. Press, New York, 1981.
12. Kac, M., *Probability and Related Topics in Physical Sciences,* Interscience Publ., London, 1957.
13. Kleinrock, L., *Queueing Systems,* Wiley, New York, 1975.
14. Kleinrock, L., *Computer Applications,* Wiley, New York, 1976.
15. Levin, B. R., *Theoretical Foundations of the Statistical Radio Engineering,* Sov. Radio, Moscow, 1968 (in Russian).
16. Pugachev, V. S., *The Theory of Random Functions and Its Application to Problems of Automatic Control,* Fizmatgiz, Moscow, 1962 (in Russian).
17. Ross, S. M., *Applied Probability Models with Optimization Applications,* Holden-Day, San Francisco, 1970.
18. Shannon, C. E., *The Mathematical Theory of Communication,* Univ. of Ill., Urbana, 1972.
19. Sharakshane, A. S., Zheleznov, I. G., Ivnitsky V. A., *Complex Systems,* Vysshaya Shkola, Moscow, 1977 (in Russian).

20. Tikhonov, V. I., Mironov, M. A., *Markov Processes,* Sov. Radio, Moscow, 1977 (in Russian).
21. Ventzel, E. S., *Operations Research: Problems, Principles, Methodology,* Nauka, Moscow, 1980 (in Russian).

Chapter 1

Definitions and General Properties of Random Processes

1.1 Definition of a Random Process Via a Probability Measure over the Path Function Space

Let I be a set and a random variable $\xi(t)$ be defined for any $t \in I$. Then the set of random variables $\{\xi(t),\ t \in I\}$ is called **a random process;** each $t \in I$ is called **a parameter;** I is called **the parametric set.** In most cases I is an interval of the real line $\mathbb{R}$. The interval may be bounded or unbounded from one or both sides; it may be closed as well as open or half-open. In all such cases t is considered as a time variable, and $\xi(t)$ is called **a value** of the random process at time t.

As a rule, the set X of possible values of random variables $\xi(t)$ is the same for each $t \in I$. In most cases X is $\mathbb{R}$ or $\mathbb{R}^m$. If $X = \mathbb{R}^m$ then the random process is called **an** m-dimensional random process. The set X is also called **the phase-state** of the random process.

Any random variable ξ can be represented by a function of an elementary event ω of a sample space Ω with a probability measure **P**, i.e. $\xi = \xi(\omega)$, $\omega \in \Omega$. Then

$$\xi(t) = \xi(t, \omega), \quad t \in I,\ \ \omega \in \Omega. \tag{1.1}$$

It is essential that all the $\xi(t)$, $t \in I$, should be defined on the same elementary event ω. The time function $\xi(t, \omega)$ at a given ω is said to be **a path**

function (realization or **a trace)** of the random process. The symbol $\{\xi(t),\ t \in I\}$ is usually reduced to a simpler one $\xi(t)$, and the shorter term "process" is preferred to "random process".

Example 1.1. Operating parameters of some electronic components follow linear laws of behavior. Both the initial value α and the velocity β of its change in time are random. Then for a path function $\xi(t)$ the following equation is satisfied: $\xi(t) = \alpha + \beta t,\ \ t \geq 0$. A pair (α, β) can be considered as an elementary event. The probability measure $\mathbf{P}$ can be defined by the joint probability density function $f_{\alpha\beta}(x, y)$ as follows:

$$\mathbf{P}\{(\alpha, \beta) \in A\} = \int\int_A f_{\alpha\beta}(x, y)\, dx dy$$

as soon as the right-hand side is defined as a Lebesgue integral for a given set $A \subset \Omega$.

Example 1.2. A harmonic oscillation with random amplitude and phase $\xi(t) = A \sin(\lambda t - \varphi)$ can be considered as a random process. In this case $\omega = (A, \varphi)$. If λ is random too then $\omega = (\lambda, A, \varphi)$.

Random processes are used as models of a wide range of real phenomena. A trace of a diffusive particle, the congestion of vehicles at a crossing, the growth of a biological population are sketch examples. The value of a random process at time t is often spoken of as "system state" or a "particle coordinate". In a mathematical context such terms as "system", "particle" and so on are used for comprehension and cannot be defined rigorously.

The structure of the sample space Ω may be very sophisticated. If ω is a multidimensional random variable or a sequence $\omega = \{\omega_n\}$ of independent random variables and an algorithm is available for the computation of $\xi(t, \omega)$ for each $t \in I$ and $\omega \in \Omega$ then the random process $\xi(t)$ can be considered as **a constructively defined process.** This definition is not rigorous because the algorithm can be defined in various manners, but it is of interest for practical applications. Since a modern computer admits obtaining realizations of independent random variables ω_n we can therefore obtain realizations of a constructively defined random process. This possibility is the basis for the simulation (statistical modelling) of random processes describing the dynamic behavior of a large variety of real things imposed by the influence of a chance.

Probability measure $\mathbf{P}$ defined over a measurable space Ω induces a measure over the space of path functions of a random process $\xi(t)$. Indeed each measurable set $A \subset \Omega$ can be associated with the set A^* of space functions $\{\xi(\cdot, \omega),\ \omega \in A\}$. In the example 1.1 the set $\{\alpha > 0,\ \ \beta < 0\}$ is associated with the set of space functions intersecting the abscissa $\{t > 0\}$ from above to below. One can set $\mathbf{P}\{A^*\} = \mathbf{P}\{A\}$. This equation defines the probability measure over the space of path functions. The realization of such a simple principle has, though, many nontrivial features when considering special cases.

1.2 Definition of a Random Process Via Multidimensional Distributions

Let $n > 1$; $A_1, ..., A_n$ be any measurable sets of the space X; $t_1, ..., t_n$ be time instants from the set I. The function

$$P_{t_1,...,t_n}(A_1, ..., A_n) = \mathbf{P}\{\xi(t_1) \in A_1, ..., \xi(t_n) \in A_n\}$$

is called **an n-dimensional distribution** of the process $\xi(t)$. The set of all such distributions for $n = 1, 2, ...$ is called **the family of finite-dimensional distributions** of the process.

If $X = \mathbb{R}$ then the definition of the family of finite-dimensional distribution can be made via the family of multidimensional distribution functions

$$F_{t_1,...,t_n}(x_1, ..., x_n) = \mathbf{P}\{\xi(t_1) \le x_1, ..., \xi(t_n) \le x_n\}.$$

In the case of $\mathbb{R}^m$ then $x_i = (x_{i1}, ..., x_{im})$, $\xi(t) = (\xi_1(t), ..., \xi_m(t))$ where $\xi_j(t)$ are one-dimensional random variables. The last formula is still valid if one uses the symbol $\{\xi(t_i) < x_i\}$ instead of $\{\xi_j(t_i) < x_{ij},\ 1 \le j \le m\}$.

Let X be a topological space and X^n be its n-product, i.e., the set of the points of the form $(x_1, ..., x_n)$, $x_i \in X$. Denote by $\mathfrak{B}^n$ the class of Borel sets in X^n defined as the minimal σ-algebra containing all the sets of the form $\{x_i \in C\}$, C being any open subset of X.

A cylindrical set of the space of functions $f(t) : I \to X$ means an arbitrary set B defined by the condition

$$(f(t_1)\ , ...,\ f(t_n)) \in A, \quad A \in \mathfrak{B}^n, \ \ t_i \in I, \ \ 1 \le i \le n.$$

In the case $X = \mathbb{R}$ any set of the form

$$\{f(t),\ t \in I :\ f(t_1) < x_1\ , ...,\ f(t_n) < x_n\}$$

is a cylindrical set.

Finite-dimensional distribution defines the probability measure for any cylindrical set of the path function space of a random process[1,2]:

$$\mathbf{P}\{B\} \ = \mathbf{P}\{(\xi(t_1)\ , ...,\ \xi(t_n)) \in A\} \ = \int_{(x_1, \ldots, x_n) \in A} \cdots \int P_{t_1,...,t_n}(dx_1, ..., dx_n).$$

In the case of $A = A_1 \times \ ... \ \times A_n$ one obtains

$$\mathbf{P}\{B\} = \mathbf{P}\{(\xi(t),\, t \in I\} \in B\} = P_{t_1,\dots,t_n}(A_1, \dots, A_n).$$

For sets of a more complex structure than cylindrical ones the probability measure is usually defined via a monotone limiting procedure. Some general properties are required concerning the space functions; for example they are assumed to be deprived of discontinuity of the second kind.

A family of multidimensional distributions is called **compatible** if the following **compatibility conditions** are satisfied.

1. If $(i_1, \dots, i_n)$ is any permutation of numbers $1, \dots, n$ then

$$P_{t_{i_1},\dots,t_{i_n}}(A_{i_1}, \dots, A_{i_n}) = P_{t_1,\dots,t_n}(A_1, \dots, A_n).$$

2. $P_{t_1,\dots,t_n}(A_1, \dots, A_n, X) = P_{t_1,\dots,t_{n-1}}(A_1, \dots, A_{n-1})$.

Kolmogorov Theorem *states that if X is a complete separable metric space (for example $\mathbb{R}$ or $\mathbb{R}^m$) and a compatible family of multidimensional distributions is defined over X then a random process $\xi(t) = \xi(t, \omega)$ can be defined in such a way that its multidimensional distributions coincide with prescribed ones from the initial family.*

Let $\mathfrak{X}$ be the path space of a random process $\xi(t)$. A separate path function will be denoted by $x(\cdot)$. A probability measure $\mathbf{P}$ is defined over $\mathfrak{X}$; see above. Then one can also define a probability measure over $\mathfrak{X}^n$, i.e. cartesian product of $\mathfrak{X}$, by the equation

$$\mathbf{P}^{(n)}\{A_1 \times \ ... \ \times A_n\} = \mathbf{P}\{A_1\} \dots \mathbf{P}\{A_n\}. \tag{1.2}$$

If an element $(x_1(\cdot), \ \dots \ , x_n(\cdot))$ is being chosen from $\mathfrak{X}^n$ in accordance with the measure $\mathbf{P}^{(n)}$ then $x_1(\cdot), \ \dots \ , x_n(\cdot)$ are called **independent path functions** (or **realizations**) of the random process $\xi(t)$. Particularly, multidimensional distributions are factorizing as

$$\mathbf{P}\{x_i(t_i) \in A_i,\ 1 \le i \le n\} = \prod_{i=1}^{n} P_{t_i}(A_i).$$

An infinite sequence of independent path functions can be defined in a similar way.

Note. Let $\xi(t, \omega)$, $\omega \in \Omega$, and $\mathbf{P}(A)$ is a probability measure over Ω. Define a probability measure $\mathbf{P}^{(n)}$ over cartesian product Ω^n by the equation (1.2). If $\omega = (\omega_1, \ \dots, \ \omega_n) \in \Omega^n$ is being chosen in accordance with the measure $\mathbf{P}^{(n)}$ then $\xi(t, \omega_i)$, $1 \le i \le n$, are independent path functions of $\xi(t)$.

Multidimensional distributions of the processes observed in reality can be estimated as accurately as desired via the observation of independent path functions, as well as each probability can be computed via the limit of a sequence of frequencies. Thus the definition of a random process via the family of multidimensional distributions enables the justification of a stochastic model of a real process.

1.3 Equivalence of Random Processes. Measurability. Separability

A random process with prescribed multidimensional distributions can be defined in various ways. It is natural to identify all such constructions in a probabilistic sense. Two random processes $\{\xi(t),\ t \in I\}$ and $\{\eta(t),\ t \in I\}$ are called **stochastically equivalent**[1,2] if their multidimensional distributions coincide.

Path functions of a random process are often subjected to transforms of various kinds. At least the measurability must be assumed to justify such transformations. Let $\xi(t) = \xi(t, \omega)$ be a random process defined over a measurable space $(X, \mathfrak{S})$ where $\mathfrak{S}$ is a σ-algebra of subsets of the space X. Let X be a topological space, I be a Lebesgue-measurable subset of X and μ is the measure over the space $I \times \Omega$ defined by the product of Lebesgue measure and $\mathbf{P}$. (That is, for each B of the form $(t,\ \omega) \in I \times \Omega$ defined by the conditions $t \in I_0$, $\omega \in \Omega_0$, I_0 being a Lebesgue-measurable subset of I, $\Omega_0 \in \mathfrak{A}$ then $\mu(B) = |I_0|\, \mathbf{P}\{\Omega_0\}$, $|I_0|$ denoting Lebesgue measure of I_0.)

A random process $\xi(t) = \xi(t, \omega)$, $t \in I$, $\omega \in \Omega$ in X is said to be **measurable** if the function $\xi(t, \omega)$ is measurable in accordance with the above defined measure over the space of pairs (t, ω). In other words, for an arbitrary Borel set C of the space X the two-dimensional set $\{(t, \omega) : \xi(t, \omega) \in C\}$ is μ-measurable.

If a random process $\xi(t)$ is a measurable one then almost every path function is Borel-measurable. Therefore one can choose such set $\Omega' \subset \Omega$, $\mathbf{P}\{\Omega'\} = 1$ that for arbitrary $\omega_0 \in \Omega'$ and Borel set $C \subset X$ the set of $t \in I$ for which $\xi(t, \omega_0) \in C$ is a Lebesgue-measurable.

Let Γ be a class of subsets of X. A random process $\xi(t)$ is said to be **separable** with respect to Γ if a finite or countable set $\Lambda \subset I$ and an "exceptional" set V of a zero probability exist such that the following property is satisfied. Let $\Delta = I \cap (a, b)$, $-\infty < a < b < +\infty$, $A \in \Gamma$. Then for any $\omega \notin V$ the inclusions $\{\xi(t, \omega) \in A,\ t \in \Delta\}$ and $\{\xi(t, \omega) \in A,\ t \in \Delta \cap \Lambda\}$ either both are true or both are false. The set Λ is called

a separability set.

Example 1.3. Let a path function of a random process $\xi(t)$ be continuous almost surely. Since any continuous function can be defined uniquely by its values over a countable everywhere dense set (e.g., the set of rational numbers) each such set can be considered as a separability set, and the process $\xi(t)$ is separable.

While considering a one-dimensional or multidimensional process the set Γ is usually defined as the class of all the closed sets. Particularly, for a separable multidimensional process $\xi(t) = (\xi_1(t), \ldots, \xi_m(t))$ with Λ defined as the set of rational points of I, the inclusion $\xi_1^2(t) + \cdots + \xi_m^2(t) \leq \leq R^2$ being satisfied for rational points of an interval (t_0, t_1) assures the same inclusion over the whole interval (t_0, t_1).

A one-dimensional random process $\xi(t)$ is separable with respect to the class of all closed sets iff for each interval (a, b) the following equation holds true for a countable set Λ:

$$\sup_{t \in I \cap (a,b)} \xi(t) = \sup_{t \in I \cap (a,b) \cap \Lambda} \xi(t), \qquad \inf_{t \in I \cap (a,b)} \xi(t) = \inf_{t \in I \cap (a,b) \cap \Lambda} \xi(t).$$

It is important to know if there exists a separable random process equivalent to a given random process. Let X be a compact metric space (for example, a bounded closed set of $\mathbb{R}$ or $\mathbb{R}^m$), I be a finite or infinite interval. Then, given an X-valued random process $\{\xi(t), \ t \in I\}$, there exists a separable X-valued process equivalent to the original one.

Example 1.4. Let α_n be independent random variables uniformly distributed over the interval $(0, 1)$. Define $\xi(\alpha_n) = 1$ and $\xi(t) = 0$ for $t \in [0, 1] \setminus \{\alpha_n\}$. Such a random process is not separable, but there exists a separable random process $\eta(t) = 1, \ 0 \leq t \leq 1$, equivalent to $\xi(t)$.

1.4 Stochastic Continuity

A sequence $\{\xi_n\}$ of one-dimensional random variables is said to converge to a random variable η **in probability** as $n \to \infty$ iff $\mathbf{P}\{|\xi_n - \eta| > \varepsilon\} \underset{n \to \infty}{\longrightarrow} 0$ for every $\varepsilon > 0$. A family of random variables $\{\xi_t\}$ converges to a random variable η **in probability** as $t \to t_0$ iff $\mathbf{P}\{|\xi_t - \eta| > \varepsilon\} \underset{t \to t_0}{\longrightarrow} 0$ for every $\varepsilon > 0$. The defined property is usually designed as follows:

$$\xi_n \xrightarrow[n \to \infty]{p} \eta, \quad \xi_t \xrightarrow[t \to t_0]{p} \eta \quad \text{or} \quad p \lim_{n \to \infty} \xi_n = \eta, \quad p \lim_{t \to t_0} \xi_t = \eta.$$

Making use of the convergence in probability one can call a one-dimensional random process $\xi(t)$ **stochastically continuous**[3] **at a point** t_0 iff

$\xi(t) \xrightarrow[t\to t_0]{p} \xi(t_0)$. A random process $\xi(t)$ is said to be **stochastically continuous** iff it is stochastically continuous at each point $t \in I$.

The stochastic continuity does not imply the continuity of its path functions. Indeed, let $I = (0, 1)$, ω be a uniform random variable over the interval (0, 1). Define a random process $\xi(t)$, $0 \le t \le 1$, by the formula

$$\xi(t) = \begin{cases} 0 & \text{if } t < \omega \\ 1 & \text{if } t \ge \omega. \end{cases}$$

Then the path functions are discontinuous whereas $\xi(t)$ is stochastically continuous due to the estimate satisfying for each $t, t_0 \in [0, 1]$:

$$\mathbf{P}\{|\xi(t) - \xi(t_0)| > \varepsilon\} \le \mathbf{P}\{\xi(t) \ne \xi(t_0)\} =$$

$$= \mathbf{P}\{\xi(t) = 0,\ \xi(t_0) = 1 \text{ or } \xi(t_0) = 0,\ \xi(t) = 1\} = |t - t_0| \to 0 \text{ as } t \to t_0.$$

The stochastic continuity is defined by appropriate properties of two-dimensional distributions $F_{t_0,t}(x_0, x) = \mathbf{P}\{\xi(t_0) < x_0,\ \xi(t) < x\}$ of the process $\xi(t)$.

If X is a metric space, $\varrho(x, y)$ being the metrics then the convergence in probability can be defined similarly to the one-dimensional case; the inequalities $|\xi_n - \eta| > \varepsilon$, $|\xi_t - \eta| > \varepsilon$ should be changed to $\varrho(\xi_n, \eta) > \varepsilon$, $\varrho(\xi_t, \eta) > \varepsilon$. Therefore the above formulated property of the stochastic continuity is obviously expanded to cover the processes in a metric space as well. Particularly, an m-dimensional process $\xi(t) = (\xi_1(t), \ldots, \xi_m(t))$ is stochastically continuous at a point t_0 iff each of the following conditions is satisfied:

1. $\mathbf{P}\left\{\sum_{i=1}^{m} (\xi_i(t) - \xi_i(t_0))^2 \ge \varepsilon^2\right\} \xrightarrow[t\to t_0]{} 0,\ \ \varepsilon > 0.$
2. $\mathbf{P}\{\max_{1\le i\le m} |\xi_i(t) - \xi_i(t_0)| > \varepsilon\} \xrightarrow[t\to t_0]{} 0,\ \ \varepsilon > 0.$
3. $\mathbf{P}\{|\xi_i(t) - \xi_i(t_0)| > \varepsilon\} \xrightarrow[t\to t_0]{} 0,\ \ 1 \le i \le m,\ \ \varepsilon > 0.$

A complex-valued random process is defined as the sum

$$\xi(t) = \alpha(t) + i\beta(t)$$

where $\alpha(t)$ and β are one-dimensional random processes called **the real component** and the **imaginary component** of the $\xi(t)$ respectively. Such a process is stochastically continuous at a point t_0 iff both its components $\alpha(t)$ and $\beta(t)$ possess this property.

1.5 Definition of a Random Process Via Mean and Covariance Functions

Let $\xi(t)$ be a one-dimensional random process possessing the property that $\mathbf{E}\,\xi^2(t) < \infty$, $t \in I$. Let us set

$$m(t) = \mathbf{E}\,\xi(t), \quad R(t_1, t_2) = \mathbf{E}\,\xi^0(t_1)\,\xi^0(t_2),$$

where $\xi^0(t) = \xi(t) - \mathbf{E}\,\xi(t)$. The function $m(t)$ is called **mean function** of the random process; the function $R(t_1, t_2)$ is called **the covariance function** of the random process. The function $D(t) = \mathbf{Var}\,\xi(t) = R(t,t)$ is called **the variance function** of the random process. The function $r(t_1, t_2) = R(t_1, t_2)/(\mathbf{Var}\,\xi(t_1)\,\mathbf{Var}\,\xi(t_2))^{1/2}$ is called **the normed covariance function** of the random process.

Let $\xi(t) = (\xi_1(t), \ldots, \xi_m(t))$ be an m-dimensional random process. The vector function $m(t) = (m_1(t), \ldots, m_m(t))$ with $m_i(t) = \mathbf{E}\,\xi_i(t)$ is called **the mean function** of the random process. The matrix function $R(t,\tau) = (\mathbf{cov}(\xi_i(t),\ \xi_j(\tau)))$ is called **the covariance matrix** of the process. Its diagonal elements $\{R_{ii}(t,\tau)\}$ are **the covariance functions** of the processes $\{\xi_i(t)\}$ considered separately whereas a nondiagonal element $R_{ij}(t,\tau) = \mathbf{cov}(\xi_i(t),\ \xi_j(\tau))$ is called **the joint covariance function** of the processes $\xi_i(t)$ and $\xi_j(t)$. The matrix function

$$r(t,\tau) = \left(\frac{\mathbf{cov}(\xi_i(t),\ \xi_j(\tau))}{\sigma_i(t)\,\sigma_i(\tau)} \right)$$

with $\sigma_i(t) = \sqrt{\mathbf{Var}\,\xi(t)}$ is called **the normed covariance function** of the process $\xi(t)$.

Let $\xi(t) = \alpha(t) + i\,\beta(t)$ be a complex-valued random process; $\bar{x}$ denotes a complex conjugate to x. The complex function $m(t) = m_1(t) + i\,m_2(t)$ with $m_1(t) = \mathbf{E}\,\alpha(t)$, $m_2(t) = \mathbf{E}\,\beta(t)$ is called **the mean function** of the process $\xi(t)$. The function $R(t_1, t_2) = \mathbf{E}\,\{\xi^0(t_1)\,\overline{\xi^0(t_2)}\}$ with $\xi^0(t) = \xi(t) - m(t)$ is called **the covariance function** of the complex-valued process $\xi(t)$. It can be expressed via the covariance function and the joint covariance function of the processes $\alpha(t)$ and $\beta(t)$:

$$R(t_1, t_2) = R_{11}(t_1, t_2) + R_{22}(t_1, t_2) + i\,(R_{21}(t_1, t_2) - R_{12}(t_1, t_2)),$$

$R(t_1, t_2) = (R_{ij}(t_1, t_2))$ being the covariance matrix of the two-dimensional process $(\alpha(t),\ \beta(t))$.

The covariance function of a complex-valued (and also a real-valued) random process is a nonnegatively defined function so that

$$\sum_{i,j=1}^{n} z_i\,\bar{z}_j\,R(t_i, t_j) \geq 0$$

for any $t_i \in I$ and any complex numbers $\{z_i\}$. If $R(t_1, t_2)$ is continuous over a square $a \leq t_1, t_2 \leq b$ then for any continuous function $z(t)$ the inequality

$$\int_a^b \int_a^b z(t)\, \overline{z(\tau)}\, R(t, \tau)\, dt\, d\tau \geq 0$$

holds true.

1.6 Mean-Square Continuity

Let ξ and η be one-dimensional random variables possessing finite variances. Let us define a norm of their mutual deviation

$$\|\xi - \eta\| = (\mathbf{E}\,(\xi - \eta)^2)^{1/2}.$$

The scalar product of ξ and η is defined as

$$(\xi,\ \eta) = \mathbf{E}\,(\xi\, \eta).$$

The scalar product of complex-valued ξ and η is defined as

$$(\xi,\ \eta) = \mathbf{E}\,(\xi\, \bar{\eta}).$$

In both cases $|(\xi,\ \eta)|^2 \leq \mathbf{E}\,|\xi|^2\, \mathbf{E}\,|\eta|^2$.

If $\{\xi_n\}$ is a sequence of random variables and η is a random variable with the finite variance then ξ_n **converges to η in the mean square**[4–6] if $\|\xi_n - \eta\|^2 \underset{n\to\infty}{\longrightarrow} 0$.

For complex random variables ξ, η a norm of their mutual deviation is defined as

$$\|\xi_n - \eta\| = (\mathbf{E}\,|\xi - \eta|^2)^{1/2} = (\mathbf{E}\,(\xi - \eta)\, \overline{(\xi - \eta)})^{1/2}.$$

The notion of the convergence is defined in a similar way.

If $\xi_n = \alpha_n + i\,\beta_n$, $\eta = \alpha_0 + i\,\beta_0$ are complex random variables then the mean square convergence $\xi_n \to \eta$ is equivalent to the mean-square convergence of the real and imaginary parts of these variables separately. The convergence of a family of random variables $\{\xi_t\}$ as $t \to t_0$ or $t \to \infty$ can be defined in a similar way.

Let $\xi = \alpha + i\,\beta$ be a complex-valued random variable. The variance of ξ is defined as $\mathbf{Var}\,\xi = \mathbf{Var}\,\alpha + \mathbf{Var}\,\beta$. A sequence of complex or real random variables of finite variances is called **mean-square fundamental** iff $\|\xi_n - \xi_m\| \to 0$ as $n, m \to \infty$.

An identity

$$\|\xi - \eta\|^2 = |\mathbf{E}\,\xi - \mathbf{E}\,\eta|^2 + \mathbf{E}\,((\xi^0 - \eta^0)\,\overline{(\xi^0 - \eta^0)})$$

is often used for the justification of the fundamentality of concrete sequences of random variables.

The Completeness Theorem. *Let $\{\xi_n\}$ be a sequence of complex-valued or real random variables with finite variances. If this sequence is mean-square fundamental then a mean-square limit η of the sequence $\{\xi_n\}$ exists such that $\|\eta\| < \infty$ and $(\xi_n,\, \zeta) \underset{n\to\infty}{\longrightarrow} (\eta,\, \zeta)$ for an arbitrary random variable $\zeta:\ \|\zeta\| < \infty$.*

If $\xi_n \to \eta$ in the mean square then $\xi_n \to \eta$ in probability.

Let $\xi(t)$ be a complex-valued or real random process and $\mathbf{Var}\,\xi(t) < \infty$, $t \in I$. The process is called **mean-square continuous at a point** t_0 iff $\|\xi(t) - \xi(t_0)\| \underset{t\to t_0}{\longrightarrow} 0$. The process is **mean-square continuous** iff the stated property is satisfied for each $t_0 \in I$.

The process $\xi(t)$ with mean function $m(t)$ and covariance function $R(t, \tau)$ is **mean-square continuous at a point** t_0 iff $m(t)$ is continuous at this point and $R(t, \tau)$ is continuous at the point $(t_0,\, t_0)$. The mean-square continuity implies the stochastic continuity.

Example 1.5. Let $\{\zeta_i\}$ be complex-valued random variables, $\mathbf{Var}\zeta_i < \infty$. Let, then, $f_i(t)$ be continuous functions. Consider a random process

$$\xi(t) = \sum_{i=1}^{n} \zeta_i\, f_i(t).$$

It is mean-square continuous since

$$\|\xi(t) - \xi(t_0)\|^2 = \sum_{i,j=1}^{n} (f_i(t) - f_i(t_0))\,\overline{(f_j(t) - f_j(t_0))}\,\mathbf{E}\,(\zeta_i\,\bar{\zeta}_j).$$

The mean function and the covariance function (matrix) of a stochastic process are called its **characteristics of the first and second orders**. Many properties of the process can be defined via these functions.

1.7 Random Processes with Continuous Path Functions

Let $\{\xi(t),\ t \in I\}$ be a random process, the I being a finite or infinite interval (closed, open or half-closed). The set X of possible values of the

$\xi(t)$ is assumed to be a topological space so that a continuous mapping $I \to X$ is of a sense. Particularly the cases $X = \mathbb{R}$ or $X = \mathbb{R}^m$ may occur. For a given $\omega \in \Omega$ the path function $\xi(t, \omega)$, $\omega \in I$ may be continuous or discontinuous. Let Ω_0 be the set of ω for which it is continuous. The process $\xi(t)$ is called **a process with continuous path functions**[1] iff $\mathbf{P}\{\Omega_0\} = 1$.

Example 1.6. Let Ω be the interval (0, 1); the probability measure is defined through the uniform distribution over this interval. Set $\xi(t, \omega) = 0$, $0 < t < 1$, if ω is an irrational number. If ω is rational point $\xi(t, \omega) = 0$ for $t < \omega$, $\xi(t, \omega) = 1$ for $t \geq \omega$. The defined random process $\xi(t) = \xi(t, \omega)$ is a process with continuous path functions. The set Ω_0 is the set of irrational points of the interval (0, 1).

If a random process has continuous path functions then it is stochastically continuous.

Many real processes (the movement of a particle etc.) are continuous in their nature. Therefore the continuity of path functions is a necessary requirement while constructing a stochastic model of each of such real processes.

Assume that the finite-dimensional distributions of a real-valued random process $\{\xi(t),\ a \leq t \leq b\}$ are given. According to **Kolmogorov Theorem**, if such $\rho > 0$, $c > 0$, $\beta > 0$ exist that for arbitrary t and τ from $[a, b]$

$$\mathbf{P}\{|\xi(t) - \xi(\tau)| > \varepsilon\} < c|t - \tau|^{1+\rho}/\varepsilon^{\beta}, \quad \varepsilon > 0,$$

then $\xi(t)$ can be defined over an appropriate sample space as a process with continuous path functions. If the above inequality holds true for a separable process then it has continuous path functions. The inequality $\mathbf{E}\,|\xi(t) - \xi(\tau)|^{\beta} \leq c|t - \tau|^{1+\rho}$ is sufficient for the validity of the Kolmogorov Theorem condition.

Let us consider the most important example of a real-valued process with continuous path function. It is **the Wiener process (the process of Brownian motion)**. It can be defined via the following properties.

1. $\xi(0) = 0$ almost surely.

2. For each $n \geq 1$ and arbitrary $0 < t_1 < \cdots < t_n$ the random variables $\xi(t_1)$, $\xi(t_2) - \xi(t_1)$, ..., $\xi(t_n) - \xi(t_{n-1})$ are independent.

3. If $0 \leq \tau \leq t$ then the random variable $\xi(t) - \xi(\tau)$ has Gaussian distribution with the mean $a(t-\tau)$ and the variance $\sigma^2(t-\tau)$. The constants a and σ^2 are called **the shift parameter** and **the diffusion parameter** of the Wiener process.

We have $\mathbf{E}\,(\xi(t) - \xi(\tau))^4 = O((t - \tau)^4)$; thus the condition of the Kolmogorov Theorem is satisfied for $\beta = 4$ and $\rho = 1$.

1.8 Random Processes without Discontinuities of the Second Kind

Let $[a, b] \subset \mathbb{R}$, X be a metric space. A function $f(t) : [a, b] \to X$ is called a function without discontinuities of the second kind if finite limits

$$f(t+0) = \lim_{\varepsilon \downarrow 0} f(t+\varepsilon) \quad \text{and} \quad f(\tau - 0) = \lim_{\varepsilon \downarrow 0} f(\tau - \varepsilon)$$

exist for each t and $\tau : a \le t \le b, \ a < \tau \le b$. Denote by $D = D_{[a,b]}$ the class of real-valued functions over the interval $[a, b]$ without discontinuities of the second kind.

A X-valued random process $\{\xi(t), \ a \le t \le b\}$ is called **a random process without discontinuities of the second kind**[1,7] iff its path function $\xi(t) = \xi(t, \omega)$ belongs to D almost surely. The definition of the class of random processes and the development of their theory was done by A.V. Skorokhod.

Many important classes of random processes are included in the class of processes without discontinuities of the second kind. The developed theory led to the unification of approaches to each such class analysis as well as enabling the achievement of many new results.

Let $u(x, y)$ be the distance between points x and y of the space X. According to **Chentsov Theorem** the inequality

$$\mathbf{E}\,\{u^p(\xi(t_1), \xi(t_2))\, u^q(\xi(t_2), \xi(t_3))\} \le c\,|t_3 - t_1|^{1+r}$$

for some $p > 0, \ q > 0, \ r > 0, \ c > 0$ is sufficient for a separable random process to be a process without discontinuities of the second kind. If a compatible family of finite-dimensional distributions is given and Chentsov's inequality is satisfied then it is possible to construct a random process without discontinuities of the second kind with prescribed distributions.

In applications it is usually possible to equate the two path functions $x(t)$ and $y(t)$ if $x(t \pm 0) = y(t \pm 0)$ for arbitrary $t \in I$. Thus, for definiteness, the continuity from the right can be assumed for path functions so that

$$\mathbf{P}\{\xi(t) = \xi(t+0), \ a \le t < b\} = 1.$$

Features of processes without discontinuities of the second kind are connected to the features of the space $D = D_{[a,b]}$. Let x and y be any functions belonging to D. Let us set

$$u_\varepsilon(t) = \inf_{|\tau - t| < \varepsilon} |x(t) - y(\tau)|, \quad u_\varepsilon = \sup_{a \le t \le b} \{u_\varepsilon(t) + \varepsilon\}.$$

Define the distance $\rho(x,y)$ between x and y by the value $\inf_{\varepsilon>0} u_\varepsilon$. (It can be shown that all the axioms of the metrics are satisfied.) A function $x \in D$ can be approximated by a piecewise constant function z so that $\rho(x,y) < \varepsilon$.

1.9 Convergence of Random Processes

In theory as well as in application it is often required to consider a process instead of another one, e.g. for analytic simplification. Hence an important role is played by limit theorems enabling the estimation of the limiting behavior of a random process under some asymptotic assumptions. We must first define a notion of the convergence of random processes[1,7,8].

Let X be a metric space and let $\mathbf{P}_n$, $\mathbf{P}$ be probability measures over X. The sequence of measures $\mathbf{P}_n$ is said **to converge to P in the weak sense** if for any bounded continuous function $f : X \to R$ the equality

$$\int_X f(x)\,\mathbf{P}_n(x) \underset{n\to\infty}{\longrightarrow} \int_X f(x)\,\mathbf{P}(x)$$

holds true. The weak convergence is denoted symbolically by

$$\mathbf{P}_n \xrightarrow[n\to\infty]{w} \mathbf{P}.$$

In the case $X = \mathbb{R}$ denote by $F_n(x)$, $F(x)$ the distribution functions associated with the measures $\mathbf{P}_n$, respectively $\mathbf{P}$. Then the weak convergence means the convergence of $F_n(x) \underset{n\to\infty}{\longrightarrow} F(x)$ at every continuity point x of the function $F(x)$.

A sequence of random processes $\{\xi_n(t),\ t \in I\}$ is said **to converge to the process** $\{\xi(t),\ t \in I\}$ **in the weak sense** if finite-dimensional distributions of $\xi_n(t)$ converge to those of $\xi(t)$ as $n \to \infty$ in the weak sense. Such a convergence is denoted as follows: $\xi_n \xrightarrow{w} \xi$.

Example 1.7. Let $\xi(t)$ be the Wiener process with a zero shift parameter and a unit diffusion parameter. For each $n \geq 1$ consider independent Gaussian random variables η_{nk}, $\mathbf{E}\,\eta_{nk} = 0$, $\mathbf{Var}\,\eta_{nk} = 1/n$. Set $\xi_n(0) = 0$, $\xi_n(k/n) = \eta_{n1} + \cdots + \eta_{nk}$. In each interval $((k-1)/n,\ k/n)$ define $\xi(t)$ by the linear interpolation. Then it can be shown that $\xi_n(t) \xrightarrow[n\to\infty]{w} \xi(t)$.

The weak convergence is a property concerning finite-dimensional distributions but not path functions of the processes. In a widely used case of the definition of both $\xi_n(t)$ and $\xi(t)$ over the same probability space so that

$$\xi_n(t) = \xi_n(t,\omega), \quad \xi(t) = \xi(t,\omega), \quad \omega \in \Omega,$$

it is important to investigate the pathwise deviation of $\xi_n(t)$ from $\xi(t)$.

The two following definitions of the pathwise convergence are of the most interest:

(i) **the uniform convergence in probability** consists of the equality

$$\sup_{t\in I} \rho(\xi_n(t),\ \xi(t)) \xrightarrow[n\to\infty]{w} 0,$$

that is

$$\mathbf{P}\left\{\rho(\xi_n(t,\omega),\ \xi(t,\omega)) < \varepsilon,\ \ t \in I\right\} \to 1,\ \ n \to \infty,$$

holds true for any $\varepsilon > 0$, ρ being the distance in the space X;

(ii) **the convergence in the Skorokhod J-topology** requiring that the following property be satisfied almost surely:

$$\rho(\xi_n(\cdot,\omega),\ \xi(\cdot,\omega)) \xrightarrow[n\to\infty]{w} 0$$

where ρ is the J-topology distance. The denotations $\xi_n(\cdot,\omega),\ \xi_n(\cdot,\omega)$ mean the consideration of the entire path function as a point of the functional space.

Example 1.8. Consider a device for the transmission of quantified signals. An input message $\xi(t)$ is received as an output message $\eta(t)$. Assume that $\xi(t) = 1$ in time intervals $(a_i,\ b_i)$, $\xi(t) = 0$ otherwise, and $\eta(t) = 1$ for $(A_i,\ B_i)$, $\eta(t) = 0$ otherwise. If $A_i - a_i,\ B_i - b_i$ are small but not vanishing variables then the J-topology distance between $\xi(t)$ and $\eta(t)$ is small too though the uniform distance equals 1. If it were not errors in the phase, so that $A_i = a_i,\ \ B_i = b_i$ and a small amplitude error were to exist, a small uniform deviation would be observed.

1.10 Invariance Principle

A researcher is not usually interested in a random process describing a phenomenon being investigated; he or she is rather interested in a real-valued functional of the process, e.g., the efficiency parameter of an engineering system. The question naturally arises to estimate the functional deviation under small deviations of path functions. (**A functional** of the process is a real-valued random variable defined by the path function.)

Let $\xi_n(t)$ and $\xi(t)$ be real-valued random processes, $f(x(\cdot))$ be a functional, $x(\cdot)$ be a symbol for a path function. Assume that $\xi_n \xrightarrow[n\to\infty]{w} \xi$. The variety of theorems asserting the statement $f(\xi_n(\cdot)) \xrightarrow[n\to\infty]{w} f(\xi(\cdot))$ under some additional conditions is called **the invariance principle**[1,8]. The following theorem by Yu.V. Prokhorov is one of the most known cases of the invariance principle.

Prokhorov Theorem. *Let $\xi_n(t)$ and $\xi(t)$ be real-valued random processes with continuous path functions defined over $I = [a,\ b]$. Assume that $\xi_n \xrightarrow[n\to\infty]{w} \xi$ and the condition*

$$\lim_{\Delta\to 0} \sup_{n>n_0} \mathbf{P} \left\{ \sup_{|t'-t''|<\Delta} |\xi_n(t') - \xi_n(t'')| > \varepsilon \right\} = 0 \tag{1.3}$$

is satisfied for any $\varepsilon > 0$. Let $f(\xi(\cdot))$ be such a functional of a path function that for an arbitrary continuous function $x(t)$ the relation

$$\sup_{\rho(y(\cdot),\, x(\cdot))\le\gamma} |f(y(\cdot)) - f(x(\cdot))| \underset{\gamma\to 0}{\longrightarrow} 0$$

holds true with $\rho(y(\cdot),\, x(\cdot)) = \max\limits_{a\le t\le b} |y(t) - x(t)|$, $y(t)$ being any continuous function over $[a,\ b]$. Under the stated conditions the sequence of distributions of the functionals $f(\xi_n(\cdot))$ converges to the distribution of the functional $f(\xi(\cdot))$ in the weak sense.

In other words, a qualitative sense of the additional conditions consists of the absence of valuable variations of a path function in all the time intervals of a length Δ and besides the continuity f over the set C of continuous $[a,\ b] \to \mathbb{R}$ functions.

According to **Prokhorov Theorem** the condition (1.3) is implied by a simpler condition

$$\mathbf{E}\ |\xi_n(t') - \xi_n(t'')|^{\delta} \le c\,|t' - t''|^{1+\rho},$$

c, ρ, δ being positive constants, t' and t'' being arbitrary points of the I.

Example 1.9. Let us consider the sequence of "random broken lines" $\xi_n(t)$ converging to the Wiener process; see example 1.7. Assume that $f(\xi(\cdot)) = \max\limits_{0\le t\le 1} \xi(t)$ and $f(\xi_n(\cdot)) = \max\limits_{0\le t\le 1} \xi_n(t)$. Let the inequality $|x(t)- -y(t)| \le \gamma$ be satisfied for $t \in [a,\ b]$. Set, for the definiteness, $a = x(c) = = \max x(t) = \max y(t) = b$. Then $y(c) \ge a - \gamma$ so that $b \ge a - \gamma$, hence $|f(x(\cdot)) - f(y(\cdot))| \le \gamma$. Therefore f is a uniformly continuous functional. Since $\xi_n(t') - \xi_n(t'')$ is a Gaussian random variable with a zero mean and a variance bounded by $|t' - t''|$ from above, an estimate $\mathbf{E}\ |\xi_n(t') - \xi_n(t'')|^4 = = O((t' - t'')^2)$ can be stated, similarly to one stated in section 1.7. According to Prokhorov Theorem the distribution of $\max\limits_{0\le t\le 1} \xi_n(t)$ satisfies the conditions of the invariance principle.

1.11 Ergodicity

The ergodicity of a random process[9,10] consists of the independence of some means related to the process of a chance. Let $\{\xi_n,\ n \geq 1\}$ be a process with discrete time, f be a real-valued function. The process is said to be **ergodic** with respect to the function f if for some set $\Omega' \subset \Omega$, $\mathbf{P}\{\Omega'\} = 1$,

$$\lim_{n\to\infty} \frac{1}{n} \sum_{k=1}^{n} f(\xi_k) = \bar{f}, \quad \omega \in \Omega',$$

$\bar{f}$ being a constant.

Example 1.10. Let $\{\xi_n\}$ be a sequence of independent identically distributed random variables, f being any real-valued function such that $\bar{f} = \mathbf{E}\, f(\xi_1)$ is finite. Then $\{\xi_n\}$ is ergodic with respect to the function f, due to the law of large numbers.

Let $\{\xi(t),\ t \geq 0\}$ be a random process in continuous time and let f be a real-valued function for which the integral $\int_0^T f(\xi(t))\, dt$ exists almost surely. The process $\xi(t)$ is said to be **ergodic** with respect to the function f if for a set $\Omega' \subset \Omega$ of a probability 1 the relation

$$\lim_{T\to\infty} \frac{1}{T} \int_0^T f(\xi(t))\, dt = \bar{f}$$

holds true with $\bar{f}$ being a constant. Let, particularly, A be a subset of X and $f_A(\cdot)$ be its indicator function, i.e. $f_A(x) = 1$ if $x \in A$, $f_A(x) = 0$ otherwise. Then the integral $\int_0^T f_A(\xi(t))\, dt$ equals the sojourn time in the set A within an interval $(0, T)$. If the process $\xi(t)$ is ergodic with respect to the function f_A then the proportion of time spent by a path function in the subset $A \subset X$ tends to a definite limit as $T \to \infty$.

Example 1.11. Let $\eta_1, \ldots, \eta_m$ be real-valued random variables and let $\lambda_1, \ldots, \lambda_m$ be distinct positive numbers. Let us define $\xi(t)$ by the equation $\xi(t) = \sum_{k=1}^{m} \eta_k \cos(\lambda_k t)$. Then

$$\frac{1}{T} \int_0^T \xi(t)\, dt = \frac{1}{T} \sum_{k=1}^{m} \frac{\eta_k}{\lambda_k} \sin(\lambda_k T) \underset{T\to\infty}{\longrightarrow} 0$$

holds true. Hence the process $\xi(t)$ is ergodic with respect to the function $f(\xi) = \xi$.

Example 1.12. Let $\xi(t)$ be the same as in the previous example and $f(\xi) = \xi^2$. We have

$$\frac{1}{T}\int_0^T \xi^2(t)\,dt = \frac{1}{2T}\sum_{k,j=1}^{m} \eta_k\eta_j \int_0^T (\cos(\lambda_k - \lambda_j)\,t +$$

$$+ \cos(\lambda_k + \lambda_j)\,t)\,dt \underset{T\to\infty}{\longrightarrow} \frac{1}{2}\sum_{k=1}^{m} \eta_k^2.$$

Hence the process $\xi(t)$ is ergodic with respect to the function $f(\xi) = \xi^2$ iff $\eta_1^2 + \cdots + \eta_m^2$ is a definite constant.

Example 1.13. Let θ be an irrational number and ω be a random variable uniformly distributed over the interval (0, 1). Define ξ_n by the equation

$$\xi_n = \{\omega + n\,\theta\}$$

where the braces denote the fractional part of a number. Set $f_a(x) = 1$ if $0 \le x < a$, $f_a(x) = 0$ if $x \ge a$. Then

$$\lim_{n\to\infty} \frac{1}{n}\,(f_a(\xi_1) + \cdots + f_a(\xi_n)) = a$$

holds true almost surely whichever a is.

The ergodicity is usually considered while investigating stationary processes; see sections 2.2 and 2.3. In this case $\mathbf{E}\,f(\xi(t))$ is independent of t, and $\bar{f} = \mathbf{E}\,f(\xi(t))$. The $\bar{f}$ is called **the average over the ensemble of realizations** of the random process. It can be estimated via a sequence of independent realizations of the process, i.e.

$$\bar{f} = p\lim_{n\to\infty} \frac{1}{n}\sum_{k=1}^{m} f(\xi_k(t)).$$

The ergodicity enables the computation of the same mean via an alone though infinitely durable realization, in other words via **the time average**

$$\lim_{T\to\infty} \frac{1}{T}\int_0^T f(\xi(t))\,dt.$$

The coincidence of both averages is the basis for many applied procedures.

Mathematical theorems stating the ergodicity of random processes under various conditions are spoken of as **ergodic theorems**. They form **the crgodicity theory**.

A stronger property is also investigated in the framework of the ergodicity theory. This property consists of the limiting independence of the process states at time t and at time τ as $t-\tau \to \infty$. Such an ergodic property is usually considered while investigating Markov and related random processes (Ch. 3 and subsequent ones). Let $\xi(t)$ be a random process in a discrete or continuous time. Let us set

$$P_t(A\,|\,x) = \mathbf{P}\,\{\xi(t) \in A \mid \xi(0) = x\}.$$

The ergodicity consists of the relation

$$P_t(A|x) \underset{t\to\infty}{\longrightarrow} \pi(A),\ x \in X,$$

being satisfied for each measurable set $A \subset X$; $\pi(A)$ is assumed to be a probability measure over X called **the ergodic distribution** of the process $\xi(t)$.

Example 1.14. Let $\{\eta_n,\ -\infty < n < \infty\}$ be a sequence of independent identically distributed random variables with a finite mean and a characteristic function $\psi(t)$, $\xi_n = \sum\limits_{k=0}^{\infty} \rho^k\, \eta_{n-k}$, where $|\rho| < 1$. Assuming that $\xi_0 = x$ we obtain

$$\xi_n = \sum_{k=0}^{n-1} \rho^k\, \eta_{n-k} + \rho^n x.$$

Therefore the dependence of ξ_n upon x vanishes as $n \to \infty$. For each Borel set A

$$\mathbf{P}\,\{\xi_n \in A\} \underset{n\to\infty}{\longrightarrow} \pi(A),$$

where $\pi(A)$ can be defined through the characteristic function

$$\varphi(\lambda) = \mathbf{E}\, e^{i\lambda\xi_n} = \prod_{k=0}^{n-1} \psi(\rho^k t).$$

Example 1.15. Let η_n be independent random variables; $\mathbf{P}\,\{\eta_n = 1\} = p$, $\mathbf{P}\,\{\eta_n = 0\} = q = 1-p$. Set $\xi_n = 0$ if $\eta_1 + \cdots + \eta_n$ is even, $\xi_n = 1$ if $\eta_1 + \cdots + \eta_n$ is odd. We have

$$\mathbf{P}\,\{\xi_n = 0 \mid \xi_0 = 0\} = \frac{1}{2}(1+\Delta^n),\quad \mathbf{P}\,\{\xi_n = 1 \mid \xi_0 = 0\} = \frac{1}{2}(1-\Delta^n),$$

$$\mathbf{P}\,\{\xi_n = 0 \mid \xi_0 = 1\} = \frac{1}{2}(1-\Delta^n),\quad \mathbf{P}\,\{\xi_n = 1 \mid \xi_0 = 1\} = \frac{1}{2}(1+\Delta^n)$$

where $\Delta = q - p$. We observe that the ergodicity holds true in each case with the exception of two cases: $p = 1$ and $p = 0$.

1.12 References

1. Gihman, I. I., Skorohod, A. V., *Introduction to the Theory of Random Processes,* Saunders, Philadelphia, 1969.
2. Doob, J. L., *Stochastic Processes,* Wiley, New York, 1967.
3. Hoel, P. G., Port, S. C., Stone, C. J., *Introduction to Stochastic Processes,* Houghton Mifflin, Boston, 1972.
4. Kelly, D. G., *Introduction to Probability,* Macmillan Publ., New York, 1994.
5. Ross, S. M., *A First Course in Probability,* Macmillan Publ., New York, 1976.
6. Ross, S. M., *Introduction to Probability Models,* Academic Press, London, 1985.
7. Skorohod, A. V., *Random Processes with Independent Increments,* Kluwer Acad. Publ., Dordrecht, 1991.
8. Billingsley, P., *Convergence of Probability Measures,* Wiley, New York, 1968.
9. Gnedenko, B. V., *The Theory of Probability,* Chelsea Publ., New York, 1968.
10. Širjaev, A. N., *Probability,* Springer, New York, 1984.

Chapter 2

Classification of Random Processes

2.1 State Space and a Parametric Set

A random process can be considered as a random mapping: $I \to X$ so that each $\xi(t)$, $t \in I$, can be realized as a point of the set X. If X is either finite or countable then the process is called **a discrete (finite or countable) random process**. If X is neither finite nor countable then $\xi(t)$ is called **a process with a continuous state space**. If $I = \{t_n\}$ then the random process is called **a process with discrete time (parameter)**. If I is an interval of the time axis $\mathbb{R}$ then $\xi(t)$ is called **a random process with continuous time (parameter)**. If I is an r-dimensional set with $r \geq 2$ then $\xi(t)$ is called **a random field**. An example of a random field: the temperature at a point (λ, φ) of the earth surface at time t; in this case $r = 3$. The processes with a discrete parameter are used in the two main situations: (i) if time is discrete in its nature as, for example, if ξ_n is number of descendants of an individual in the n-th generation, and (ii) if it is convenient to consider a discrete sample of a process with continuous time; for example, the temperature in a controlled technological device is measured each τ seconds.

2.2 Stationary in the Broad Sense Random Processes

In this section we shall consider random processes defined over a set $I \in \mathbb{R}$. A real-valued random process $\{\xi(t),\ t \in I\}$ is said to be **stationary in the broad sense** if

$$\mathbf{Var}\,\xi(t) < \infty,\ \ \mathbf{E}\,\xi(t) = m(t) = a,\ \ \mathbf{cov}\,(\xi(t), \xi(\tau)) = R(t-\tau),\ t \in I,\ \tau \in I.$$

In other words, the mean function is constant and the covariance function depends only on the difference $t - \tau$. The constant a is called **mean value** of the stationary process. The function $R(t)$ is called **the covariance function** of the stationary process. (The words "in the broad sense" are often omitted in cases where no confusion can occur.)

A complex-valued random process $\{\xi(t),\ t \in I\}$ is said to be **stationary in the broad sense** if

$$\mathbf{Var}\,\xi(t) < \infty, \quad \mathbf{E}\,\xi(t) = m(t) = a,\ t \in I,$$

$$\mathbf{E}\,(\xi(t) - a)\,\overline{(\xi(\tau)) - a)} = R(t - \tau),\ t \in I,\ \tau \in I.$$

Example 2.1. Let ξ_k, η_k be pairwise zero-correlated random variables, $\mathbf{E}\,\xi_k = \mathbf{E}\,\eta_k = 0$, $\mathbf{Var}\,\xi_k = \mathbf{Var}\,\eta_k = b_k^2$. Define a random process $\xi(t)$ via the expression

$$\xi(t) = \sum_{k=1}^{n} (\xi_k \cos \lambda_k t + \eta_k \sin \lambda_k t), \quad -\infty < t < \infty,$$

where λ_k are distinct positive numbers. We have

$$\mathbf{E}\,\xi(t) = 0, \quad \xi^0(t)\,\xi^0(\tau) = \xi(t)\,\xi(\tau) =$$

$$= \sum_{k=1}^{n} (\xi_k^2 \cos \lambda_k t \cos \lambda_k \tau + \eta_k^2 \sin \lambda_k t \sin \lambda_k \tau) + \psi(t, \tau),$$

where $\psi(t, \tau)$ is a sum of products of zero-correlated variables $\xi_k \xi_j$, $\eta_k \eta_j$, $\xi_k \eta_j$. Hence $\mathbf{E}\,\psi(t, \tau) = 0$ and

$$R(t, \tau) = \mathbf{E}\,\xi^0(t)\,\xi^0(\tau) = \sum_{k=1}^{n} b_k^2 \cos \lambda_k (t - \tau) = R(t - \tau).$$

Therefore the process $\xi(t)$ is stationary.

Example 2.2. Let $\zeta_1, \ldots, \zeta_n$ be complex-valued random variables with finite variances, and

$$\mathbf{E}\,\zeta_k = 0, \quad 1 \le k \le n, \quad \mathbf{E}\,(\zeta_k\,\bar{\zeta}_j) = 0 \ \text{ for } \ j \ne k,$$

$$\mathbf{E}\,|\zeta_k|^2 = \mathbf{E}\,(\zeta_k\,\bar{\zeta}_k) = b_k^2, \quad 1 \le k \le n.$$

Define a complex-valued random process

$$\xi(t) = \sum_{k=1}^{n} \zeta_k\, e^{i\lambda_k t}, \quad -\infty < t < \infty,$$

where $\{\lambda_k\}$ are distinct real numbers. Then

$$\mathbf{E}\,\xi(t) = \sum_{k=1}^{n} \mathbf{E}\,\zeta_k\, e^{i\lambda_k t} = 0, \quad R(t,\tau) = \mathbf{E}\,\xi^0(t)\,\overline{\xi^0(\tau)} = \mathbf{E}\,\xi(t)\,\overline{\xi(\tau)} =$$

$$= \sum_{k=1}^{n}\sum_{j=1}^{n} \mathbf{E}\,\zeta_k\,\bar{\zeta}_j\; e^{i(\lambda_k t - \lambda_j \tau)} = \sum_{k=1}^{n} \mathbf{E}\,\zeta_k\,\bar{\zeta}_k\; e^{i\lambda_k(t-\tau)} = \sum_{k=1}^{n} b_k^2\, e^{i\lambda_k(t-\tau)}.$$

The process is stationary.

Example 2.3. Let ω be a real-valued random variable with the distribution function $F(x)$, and φ be a random variable uniformly distributed over the interval $(0,\ \pi)$. We assume that ω and φ are independent. Define a random process

$$\xi(t) = \sigma\, e^{i(\omega t - \varphi)}, \quad -\infty < t < \infty.$$

Then

$$\mathbf{E}\,\xi(t) = \frac{\sigma}{2\pi}\int_{-\infty}^{\infty} e^{ixt}\,dF(x)\int_{0}^{2\pi} e^{-iy}\,dy = 0,$$

$$\mathbf{E}\,\xi(t)\,\overline{\xi(\tau)} = \sigma^2\,\mathbf{E}\,e^{i\omega(t-\tau)} = \sigma^2\,\psi(t-\tau),$$

where $\psi(t)$ is the characteristic function of the variable ω. We obtain a stationary process.

A multidimensional random process $\xi(t) = (\xi_1(t),\ \ldots,\ \xi_m(t))$ with $\mathbf{Var}\,\xi_i(t) < \infty$, $t \in I$, is said to be **stationary in the broad sense** if $\mathbf{E}\,\xi(t) = a = (a_1, \ldots, a_m)$ and $\mathbf{cov}\,(\xi_i(t), \xi_j(\tau)) = R_{ij}(t-\tau)$ for $t \in I$, $\tau \in I$. The matrix function $R(t) = (R_{ij}(t))$ is called **the covariance matrix function** of the multidimensional random process $\xi(t)$. Real random processes $\xi(t)$ and $\eta(t)$ stationary in the broad sense are said to be **stationary connected** if $\mathbf{cov}\,(\xi(t), \eta(\tau)) = R_{\xi\eta}(t-\tau)$. The function $R_{\zeta\eta}(\tau)$ is called **joint covariance function** of $\xi(t)$ and $\eta(t)$. Hence the components of a multidimensional stationary process are stationary connected. A multidimensional complex-valued random process stationary in the broad sense is defined in a similar way.

2.3 Stationary Random Processes and Random Processes with Stationary Increments

Let $\{\xi(t),\ t \in I\}$ be a random process with a measurable phase-state X, $I \subset \mathbb{R}$. The process is said to be **stationary in the strict sense** if its finite-dimensional distributions are invariant of time shift, so that for any $n \geq 1$ and arbitrary measurable sets $A_i \subset X$

$$\mathbf{P}\{\xi(t_1+\tau) \in A_1,\, \dots,\, \xi(t_n+\tau) \in A_n\} = \mathbf{P}\{\xi(t_1) \in A_1,\, \dots,\, \xi(t_n) \in A_n\}$$

if $t_i \in I,\ t_i+\tau \in I,\ 1 \leq i \leq n$. In the case where no confusion can occur we use a shorter term "stationary process" omitting the words "in the strict sense".

If $X \subset \mathbb{R}$ or $X \subset \mathbb{R}^m$ then the strict stationarity condition can be expressed via finite-dimensional distributions as follows:

$$F_{t_1,\dots,t_n}(x_1,\, \dots,\, x_n) = F_{t_1+\tau,\dots,t_n+\tau}(x_1,\, \dots,\, x_n).$$

If $\{\xi(t),\ t \in I\}$ is a one-dimensional (m-dimensional) random process $\xi(t)$ stationary in the strict sense and moreover $\mathbf{Var}\,\xi(t) < \infty$, $t \in I$ (the variances of the components of the $\xi(t)$ are finite for $t \in I$) then $\xi(t)$ is stationary in the broad sense also. At the same time, the stationary in the broad sense (concerning only two-dimensional distributions) does not imply the stationarity in the strict sense.

If $I = \{n\,\Delta + a,\ n \geq 0\}$ or $I = \{n\,\Delta,\ -\infty < n < \infty\}$ then the stationary random process $\{\xi(t),\ t \in I\}$ is called **a stationary random sequence** (in the strict sense).

Example 2.4. Let $\{\eta_n\}$ be a sequence of independent random variables, $\mathbf{P}\{\eta_n = 1\} = p$, $\mathbf{P}\{\eta_n = 0\} = q = 1 - p$. Set $\xi_n = \eta_{n-1}\eta_n$. The defined sequence $\{\xi_n\}$ is stationary. Indeed, a joint probability $\mathbf{P}\{\xi_{i_1} = j_1,\, \dots,\, \xi_{i_n} = j_n\}$ equals the probability of some event concerning $(\eta_{i_1-1}, \eta_{i_1},\, \dots,\, \eta_{i_n-1}, \eta_{i_n})$. Obviously this probability is independent of time shifting.

Let $\xi(t)$ be a one-dimensional or m-dimensional random process. Consider any $n \geq 1$ and arbitrary $t_0 < t_1 < \cdots < t_n$ and set $\eta_k = \xi(t_k) - \xi(t_{k-1})$, $1 \leq k \leq n$. If the distributions of random variables $(\eta_1,\, \dots,\, \eta_n)$

are invariant of a transformation $(t_0, t_1, \dots, t_n) \to (t_0+\tau, t_1+\tau, \dots, t_n+\tau)$ then the process is said to be **a random process with stationary increments.**

Example 2.5. Let $\zeta(t)$ be a real-valued random process stationary in the strict sense with Rieman-integrable path functions. Consider a new process

$$\xi(t) = \int_0^t \zeta(x)\, dx.$$

The $\xi(t)$ is a process with stationary increments. Indeed its increments $\eta_k(\tau) = \int_{t_{k-1}+\tau}^{t_k+\tau} \zeta(x)\, dx$ can be expressed as the limits of integral sums of the form $\Delta \sum_i \zeta(t_{k-1} + \tau + i\Delta)$ over appropriate intervals assuming Δ to tend to zero. The integral sums are distributed independently of τ; so are the limits in probability.

A complex random process $\{\xi(t),\ t \in I\}$ possessing a finite variance is said to be **a random process with orthogonal increments** if for arbitrary $t_i \in I$, $t_0 \le t_1 \le t_2 \le t_3$

$$\mathbf{E}\, (\xi(t_1) - \xi(t_0))\, \overline{(\xi(t_3) - \xi(t_2))} = 0.$$

2.4 Random Processes with Independent Increments

Let $\{\xi(t),\ t \in I\}$ be a one-dimensional or m-dimensional random process. Consider its increments $\eta_1 = \xi(t_1) - \xi(t_0), \dots, \eta_n = \xi(t_n) - \xi(t_{n-1})$. The process $\xi(t)$ is called **a random process with independent increments** if the random variables $\xi(t_0), \eta_1, \dots, \eta_n$ are independent for arbitrary $n \ge 0$ and $0 \le t_0 \le t_1 \le \dots \le t_n$.

Example 2.6. The Wiener process defined in section 1.7 is a process with independent increments.

Example 2.7. A random process $\{\xi(t),\ t \in I\}$ with independent increments is said to be **a Poisson process** if $\xi(0) = 0$ and an increment $\xi(t) - \xi(\tau)$ is Poisson distributed with parameter $\Lambda(t) - \Lambda(\tau)$ for arbitrary $0 \le \tau \le t$. The nondecreasing function $\Lambda(t)$ is called **the leading function** of the Poisson process.

The most widely used **stationary Poisson process** is characterized by the identity $\Lambda(t) - \Lambda(\tau) = \lambda(t - \tau)$; a positive parameter λ is called

the parameter of the stationary Poisson process. Thus we have

$$\mathbf{P}\{\xi(t_0+t)-\xi(t_0)=k\}=e^{-\lambda t}(\lambda t)^k/k!,\quad k\geq 0,$$

for arbitrary $t_0 \geq 0$ and $t \geq 0$.

2.5 Point Processes. Memoryless Property

Let $(\Omega,\ \mathfrak{A},\ \mathbf{P})$ be a probability space, and $x_n = x_n(\omega)$ be a random sequence of points $x_n \in \mathbb{R}^m$. For each subset $A \subset \mathbb{R}^m$ denote by $N(A)$ the number of points $x_n \in A$.

If $N(A)$ is finite almost surely for an arbitrary bounded set A, then the random sequence $\{x_n\}$ is called **a point process** in the space $\mathbb{R}^m$.

Instead of the sequence $\{x_n\}$ the point process can be considered as **the random measure** $\{N(A),\ A \subset \mathbb{R}^m\}$, where $N(A) = N(A,\omega) = \sum\limits_{n:\, x_n \in A} 1$. The random measure is integer-valued, nonnegative and σ-additive so that

$$N\Big(\bigcup_{k=1}^{\infty} A_k\Big) = \sum_{k=1}^{\infty} N(A_k)$$

for any disjoint sets $A_k \subset \mathbb{R}^m$ almost surely.

Consider hereinafter point processes in $\mathbb{R}$ or $\mathbb{R}^+ = (0,\ \infty)$. Such point processes are also called **random flows of events**. Each such point process can be defined in each of the following ways:

(i) as a random sequence $\{t_n\}$ usually ordered so that $t_n \leq t_{n+1}$ almost surely,

(ii) as a counting process

$$N(t) = \begin{cases} \sum\limits_{n:\, 0\leq t_n<t} 1 & \text{if } t \geq 0 \\ -\sum\limits_{n:\, t\leq t_n<0} 1 & \text{if } t < 0. \end{cases}$$

The definition of the point process implies the finiteness of the $N(t)$ almost surely. With each t_n one can associate an "event" of the point process at time t. The number of events in a half-interval $[a,\ b)$ equals $N(b) - N(a)$. If $t_n = t$ we shall say that at time t the n-th event of the random flow occurs. A case of r simultaneous events at time t may also happen, so that $N(t+0) - N(t) = r$.

Let $T_0 < T_1 < \cdots < T_n$ be fixed time epochs. Consider random variables $\eta_1 = N(T_1) - N(T_0)$, ..., $\eta_n = N(T_n) - N(T_{n-1})$. These variables equal numbers of events in half-intervals $[T_0, T_1)$, ..., $[T_{n-1}, T_n)$.

A point process is a well-defined random process. It can be defined in the probabilistic sense via each of the following parameters.

(i) The family of finite-dimensional distributions of the counting process $N(t)$ defined via

$$\mathbf{P}\{N(T_1) = k_1, \ldots, N(T_n) = k_n\}.$$

(ii) The family of the finite-dimensional distributions of the process $N(t)$ increments via

$$\mathbf{P}\{N(T_i) - N(T_{i-1}) = j_i, \ 1 \le i \le n\}.$$

(iii) The family of the finite-dimensional distributions of the sequence $\{t_n\}$. For example, in the case $\{n\} = \{1, 2, \ldots\}$ it suffices to define joint distribution functions

$$F_n(T_1, ..., T_n) = \mathbf{P}\{t_k \le T_k, \ 1 \le k \le n\}.$$

(iv) The family of interevent distributions. For the case $\{n\} = \{1, 2, \ldots\}$, $t_1 \le t_2 \le ...$ it suffices to define joint distribution functions

$$\mathbf{P}\{z_k \le T_k, \ 1 \le k \le n\}$$

where $z_1 = t_1$, $z_k = t_k - t_{k-1}$, $k \ge 2$.

There also exist some other ways of a point process definition.

A point process $\{t_n\}$ in $\mathbb{R}$ is said to be **stationary** iff $N(t)$ is a random process with stationary increments so that a multidimensional random variable

$$(N(T_k + \tau) - N(T_{k-1} + \tau), \ 1 \le k \le n)$$

is distributed independently of τ.

If $(N(T_k) - N(T_{k-1}), \ 1 \le k \le n)$ are independent for arbitrary $n \ge 2$ and $T_1 \le T_2 \le ... \le T_n$ then the point process is said to be **a memoryless point process**. In other words, in this case $N(t)$ is a process with independent increments.

A point process $\{t_n\}$ in $\mathbb{R}^m$ is called **simple** if t_n are disjoint points almost surely.

Consider an interval (a, b) and divide it into n subintervals $(a, a + \Delta)$, $[a + \Delta, \ a + 2\Delta)$, ..., $[a + (n-1)\Delta, \ b]$, where $\Delta = (b - a)/n$. Define by $\psi_2(n)$ the probability of such a complex event: {At least one of the n

subintervals contains at least two moments of the point process events}. The point process is simple iff $\psi_2(n) \underset{n\to\infty}{\longrightarrow} 0$ for arbitrary a and b.

Let $N(t)$, $t \geq 0$ be a stationary Poisson process with the parameter λ. The $N(t)$ can be considered as the counting process for a point process being called **the simplest (or stationary Poisson) point process (or flow)** with parameter λ.

The following properties hold true for the simplest point process[1]:

(i) the random variables t_1, $t_2 - t_1$, ..., $t_n - t_{n-1}$, ... are independent, their common probability density function is $\lambda e^{-\lambda t}$, $t \geq 0$,

(ii) the point process is memoryless,

(iii) the number of events in an interval (t_0, t) is a Poisson random variable with the parameter $\lambda(t - t_0)$ for $0 \leq t_0 \leq t$,

(iv) the mean number of events in an interval (t_0, t) equals $\lambda(t - t_0)$,

(v) the probability of an event occuring within the interval $(t, t+h)$ equals $\lambda h + o(h)$, $h \to 0$,

(vi) the probability of at least two events within the same interval equals $o(h)$,

(vii) if T is a fixed time epoch, $T \geq 0$ then the time from T to the next event epoch is exponentially distributed with the parameter λ.

2.6 Markov Processes

Let $(\xi_1, ..., \xi_n)$ be a multidimensional random variable, ξ_i being one-dimensional or multidimensional variables, in their turn. One can write:

$$\mathbf{P}\{\xi_1 \in A_1, \xi_2 \in A_2\} = \int_{A_1} dF_1(x_1) \int_{A_2} dF_2(x_2|x_1),$$

$$\mathbf{P}\{\xi_1 \in A_1, \xi_2 \in A_2, \xi_3 \in A_3\} =$$

$$= \int_{A_1} dF_1(x_1) \int_{A_2} dF_2(x_2|x_1) \int_{A_3} dF_3(x_3|x_1, x_2),$$

. . . .

$$\mathbf{P}\{\xi_1 \in A_1, \ldots, \xi_n \in A_n\} = \int_{A_1} dF_1(x_1) \int_{A_2} dF_2(x_2|x_1) \quad \ldots$$

$$\ldots \int_{A_n} dF_n(x_n|x_1, \ldots, x_{n-1})$$

for arbitrary Borel sets $A_1, \dots, A_n$ where $F_1(x_1)$ is the distribution function of the random variable ξ_1; $F_k(x_k, |x_1, \dots, x_k)$ is the conditional distribution function of ξ_k given $\xi_1 = x_1, \dots, \xi_{k-1} = x_{k-1}$. The random variables $\xi_1, \dots, \xi_n$ are said to be tied in **a Markov chain** if the following condition is satisfied:

$$\mathbf{P}\{\xi_1 \in A_1, \dots, \xi_n \in A_n\} = \int_{A_1} dF_1(x_1) \int_{A_2} dF_2(x_2|x_1) \dots \int_{A_n} dF_n(x_n|x_{n-1})$$

where $F_k(x_k|x_{k-1})$ denotes the conditional distribution function of ξ_k given $\xi_{k-1} = x_{k-1}$. The condition of the markovity can be also expressed via equation

$$F_k(x_k|x_1, \dots, x_{k-1}) = F_k(x_k|x_{k-1}), \qquad 2 \le k \le n.$$

A random process $\xi(t)$, $t \in I$, defined over a set $I \subset \mathbb{R}$ is called **a Markov process**[1−3] if for arbitrary $t \ge 2$ and $t_i \in I$, $t_1 < t_2 < \dots < t_n$ the random variables $\xi(t_1)$, $\xi(t_2)$, ..., $\xi(t_n)$ are tied in a Markov chain. The function $P(t_0, x; t, y) = \mathbf{P}\{\xi(t) \in A \mid \xi(t_0) = x\}$ defined for $t \ge t_0$, $t \in I$, $t_0 \in I$, $x \in X$, where A is an arbitrary Borel set, is called **the Markov transition probability function**. This term is also kept for the conditional distribution. For arbitrary Borel sets A_k the following equation holds true:

$$\mathbf{P}\{\, \xi(t_1) \in A_1, \dots, \xi(t_n) \in A_n \,\} = \int_{A_1} dF_{t_1}(x_1) \int_{A_2} dF(t_1, x_1; t_2, x_2) \dots$$

$$\dots \int_{A_n} dF(t_{n-1}, x_{n-1}; t_n, x_n).$$

In the last equation the differential always concerns the last argument of the function.

A Markov process with a finite or countable state space X is called **a Markov chain**. As concerns the parametric set I, two cases are usually considered: **discrete time Markov chains** for which $I = \{0, 1, 2, \dots\}$ or, sometimes, $I = \{0, \Delta, 2\Delta, \dots\}$, and **continuous time Markov chains** with $I = \{t \ge 0\}$.

The definition of a Markov chain was developed by A.A.Markov. Traditionally a Markov chain was defined for outcomes of trials rather than random variables. Let us consider a sequence of trials $U_0, U_1, \dots$ with possible outcomes $i, j, \dots$ from a finite or countable set X. The outcomes of the trials are said to be tied in a Markov chain if the following property (called **the Markov property**, or **the property of markovity**) is satisfied: given the outcomes $\omega_0, \omega_1, \ldots, \omega_{n-1}$ of trials $U_0, U_1, \ldots, U_{n-1}$ the outcome of trial U_n depends only on ω_{n-1} but not on $\omega_0, \ldots \omega_{n-2}$. The same property can be expressed in an equivalent form as the independence of the past and the future if the present is known. Thus, let A and B be arbitrary events defined by outcomes of the trials $U_0, \ldots U_{n-1}$ and, respectively, of the trials $U_{n+1}, U_{n+2}, \ldots$. Then given an outcome i of the trial U_n we have

$$\mathbf{P}\{AB \,|\, i\} = \mathbf{P}\{A \,|\, i\} \, \mathbf{P}\{B \,|\, i\}.$$

A Markov process is called **time-homogeneous** if $P(t_0, x; t, A) = P_{t-t_0}(x; A)$. A discrete time Markov chain is called **time-homogeneous** if $\mathbf{P}\{\xi_{n+m} = j \mid \xi_m = i\} = p_{ij}^{(n)}$. In other words, the dependence on two time arguments $n \geq 0$ and $m \geq 0$ can be reduced to n only. A continuous time Markov chain is called **time-homogeneous** if

$$\mathbf{P}\{\xi(t_0 + t) = j \mid \xi(t_0) = i\} = p_{ij}(t)$$

for each $t_0 \geq 0$ and $t \geq 0$.

A Markov process $\xi(t)$ with the phase-space $\mathbb{R}^m$ is called **space-homogeneous** if $F(t_0, x; t, y) = F(t_0, x + z; t, y + z)$ for arbitrary $z \in \mathbb{R}^m$.

A space-homogeneous Markov chain is called **a random walk**. If moreover the Markov chain is time-homogeneous the random walk is called **a time-homogeneous random walk**. If $\{\eta_n\}$ is a sequence of independent identically distributed random variables then the sequence of sums $s_n = \eta_1 + \cdots + \eta_n$ is a time homogeneous random walk.

The family of finite-dimensional distributions of a Markov process is uniquely defined via the Markov transition probability function $P(t_0, x; t, a)$ and **the initial distribution** $P_0(A) = \mathbf{P}\{\xi(0) \in A\}$. This distribution can be defined via the distribution function $F_0(x) = \mathbf{P}\{\xi(0) < x\}$. In the case of a Markov chain it suffices to define **initial probabilities** $p_i^{(0)} = \mathbf{P}\{\xi(0) = i\}$.

2.7 Semi-Markov Processes

The concept of a semi-Markov process is not more general then the concept of a Markov process but it is convenient in many applied problems. The concept is due to P. Lévy.

Consider a system which can possess a state $\nu(t) \in X$ for any $t \geq 0$. The X is assumed to be a nonempty measurable set. A path function of the process $\nu(t)$ which is defined in what follows is assumed to be a stepwise right-continuous function almost surely. The construction of such a function means the definition of two sequences $\{\nu_n, n \geq 0\}$ and $\{z_n, n \geq 0\}$ where ν_0 is the initial state; ν_n is the state after the nth transmission ("jump") for $n \geq 1$; z_0 is the sojourn time in the initial state and z_n is the time between the nth and the $(n+1)$th transmissions for $n \geq 0$. The case $\nu_n = \nu_{n+1}$ is not excluded, so that a "jump" may be a zero value.

A stepwise-constant right-continuous random process $\{\nu(t),\ t \geq 0\}$ with a measurable phase-space X is called **a semi-Markov process (SMP)** if the following property holds true. Let the nth transmission occur at time t and the system state is known to be $i \in X$ after it. Let moreover an additional information concerning the process path function in the time interval $[0,\ t)$ be available. Then an event $\{z_n \leq z,\ \nu_{n+1} \in A\}$ can occur with the probability $Q_i(z, A)$ for an arbitrary measurable set $A \subset X$. Nonnegative nondecreasing functions $Q_i(z, A)$ satisfy the equalities

$$Q_i(\infty, A) = 1, \quad i \in X.$$

Having in mind the majority of applications we confine ourselves to considering a SMP with a finite or countable phase-space X. In this case we shall denote the probability of the event $\{z_n \leq z,\ \nu_{n+1} = j\}$ by $Q_{ij}(z)$ under the above formulated conditions. The matrix function $Q(z) = Q_{ij}(z)$ is called **the Q-matrix** of the SMP.

An alternative approach leads to the same concept of the SMP. Assume that some factors control the behavior of the process $\nu(t)$. As soon as the process enters a state i at time τ, some operations start. An operation j terminates in an interval $(\tau + t,\ \tau + t + dt)$ with a probability $\lambda_{ij}(t)dt$ if no operations terminate in the interval $(\tau,\ t+\tau)$. In the case the operation j terminates the process jumps to the state j. Thus there is a competition of operations, and the winner carries the system away making it jump to its state. The following equations relate $\lambda_{ij}(t)$ to $Q_{ij}(t)$. Set

$$\lambda_i(t) = \sum_j \lambda_{ij}(t).$$

Then

$$dQ_{ij}(t) = \lambda_{ij}(t)\ \exp\{-\int_0^t \lambda_i(x)dx\}.$$

Conversely, having denoted

$$Q_i(t) = \sum_j Q_{ij}(t)$$

one obtains

$$\lambda_{ij}(t)dt = \frac{dQ_{ij}(t)}{1 - Q_i(t)}$$

provided that $Q_{ij}(t)$ is absolutely continuous.

Example 2.8. A component of an engineering system is subjected to a random failure. The time to failure ξ has a distribution function $F(x)$. A failed component is being repaired during a random time η with a distribution function $H(x)$. If a random time ζ has elapsed after switching a component in and the component has not failed within this time, a planned change of the component to a new one is done. Denote by $G(x)$ the distribution function of ζ.

A random process describing the behavior of the system has two states:
State 0: a component is operating,
State 1: the repair is being done.
We have:

$$Q_{10}(x) = H(x), \quad Q_{11}(x) = 0,$$

$$Q_{01}(x) = \int_0^x (1 - G(t))dF(t), \quad Q_{00}(x) = \int_0^x (1 - F(t))dG(t).$$

As has been stated, the behavior of an SMP after its first transition can be characterized by $\{Q_{ij}(x)\}$ or by $\{\lambda_{ij}(x)\}$. Its behavior up to the first transition can be characterized by constants $\{p_i^{(0)}\}$ and functions $\{Q_{ij}^{(0)}(x)\}$. They can be explained in such a way. The initial state of SMP at time 0 is chosen at random; this state is i with the probability $p_i^{(0)}$. If this event really happens then the next transition will happen in a time interval $[0, t]$ and the next state will be j with the probability $Q_{ij}^{(0)}(x)$.

The Markov chain $\{\nu_n\}$ having appeared in the definition of an SMP is called its **embedded Markov chain**. The concept of embedded Markov chains has been established by D. Kendall who has also developed some approaches to analyze queueing systems by this method.

2.8 Renewal Processes (Recurrent Point Processes)

Let $\{\xi_n, n \geq 1\}$ be a sequence of independent nonnegative random variables. Assume that ξ_n, $n \geq 2$, has a distribution function $F(x)$ and ξ_0 has a distribution function $F_0(x)$. Define the sequence of sums $s_n = \xi_1 + \cdots + \xi_n$. Such a sequence $\{s_n, n \geq 1\}$ is called **a renewal process**[4,5] **(recurrent point process, recurrent flow)**. If $F_0(x) \neq F(x)$ the renewal (recurrent point) process is called **delayed**. The formation of this concept is very fruitful because the renovation is pertinent to many real processes. For example, the process of failures of a tool is recurrent if each failed tool is immediately changed by a new one similar to it in the probabilistic sense. A special role of $F_0(x)$ can be explained with the fact that the origin may be either synchronized or nonsynchronized with the renewal process.

A recurrent process is **ordinary** if $F_0(0) = F(0) = 0$.

Sometimes it is convenient to assume a positive probability for the event $\xi_n = \infty$. In such a case the renewal process $\{s_n\}$ consists of a finite random number of points.

Such a process is called **a generalized (broken) renewal process**. The random variable $s^* = \max\{s_n : s_n < \infty\}$ is called **the break point**.

A path function $\{s_n,\ n \geq 0\}$ defines a continuous time process $N(t)$ where $N(t)$ is the number of renewals in a half interval $[0, t)$.

A stationary simple recurrent process is characterized by the following conditions:

(i) $F(0) = 0$,

(ii) $\tau = \int\limits_0^\infty x dF(x) \ < \ \infty$,

(iii) $dF_0(x) = \frac{1}{\tau}(1 - F(x))dx$.

2.9 Regenerative Processes

Define first a breaking random process, that is, a process with a chance-dependent duration. Let $(\Omega,\ \mathfrak{A},\ \mathbf{P})$ be a probability space. Define a pair $(\tau(\omega),\ f(t,\omega),\ 0 \leq t < \tau(\omega))$, i.e., the duration of the process and its path function. It is assumed that $\tau(\omega)$ is a positive random variable. As to the function $f(t,\omega)$, we assume that its values belong to a measurable space $(X,\ \mathfrak{S})$. For any $A \in \mathfrak{S},\ t \geq 0$, the event $\{\tau(\omega) > t,\ f(t,\omega) \in A\}$ belongs to the main σ-algebra $\mathfrak{A}$ of events for which a probability is defined. This assumptional distribution functions

$$F_{t_1,\dots,t_n}(A_1,\dots,A_n) = \mathbf{P}\{\tau(\omega) > \max_{1\leq i\leq n}\{t_i\};\ f(t_i,\omega) \in A_i,\ 1 \leq i \leq n\}.$$

A breaking random process is said to be **separable** with respect to a system of sets $\mathfrak{B}$ if there exists such a countable set Λ called **the separability set** that for arbitrary $A \in \mathfrak{B}$ the relations $\{\tau(\omega) > b,\ f(t,\omega) \in A,\ a \leq t \leq \leq b\}$ and $\{\tau(\omega) > b,\ f(t,\omega) \in A,\ t \in \Lambda,\ a \leq t \leq b\}$ either both are true or both are false.

Assume that two breaking random processes are defined: $(\tau_0(\omega), f_0(t,\omega), 0 \leq t < \tau_0(\omega))$ and $(\tau(\omega),\ f(t,\omega),\ 0 \leq t < \tau(\omega))$.

Consider a sequence of independent trials with outcomes ω_n, $n \geq 0$, each following the probability measure $\mathbf{P}$ over Ω, and observe the path functions of the breaking random processes defined by these ω_n: $\tau_0 = \tau_0(\omega_0)$, $f_0(t) = f_0(t, \omega_0)$, $0 \leq t < \tau_0$; $\tau_n = \tau(\omega_n)$, $f_n(t) = f(t, \omega_n)$, $0 \leq t < \tau_n$, $n \geq 1$. We paste together the path functions. In other words we set $\tilde{\omega} = (\omega_0, \omega_1, \omega_2, \ldots)$, $t_n = \tau_0 + \cdots + \tau_n$, $\xi(t, \tilde{\omega}) = f_0(t)$, $t < \tau_0$, $\xi(t, \tilde{\omega}) = f_n(t - t_{n-1})$ if $t_{n-1} \leq t < t_n$. The defined random process $\xi(t) = \xi(t, \tilde{\omega})$ is called **a regenerative process**. The sequence of points $\{t_n\}$ is a recurrent point process. It is called **an embedded recurrent point process** of the regenerative process $\xi(t)$.

Example 2.9. A device state at time t is characterized by some parameters $x_1(t), \ldots, x_m(t)$. The parameters are being measured continuously. As soon as a parameter achieves a critical level, the device adjustment starts for returning the parameters to original levels $x_1^0, \ldots, x_m^0$. Denote by t_n the nth time of terminating the adjustment. If the system behavior in an interval (t_{n-1}, t_n) follows a time-homogeneous Markov process then the process $x_i(t)$ is a regenerative process $\{t_n\}$ being its embedded recurrent point process.

Since the theory of Markov processes is a highly developed branch of the random processes theory, it is often convenient to reduce constructive random processes to Markov ones. As to regenerative processes, such a reduction can be done as follows. Consider a random process $(\alpha(t), \omega(t), \gamma(t))$ where

$$\alpha(t) = 0 \text{ if } t < \tau_0, \quad \alpha(t) = 1 \text{ if } t \geq \tau_0,$$

so that $\alpha(t)$ indicates if the initial cycle is over before time t. If $\alpha(t) = 0$ then $\omega(t) = \omega_0$, $\gamma(t) = t$; if $\alpha(t) = 1$ and $t_{n-1} \leq t < t_n$ then $\omega(t) = \omega_n$, $\gamma(t) = t - t_{n-1}$. This process is a time-homogeneous Markov process. A value of the regenerative process can be reconstructed uniquely via the constructed process. Indeed,

$$\xi(t) = f_0(\gamma(t), \omega(t)) \text{ if } \alpha(t) = 0,$$

$$\xi(t) = f_1(\gamma(t), \omega(t)) \text{ if } \alpha(t) = 1,$$

$$\text{or } \xi(t) = f_{\alpha(t)}(\gamma(t), \omega(t)).$$

An interesting special case of a regenerative process is **a strongly regenerative process** (see also Section 8.3). Its characteristic property consists of the following. The $\tau(\omega)$ equals the sum of independent random variables $\tau'(\omega)$ and $\tau''(\omega)$, the latter being distributed exponentially. In the interval of the length of $\tau''(\omega)$ the function $f(t, \omega)$ usually vanishes.

Example 2.10. Consider a device subjected to random failures. A time to failure is exponentially distributed. A repair time is a random variable with some distribution. The device capacity equals ω_0 at a time when it is in the up-state; otherwise the capacity is defined by the time elapsed from the start of a current repair. Denote by $\omega(t)$ the capacity of the device at the time t. The process $\omega(t)$ is a strongly regenerative process.

2.10 Gaussian Proccsses

Gaussian (or normal) distribution is the distribution of an m-dimensional random variable $\xi = (\xi_1, \ldots, \xi_n)$ possessing the characteristic function of the form

$$\varphi_\xi(\lambda) = \varphi_\xi(\lambda_1, \ldots, \lambda_m) = \mathbf{E}\exp\{i\sum_{k=1}^{m}\lambda_k\xi_k\} = \exp\{i(\lambda, \mathbf{a}) - \frac{1}{2}(\mathbf{A}\lambda, \lambda)\}$$

where λ and $\mathbf{a}$ are column vectors, $(\mathbf{a}, \mathbf{b})$ denotes the scalar product of vectors $\mathbf{a}$ and $\mathbf{b}$, and $\mathbf{A}$ is a symmetrical nonnegatively defined matrix. If a_j is the jth component of the vector $\mathbf{a}$ then $a_j = \mathbf{E}\xi_j$, $1 \le j \le m$. The matrix $\mathbf{A}$ is the covariance matrix of the random variable ξ, i.e.,

$$\mathbf{A} = (\mathbf{cov}(\xi_i, \xi_j)).$$

Let $\xi(t)$ be a random process for which a random variable $(\xi(t_1), \ldots \ldots, \xi(t_n))$ has a Gaussian distribution whatever n and $t_k \in I$ be. Such a random process is called **a Gaussian (or normal) process**[5]. The family of finite-dimensional distribution of a Gaussian distribution is defined uniquely by the vector of means and the covariance function. Thus an exhaustive probabilistic description of a Gaussian process (in the framework of finite-dimensional distributions) can be achieved via considering its mean function $m(t)$ and covariance function $R(t, \tau)$.

2.11 Martingales and Semi-Martingales

Let $I \subset \mathbb{R}$. An m-dimensional random process $\{\xi(t),\ t \in I\}$ with finite $m(t) = \mathbf{E}\,\xi(t)$, $t \in I$, is said to be **a martingale** if the following property holds true. Choose arbitrary points $t_1 < \cdots < t_{n+1}$ of the set I. Then

$$\mathbf{E}\,(\,\xi(t_{n+1}) \mid \xi(t_k) = x_k,\ 1 \le k \le n) = x_n$$

for arbitrary $x_k \in \mathbb{R}^m$, $1 \le k \le n$. It is useful to express the martingale property in another equivalent form. Let $\mathbf{\Lambda}$ be a $\mathbf{P}$-measurable set defined by conditions concerning values of the variables $\xi(t_1), \ldots, \xi(t_n)$. In other words, $\mathbf{\Lambda} = \{\omega : \ (\xi(t_1), \ \dots, \ \xi(t_n)) \in A\}$ for some set $A \subset \mathbb{R}^{mn}$. Let $I_{\mathbf{\Lambda}}$ be the indicator function of the set $\mathbf{\Lambda}$, i.e., $I_{\mathbf{\Lambda}} = 1$ if $\omega \in \mathbf{\Lambda}$, $I_{\mathbf{\Lambda}} = 0$ otherwise. Then

$$\mathbf{E}\, \xi(s)\, I_{\mathbf{\Lambda}} = \mathbf{E}\, \xi(t)\, I_{\mathbf{\Lambda}},$$

or, equivalently,

$$\int\limits_{\mathbf{\Lambda}} \xi(s)\, d\mathbf{P} = \int\limits_{\mathbf{\Lambda}} \xi(t)\, d\mathbf{P}$$

as soon as $t_1 < \cdots < t_n \leq s < t$.

The most valuable application of a martingale is connected to the estimation of a random variable via observing a random process related to it. Let ξ and $\{\eta(t), \ t \in I\}$ be defined over the same probability space, $I \subset \mathbb{R}$. Denote by $\xi(t)$ the conditional expectation of ζ given the path function $\{\eta(u), \ u \leq t, \ u \in I\}$. Then the random process $\{\xi(t), \ t \in I\}$ is a martingale. Let us set $I^* = I \cup \infty$, $\xi(\infty) = \zeta$. Then $\{\xi(t), \ t \in I^*\}$ is a martingale too.

Example 2.11. Independent random variables $\eta_1, \ \eta_2, \ \dots$ are successively observed. Under a hypothesis H_0 the variables $\{\eta_k\}$ have a probability density function $p_0(x)$ whereas under an alternative hypothesis H_1 they have a density $p_1(x)$. Let α_0 be a priori probability of the hypothesis H_0. The posteriorly probability of the hypothesis H_0 given the observations $\eta_1, \ \dots, \ \eta_n$ is defined by the expression

$$\xi_n = \frac{\alpha_0\, p_0(\eta_1) \ \dots \ p_0(\eta_n)}{\alpha_0\, p_0(\eta_1) \ \dots \ p_0(\eta_n) + (1 - \alpha_0)\, p_1(\eta_1) \ \dots \ p_1(\eta_n)}.$$

Then the sequence $\{\alpha_0, \ \xi_1, \ \xi_2, \ \dots\}$ is a martingale with a discrete time parameter.

Consider a one-dimensional random process $\{\xi(t), \ t \in I\}$ with $I \in \mathbb{R}$. Assume that the following property is satisfied. If $n \geq 1$, $t_1 < \cdots < t_n \leq$ $\leq s < t$, and $\mathbf{\Lambda}$ is an arbitrary measurable subset of Ω defined by a condition concerning $\xi(t_1), \ \dots, \ \xi(t_n)$. Then

$$\mathbf{E}\, \xi(s)\, I_{\mathbf{\Lambda}} \leq \mathbf{E}\, \xi(t)\, I_{\mathbf{\Lambda}}.$$

In such a case the random process $\{\xi(t), \ t \in I) \}$ is called **a submartingale**.

Example 2.12. Let $\{\eta_n\}$ be a sequence of one-dimensional independent random variables possessing nonnegative means. Then the sequence of sums $\{s_n\}$, $s_n = \xi_1 + \cdots + \xi_n$, is a submartingale. If we change the sign of the inequality in the definition of a submartingale, we obtain the definition of **a supermartingale**. Hence, for a supermartingale,

$$\mathbf{E}\, \xi(s)\, I_{\mathbf{\Lambda}} \geq \mathbf{E}\, \xi(t)\, I_{\mathbf{\Lambda}}$$

under the above stated conditions. Submartingales and supermartingales are called **semi-martingales**. Sometimes this notion is used instead of the term "submartingales".

2.12 References

1. Cox, D. R., Miller, H. D., *The Theory of Stochastic Processes,* Chapman and Hall, London, 1972.
2. Bailey, N. T. J., *The Elements of Stochastic Processes,* Wiley, New York, 1964.
3. Hoel, P. G., Port, S. C., Stone, C. J., *Introduction to Stochastic Processes,* Houghton Mifflin, Boston, 1972.
4. Haight, F. A., *Applied Probability,* Plenum Press, New York, 1981.
5. Grimmett, G., Stizzaker, D., *Probability and Random Processes,* Clarendon Press, Oxford, 1982.

Chapter 3

Discrete-Time Markov Chains

3.1 Definitions and Elementary Relations

Let $\{S_n,\ n \geq 0\}$ be a sequence of trials each possessing a finite or countable set of states, and let ε_n be an outcome of the nth trial. If the conditional distribution of ε_{n+1} given $\varepsilon_0, \ldots, \varepsilon_n$ does not depend on $\varepsilon_{n-1}, \varepsilon_{n-2}, \ldots$, then the sequence $\{\varepsilon_n,\ n \geq 0)$ is called **a discrete-time Markov chain** (or, simply, a **Markov chain**[1−7]). More precisely, a Markov chain is defined through the following equation called **the Markov property**: for any $n,\ x_0, \ldots, x_{n-1}, x_n, y$

$$\mathbf{P}\{\varepsilon_{n+1} = y \,|\, \varepsilon_n = x;\ \varepsilon_0 = x_0, \ldots, \varepsilon_{n-1} = x_{n-1}\} = \mathbf{P}\{\varepsilon_{n+1} = y \,|\, \varepsilon_n = x\}.$$

The property is equivalent to the equation

$$\mathbf{P}\{\varepsilon_0 = x_0, \ldots, \varepsilon_n = x_n\} = \mathbf{P}\{\varepsilon_0 = x_0\}\,\mathbf{P}\{\varepsilon_1 = x_1 \mid \varepsilon_0 = x_0\} \ldots$$
$$\ldots \mathbf{P}\{\varepsilon_n = x_n \mid \varepsilon_{n-1} = x_{n-1}\}$$

and also to the equation

$$\mathbf{P}\{AB \mid \varepsilon_n = x\} = \mathbf{P}\{A \mid \varepsilon_n = x\}\,\mathbf{P}\{B \mid \varepsilon_n = x\} \qquad (3.1)$$

where A is an arbitrary event defined by outcomes of trials $0, 1, \ldots, n-1$, and B is an event defined by outcomes of trials $n+1,\ n+2, \ldots$. The equation (3.1) is often described as "the independence of future and past if the current state is known".

Let $\{\xi_n,\ n \geq 0\}$ be a sequence of one-dimensional or multidimensional random variables defined over the same probability space, so that $\xi_n = f_n(\omega)$ where ω is an elementary event of the probability space. Then the sequence $\{\xi_n,\ n \geq 0\}$ is called **a Markov chain** if the equation

$$\mathbf{P}\{\xi_0 \in A_0,\ \ldots,\ \xi_n \in A_n\} = \iint\limits_{x_i \in A_i,\, 0 \leq i \leq n} \mu_0(dx_0)\, \mu_1(x_0\,, dx_1)\ \ldots\ \mu_n(x_{n-1},\ dx_n)$$

holds true for every $n \geq 1$ and arbitrary Borel sets $A_0, \ldots, A_n$ where $\mu_0(A)$ is a probability measure and $\mu_i(x, A)$ are probability measures in A given x. Very often ξ_n is called the position of a particle or state of a system at time n (or at step n). The μ_0 is called **the initial distribution** (the distribution of the initial state) whereas $\mu_n(x,\ A)$ is the conditional distribution of the state at time n (at step n) given a state x at time (step) $n-1$. If multidimensional random variables $(\xi_0,\ \ldots,\ \xi_n)$ are continuous then the joint densities of these variables can be defined by the equation

$$p_n(x_0,\ \ldots,\ x_n) = \int \cdots \int p_0(x_0)\, p_1(x_0,\ x_1)\ \ldots\ p_n(x_{n-1},\ x_n)\, dx_0\ \ldots\ dx_n,$$

where $p_0(x)$ is the probability density function of ξ_0, and $p_n(x_{n-1},\ x_n)$ is the conditional density of ξ_n given $\xi_{n-1} = x_{n-1}$.

A Markov chain (ξ_n) is called **time-homogeneous** if $\mu_n(x,\ A)$ does not depend on n.

Below we confine ourselves to the consideration of time homogeneous Markov chains with a finite or countable set X of possible values of each variable ξ_n. The set X is usually called **the phase space** of a Markov chain. States of ξ_n will be denoted by letters i, j and so on. In a case where the set of summation is not indicated the sum over the entire set X is mentioned.

Denote by ξ_n the state of the Markov chain at step n, by $p_i^{(n)} = \mathbf{P}\{\xi_n = i\}$, $p_{ij} = \mathbf{P}\{\xi_n = j \mid \xi_{n-1} = i\}$, $n \geq 1$, **the one-step transition probabilities**, and by $p_{ij}^{(n)} = \mathbf{P}\{\xi_{m+n} = j \mid \xi_m = i\}$, **the n-step transition probabilities**. The matrix $P = (p_{ij})$ is called **the transition matrix** of the Markov chain whereas $P^{(n)} = (p_{ij}^{(n)})$ is called **the n-step transition matrix** of the Markov chain. The equation

$$p_{ij}^{(n+m)} = \sum_k p_{ik}^{(n)}\, p_{kj}^{(m)},\quad n \geq 0,\ m \geq 0,$$

holds. It is called **the Markov equation**. It can be written as

$$P^{(m+n)} = P^{(m)}\,P^{(n)}$$

in a matrix form.

We have $P^{(n)} = P^n$; thus one-step transition probabilities define $p_{ij}^{(n)}$ for any n. The absolute probabilities satisfy the equation

$$p_j^{(m+n)} = \sum_i p_i^{(m)}\,p_{ij}^{(n)}, \quad n \geq 0,\ m \geq 0$$

which is called **the Markov equation** as well. A special case $m = 0$ can be written as $p_j^{(n)} = \sum\limits_i p_i^{(0)}\,p_{ij}^{(n)}$ so that the absolute probabilities can be expressed via initial probabilities and n-step transition probabilities.

The probability of a chain of states is defined as follows:

$$\mathbf{P}\{\xi_m = i_m,\ \xi_{m+1} = i_{m+1},\ \ldots,\ \xi_n = i_n\} = p_{i_m}^{(m)}\,p_{i_m i_{m+1}}\,\cdots\,p_{i_{n-1} i_n}.$$

Let A be a given subset of X. The probability $P_{ij}^{(n)}(A)$ of the event $\{\xi_n = j;\ \xi_k \in A,\ 1 \leq k \leq n-1\}$ given $\xi_0 = i \in A$ can be defined by the expression

$$P_{ij}^{(n)}(A) = \sum_{i_1,\ldots,i_{n-1} \in A} p_{ii_1}\,p_{i_1 i_2}\,\cdots\,p_{i_{n-1} j} \qquad (3.2)$$

and can be also computed by the recurrent formula

$$P_{ij}^{(n)}(A) = \sum_{k \in A} p_{ik}^{(n-1)}(A)\,p_{kj}, \quad n \geq 1; \quad P_{ij}^{(0)}(A) = \delta_{ij}$$

where δ_{ij} is Kronecker delta.

Let $f_{ij}^{(n)}$ denote the probability of the event $\{\xi_n = j;\ \xi_k \neq j,\ 1 \leq j \leq \leq n-1\}$ given $\xi_0 = i$. We have

$$f_{ij}^{(n)} = p_{ij}^{(n)} - \sum_{m=1}^{n-1} f_{ij}^{(m)}\,p_{jj}^{(n-m)}, \quad n \geq 1.$$

The probability $f_i^{(n)} = f_{ii}^{(n)}$ is called the probability of **the first return** of the Markov chain to the state i in n steps.

Let j be a given state and τ_{mn} be the number of such steps t, $m \leq t \leq n$, at which $\xi_t = j$. We have

$$\mathbf{E}\,\tau_{mn} = \sum_{k=m}^{n} p_j^{(k)}, \quad \mathbf{E}\{\tau_{mn} \mid \xi_0 = i\} = \sum_{k=m}^{n} p_{ij}^{(k)},$$

$$\mathbf{Var}\,\tau_{mn} = 2\sum_{m\le r<s\le n} p_j^{(r)}\,(p_{ij}^{(s-r)} - p_j^{(s)}) + \sum_{k=m}^{n} p_j^{(r)}\,(1-p_j^{(r)}),$$

$$\mathbf{Var}\,\{\tau_{mn}\mid \xi_0=i\} = 2\sum_{m\le r<s\le n} p_{ij}^{(r)}\,(p_{ij}^{(s-r)} - p_{ij}^{(s)}) + \sum_{k=m}^{n} p_{ij}^{(r)}\,(1-p_{ij}^{(r)}).$$

3.2 Classification of States of a Markov Chain

A state j is called **accessible** from a state i if $p_{ij}^{(n)} > 0$ for some $n \ge 0$. If j is accessible from j and k is accessible from j then k is accessible from i. Denote by $R = (r_{ij})$ the 0, 1-matrix defined as follows: $r_{ij} = 1$ if j is accessible from i, $r_{ij} = 1$ otherwise. We have

$$R = \lim_{n\to\infty} R^{(n)}$$

where the matrixes $R^{(n)} = (r_{ij}^{(n)})$ can be computed by the recurrent formula

$$r_{ij}^{(n)} = \begin{cases} 1 & \text{if } r_{ij}^{(n-1)} = 1 \\ 1 & \text{if } \sum\limits_k r_{ik}^{(n-1)}\,p_{jk} > 0 \\ 0 & \text{otherwise} \end{cases}$$

where $r_{ij}^{(0)} = \delta_{ij}$. If $\mid X \mid = m$ then $R = R^{(m-1)}$.

States i and j are called **communicating** if each is accessible from the other, i.e., $r_{ij} = r_{ji} = 1$. We call **a class of communicating states** such a set $A \subset X$ for which the following conditions are satisfied: (i) each states i and j belonging to A are communicating, and (ii) no state j outside A is accessible from a state i belonging to A.

A class K of states of a Markov chain is called **closed** if $p_{ij} = 0$ as soon as $i \in K$ and $j \notin K$. Therefore given $\xi_n \in K$ we have $\xi_m \in K$ for every $m \ge n$.

Each state i belongs to a class of communicating states. Hence X is divided into classes of communicating states. Some of the classes are closed. Let us denote the closed classes by $C_1, C_2, \ldots$ and also denote by C_0 the set of states belonging to no closed class. Each state from C_0 is called **transient**.

Set $f_{ij} = \sum\limits_{n=1}^{\infty} f_{ij}^{(n)}$, $f_i = \sum\limits_{n=1}^{\infty} f_i^{(n)}$. Then f_{ij} is the probability of **the reaching** of state j from state i whereas $f_i = f_{ii}$ is the probability of **the return** to the state i.

Let $\xi_0 = i \in C_0$. Then with a positive probability $1 - f_i$ the Markov chain never returns to the state i. Reaching probabilities f_{ij} can be defined by the system of equations

$$f_{ij} = p_{ij} + \sum_{k \neq j} p_{ik} f_{kj}, \quad i, j \in X.$$

If $i \in C_r$, $j \notin C_r$ for some $r > 0$, then $f_{ij} = 0$. For $i \in C_0$, $j \in C_0$ the following equation hold true:

$$f_{ij} = p_{ij} + \sum_{j \neq k \in C_0} p_{ik} f_{kj}.$$

Let f_{irk} be the probability of the following event: given $\xi_0 = i \in C_0$ the Markov chain enters the class C_r, $r \geq 1$, through the state k. Then

$$f_{irk} = \mathbf{P}\left\{ \bigcup_{n=1}^{\infty} \{\xi_n = k;\ \xi_m \notin C_r,\ m < n\} \,|\, \xi_0 = i \right\}.$$

The probabilities being considered satisfy the system of equations

$$f_{irk} = p_{ik} + \sum_{i \neq j \in C_0} p_{ij} f_{jrk}, \ i \in C_0 \ \ (r \geq 1,\, k \in C_r).$$

Let $p_{ik}^{(n)}(C_0)$, $i \in C_0$, $k \in C_0$, be defined accordingly to (3.2). Then the equation

$$p_{ij}^{(n)} = \sum_{k \in C_r} p_{ik} p_{kj}^{(n-1)} + \sum_{m=1}^{n} \sum_{k \in C_0,\, l \in C_r} p_{ik}^{(m)}(C_0)\, p_{kl}\, p_{lj}^{(n-m-1)}, \quad i \in C_0,$$

holds for $j \in C_r$, $r > 0$, $n \geq 1$.

A time-homogeneous Markov chain $\{\xi_n,\, n \geq 0\}$ is called **irreducible** if its phase-space is a closed class of communicating states. Let $\{\xi_n,\, n \geq 0\}$ be a time-homogeneous Markov chain and C_r be a closed class of its communicating states. The chain $\{\xi_n\}$ being considered in a reduced phase-space C_r instead of X is irreducible. Therefore irreducible Markov chains are the main interest.

Let $\{\xi_n,\, n \geq 0\}$ be an irreducible time-homogeneous Markov chain. A state i is called **transient** if $f_i < 1$ and **recurrent** if $f_i = 1$.

Assume that $\xi_n = i$ and $\xi_{n+\theta_i} = i$ for the first time. Then the random variable θ_i is called **the return time** (to the state i). In a case where $\xi_m \neq i$ for every $m > n$ we set $\theta_i = \infty$. We have $\mathbf{P}\{\theta_i < \infty\} = 1$ for a recurrent state i. Set $\tau_i = \mathbf{E}\,\theta_i$. A state i is called **a recurrent zero state** if it is recurrent and $\tau_i = \infty$. A state i is called **positive** if it is recurrent and $\tau_i < \infty$.

Theorem 3.1. *Only three possibilities exist: (i) all the states $i \in X$ are transient, (ii) all the states $i \in X$ are recurrent zero, (iii) all the states $i \in X$ are positive. If X is finite then the case (ii) is impossible.*

Theorem 3.2. *A state i is transient if $\sum\limits_n p_{ii}^{(n)} < \infty$; it is recurrent if $\sum\limits_n p_{ii}^{(n)} = \infty$. A recurrent state is zero if $p_{ii}^{(n)} \underset{n\to\infty}{\longrightarrow} 0$ or positive if $\overline{\lim\limits_{n\to\infty}}\, p_{ij}^{(n)} > 0$.*

Let i be a state. Denote by t_i the greatest common divisor of such positive n for which $p_{ii}^{(n)} > 0$. A state i is called **periodic**, l being its **period** if $t_i = l > 1$. A state i is **aperiodic** if $t_i = 1$.

Theorem 3.3. *All the states of an irreducible time-homogeneous Markov chain are either aperiodic or periodic with the same period l.*

In the former case the Markov chain is called **ergodic**; in the latter case it is called **periodic** with the period l.

3.3 Ergodic Theorems

The most interesting is the behavior of transition probabilities $p_{ij}^{(n)}$ as n goes against infinity, in the case of an ergodic Markov chain[1−9]. We have

$$p_{ij}^{(n)} \underset{n\to\infty}{\longrightarrow} 1/\tau_j.$$

The phase space of a positive periodic Markov chain with a period $l > 1$ is partitioned onto l nonempty sets $X_0, X_1, \ldots, X_{l-1}$ according to the following principle. If $\xi_0 \in X_r$ then the event $\{\xi_n \in X_s\}$ occurs at any step n comparable with $s - r$ modulo l. Such an event is impossible for n which does not satisfy this condition. The partition X onto classes X_k is unique.

Theorem 3.4. *Let $p_{ij}^{(n)}$ be n-step transition probabilities of a periodic positive Markov chain. Then*

$$p_{ij}^{(n)} = 0 \quad \text{if } i \in X_r,\ j \in X_s,\ n \notin \{kt + s - r\};$$

$$p_{ij}^{(nt+s-r)} \underset{n\to\infty}{\longrightarrow} t/\tau_j \quad \text{if } i \in C_r,\ j \in C_s.$$

If a Markov chain is irreducible and positive then mean reaching times τ_{ij} are finite for each i and j.

Example 3.1. Consider a random walk $\{\xi_n\}$ on a half-line $\mathbb{R}^+$ defined as follows. If $\xi_n = i > 0$ then $\xi_{n+1} = i+1$ with a probability $p = (1+\triangle)/2$. If $\xi_n = 0$ then the two probabilities exist: either $\xi_{n+1} = 1$ with probability p or $\xi_{n+1} = 0$ with probability q. The transition matrix of the Markov chain has the form

$$\begin{pmatrix} q & p & 0 & 0 & \dots \\ q & 0 & p & 0 & \dots \\ 0 & q & 0 & p & \dots \\ . & . & . & . & \dots \end{pmatrix}$$

If $\triangle > 0$ then the event $\{\xi_n$ never returns to state $i\}$ has a positive probability.

The system of equations

$$p_j = \sum_i p_i \, p_{ij}, \quad j \in X,$$

under the normalizing condition $\sum\limits_j p_j = 1$ is called **the stationarity equation** of a Markov chain. If a Markov chain is ergodic then the stationarity equation has a unique solution provided that $\sum\limits_j |p_j| < \infty$. This solution is called **the ergodic distribution** of the Markov chain.

Consider an irreducible aperiodic Markov chain. Suppose that a solution $\{u_j\}$ of the stationarity equation satisfies the condition $\sum\limits_j |u_j| < \infty$; then the Markov chain is ergodic, the $\{u_j\}$ being its ergodic distribution.

Consider an irreducible periodic Markov chain with a period t. Consider the distributions

$$u_j^{(r)} = \begin{cases} t u_j / \sum\limits_i u_i & \text{if } j \in X_r \\ 0 & \text{if } j \notin X_r, \end{cases}$$

$0 \le r \le t-1$. Then $p_{ij}^{(nt+s-r)} \underset{n\to\infty}{\longrightarrow} u_j^{(s)}$ for $i \in X_r$.

Let $X = \{0, 1, 2, \dots\}$, $U(z) = \sum\limits_{j=0}^{\infty} u_j z^j$ and $|U(z)| < \infty$ for some z, $|z| > 1$. Then $\sum\limits_j |u_j| < \infty$. This property is often used while solving the stationarity equation by the method of generating functions.

Bernstein Theorem. *Suppose that such $i_0 \in X$ and $\lambda > 0$ exist that $p_{ii_0} \ge \lambda$ for each $i \in X$. Then the ergodic distribution $\{\pi_j\}$ exists so that*

$p_{ij}^{(n)} \underset{n\to\infty}{\longrightarrow} \pi_j$ *for each* $i \in X$. *Moreover*

$$\sum_j |p_{ij}^{(n)} - p_{kj}^{(n)}| \le 2(1-\lambda)^n, \quad n \ge 0,$$

and

$$\sum_j |p_{ij}^{(n)} - \pi_j| \le 2(1-\lambda)^n, \quad n \ge 0,$$

for arbitrary i *and* k.

Consider an irreducible aperiodic Markov chain $\{\xi_n\}$. Suppose that such a nonnegative sequence $\{f_i\}$ exists that the following two conditions are satisfied:

(i) $\sum_j f_j p_{ij} < \infty$ for every $i \in X$,

(ii) $\sum_j f_{ij} p_{ij} - f_i \le -\varepsilon$ for some $\varepsilon > 0$ and every i except a finite number of states.

Then the Markov chain $\{\xi_n\}$ is ergodic.

Example 3.2. Consider a one-channel queueing system. Denote by η_n the service time of customer n. The η_n are considered to be independent random variables with a distribution function $B(t)$ and a finite mean τ. Assume that the incoming flow is Poisson with a parameter λ (see Section 2.5). In such a case the number of customers arriving within an interval x has a Poisson distribution with the parameter λx. Denote by ξ_n the number of customers in the queue at the time when customer n has been just serviced. Then

$$\xi_n = \begin{cases} \xi_{n-1} + \gamma_n - 1 & \text{if} \quad \xi_{n-1} \ge 1 \\ \gamma_n & \text{if} \quad \xi_{n-1} = 0 \end{cases}$$

where γ_n denotes the number of customers arriving within the service time of customer n. Then we have

$$\sum_j f_j p_{0j} = \mathbf{E}\gamma_n = \int_0^\infty \lambda x \, dB(x) = \lambda\tau,$$

$$\sum_j f_j p_{ij} - f_i p_i = \mathbf{E}\gamma_n - 1 = \lambda\tau - 1, \quad n \ge 1.$$

The above mentioned ergodicity conditions are satisfied for $\varepsilon = 1 - \lambda\tau$ if $\lambda\tau < 1$. (The irreducibility follows from the condition $\mathbf{P}\{\gamma_n = 0\} = = \int_0^\infty e^{-\lambda x} dF(x) > 0$, so that the state $i-1$ is accessible from a state $i > 0$;

at the same time $\mathbf{P}\{\gamma_n = k\} > 0$ for each $k = 1, 2, \dots$. Hence, any $j > i$ is accessible from a state i. The aperiodicity follows from the fact that the event $\{\xi_{n-1} = 0,\ \gamma_n = 0\}$ implies $\{\xi_n = 0\}$ so that the return in time 1 is possible.)

Example 3.3. Let ξ_n be the number of descendants of a biologic particle in generation n; $\xi_0 = 1$. Each particle generates a random number η of particles of the next generation; $g_k = \mathbf{P}\{\eta = k\}$, $g_0 + g_1 + \cdots = 1$. Suppose that the birth processes are independent among particles and generations. Suppose that a force prevents the degeneration of the population so that $\xi_{n+1} = 1$ as soon as $\xi_n = 0$. If $\xi_n = i > 0$ then $\xi_{n+1} = \eta_1 + \cdots + \eta_i$ where η_i are independent realizations of the random variable η. Hence, for a given $i > 0$ $\mathbf{E}\{\xi_{n+1} \,|\, \xi_n = i\} = i\,\tau$ where $\tau = \mathbf{E}\eta = \sum_j j\,g_j$. Setting $f_i = i$ one obtains

$$\sum_j f_j\,p_{ij} - f_i\,p_i = \begin{cases} i(\tau - 1) & \text{if} \quad i > 0 \\ 1 & \text{if} \quad i = 0. \end{cases}$$

This expression does not exceed $-\varepsilon = \tau - 1$ as soon as $\tau < 1$ and $i > 0$. The irreducibility and aperiodicity are satisfied if $g_0 + g_1 < 1$. Therefore the conditions $\tau < 1$ and $g_0 + g_1 < 1$ imply the ergodicity of the Markov chain $\{\xi_n\}$. In particular, the mean reaching time of the state 0 from a given state i is finite.

Consider a Markov chain $\{\zeta_n\}$ of a similar structure, with the distinction that the state 0 is absorbing for ζ_n: the event $\{\zeta_n = 0\}$ implies $\{\zeta_{n+1} = 0\}$. The absorbing time for the sequence $\{\zeta_n\}$ equals the reaching time for $\{\xi_n\}$. Therefore the mean absorbing time is finite. It can be treated as the time of the degeneration of a biological population.

3.4 Method of Generating Functions

Introduce **the generating functions**[3,4,10–12]

$$P_{ij}(z) = \sum_{n=1}^{\infty} p_{ij}^{(n)}\, z^{n-1}, \quad |z| \le 1.$$

The system of equations

$$z \sum_j P_{jk}(z)\, p_{ij} = P_{ik}(z) - p_{ik}$$

holds true.

Let $X = \{1, 2, \ldots, a\}$. Denote by $D(z)$ the determinant of the last system of equations.

In the most typical case if no multiple root exists the solution of the system can be reduced to the sum of simple fractions

$$p_{ik}^{(n)} = \frac{\rho_{ik}^{(1)}}{z_1^n} + \cdots + \frac{\rho_{ik}^{(a)}}{z_a^n}$$

where $z_1, \ldots, z_a$ are the roots of the equation $D(z) = 0$. The coefficients satisfy the systems of equations

$$\rho_{jk}^{(r)} = z_r \sum_{i=1}^{a} p_{ji}\, \rho_{ik}^{(r)}, \quad 1 \le r \le a$$

and

$$\rho_{jm}^{(r)} = z_r \sum_{k=1}^{a} \rho_{jk}^{(r)}\, p_{km}, \quad 1 \le r \le a.$$

Consider systems of equations

$$x_j = z \sum_{i=1}^{a} p_{ji}\, x_i; \quad y_m = z \sum_{i=1}^{a} y_k\, p_{km}.$$

At the points $z = z_r$ they have non-trivial solutions $(x_1^{(r)}, \ldots, x_a^{(r)})$, $(y_1^{(r)}, \ldots$ $\ldots, y_a^{(r)})$, which are uniquely defined but constant factors. Then

$$\rho_{jk}^{(r)} = x_j^{(r)}\, y_k^{(r)} \Big/ \sum_{i=1}^{a} x_i^{(r)}\, y_i^{(r)}.$$

Example 3.4. Consider a Markov chain with two states and the transition matrix

$$P = \begin{pmatrix} \alpha & 1-\alpha \\ \beta & 1-\beta \end{pmatrix}.$$

Then

$$D(z) = \begin{vmatrix} \alpha z - 1 & (1-\alpha)z \\ \beta z & (1-\beta)z - 1 \end{vmatrix} = (z-1)\,((\alpha-\beta)\,z - 1).$$

If $|\alpha - \beta| = 1$ then $\{\xi_n\}$ is a deterministic sequence given ξ_0. If $|\alpha - \beta| < 1$ then the equation $D(z) = 0$ has two distinct roots: $z_1 = 1,\ z_2 = 1/(\alpha - \beta)$. One obtains

$$P^n = \begin{pmatrix} p_{11}^{(n)} & p_{12}^{(n)} \\ p_{21}^{(n)} & p_{22}^{(n)} \end{pmatrix} = \begin{pmatrix} \pi_1 + \pi_2\,(\alpha-\beta)^n & \pi_2 - \pi_2\,(\alpha-\beta)^n \\ \pi_1 - \pi_1\,(\alpha-\beta)^n & \pi_2 + \pi_1\,(\alpha-\beta)^n \end{pmatrix}$$

where $\pi_1 = \beta/(1-\alpha+\beta)$, $\pi_2 = (1-\alpha)/(1-\alpha+\beta)$.

Example 3.5. Consider a Markov chain with three states 1, 2 and 3 possessing the transition matrix

$$\begin{pmatrix} 1-2b & 2b & 0 \\ 0 & 1-b & b \\ 0 & 0 & 1 \end{pmatrix}$$

with $0 \le b \le 1/2$. It is required to get some expressions for $p_{13}^{(n)}$. One obtains

$$D(z) = \begin{pmatrix} (1-2b)z-1 & 2bz & 0 \\ 0 & (1-b)z-1 & bz \\ 0 & 0 & z-1 \end{pmatrix} =$$

$$= (z-1)\,((1-b)\,z-1)\,((1-2b)\,z-1),$$

hence $z_1 = 1$, $z_2 = 1/(1-b)$, $z_3 = 1/(1-2b)$; $p_{13}^{(n)} = 1-2\,(1-b)^n+(1-2\,b)^n$.

The method of generating functions enables the discovering of the rate of convergence of $p_{ij}^{(n)}$, $p_j^{(n)}$ to their limits as $n \to \infty$. Without a loss of generality, suppose that the roots of the equation $D(z) = 0$ satisfy the condition $|z_1| \le |z_2| \le \cdots \le |z_a|$. If $|z_1| = 1 < |z_2| = \cdots = |z_m| < |z_{m+1}|$ and all roots $z_2, \ldots, z_m$ are simple, then $p_{ij}^{(n)} = \tau_j + O(|z_2|^{-n})$.

3.5 Unbounded Random Walk

We call **an** m**-dimensional (unbounded) random walk**[11] a time-homogeneous Markov chain $\{\xi_n,\ n \ge 0\}$ with the phase space $\mathbb{Z}^m$ and transition probabilities satisfying the condition $p_{ij} = p_{0,j-1}$, or $p_{(i_1,\ldots,i_m)\,(j_1,\ldots,j_m)} = p_{(0,\ldots,0)\,(j_1-i_1,\ldots,j_m-i_m)}$. The condition implies the relation $p_{ij}^{(n)} = p_{0,j-i}^{(n)}$ for any $n \ge 0$.

A simple random walk is an m-dimensional random walk for which $p_{0j} = 1/(2m)$ for all the vectors j of the form $(0,\ldots,0,\pm 1,0,\ldots,0)$. For example, in the case $m = 2$ the particle can move to each of the states $(1,0)$, $(-1,0)$, $(0,1)$, $(0,-1)$ from the state $(0,0)$ in a step with probability $1/4$. An unbounded one-dimensional random walk is said to be **bernoullian** if the transition probabilities $p_{01} = p$, $p_{0,-1} = q = 1-p$.

Denote by $f_{ij}^{(n)}$ the probability of the first passage to a state j from a state i occurring at step n; set $f_{ij}^{(0)} = 0$. We have

$$f_{ij} = \sum_{n-1}^{\infty} f_{ij}^{(n)}, \qquad G_{ij}^{(n)} = \sum_{k-0}^{n} p_{ij}^{(k)}, \qquad G_{ij} = \sum_{k=0}^{\infty} p_{ij}^{(k)},$$

$$G_{00}^{(n)} = G^{(n)}, \quad G_{00} = G, \quad f_{00}^{(n)} = f^{(n)}, \quad f_{00} = f.$$

A random walk is called **recurrent** if $f = 1$ and **transient** if $f < 1$. The equation $G = f/(1-f)$ holds ($G = \infty$ if $f = 1$). The recurrence of the random walk is the same as that of the recurrence of the Markov chain $\{\xi_n\}$. For a random walk the equation $\lim_{n\to\infty} G_{ij}^{(0)}/G^{(n)} = f_{ij}$ holds true.

In the case of a bernoullian random walk with $p > q$ we have $f_{0i} = 1$ for any $j > 0$; $f_{0i} = (p/q)^j$ if $j \notin X_0$. If $p = q$ then a bernoullian random walk is recurrent.

Denote by X_0 the set of states accessible from the state 0. Then $G_{0j} = = f_{0j} = 0$ if $j \notin X_0$. For a transient random walk $G_{0j} < \infty$ for an arbitrary j. For a recurrent random walk $G_{0j} = \infty$ for any j. A random walk is said to be **aperiodic** if each point of the set X can be expressed as the difference $x - y$ where $x \in X_0$, $y \in X_0$. For an aperiodic recurrent random walk the relation $\lim_{n\to\infty} G_{0j}^{(n)}/G^{(n)} = 1$ is satisfied for every $j \in X$. If a random walk is aperiodic then either $G_{ij} = \infty$ for any i and j and the random walk is recurrent, or $G_{ij} < \infty$ for any i and j and the random walk is transient.

A one-dimensional random walk is said to be **left-semicontinuous** if $p_{0\,-1} > 0$, $p_{0\,j} = 0$ for every $j \geq 2$. A random walk is said to be **right-semicontinuous** if $p_{0\,1} > 0$, $p_{0\,j} = 0$ for every $j \geq 2$. If a random walk is aperiodic, transient and right-semicontinuous and, moreover, $0 < \sum_{j=-\infty}^{\infty} j\,p_{0j} < \infty$, then $f_{0j} = 1$ for every $j > 0$, $f_{0j} < 1$ for every $j \leq 0$. If a random walk is aperiodic, transient and left-semicontinuous and, moreover, $-\infty < \sum_{j=-\infty}^{\infty} j\,p_{0j} < \infty$, then $f_{0j} = 1$ for every $j < 0$, $f_{0j} < 1$ for every $j \geq 0$. In all the rest of cases of an aperiodic transient random walk we have $f_{ij} < 1$ for each i and j.

Suppose that $\sum_j |j|\,p_{0j} < \infty$. Set $\mu = \sum_j j\,p_{0j}$. If $\mu = 0$ then the random walk is recurrent; if $\mu \neq 0$ the random walk is transient. The number R_n of distinct points of the set $\{\xi_0, \xi_1, \ldots, \xi_n\}$ is called **the range** of the random walk. We have

$$R_n/n \xrightarrow[n\to\infty]{p} 1 - f > 0$$

for a transient random walk and

$$R_n/n \xrightarrow[n\to\infty]{p} 0$$

for a recurrent one. We have also

$$\delta_{i0} + \sum_{n=1}^{\infty} (p_{00}^{(n)} - p_{i0}^{(n)}) = \lim_{n\to\infty} \left(\sum_{k=n}^{\infty} f_{i0}^{(k)} \Big/ \sum_{k=n}^{\infty} f^{(k)} \right)$$

for an aperiodic random walk.

Set

$$\psi(\lambda) = \sum_j e^{i(\lambda, j)} p_{0j}$$

where

$$\lambda = (\lambda_1, \ldots, \lambda_m), \; j = (j_1, \ldots, j_m), \; (\lambda, j) = \lambda_1 j_1 + \cdots + \lambda_m j_m.$$

Then the characteristic function $\varphi_n(\lambda)$ of the random vector $\xi_n - \xi_0$ is given by the relation $\varphi(\lambda) = \psi^n(\lambda)$.

Some criteria of the recurrence of a random walk are expressed via the characteristic function $\varphi(\lambda)$. Denote by C the domain of $\lambda = (\lambda_1, \ldots, \lambda_m)$ of the form $\{-\pi < \lambda_i < \pi, \; 1 \le i \le m\}$. A random walk is transient iff

$$\lim_{t \uparrow 1} \int_C \mathrm{Re}\left\{\frac{1}{1 - t\,\psi(\lambda)}\right\} d\lambda < \infty.$$

3.6 Bounded Random Walk

In various applications a random walk with one-sided restrictions is of great interest[11,13,14]. For example, in a one-dimensional case $p_{ij} = p_{i-j}$ for $i \ge i_0$. Usually the system behavior is described by a multidimensional random walk $\{\xi_n\}$ and the components of the vector ξ_n are nonnegative, according to a physical sense. For example, each of them is defined as the number of some object (queue length, failed channels etc.).

A random walk may be non-truncated as well as truncated (stopped) in time. In the former case a steady-state (stationary) distribution is of the most interest. In most cases it is the ergodic distribution. In the latter case the stopping time as well as the position of a walking particle at this moment (and also other related parameters) is of interest.

In both cases the method of generating functions is the most powerful tool for the investigation of random walks. We shall consider some important examples.

Example 3.6. Let $\{\xi_n, \; n \ge 0\}$ be a random walk of the Example 3.2 so that $\xi_n = \xi_{n-1} + \gamma_n - 1$ if $\xi_{n-1} > 0$ and $\xi_n = \gamma_n$ if $\xi_{n-1} = 0$, the γ_n being independent random variables with a generating function $g(z)$. The existence of the ergodic distribution is proved in Section 3.3 for the case $\mathbf{E}\,\gamma_n < 1$. We have $p_{ij} = g_{j-i+1}$ for $i > 0$, $p_{0j} = g_j$ where $g_j = \mathbf{P}\{\gamma_n = j\}$, $j \ge 0$. Hence the ergodic distribution $\{\pi_j\}$ satisfies the system of equations

$$\pi_j = \pi_0 g_j + \sum_{i=1}^{j+1} \pi_i g_{j-i+1}.$$

Multiplying both sides by z^j and summing up in j over $\mathbb{R}^+$ one obtains the equation

$$\pi(z) = \pi_0 g(z) + (\pi(z) - \pi_0) g(z)/z.$$

Hence

$$\pi(z) = \pi_0 (1-z) g(z)/(g(z) - z).$$

In a neighborhood of the point $z = 1$ the expansion

$$g(z) = g(1) + g'(z)(z-1) + o(z-1) = 1 + \mathbf{E}\gamma(z-1) + o(z-1)$$

holds true. At the same time $\pi(1) = 1$. Thus

$$\pi_0(1-z) \sim (1-z)(1-\mathbf{E}\gamma), \quad z \to 1,$$

therefore $\pi_0 = 1 - \mathbf{E}\gamma$. Finally

$$\pi(z) = (1-\mathbf{E}\gamma)(1-z)\, g(z)/(g(z)-z), \quad |z| \le 1,$$

as soon as $\mathbf{E}\gamma < 1$.

Assume, similarly to Section 3.3, that γ_n is the number of events in a random interval of a length η_n, $\mathbf{P}\{\eta_n < x\} = B(x)$. γ_n is a Poisson random variable with the parameter λx given $\eta_n = x$. Then the equality

$$g(z) = \mathbf{E}\, z^{\gamma_n} = \int_0^\infty \mathbf{E}\{z^{\gamma_n} \,|\, \eta = x\}\, dB(x) = \int_0^\infty e^{-\lambda_x (1-x)}\, dB(x) = \psi(\lambda(1-z))$$

holds where $\psi(s) = \int_0^\infty e^{-sx}\, dB(x)$. Moreover $\mathbf{E}\gamma = \lambda\tau = \rho$ where $\tau = \int_0^\infty x\, dB(x)$. A well known **Pollaczek-Chinčin formula**

$$\pi(z) = (1-\rho)(1-z)\,\psi(\lambda(1-z))/(\psi(\lambda(1-z)) - z), \quad |z| \le 1, \ \rho < 1$$

holds true.

Example 3.7. Consider a left-semicontinuous random walk $\{\xi_n, n \ge 0\}$. Suppose that $\xi_0 = 1$ and that the origin is the absorbing point. The transition matrix is of the form

$$P = \begin{pmatrix} 1 & 0 & 0 & 0 & \dots \\ g_{-1} & g_0 & g_1 & g_2 & \dots \\ 0 & g_{-1} & g_0 & g_1 & \dots \\ 0 & 0 & g_{-1} & g_0 & \dots \\ \cdot & \cdot & \cdot & \cdot & \dots \end{pmatrix}$$

where $\{g_j\}$ is the distribution of a step of the random walk from a nonzero state. Denote $g(z) = \sum_{k=-1}^{\infty} g_k\, z^{k+1}$, $|z| \le 1$ and set $\tau = \min\{n : \xi_n = 0\}$ where τ is the absorbing time. Let $\eta = \xi_1 - \xi_0$, $\xi_0 = 1$. If $\eta = -1$ then $\tau = 1$; if $\eta = k \ge 0$ it is necessary to reach the points $k,\ k-1, \dots, 1, 0$ for a particle to be absorbed. But the reaching time of the state i from $i+1$ is distributed similarly to the reaching time of 0 from 1. The corresponding reaching times are independent. Thus we have

$$\varphi(z) = g_{-1}\, z + \sum_{k=0}^{\infty} g_k\, z\, \varphi^{k+1}(z) = z\, g(\varphi(z))$$

for

$$\varphi(z) = \mathbf{E}\{z^{\tau} \,|\, \xi_0 = 1\}.$$

There exists an inverse function $\alpha(z)$ for the function $\varphi(z)$ in the interval $0 \le z \le 1$. Setting $\varphi(z) = t$ one obtains the equation $t = \alpha(t)\, g(t)$. Hence $\alpha(t) = t/g(t)$, $0 \le t \le 1$.

Therefore the function $\varphi(z)$ is defined within the interval $[0, 1]$ as the inverse to the function $z/g(z)$. Since $g_{-1} = p$, $g_0 = q = 1 - p$ we obtain the equation $g(z) = p + q\, z$, $\varphi(z) = z\,(p + q\, \varphi(z))$. Hence

$$\varphi(z) = p\, z/(1 - q\, z) = \sum_{k=1}^{\infty} p q^{k-1} z^k, \quad \mathbf{P}\{\tau = k \,|\, \xi_0 = 1\} = p\, q^{k-1}, \quad k \ge 1.$$

Let τ_i be the absorption time given an initial state i. Then τ_i equals the sum of i independent random variables with the generating function $\varphi(z)$. Hence $\mathbf{E}\, z^{\tau_i} = \varphi^i(z)$, $i \ge 1$.

The following interpretation may illustrate possible applications. Consider a model of the behavior of a bunker as a part of a technological line. An item is taken from the bunker each time unit. The bunker is supplied according to a random procedure; the probability of k units to be supplied within a unit time equals g_{k-1}, independently for each unit time. Then the τ_i above equals the first time the bunker is empty given i items in it initially.

Example 3.8. Consider a Markov chain $\{\xi_n,\ n \geq 0\}$ possessing values $(i,\ j)$ where i and j belong to $\mathbb{N}^+$. The following kinds of one-step transitions are the only possibilities: (i) from $(i,\ j) \neq (0,\ 0)$ to $(i+1,\ j)$ with the probability a, (ii) from $(0,\ 0)$ to $(1,\ 0)$ with probability $a/(a+b)$, (iii) from $(i,\ j) \neq (0,\ 0)$ to $(i,\ j+1)$ with the probability b, (iv) from $(0,\ 0)$ to $(0,\ 1)$ with the probability $b/(a+b)$, (v) from $(i,\ j)$ to $(i-1,\ j)$ with the probability $c = 1-a-b$ if $i > 0$, (vi) from $(i,\ j)$ to $(i,\ j-1)$ with the probability c if $i = 0, j > 0$. The stationary equation has the form

$$p_{ij} = c p_{i+1\,j} + c p_{i\,j+1}\, \delta_{i0} + a \left(1 + \delta_{i1}\,\delta_{j0}\,\frac{c}{a+b}\right) p_{i-1\,j} +$$

$$+\, b \left(1 + \delta_{i1}\,\delta_{j0}\,\frac{c}{a+b}\right) p_{i\,j-1},\quad i \geq 0,\ j \geq 0.$$

Let us introduce a two-dimensional generating function

$$A(x,y) = \sum_{i,j=1}^{\infty} p_{ij}\, x^i\, y^j$$

and a one-dimensional generating function

$$B(y) = \sum_{j=0}^{\infty} p_{0j}\, y^j.$$

Then the stationarity equation is reduced to the following equation:

$$\omega(x,y)\, A(x,y) = c\,(x-y)\, B(y) + p_{00}\, \sigma(x,y)$$

where

$$\omega(x,y) = [x\,(1-ax-by)-c]\, y,\quad \sigma(x,y) = \frac{cxy}{a+b}\,(ax+by) - cx.$$

The equation involves two unknown functions $A(x,y)$ and $B(y)$. Since both the functions are analytic in the domain $|x| \leq 1,\ \ |y| \leq 1$ of complex variables x and y the right-hand side must vanish as soon as $\omega(x,y) = 0$. Denote by $f(y)$ the root of the equation $1 - a\,x - b\,y = c/x$. We obtain the equation $c\,(f(y)-y)\,B(y) + p_{00}\,\sigma(f(y),\,y) = 0$ from which the $B(y)$ can be uniquely obtained with a constant factor p_{00}. Now we can substitute the known function $B(y)$ into the equation defining $A(x,y)$. The function is well defined, taking into account the normalizing condition $A(1,1) = 1$.

The theory of one-dimensional random walks has been developed considerably[11,13,14], especially concerning the investigation of the reaching time of a boundary and the particle position after the boundary crossing.

The following reasoning illustrates mathematical tools for the solution of such problems.

Let $\{\xi_n\}$ be an unbounded one-dimensional random walk; $\xi_0 = 0, T = \min\{ n : \xi_n > 0 \}$. Then the following well-known **Spitzer identity** holds true:

$$1 - \mathbf{E}\{t^T z^{\xi_T}\} = \exp\left\{ - \sum_{k=1}^{\infty} \frac{1}{k} t^k \mathbf{E}\{ z^{\xi_k} I_k \} \right\}$$

where $I_k = 1$ if $\xi_k > 0, otherwise\ I_k = 0$. The stated identity and many related ones are called **factorization identities**. All of them are derived by the method of **factorization.**

Let $\varphi(z)$ be a continuous function defined on the unit circle $|z| = 1$ of a complex plane. The factorization means that one finds such functions $\varphi_1(z)$ and $\varphi_2(t)$ that

(i) $\varphi(z) = \varphi_1(z)\,\varphi_2(z)$ for $|z| = 1$,

(ii) $\varphi_1(z)$ admits an analytical prolongation inside the circle, i.e. to the domain $|z| < 1$,

(iii) $\varphi_2(z)$ admits a bounded analytical prolongation outside the circle, i.e. to the domain $|z| > 1$.

The approaches to the solution of such problems involve boundary problems for analytical functions and singular integral equations. For right-semicontinuous as well as left-semicontinuous random walks (see Examples 3.6 to 3.8) there exist many explicit expressions for the parameters related to these random walks.

3.7 References

1. Bailey, N. T. J., *The Elements of Stochastic Processes,* Wiley, New York, 1964.
2. Chung K. L., *Markov Chains: With Stationary Transition Probabilities,* Springer, Berlin, 1967.
3. Feller, W., *An Introduction to Probability Theory and Its Applications,* 1, Wiley, New York, 1968.
4. Feller, W., *An Introduction to Probability Theory and Its Applications,* 2, Wiley, New York, 1971.
5. Cox, D. R., Miller, H. D., *The Theory of Stochastic Processes,* Chapman and Hall, London, 1972.
6. Hoel, P. G., Port, S. C., Stone, C. J., *Introduction to Stochastic Processes,* Houghton Mifflin, Boston, 1972.
7. Grimmett, G., Stizzaker, D., *Probability and Random Processes,* Clarendon Press, Oxford, 1982.

8. Sarymsakov, T. A., *Foundations of the Theory of Markov Chains,* Gostekhizdat, Moscow, 1954 (in Russian).
9. Širjaev, A. N., *Probability,* Springer, New York, 1984.
10. Klimov, G. P., *Bedienungsprozesse,* Birkhäuser, Basel, 1979.
11. Spitzer, F., *Principles of Random Walk,* Springer, New York, 1976.
12. Kemeny, J. G., Snell J. L., *Finite Markov Chains,* Springer, Heidelberg, 1976.
13. Malyshev, V. A., *Random Walks. Wiener-Hopf Equations in a Quarter of a Plain. Galois Automorphisms,* Moscow Univ. Publ., Moscow, 1970 (in Russian).
14. Borovkov, A. A., *Probability Theory,* Nauka, Moscow, 1976 (in Russian).

Chapter 4

Main Classes of Constructively Defined Random Processes

4.1 Poisson Process

We call such a random process $\{\xi(t),\ t \geq 0\}$ **Poisson** which satisfies the following conditions:

(i) $\xi(0) = 0$ with probability 1,

(ii) $\xi(t)$ is a separable process with independent increments (for convenience it is assumed that a path function of $\xi(t)$ is right-continuous with probability 1),

(iii) for each t and $\tau\ :\ 0 \leq t \leq \tau$ the increment $\xi(\tau) - \xi(t)$ is a Poisson random variable with the parameter $\Lambda(\tau) - \Lambda(t)$ where $\Lambda(t)$ is a nondecreasing function called **the leading function** of the process $\xi(t)$; without a loss of the generality assume that $\Lambda(0) = 0$ and that Λ is right-continuous.

A Poisson process has stepwise path functions with unit steps with probability 1. The function $\Lambda(t)$ may possess a finite or countable set of discontinuity points $\{a_n\}$.

A Poisson process is called **regular** if it is stochastically continuous. A Poisson process is regular iff its leading function is continuous on $\mathbb{R}^+$. A Poisson process is called **singular** if the discontinuity points of its path function belong to a finite or countable set A. A Poisson process is singular iff its path function is stepwise and $A = \{ a_n \}$. In such a case

$$\xi(t) = \sum_{n:\, a_n \le t} \eta_n$$

where η_n are independent Poisson random variables with parameters $\lambda_n = \Lambda(a_n) - \Lambda(a_n - 0)$.

Let $\xi(t)$ be a Poisson process with a leading function $\Lambda(t)$ which possesses a set $\{a_n\}$ of discontinuity points; set $\lambda_n = \Lambda(a_n) - \Lambda(a_n - 0)$. Then $\xi(t)$ can be expressed as the sum of two independent Poisson processes: $\xi(t) = \xi_0(t) + \xi_1(t)$ in such a way that $\xi_0(t)$ is a regular Poisson process and $\xi_1(t)$ is a singular Poisson process. The leading functions $\Lambda_0(t)$ and $\Lambda(t)$ of these processes are defined as follows:

$$\Lambda_1(t) = \sum_{n:\, a_n \le t} \lambda_n,$$

$$\Lambda_0(t) = \Lambda(t) - \Lambda_1(t).$$

Moreover, if the leading function $\Lambda(t)$ of a Poisson process $\xi(t)$ can be expressed as the sum of nondecreasing right-continuous functions $\Lambda_k(t)$, then the process itself can be expressed as the sum

$$\xi(t) = \sum_k \xi_k(t)$$

where $\xi_k(t)$ are independent Poisson processes with leading functions $\Lambda_k(t)$.

A path function $\xi(t)$ of a regular Poisson process with a leading function $\Lambda(t)$ can be defined via the sequence of its discontinuity points $\{t_n,\, n \ge 1\}$. The t_n are random variables distributed as follows:

$$\mathbf{P}\{t_1 \ge t\} = 1 - \exp\{-\Lambda(t)\}, \quad t \ge 0.$$

For $n \ge 2$ the conditional distribution of t_n given $t_1, \dots, t_{n-1}$ is defined by the equality

$$\mathbf{P}\{t_n \le t \,|\, t_1, \dots, t_{n-1}\} = 1 - \exp\{-\Lambda(t) + \Lambda(t_n)\}, \quad t \ge t_n.$$

The following property of a Poisson process is valuable. Assume that $\xi(T) = n$. Then

$$\xi(t) = \sum_{k:\, \tau_k \le t} 1$$

where $\tau_1, \dots, \tau_n$ are independent random variables distributed as

$$\mathbf{P}\{\tau_k \le t\} = \Lambda(t) \,/\, \Lambda(T), \qquad 0 \le t \le T.$$

Time-homogeneous (stationary) Poisson process (see Section 2.5) is characterized by the condition $\Lambda(t) = \lambda t$, $t \geq 0$. A positive constant λ is called **the parameter** of the process.

If $\xi(t)$ is a Poisson process then a flow of homogeneous events can be defined in such a way. The number of events at each time t equals $\xi(t) - \xi(t-0)$. Such a flow is called **a Poisson flow** of homogeneous events. In particular, we obtain the simplest (stationary Poisson) flow; see Section 2.5.

A Poisson process as well as a Poisson flow of homogeneous events can be defined not only over $\mathbb{R}$, but also over $\mathbb{R}^m$ as well as over an arbitrary measurable space $(R, \mathfrak{L})$ where $\mathfrak{L}$ is a σ-algebra of measurable subsets of R. Let $(\Omega, A, \mathbf{P})$ be the main probability space and each $\omega \in \Omega$ defines an integer-valued measure μ over (R, L), i.e. an integer-valued set function satisfying the σ-additivity condition with probability 1 so that $\mu(A) = \sum_n \mu(A_n)$ as soon as $\{A_n\}$ is a sequence of distinct sets. (The case $\mu = \infty$ is not excluded.) An integer-valued random measure is also called **a point process**.

An integer-valued random measure μ is called **Poisson** if a measure $\Lambda(A)$ exists such that $\mu(A_n)$ are independent random variables with parameters $\Lambda(A_n)$ for a sequence $\{A_n\}$ of disjoint measurable subsets of R. The Λ is called **the leading measure** of the point process.

An integer-valued random measure defines a flow of homogeneous events in the space R if the σ-algebra $\mathfrak{L}$ includes all the one-point sets. A flow of homogeneous events associated with a Poisson random measure is called **a Poisson flow** in the R.

Let $\mu(A)$ be an integer-valued measure in $\mathbb{R}^m$, $A \in L$, the class of Borel sets of $\mathbb{R}^m$ and the following conditions are satisfied.

1. For any $n \geq 1$ and disjoint sets $A_1, \ldots, A_n$ the random variables $\mu(A_1), \ldots, \mu(A_n)$ are independent.

2. A nonnegative continuous function $\lambda(x)$, $x \in \mathbb{R}^m$, exists so that for a small neighborhood u_x of a point x

$$\mathbf{P}\{\mu(u_x) > 0\} = \lambda(x)\,|u_x| + o(|u_x|)$$

where $\mu(u_x)$ is the Lebesgue measure of the set u_x.

3. Under the same conditions

$$\mathbf{P}\{\mu(u_x) \geq 2\} = o(|u_x|).$$

It is also assumed that the diameter of u_x tends to zero and that the expressions $o(|u_x|)$ in the conditions 2 and 3 are uniform in x as soon as x belongs to a bounded subset of R.

Under the stated condition the random measure μ is Poisson, and its leading measure is

$$\Lambda(A) = \int_A \lambda(x)\, dx.$$

If $\lambda(x) = \lambda = const$ then the Poisson measure is called **stationary** and the associated flow is called **the simplest (stationary Poisson)** flow (point process) in the space R.

Some examples of real point processes close to Poisson ones are: bubbles in a metal pattern, defects in textile, rolled metal, paper etc., worms in a sack of flour.

Consider a measurable mapping $f : R \longrightarrow R'$ where (R', L') is a measurable space. Then a Poisson random measure μ with a leading function $\Lambda(A)$ is transformed into another Poisson random measure μ' defined by the equality

$$\mu'(f(A)) = \mu(A), \qquad f(A) \in \mathfrak{L}'.$$

The leading function of the new measure is defined by the equality

$$\Lambda(f(A)) = \Lambda(A), \qquad f(A) \in \mathfrak{L}'.$$

The "times" t_n of spatial events of the initial point process transfer to "times" $t_n' = f(t_n) \in R'$ of the new one.

In particular, the simplest flow $\{t_n\}$ with the parameter λ can be transformed into a Poisson flow $\{t_n'\}$ by a transformation $\{t_n'\} = f(t_n)$ where f is a nondecreasing function. The leading function $\Lambda'(t)$ of the new flow can be defined as follows: $\Lambda'(t) = \lambda f^{-1}(t)$.

4.2 Continuous-Time Markov Chains

A continuous-time Markov chain[1–9] is defined as a time-homogeneous Markov process $\{\xi(t), t \geq 0\}$ in a finite or denumerable phase space X. We shall use the term "Markov chain" in this section omitting the word "continuous-time."

A Markov chain is characterized by its markovian transition function

$$p_{ij}(t) = \mathbf{P}\{\xi(t_0 + t) = j \,|\, \xi(t_0) = i\}, \qquad i, j \in X.$$

This function satisfies **Kolmogorov-Chapman equations**

$$p_{ij}(t + \tau) = \sum_k p_{ij}(t)\, p_{kj}(\tau), \qquad t \geq 0,\ \tau \geq 0.$$

Absolute probabilities $p_j(t) = \mathbf{P}\{\xi(t) = j\}$ satisfy the equations

$$p_j(t) = \sum_i p_i(s)\, p_{ij}(t-s)$$

as soon as $0 \le s \le t$.

The matrix function $P(t) = (p_{ij}(t)),\ t \ge 0,$ is called **the transition matrix** of the Markov chain.

The expression $\sum_j |p_{ij}(t+h) - p_{ij}(t)|$ tends to zero as $h \to 0$ uniformly in any interval $0 < \delta \le t < \infty$. In particular, a function $p_{ij}(t)$ is uniformly continuous in an interval $[\delta, \infty)$. Moreover, either it vanishes for any $t > 0$ or it is positive for any $t > 0$.

For a stochastically continuous Markov chain $\xi(t)$ the following properties hold true.

The limit

$$\lim_{t\to 0} p_{ij}(t) = p_{ij}(0) = \delta_{ij}$$

exists where δ_{ij} is Kronecker delta. A function $p_{ij}(t)$ is uniformly continuous in the interval $\mathbb{R}^+$. For any i

$$\lim_{h\to 0} [1 - p_{ij}(h)/h] = -p'_{ii}(0)$$

exists (the right-hand side derivative is meant). The case of infinite value of the limit is not excluded. The limit is called **the exit rate** of the Markov chain and is denoted by $\lambda_i = -\lambda_{ii}$. For any states $i \ne j$ a finite limit $\lambda_{ij} = \lim\limits_{h\to 0} p_{ij}(h)/h$ exists; it is called **the transition rate** from state i to state j.

Consider the matrix $Q = (\lambda_{ij}) = (p'_{ij}(0))$ called **the Q-matrix** of the matrix function $P(t)$. If $\lambda_i < \infty$ then $p'_{ij}(t)$ exists and is continuous on $\mathbb{R}^+$. In such a case

$$|p_{ij}(t+h) - p_{ij}(t)| \le \lambda_i h, \quad \sum_j p'_{ij}(t) = 0$$

for any $t \ge 0$ and

$$\sum_j |p'_{ij}(t)| \le 2\lambda_i, \quad t \ge 0; \quad p'_{ij}(s+t) = \sum_k p'_{ik}(s)\, p_{kj}(t), \quad s > 0,\ t \ge 0.$$

If $\lambda_j < \infty$ then $p_{ij}(t)$ have the derivative continuous for $t \ge 0$ and moreover

$$p'_{ij}(t+s) = \sum_k p_{ik}(t)\, p'_{kj}(s), \quad t \ge 0,\ s > 0.$$

A state i is called **stable** if $\lambda_i < \infty$ and **instantaneous** if $\lambda_i = \infty$. A Markov chain is called **locally regular** if $\lambda_i = \sum_{j \neq i} \lambda_{ij} < \infty$.

According to the very definition of exit and transition rates we have

$$\mathbf{P}\{\xi(t+h) \neq i \,|\, \xi(t) = i\} = \lambda_i h + o(h), \quad h \to 0,$$

$$\mathbf{P}\{\xi(t+h) = j \,|\, \xi(t) = i\} = \lambda_{ij} h + o(h), \quad h \to 0, \quad j \neq i.$$

If all the states are stable and moreover in the expression $\lambda_{ij} h + o(h)$ the remainder term $o(h)$ is uniform in i given j then the Markov chain $\xi(t)$ is called **regular**.

The definition of a Markov chain via the Q-matrix is said to be **the infinitesimal definition** of the Markov chain. This term means that its global behavior is defined (in the probabilistic sense) via its behavior in infinitely-small time intervals.

There exists another **constructive definition** of a Markov chain.

We assume that a path function $\xi(t)$ is right-continuous step function with probability 1. Hence it is sufficient to define the sequence of its states $\xi_0, \xi_1, \xi_2, \ldots$ and the duration of the time intervals $\tau_0, \tau_1, \tau_2, \ldots$ within which these states are valid. Thus we have

$$\mathbf{P}\{\xi_n = j \mid \xi_0, \ldots, \xi_{n-1}; \tau_0, \ldots, \tau_{n-1}\} = \lambda_{\xi_{n-1} j} / \lambda_{\xi_n}, \tag{4.1}$$

$$\mathbf{P}\{\xi_1 = j \mid \xi_0; \tau_0\} = \lambda_{\xi_0 j} / \lambda_{\xi_0}. \tag{4.2}$$

The value ξ_0 is chosen at random, according to the distribution $\{p_i(0)\}$. If ξ_0 is known then τ_0 is realized as an exponential random variable with the parameter λ_{ξ_0}. Then ξ_1 is chosen according to the distribution (4.2). In general, if values of $\xi_0, \ldots, \xi_{n-1}, \tau_0, \ldots, \tau_{n-1}$ are already chosen then ξ_n is chosen according to the distribution (4.1).

If ξ_n is known then τ_n is realized as an exponential random variable with the parameter λ_{ξ_n}.

The described recurrent procedure enables constructing the path function $\xi(t)$ in a random interval $[0, t^*)$ where $t^* = \sum_{n=0}^{\infty} \tau_n$. If $t^* = \infty$ then the process is defined over the entire half-line $\mathbb{R}^+$. If $t^* < \infty$ then the process is **breaking**, t^* being **the breaking point**.

A Markov chain with **bounded exit rates** so that $\lambda_i \leq c$ for all the $i \in X$ possesses the property that $t^* = \infty$, hence there is a finite number of discontinuities of a path function in a finite time interval with probability 1.

If $\xi(t)$ is a regular separable right-continuous Markov chain with bounded exit rates then its definition via the Q-matrix is equivalent to the constructive definition. In such a case we have $\sum_j p_{ij} = 1,\ t \geq 0$. If exit rates are not bounded then in some cases $t^* < \infty$, or, equivalently, $\sum_j p_{ij} < 1$.

A state j of the Markov chain is **accessible** from a state i if there exists such a chain of states $i = i_0,\ i_1,\ \ldots,\ i_n = j$ that $\lambda_{i_0 i_1} > 0,\ \ldots, \lambda_{i_{n-1} i_n} > 0$. If i and j are mutually accessible then they are called **communicating states** .

Let $\xi(t)$ be a regular Markov chain, all the states being communicating. Assume also that the exit rates are bounded. Then the following statements hold true.

1. If the Markov chain $\{\xi(nh + \Delta),\ n \in \mathrm{N}^+\}$ possesses an ergodic distribution $\{\pi_j\}$ for some $h > 0$ and $\Delta \geq 0$ then the Markov chain $\{\xi(t),\ t \geq 0\}$ has the same **ergodic distribution**. This means that for each $j \in X$:

(i) $p_j(t) \underset{t\to\infty}{\longrightarrow} \pi_j$ independently of $\{p_i(0)\}$;

(ii) $p_{ij}(t) \underset{t\to\infty}{\longrightarrow} \pi_j,\ i \in X$.

2. Given $s > 0$ and i, the system of equation

$$(\lambda_i + s)\, x_i = 1 + \sum_{k\neq j,\, k\neq i} \lambda_{jk}\, x_k, \quad i \neq j,$$

has a unique solution $x_j = x_j(s),\ j \neq i$. Then the expression

$$\rho_i = \frac{1}{\lambda_i}\,[1 + \sum_{j\neq i} \lambda_{ij}\, x_j(+0)]$$

equals the mean time between two consecutive visits to the state i. If $\rho_i < \infty$ at least for some i then $\xi(t)$ possesses the ergodic distribution $\{\pi_j\}$ where $\pi_j = 1/(\lambda_j \rho_j),\ j \in X$.

3. We call **a stationary (steady-state) distribution** of a Markov chain such a distribution $\{x_j\}$ that the condition

$$x_j = \sum_i x_i\, p_{ij}(t), \quad t > 0,$$

is satisfied. In other words, if the initial distribution $\{p_j(0)\}$ is stationary then the time-dependent distribution $\{p_j(t)\}$ is also stationary for any $t > 0$. For a distribution $\{x_j\}$ to be stationary it is necessary and sufficient that

$$\sum_{i\in X} x_i\, \lambda_{ij} = 0, \quad j \in X, \tag{4.3}$$

with the norming condition $\sum_{i \in X} x_i = 1$. Any solution of the equation (4.3) satisfying the condition

$$0 < \sum_j |x_j| < \infty$$

defines a stationary distribution $\{y_j\}$ where

$$y_i = x_j \Big/ \sum_i x_i, \quad j \in X.$$

4. The stationary distribution is unique and coincides with the ergodic distribution.

5. The probabilities $p_j(t) = \mathbf{P}\{\,\xi(t) = j\,\}$ satisfy the system of **forward Kolmogorov equations**

$$p_j'(t) + \lambda_j\, p_j(t) \;=\; \sum_{i \neq j} p_i(t)\, \lambda_{ij}, \quad j \in X,$$

and are defined uniquely as the bounded solution of the system under the initial conditions $p_i(+0) = p_i(0)$. In particular, the transition probabilities $p_{ij}(t)$ can be obtained as the special case of $p_j(t)$ satisfying the condition $p_k(0) = \delta_{ik},\ k \in X$.

6. The functions $p_{ij}(t)$ satisfy the system of **backward Kolmogorov equations**

$$p_{ij}'(t) + \lambda_i\, p_{ij} \;=\; \sum_{k \neq i} \lambda_{ik}\, p_{kj}(t), \quad i,\, j \in X,$$

and are defined uniquely as the solution of this system satisfying the initial condition $p_{ij}(+0) = \delta_{ij}$.

7. Let $\{t_n\}$ be discontinuity points of $\xi(t)$; $0 < t_1 < t_2 < \cdots$. Then the sequence $\xi_n = \xi(t_n)$ is a discrete-time Markov chain with transition probabilities $p_{ij} = \lambda_{ij}/\lambda_i$ for $i \neq j$, $p_{ii} = 0$. It is called **an embedded Markov chain** of the random process $\xi(t)$. If $\{\xi_n\}$ possesses the ergodic distribution f_j where

$$f_j = \lim_{n \to \infty} \mathbf{P}\{\,\xi_n = j\,\},$$

then the constants

$$\pi_j = \frac{f_j}{\lambda_j \sum_i \frac{f_i}{\lambda_i}}, \quad j \in X,$$

define the ergodic distribution of $\xi(t)$ provided that

$$\sum_i \frac{f_i}{\lambda_i} < \infty.$$

Regular Markov chains with bounded exit rates are the main applied interest, though, strictly speaking, the variety of all possible Markov chains is much wider. In particular, random processes "passing to the infinity" in a finite time are also investigated. For a logical completion of the theory it was necessary to construct a prolongation of a path function after such "explosion" times[5–9].

Some other parameters of continuous-time Markov processes, besides transition probabilities, are of applied interest. The most important parameters are outlined below.

1. The probability $p_{ij}(t, A)$ of the event $\{\xi(t) = j$ and $\xi(u) \in A$ for all $u \in [0, t]\}$ under the condition $\xi(0) = i$ is defined as the solution of the system

$$p'_{ij}(t, A) + \lambda_j p_{ij}(t, A) = \sum_{i \neq j\, k \in A} p_{ik}(t, A)\, \lambda_{kj}, \quad i \in A,\ j \in A,$$

under the initial condition $p_{ij}(+0) = \delta_{ij}$. If $i \notin A$ or $j \notin A$ then $p_{ij}(t, A) = 0$.

2. Let $\tau_i(A)$ be the sojourn time in a set A of states given an initial state i. Then

$$\mathbf{P}\{\tau_i(A) > t\} = \sum_{j \in A} p_{ij}(t, A).$$

3. Let sets A, B $(B \subset A)$ be chosen and let it be known that $\xi(0) = i \in A$. Denote by v_i the time spent by a path function in the set B up to the time of the first exit from the set A. (Not excepting the case $v_i = \infty$.) Set $\varphi_i(s) = \mathbf{E}\, e^{-s v_i}$, $Re\, s > 0$. Then $\{\varphi_i(s)\}$ satisfies the following system of equations:

$$\lambda_i \varphi_i(s) = \sum_{i \neq j \in A} \lambda_{ij} \varphi_j(s), \quad i \notin B,$$

$$(\lambda_i + s)\, \varphi_i(s) = \sum_{i \neq j \in A} \lambda_{ij} \varphi_i(s) + \sum_{i \neq j \notin A} \lambda_{ij}, \quad i \in B.$$

4. Consider n independent integer-valued Markov chains $\xi_1(t), \ldots, \xi_n(t)$ with transition rates $\lambda_{ij}^{(k)}$, $1 \leq k \leq n$, and moreover $\lambda_{ij}^{(k)} = 0$ for $|i-j| \neq 1$, R being a n-dimensional set possessing such a property: if $(i_1, \ldots, i_n) \in R$, then $(j_1, \ldots, j_n) \in R$, as soon as $j_1 \geq i_1, \ldots, j_n \geq i_n$. Let $u(t)$ be the mean time spent by $(\xi_1(t), \ldots, \xi_n(t))$ in the set R within an interval $(0, t)$. We have

$$u(t) = \sum_{(i_1, \ldots, i_n) \in R} \int_0^t p_{1 i_1}(x) \ldots p_{n i_n}(x)\, dx,$$

where $p_{kj}(t)$ denotes the probability of a state i at time t for a Markov chain $\xi_k(t)$.

5. Under the same assumptions, let $v(t)$ be the mean number of visits to the set R within an interval $(0, t)$. Denote by Γ the set of integer-valued vectors $(i_1, \ldots, i_n)$ which are not included into R and such that the addition of a unit to the kth component implies the passage into the set R. Then we have

$$v(t) = \sum_{k=1}^{n} \sum_{(i_1, \ldots, i_n) \in \Gamma_k} \lambda^{(k)}_{i_k, i_k+1} \int_0^t p_{1 i_1}(x) \ldots p_{n i_n}(x)\, dx.$$

6. Let us say that an event A_τ occurred if a path function $\xi(t)$ visited a fixed state i at time τ and spent time there exceeding Δ without interruption. Denote by $w(t)$ the mean number of events A_τ within an interval $(0, t)$. Then

$$w(t) = e^{-\lambda_i \Delta} \sum_{j \neq i} \lambda_{ji} \int_0^t p_j(x)\, dx$$

where $p_j(t) = \mathbf{P}\{\xi(t) = j\}$.

7. Let $\lambda_k(t_1, \ldots, t_k)\, dt_1, \ldots, dt_k$ be the probability of some visits of a path function $\xi(t)$ to a state i within intervals $(t_1, t_1 + dt_1), \ldots, (t_k, t_k + dt_k)$. The function $\lambda_k(t_1, \ldots, t_k)$ is called **a k-rate**[10] of the flow of visits to the state i. Then

$$\lambda_k(t_1, \ldots, t_k) = \sum_{j \neq i} \lambda_{ji} p_j(t_1)\, \psi(t_2 - t_1) \cdots \psi(t_k - t_{k-1})$$

where

$$\psi(t) = \sum_{j \neq i} \lambda_{ji} p_{ij}(t).$$

Factorial moments of the number ν of visits to the state i within time interval $(0, t)$ can be expressed via k-rate as follows:

$$\mathbf{E}\nu(\nu - 1) \ldots (\nu - k + 1) = k! \int \underset{0<t_1<\ldots<t_k<t}{\cdots} \int \lambda_k(t_1, \ldots, t_k)\, dt_1 \ldots dt_k.$$

Continuous-time Markov chains have a very wide area of applications. They adequately describe the behavior of many physical as well as technological, biological and other kinds of systems. They are also widely used in operations research. Below we give a rather general approach to the construction of a markovian model of a system's behavior.

Assume we have a system the state of which at a time $t \geq 0$ certainly belongs to a finite or denumerable set X. While staying in the state $i \in X$, some operations O_{ij} are processed; $1 \leq j \leq |i|$ where $|i|$ denotes a nonnegative integer which can be called the "weight" of a state i. It is assumed that the duration of each operation is exponentially distributed, independently of the others. In other words, if the system possesses a state i at time t then the ith operation can terminate in a small interval $(t, t+h)$ with the probability $\mu_{ij}h + o(h)$. Two or more operations may be terminated within the same time interval with a probability $o(h)$, $h \to 0$.

As a further step, one should define a probability $p_{ik}^{(j)}$ of the transfer of the system state from i to j at a time of the termination of the operation O_{ij}.

Denoting by $\xi(t)$ the state of the system at a time $t \geq 0$ we conclude that $\xi(t)$ is a continuous-time Markov chain with transition rates

$$\lambda_{ik} = \sum_j \mu_{ij} \, p_{ik}^{(j)} .$$

In engineering systems some of the operations O_{ij} have a real sense; technological operations usually called service, the repair of failed components, maintenance etc. are such real operations. Besides, some fictious operations need to be considered while constructing a stochastic model. A fictious operation usually consists of waiting for the occurrence of an event, such as that of the arrival of a customer or the failure of a component. Many such events occur according to a markovian law; hence the time to the occurrence of each of them can be conceived as an exponential random variable.

Example 4.1. There are two machines executing the servicing of customers. If both the machines are operating then the service rate equals μ_{12}; if only the first (second) one is operating then the service rate equals μ_1 (resp. μ_2). If both the machines have failed the service rate equals zero. The failure rate of the first machine equals $\alpha_{12}^{(1)}$ $(\alpha_1^{(1)})$ at a time when both machines are operating (resp. only the first one is operating). The values of the failure rate of the second machine $\alpha_{12}^{(2)}$ and $(\alpha_2^{(2)})$ are defined in a similar way. The repair rate of failed machines equals β. Customers arrive according to a Poisson flow with the intensity λ provided that there are no customers on both machines. The system's behavior can be described by the continuous time Markov chain with 8 possible states: 1, 2 - both machines are operating; 3, 4 - only the first machine is operating; 5, 6 - only the second machine is operating; 7, 8 - both machines have failed. The odd states describe the situation when there are no customers on both machines, the even states describe the situation when either one of the machines is servicing a customer or both machines have failed and a customer waits for the termination of the repair. A Markov chain describing the system's behavior has the following transition rates λ_{ij} $(i \neq j)$:

$$\lambda_{12} = \lambda, \quad \lambda_{13} = \alpha_{12}^{(1)}, \quad \lambda_{15} = \alpha_{12}^{(2)}, \quad \lambda_{21} = \mu_{12}, \quad \lambda_{24} = \alpha_{12}^{(1)}, \quad \lambda_{31} = \beta,$$

$$\lambda_{34} = \lambda, \quad \lambda_{42} = \beta, \quad \lambda_{43} = \mu_1, \quad \lambda_{48} = \alpha_1^{(1)}, \quad \lambda_{51} = \beta, \quad \lambda_{56} = \lambda,$$

$$\lambda_{57} = \alpha_2^{(2)}, \quad \lambda_{62} = \beta, \quad \lambda_{65} = \mu_2, \quad \lambda_{68} = \alpha_2^{(2)}, \quad \lambda_{73} = \beta, \quad \lambda_{75} = \beta,$$

$$\lambda_{78} = \lambda, \quad \lambda_{84} = \beta, \quad \lambda_{86} = \beta, \quad \lambda_{ij} = 0 \text{ for all other } i,\, j.$$

The solution of a system of linear algebraic equations is required for the computation of a steady-state distribution. Well known standard numerical procedures of linear algebra such as Gauss method are widely used. But a fine structure of the system usually helps to apply a simpler procedure than a universal one. Very often the matrix of the system is of a three-diagonal form. In such a case simple recursive equations exist enabling a considerable reduction of the computations. Another efficient approach exploits the possibility of a state aggregation of a Markov process.

Assume that the phase state X of a Markov chain $\xi(t)$ can be partitioned into disjoint subsets $\{X_k\}$ called macrostates in such a way that "domestic" transitions within a macrostate are non-vanishing whereas transition among macrostates are comparatively rare. Strictly speaking, this property can be formalized in such a way that all the states within a given X_k are communicating, the transition rates being equal λ_{ij}. On the other hand, if $i \in X_\alpha$ and $j \in X_\beta$ where $\alpha \neq \beta$ then

$$\lambda_{ij} = \varepsilon \, \lambda_{ij}^{(0)}$$

where ε is a small positive parameter. Introduce a merged process $\nu_\varepsilon(t)$ which equals α as soon as $\xi(t) \in X_\alpha$. Then the following property holds true at least for a finite state Markov chain $\xi(t)$. For any $n \geq 1$; $\tau_1 < \cdots < \tau_n$ and macrostates $\alpha_1, \ldots, \alpha_n$ the following relation holds true:

$$\mathbf{P}\{\eta_\varepsilon(\frac{\tau_k}{\varepsilon}) = \alpha_k, \;\; 1 \leq k \leq n\} \underset{\varepsilon \to 0}{\longrightarrow} \mathbf{P}\{\gamma(\tau_k) = \alpha_k, \;\; 1 \leq k \leq n\}$$

where $\gamma(t)$ is a stationary continuous-time Markov chain in the space of macrostates with the transition rates $\mu_{\alpha\beta}$ defined as follows.

Let $\{\pi_i^{(\alpha)}, \; i \in X_\alpha\}$ be the ergodic distribution of the process $\xi(t)$ taking values in the set X_α provided $\varepsilon = 0$. Then

$$\mu_{\alpha\beta} = \sum_{i \in X_\alpha} \sum_{j \in X_\beta} \pi_i^{(\alpha)} \lambda_{ij}^{(0)}.$$

The approach admits a lot of generalizations for much more complex processes and is widely used while analyzing networks.

Time-dependent parameters of Markov chains require more sophisticated approaches than steady-state ones, though Laplace transforms and hence the moments of distributions can be computed almost as simply as the steady-state distribution.

Introduce the Laplace transform

$$\varphi_j(s) = \int_0^\infty e^{-st}\, p_j(t)\, dt.$$

Then the system of forward Kolmogorov differential equations is reduced to the system

$$s\,\varphi_j(s) - \sum_i \varphi_i(s)\,\lambda_{ij} = p_j(0), \;\; j \in X,$$

where $\lambda_{ii} = -\lambda_i$.

Example 4.2. Let $X = \{0, 1\}$ and the Q-matrix has the form

$$\begin{pmatrix} -\lambda & \lambda \\ \mu & -\mu \end{pmatrix}.$$

Then

$$\begin{cases} (s + \lambda)\,\varphi_0(s) - \mu\,\varphi_1(s) = a \\ -\lambda\,\varphi_0(s) + (s + \mu)\,\varphi_1(s) = b \end{cases}$$

where $a = p_0(0)$, $b = p_1(0)$. The solution of the system has the form

$$\varphi_0(s) = \frac{\mu}{\lambda + \mu}\,\frac{1}{s} + \frac{a\lambda - b\mu}{\lambda + \mu}\,\frac{1}{s + \lambda + \mu},$$

$$\varphi_1(s) = \frac{\lambda}{\lambda + \mu}\,\frac{1}{s} - \frac{a\lambda - b\mu}{\lambda + \mu}\,\frac{1}{s + \lambda + \mu}.$$

The inversion of the Laplace transform gives explicit expressions

$$p_0(t) = \frac{\mu}{\lambda + \mu} + \frac{a\lambda - b\mu}{\lambda + \mu}\, e^{-(\lambda+\mu)t},$$

$$p_1(t) = \frac{\lambda}{\lambda + \mu} - \frac{a\lambda - b\mu}{\lambda + \mu}\, e^{-(\lambda+\mu)t}.$$

In particular

$$\pi_0 = \lim_{t\to\infty} p_0(t) = \frac{\mu}{\lambda+\mu}, \quad \pi_1 = \lim_{t\to\infty} p_1(t) = \frac{\lambda}{\lambda+\mu}.$$

Evidently the convergence to the limits is an exponential rate. If one sets $a = \frac{\mu}{\lambda+\mu}$, $b = \frac{\lambda}{\lambda+\mu}$ then $p_0(t) = \pi_0$, $p_1(t) = \pi_1$ for each $t \geq 0$; hence the distribution $\{\pi_0, \pi_1\}$ is a stationary one.

The main numerical approach to the investigation of markovian models of systems is the method of simulation; see Chapter 17.

Classification of states.

Let $\xi(t)$ be a Markov chain in a closed class X of communicating states. Assume that $\xi(0) = i \in X$. Define θ_i as the first return time to the state i. In the case no return ever occurs set $\theta_i = \infty$.

A state $i \in X$ is called **transient** if $\theta_i = \infty$ with a positive probability; otherwise it is called **recurrent**. A recurrent state i is called **recurrent zero** state if $\mathbf{E}\,\theta_i = \infty$; otherwise it is called an **ergodic** state.

The following statement holds true. Either all the states are transient, or all the states are recurrent zero, or all the states are ergodic. In the latter case the Markov chain is called **ergodic**. Moreover, if X is finite and all the states are communicating then a Markov chain is ergodic. The distinction with discrete-time Markov chains is that no periodic phenomena may occur. For an ergodic Markov chain $p_{ij}(t) \underset{t\to\infty}{\longrightarrow} \pi_j$ independently of i and $p_j(t) \underset{t\to\infty}{\longrightarrow} \pi_j$ independently of the initial distribution. If moreover X is finite then the convergence is of an exponential rate in both cases.

4.3 Markov Process in a Finite or Countable Phase-Space

We shall consider here a nonhomogeneous Markov process in a finite or countable phase-space. A process is said to be **regular** in a time interval $a < t < b$ if for some functions $\lambda_{ij}(t)$ and $\lambda_i(t) = \sum_{j\neq i} \lambda_{ij}(t)$ continuous on $[a, b]$ the following equalities hold true:

$$\mathbf{P}\{\xi(t+\Delta_1) \neq i \,|\, \xi(t-\Delta_0) = i\} = \lambda_i(t)\,(\Delta_0+\Delta_1) + o(\Delta_0+\Delta_1),$$

$$\mathbf{P}\{\xi(t+\Delta_1) = j \,|\, \xi(t-\Delta_0) = i\} = \lambda_{ij}(t)\,(\Delta_0+\Delta_1) + o(\Delta_0+\Delta_1),$$

where

$$j \neq i, \;\; \Delta_0 \geq 0, \;\; \Delta_1 \geq 0, \;\; \Delta_0+\Delta_1 \to 0, \;\; a \leq t-\Delta_0, \;\; t+\Delta_1 \leq b$$

and moreover the term $o(\Delta_0 + \Delta_1)$ is uniform in t, $a \le t \le b$, and in the last equality also in i for a given j. A process is said to be **regular** in an infinite interval $[t_0, \infty)$ if it is regular in an arbitrary interval $[t_0, T]$. A process is said to be **piecewise regular** in a finite interval $[a, b]$ if there exists such a finite partition of this interval into intervals $\{[a_i, b_i]\}$ that it is regular in each of them. A process is **piecewise regular** in an infinite time interval if it is piecewise regular in an arbitrary finite subinterval.

An interval in which a process is regular is called **a regularity interval** of the process. The points in which the regularity does not hold true are its **singularity points.**

The function $\lambda_{ij}(t)$ is called **the transition rate** from the state i to j if $i \neq j$; the function $\lambda_i(t) = -\lambda_{ii}(t)$ is called **the exit rate** from a state i.

Let t be a singularity point. Consider the expression

$$f_{ij}(t) = \lim_{\Delta_0,\, \Delta_1 \to 0} \mathbf{P}\{\xi(t + \Delta_1) = j \,|\, \xi(t - \Delta_0) = i\}$$

where $\Delta_0 > 0$, $\Delta_1 > 0$. It is called **the instantaneous transition probability** from state i to state j.

The definition of the Markov process via the transition rates, exit rates and instantaneous transition probabilities is called its **infinitesimal definition.**

Hereinafter we suppose that the considered Markov process is piecewise regular, has right-continuous path functions with probability 1 and that $\lambda_i(t) \le \lambda(t)$ where $\lambda(t)$ is a function bounded in any bounded time interval. Indeed the infinitesimal parameters define a family of Markov processes each of which is characterized by a special initial distribution $\{p_i(0)\}$.

There exists also **a constructive definition** of a regular Markov process which differs from the definition of a time-continuous Markov chain by only two points:

(i) if t is a time of a visit to the state i then the next discontinuity point τ of the path function $\xi(t)$ has the following distribution:

$$\mathbf{P}\{\tau \le x\} = 1 - \exp\left\{ - \int_t^x \lambda_i(u)\, du \right\},$$

(ii) if t is an exit time of the process from a state i then it transfers into a state j with the probability $\lambda_{ij}(t)/\lambda_i(t)$. A piecewise regular Markov process can also be defined constructively via its successive prolongation within regularity intervals; we will omit trivial details.

Introduce functions

$$p_{ij}(t_0, t) = \mathbf{P}\{\xi(t) = j \,|\, \xi(t_0) = i\}, \ \ t_0 < t.$$

They are called **transition probabilities** of the process $\xi(t)$. If $t_0 < t$ and both t_0 and t belong to the same regularity interval then the system of **forward Kolmogorov equations**

$$\frac{\partial p_{ij}(t_0, t)}{\partial t} = \sum_k \lambda_{kj}(t) \, p_{ik}(t_0, t)$$

as well as the system of **backward Kolmogorov equations**

$$\frac{\partial p_{ij}(t_0, t)}{\partial t_0} = -\sum_k \lambda_{ik}(t_0) \, p_{kj}(t_0, t), \quad \text{where } \lambda_{ii}(t) = -\lambda_i(t)$$

holds true. Their solutions are continuous as t_0 and t tend to the bounds of the regularity interval. They also satisfy the condition

$$p_{ij}(t_0, t) \underset{|t-t_0|\to 0}{\longrightarrow} \delta_{ij}.$$

For a singularity point t one obtains

$$p_{ij}(t_0, t) = p_{ij}(t_0, t+0) = \sum_k p_{ik}(t_0, t-0) \, f_{kj}(t).$$

Consider, for example, the case of the lone singularity point τ, $t_0 < < \tau < t_1$. We can solve the system of forward Kolmogorov equations for $p_{ij}(t_0, t)$ in an interval (t_0, t) and obtain

$$p_{ij}(t_0, \tau - 0) = \lim_{t \uparrow \tau} p_{ij}(t_0, t).$$

At the second step we jump over the point τ obtaining

$$p_{ij}(t_0, \tau + 0) = \sum_k p_{ik}(t_0, \tau - 0) \, f_{kj}(\tau).$$

We solve the system of forward Kolmogorov equations in the interval $[\tau, \tau_1]$ using $p_{ij}(t_0, \tau + 0)$ as initial values. Finally we obtain $p_{ij}(t_0, t_1)$.

Let t_0 and t belong to the same regularity interval. There exists a numerical algorithm of the solution of the system of forward Kolmogorov equations based on the following probabilistic reasoning. Denote by $\gamma(t)$ the number of jumps of the process $\xi(t)$ in an interval (t_0, t) and set

$$p_{ij}^{(m)}(t_0, t) = \mathbf{P}\{\xi(t) = j, \gamma(t) = m \,|\, \xi(t_0) = i\}.$$

Then, by the entire probability formula

$$p_{ij}(t_0, t) = \sum_{m=0}^{\infty} p_{ij}^{(m)}(t_0, t).$$

All the terms of this series are nonnegative and can be computed according to the following recurrent procedure:

$$p_{ij}^{(0)}(t_0, t) = \delta_{ij} \exp\left\{-\int_{t_0}^{t} \lambda_i(x)\, dx\right\},$$

$$p_{ij}^{(m)}(t_0, t) = \sum_k \int_{t_0}^{t} p_{ik}^{(m-1)}(t_0, x)\, \lambda_{kj}(x) \exp\left\{-\int_{x}^{t} \lambda_j(y) dy\right\} dx, \quad m \geq 1.$$

The sufficient number of terms to be kept can be estimated in such a way. Set $a = \int_{t_0}^{t} \lambda(x)\, dx$ where $\lambda(t)$ is an upper bound of $\lambda_i(t),\ i \in X$. Then

$$\sum_{m=r}^{\infty} \sum_j p_{ij}^{(m)}(t_0, t) \leq e^{-a} \sum_{m=r}^{\infty} \frac{a^m}{m!}.$$

Hence an appropriate r can be chosen using the Poisson distribution.

Markov process with periodic rates.

Let $\lambda_{ij}(t)$ be a periodic function with a period T. Consider the Markov chain $\{\xi_n\}$ where $\xi_n = \xi(nT)$. Its transition probabilities

$$p_{ij} = p_{ij}(0, T)$$

can be computed via the solution of forward Kolmogorov equations in the interval of the length T. Suppose that the ergodic distribution $\{\pi_j\}$ of $\{\xi_n\}$ exists. Then a quasi-stationary behavior is peculiar to the Markov process $\xi(t)$. This means that

$$p_{ij}(t) - A_j(t) \underset{t\to\infty}{\longrightarrow} 0, \quad p_j(t) - A_j(t) \underset{n\to\infty}{\longrightarrow} 0$$

where $A_j(t)$ are periodic functions defined as follows:

$$A_j(nT + \tau) = \sum_i \pi_i\, p_{ij}(0, \tau).$$

4.4 Birth and Death Process

Let $\{\xi(t),\ t \geq 0\}$ be a time-homogeneous Markov chain in the phase space $X = \{0, 1, 2, \ldots\}$. This Markov chain is called a **birth process** if a path function $\xi(t)$ is a nondecreasing step function with unit jumps and $\xi(0) = 0$ (both properties should hold with probability 1).

The probabilistic structure of the process is very simple. Denote by t_k the time of the kth jump of a path function. Then $\xi(t) = 0$ if $0 \le t < t_1$; $\xi(t) = k$ if $t_k \le t < t_{k+1}$ for $k \ge 1$. As concerns the points t_k, the variables $t_1, t_2 - t_1, t_3 - t_2, \dots$ are independent and distributed exponentially with parameters $\lambda_0, \lambda_1, \lambda_2, \dots$.

Set $p_k(t) = \mathbf{P}\{\xi(t) = k\}$. Then $p_0(t) = e^{-\lambda_0 t}$, $t \ge 0$, and

$$p_k(t) = \lambda_{k-1} e^{-\lambda_k t} \int_0^t e^{\lambda_k x} p_{k-1}(x)\, dx, \quad k \ge 1,\ t \ge 0.$$

The constants λ_k are called **birth parameters.** If $\xi(t) = k$ then a jump from k to $k+1$ can occur in a small interval $(t, t+dt)$ with the probability $\lambda_k\, dt$. The case $t_n \underset{n\to\infty}{\longrightarrow} t^* < \infty$ is not excluded. In such a case the path function reaches infinity in a finite time interval.

Though iff $\sum\limits_{n=0}^{\infty} \lambda_n^{-1} = \infty$ then $t_n \to \infty$ with probability one. If such a series is convergent then t^* is finite with probability one, and

$$\mathbf{E}\, t^* = \sum_{n=0}^{\infty} \lambda_n^{-1}, \quad \mathbf{Var}\, t^* = \sum_{n=0}^{\infty} \lambda_n^{-2}.$$

In the former case $\sum\limits_{k=0}^{\infty} p_k(t) = 1$ for each $t \ge 0$.

The function $p_k(t)$ satisfies the system of Kolmogorov differential equations

$$p_0'(t) = -\lambda_0\, p_0(t), \quad t > 0,$$

$$p_n'(t) = -\lambda_n\, p_n(t) + \lambda_{n-1}\, p_{n-1}(t), \quad n > 0,\ t > 0,$$

with initial conditions $p_0(0) = 1$, $p_n(0) = 0$, $n > 0$. As a special case we obtain the Poisson process with parameter λ. For such a process $\lambda_n = \lambda$, $n \ge 0$.

A birth process describes the dynamics of the growth of a biological population in which the birth rate depends on the current size of the population. The same mathematical model can also describe many real processes of another nature.

Consider an example from the reliability theory. Consider a system consisting of n components functioning independently. Each of the components can fail in a time interval $(t, t+dt)$ with a probability λdt if it has not failed up to time t. Denote by $\xi(t)$ the number of failed components at a time t. If $\xi(t) = k$ then $n - k$ components are operating at time t; hence

$$\mathbf{P}\{\xi(t+h) = k+1 \,|\, \xi(t) = k\} = (n-k)\,(1 - e^{-\lambda h})\,e^{-(n-k-1)\lambda h} =$$

$$= (n-k)\lambda h + o(h), \quad h \to 0.$$

Therefore the $\xi(t)$ is a birth process with $\lambda_k = (n-k)\lambda$ for $k < n$, $\lambda_k = 0$ for $k \geq n$.

A **death process** is a time-homogeneous Markov chain in the phase space $X = \{0, 1, 2, \dots\}$ possessing nondecreasing path function which can have only jumps of the value -1 with probability 1. Such a process is characterized by **death parameters** μ_k, $k \geq 1$, defined as follows:

$$\mathbf{P}\{\xi(t+\Delta) = k-1 \,|\, \xi(t) = k\} = \mu_k\,\Delta + o(\Delta), \quad \Delta \to 0,$$

and the initial distribution $\{p_i(0)\}$. The system of forward Kolmogorov equations has the form

$$p_0'(t) = \mu_1\,p_0(t), \quad t > 0,$$

$$p_n'(t) = -\mu_n\,p_n(t) + \mu_{n+1}\,p_{n+1}(t), \quad n > 0, \; t > 0.$$

If $\xi(0) = n$ with probability 1 then the process $\eta(t) = n - \xi(t)$ is a birth process with birth parameters $\lambda_k = \mu_{n-k}$.

Example 4.3. The service to a customer consists of n phase. The time required for the phase k is an exponential random variable η_k with parameter α_k. Assume that the service begins at time $t = 0$ and that $\{\alpha_k\}$ are independent. Denote by $\xi(t)$ the remaining number of phases at time t. Then the $\xi(t)$ is a death process with death parameters $\mu_k = \alpha_{n-k+1}$ and the initial distribution $p_k(0) = \delta_{kn}$.

We call **a birth and death process** a continuous-time Markov chain in the phase space $X = \mathbb{N}$ that a path function $\xi(t)$ has no jumps of a value apart from ± 1. Such a process can be described constructively as follows. Denote by t_n the time of the nth jump of the path function so that $0 < t_1 < t_2 < \cdots$ and ξ_k the value of the process in the kth interval, i.e. $\xi(t) = \xi_0$ if $0 \leq t < t_1$, $\xi(t) = \xi_1$ if $t_1 \leq t < t_2$ and so on. Then $\xi(0)$ is chosen at random; $\mathbf{P}\{\xi_0 = i\} = p_i(0)$. If $\xi_0 = i$ then t_1 is an exponential random variable with the parameter $\lambda_i + \mu_i$. Assume that $\xi_0, \dots, \xi_{n-1}$, $t_1, \dots, t_n$ are known for some $k \geq 1$. The ξ_n is chosen at random: if $\xi_{n-1} = i$ then $\xi_n = i+1$ with probability $f_i = = \lambda_i/(\lambda_i + \mu_i)$ or $\xi_n = i - 1$ with probability $g_i = \mu_i/(\lambda_i + \mu_i)$, independently of $\xi_0, \dots, \xi_{n-1}$, $t_1, \dots, t_n$. At the time when ξ_n is already chosen (say, $\xi_n = k$) the t_{n+1} is realized, the difference $t_{n+1} - t_n$ being independent of $\xi_0, \dots, \xi_n$, $t_1, \dots, t_n$ and distributed exponentially with the parameter $\lambda_k + \mu_k$. The constants λ_k are called **birth parameters**; the μ_k are called **death parameters** of a birth and death process.

An embedded Markov chain $\{\xi_n\}$ has the following transition matrix:

$$P = \begin{pmatrix} 0 & 1 & 0 & 0 & . & . & . \\ g_1 & 0 & f_1 & 0 & . & . & . \\ 0 & g_2 & 0 & f_2 & . & . & . \\ 0 & 0 & g_3 & 0 & . & . & . \\ . & . & . & . & . & . & . \end{pmatrix}$$

Though if $\xi_n = k$ with $\lambda_k = \mu_k = 0$ then the Markov chain breaks at that step. This means that the process $\xi(t)$ is absorbed in the state k.

State probabilities $p_j(t) = \mathbf{P}\{\xi(t) = j\}$ satisfy the system of forward Kolmogorov equations

$$p_n'(t) = -(\lambda_n + \mu_n)p_n(t) + \lambda_{n-1}p_{n-1}(t) + \mu_{n+1}p_{n+1}(t), \quad n \geq 0,$$

where $\lambda_{-1} = 0$. The equations satisfy the initial conditions $\lim_{t\to 0} p_j(t) = = p_j(0)$.

Assume that $\lambda_k > 0$ for $0 \leq k \leq N-1$, $\lambda_N = 0$, $\mu_k > 0$, $1 \leq k \leq N$, and consider the process $\xi(t)$ in the set $\{0, 1, \ldots, N\}$ only. Then the ergodic distribution of the form

$$\pi_j = \frac{\theta_j}{(\theta_0 + \cdots + \theta_N)}, \quad 0 \leq j \leq N,$$

exists where

$$\theta_j = \frac{\lambda_0 \ldots \lambda_{j-1}}{\mu_1 \ldots \mu_j}.$$

Suppose that $\lambda_k > 0$, $k > 0$ and $\mu_k > 0$, $k \geq 1$, then the ergodic distribution exists iff

$$\sum_{k=0}^{\infty} \theta_k < \infty, \quad \sum_{k=0}^{\infty} \left\{ \sum_{i=0}^{k} \frac{\theta_i}{\lambda_k \theta_k} \right\} = \infty.$$

If these conditions are satisfied then the ergodic distribution is given by the expression

$$\pi_j = \frac{\theta_j}{\sum_{k=0}^{\infty} \theta_k}, \quad j \geq 0.$$

Let $\xi(t)$ be a birth and death process with the ergodic distribution $\{\pi_j\}$. Denote by $a_n^{(r)}$, $b_n^{(r)}$ the durations of consecutive sojourn times in the sets of states $\{0, 1, \ldots, r\}$ and $\{r, r+1, \ldots\}$ resp. Then

$$\mathbf{E}\, a_n^{(r)} = \frac{\pi_0 + \cdots + \pi_r}{\lambda_r\, \pi_r}, \qquad \mathbf{E}\, b_n^{(r)} = \frac{\pi_{r+1} + \pi_{r+2} + \cdots}{\lambda_r\, \pi_r}, \quad n \geq 2.$$

Suppose that $\lambda_0 = 0$, then the state 0 is absorbing. Denote by u_m the probability of the absorption given the initial state *mand* T_m the mean time to the absorption under the same condition. Then

$$u_m = \sum_{i=m}^{\infty} \prod_{j=1}^{i} \frac{\mu_j}{\lambda_j} \Bigg/ \left(1 + \sum_{i=1}^{\infty} \prod_{j=1}^{i} \frac{\mu_j}{\lambda_j}\right) \quad \text{for} \quad \sum_{i=1}^{\infty} \prod_{j=1}^{i} \frac{\mu_j}{\lambda_j} < \infty,$$

$u_m = 1$ otherwise;

$$\lambda_0\, T_m = \sum_{i=1}^{\infty} \theta_i + \sum_{r=1}^{m-1} \left(\prod_{k=1}^{r} \frac{\mu_k}{\lambda_k}\right) \sum_{j=r+1}^{\infty} \theta_j \quad \text{for} \quad \sum_{i=1}^{\infty} \theta_i < \infty,$$

$T_m = \infty$ otherwise.

4.5 Application of Birth and Death Processes in Queueing Theory and Reliability Theory

The theory of birth and death processes enables the derivation of an elementary queueing theory[11,12]. Let a queueing system include m service channels and l waiting seats. There is the simplest (stationary Poisson) input of customers with a parameter λ called **the input rate**. Each channel can service only one customer at any time. A customer entered at a time when all the channels were engaged has to take a waiting seat, but only if any is vacant; otherwise it is lost. The service time of a customer is an exponential random variable with the parameter μ called **the service rate**. The interarrival intervals and service times of customers are independent.

The described system is usually coded as $M/M/m/l$, according to the Kendall classification of queueing systems. The system $M/M/m/0$ is called **a loss system**, the system $M/M/m/\infty$ (or, in the usual reduced notation, $M/M/m$) is called **a system with an unbounded queue**, and the system $M/M/m/l$ with $0 < l < \infty$ is **a system with a restricted queue**.

Consider a random process $\nu(t)$ equal to the number of customers (**queue length**) in the system at a time t. The $\nu(t)$ is a birth and death process in the phase space $\{0, 1, \dots, m+l\}$ if $m+l < \infty$ or $\mathbb{N}$ if $m = \infty$ or $l = \infty$. We have $\lambda_k = \lambda,\ 0 \le k < m+l,\ \ \lambda_{m+l} = 0\ \ (m+l < \infty)$ and $\mu_k = k\,\mu,\ k \le m,\ \ \mu_k = m\,\mu,\ \ k > m\ \ (m < \infty)$. Hence the ergodic distribution can be defined by the expression $\{\pi_j\}$ where

$$\pi_j = \lim_{t\to\infty} \mathbf{P}\{\nu(t) = j\}$$

for the system $M/M/m/0$. For the system with a restricted queue we have

$$\pi_k = \frac{1}{k!}\left(\frac{\lambda}{\mu}\right)^k \pi_0, \quad 0 \le k \le m,$$

$$\pi_k = \frac{1}{m!\,m^{k-m}}\left(\frac{\lambda}{\mu}\right)^k \pi_0, \quad m+1 \le k \le m+l,$$

where

$$\pi_0^{-1} = \sum_{i=0}^{m-1} \frac{1}{k!}\left(\frac{\lambda}{\mu}\right)^k + \frac{1}{m!}\left(\frac{\lambda}{\mu}\right)^m \left(1 - \left(\frac{\lambda}{m\,\mu}\right)^{l+1}\right)\left(1 - \frac{\lambda}{m\,\mu}\right)^{-1}.$$

As to the system with an unbounded queue, the ergodic distribution exists iff $\lambda < m\,\mu$. In such an event

$$\pi_k = \frac{1}{k!}\left(\frac{\lambda}{\mu}\right)^k \pi_0, \quad 0 \le k \le m, \qquad \pi_k = \frac{1}{m!\,m^{k-m}}\left(\frac{\lambda}{\mu}\right)^k \pi_0, \quad k > m;$$

$$\pi_0^{-1} = \sum_{i=0}^{m-1} \frac{1}{k!}\left(\frac{\lambda}{\mu}\right)^k + \frac{1}{m!}\left(\frac{\lambda}{\mu}\right)^m \left(1 - \frac{\lambda}{m\,\mu}\right)^{-1}.$$

A queueing system is in the steady state iff $p_i(0) = \pi_i$ for each $i \ge 0$. If $m+l < \infty$ then the loss probability equals π_{m+l}. This probability means the limit of the frequency of lost customers among the first n customers is $n \to \infty$. The mean number of lost customers in a unit time interval equals $\lambda\,\pi_{m+l}$.

Consider **the waiting time** w of a customer measured from the time of its arrival to the beginning of its servicing. We have $w = 0$ if $m = \infty$ or $l = 0$. For the $M/M/m$ queueing system

$$\mathbf{E}\,w = \left(\frac{\lambda}{\mu}\right)^m \pi_0 \left\{ m!\,m\,\mu\left(1 - \frac{\lambda}{m\,\mu}\right)^2 \right\}^{-1}$$

as soon as $\lambda < m\mu$. The mean number of engaged waiting seats equals $\lambda\,\mathbf{E}w$.

For the system $M/M/M/l$ with $0 < m + l < \infty$ the mean waiting time for a customer which is not lost is given by the expression

$$\mathbf{E}\,w' = \frac{1}{\lambda(1-\pi_{m+l})} \sum_{k=m+1}^{m+l} (k-m)\,\pi_k =$$

$$= \left(\frac{\lambda}{\mu}\right)^m \pi_0 \left(1-(l+1)\left(\frac{\lambda}{m\mu}\right)^l + l\left(\frac{\lambda}{m\mu}\right)^{l+1}\right)\left(m!\,m\,\mu\left(1-\frac{\lambda}{m\,\mu}\right)^2\right)^{-1}.$$

The approach of birth and death processes describes some other queueing phenomena as well. Consider, for example, a queueing system with impatient customers with the following peculiarities of their local behavior. If the queue length equals k at the arrival time then the customer joins the queue only with the probability b_k. A customer can also renege from a waiting seat with the rate α as well as during the servicing with the rate β. The birth and death process is then characterized by parameters $\lambda_k = b_k\,\lambda$, $k \geq 0$; $\mu_k = k\,(\mu+\beta)$, $k \leq m$; $\mu_k = m\,(\mu+\beta)+(k-m)\,\alpha$, $k > m$. The condition

$$\lim_{n\to\infty} \frac{\lambda\, b_n}{m\,(\mu+\beta)} < 1$$

is sufficient for the ergodicity of the Markov chain $\nu(t)$. The probability of a loss at the arrival time equals

$$\sum_{k=0}^{\infty} (1-b_k)\,\pi_k.$$

The probability of a loss from a waiting seat equals

$$\frac{\alpha}{\lambda}\sum_{k\geq m} (k-m)\,\pi_k$$

and the probability of a loss during the servicing equals

$$\frac{\beta}{\lambda}\left(\sum_{k=1}^{m-1} k\,\pi_k + m\sum_{k=m}^{\infty} \pi_k\right).$$

The dependence of the input on the state of the queue can also be taken into account. Consider for example a well known machine interference problem. Suppose that a system includes N machines serviced by m operators. An operating machine can fail with a probability λdt in a small interval

dt. The service time is an exponential random variable with the parameter μ. Denoting by $\nu(t)$ the number of failed machines at a time t one obtains $\lambda_k = (N-k)\,\lambda,\;\; \mu_k = k\,\mu$ if $k \le m,\;\; \mu_k = m\,\mu$ if $k > m$.

A lot of results of **the redundancy theory** are also based on the birth and death processes. It is required to assume that the time to failure as well as repair time are exponential random variables.

We describe here a model of a redundant system. Let the system consist of N identical components. If k components have failed then the regime for the rest $N-k$ components is defined as follows: n_{1k} of them are in a loaded regime, n_{2k} components are in a light regime and the rest $N-k-$ $-n_{1k}-n_{2k}$ components are used as cold standby ones. The failure rate of a component depends on the regime of its loading and equals $\lambda^{(1)}$, $\lambda^{(2)}$ or $\lambda^{(3)}$ for n_{1k}, n_{2k} and all the rest of the components, respectively. We have

$$\lambda_k = n_{1k}\,\lambda^{(1)} + n_{2k}\,\lambda^{(2)} + (N-k-n_{1k}-n_{2k})\,\lambda^{(3)};$$

$$\mu_k = k\,\mu,\;\; k \le m; \qquad \mu_k = m\,\mu,\;\; k > m.$$

There are **repairable** $(m > 0)$ as well as **unrepairable** $(m = 0)$ systems. For former ones $p_j(t) \underset{t\to\infty}{\longrightarrow} \pi_j > 0$ $(0 \le j \le N)$ whereas for latter ones $p_j(t) \underset{t\to\infty}{\longrightarrow} 0$ for $j > N$, $p_N(t) \underset{t\to\infty}{\longrightarrow} 1$ but a trivial case of vanishing a λ_k for $k < N$. Main steady-state parameters of a repairable system can be expressed via the ergodic distribution. The **availability** of the system

$$A = \lim_{t\to\infty} \mathbf{P}\{\nu(t) \le r\} = \sum_{j=0}^{r} \pi_j$$

in the case if $\{0, 1, \ldots, r\}$ is the set of the system operating states. **The mean efficiency** of the system is

$$\bar{w} = \sum_{j=0}^{N} w_j\,\pi_j$$

where w_j is the value of the efficiency given a state j. **The reliability function**

$$R(T) = \lim_{t\to\infty} \mathbf{P}\{\nu(t+u) \le r,\; 0 \le u \le T\}$$

can be obtained via the expression

$$R(T) = \sum_{j=0}^{r} \pi_j\,\varphi_{j,r+1}(T)$$

where $\varphi_{j,r+1}(T)$ is the probability of the staying of the process $\nu(t)$ in the state set $\{0, 1, \ldots, r\}$ during the time T given an initial state j. A bound

$$\varphi_{j,r+1}(T) \geq \mathbf{P}\{\gamma_j + \cdots + \gamma_r > T\}$$

holds true, the $\gamma_j, \ldots, \gamma_r$ being independent exponential random variables with parameters $\lambda_j, \ldots, \lambda_r$. **The mean sojourn time** in the set of operating states

$$T_0 = \frac{1}{\lambda_r \pi_r} (\pi_0 + \cdots + \pi_r).$$

The mean sojourn time in the set of the system failure states

$$T_1 = \frac{1}{\lambda_r \pi_r} (\pi_{r+1} + \cdots + \pi_N).$$

The main parameter of an unrepaired system is the probability $P(T)$ of the availability within a time interval $(0, T)$ given an operating state of all the components at time $t = 0$. We have

$$P(T) = \mathbf{P}\{\gamma_0 + \cdots + \gamma_r > T\}$$

where $\{\gamma_i\}$ are the same as above.

4.6 Main Relations for a Semi-Markov Process

Let $\nu(T)$ be a semi-Markov process (SMP). Set

$$Q_i(x) = \sum_j Q_{ij}(x).$$

Then $Q_i(x)$ is the distribution function of the staying of $\nu(t)$ in a state i. Let, then, $H_j(t)$ be the mean number of visits of the process to a state j within an interval $(0, t)$. For the definiteness assume that $Q_i(+0) = 0$ so that multiple jumps of the process are not admitted at a time. Then the following equation holds true:

$$dH_j(t) = \sum_i p_i^{(0)} \, dQ_{ij}^{(0)}(t) + \sum_i \int_0^t dQ_{ij}(t-x) \, dH_i(x), \quad t > 0.$$

Set

$$h_j^*(s) = \int_0^\infty e^{-st} \, dH_j(t), \; q_{ij}^*(s) = \int_0^\infty e^{-st} \, dQ_{ij}(t), \; q_{ij}^{(0)*}(s) = \int_0^\infty e^{-st} \, dQ_{ij}^{(0)}(t).$$

Then

$$h_j^*(s) = \sum_i p_i^{(0)} \, q_{ij}^{(0)*}(s) + \sum_i h_i^*(s) \, q_{ij}^*(s), \quad j \in X.$$

Assume that at least one of the following conditions is satisfied:

(i) the embedded Markov chain $\{\nu_n\}$ possesses the ergodic distribution;

(ii) $Q_i(\delta) \leq 1 - \varepsilon, \ i \in X,$ for some positive δ and ε.

Then the above written system of equations for Laplace transforms has the unique solution satisfying the condition. This solution coincides with the above defined $\{h_j^*(s)\}$.

Set $p_j(t) = \mathbf{P}\{\nu(t) = j\}$. Then

$$p_j(t) = p_j^{(0)} \, (1 - Q_j^{(0)}(t)) + \int_0^t (1 - Q_j(t - x)) \, dH_j(x), \quad t \geq 0,$$

where

$$P_j^{(0)}(t) = \sum_k Q_{jk}^{(0)}.$$

Hence

$$s \int_0^\infty e^{-st} p_j(t) dt = p_j^{(0)} \, (1 - q_j^{(0)*}(s)) + h_j^*(s) \, (1 - q_j^*(s)),$$

where

$$q_j^{(0)*}(s) = \sum_k q_{jk}^{(0)*}(s).$$

Set

$$p_j(t, T) = \mathbf{P}\{\nu(t + u) = j, \ 0 \leq u \leq T\}.$$

Then for $Q_{ii}(t) = 0, \ t \geq 0,$

$$p_{ij}(t, T) = p_j^{(0)} \, (1 - Q_j^{(0)}(t + T)) + \int_0^t (1 - Q_j(t + T - x)) \, dH_j(x), \quad t \geq 0.$$

In various applications the steady state parameter of a SMP is used most frequently.

Let the embedded Markov chain $\nu(t)$ be irreducible and moreover possess the ergodic distribution $\{\pi_j\}$ where

$$\pi_j = \lim_{n \to \infty} \mathbf{P}\{\nu_n = j\}.$$

Assume that $Q_i(\infty) = 1$, $i \in X$, and $\tau_i = \int\limits_0^\infty x\, dQ_i(x) < \infty$. (The latter constant equals the mean time of the staying in a state i.) Assume also that $\sum\limits_j \tau_j \pi_j = \tau < \infty$. Under the stated conditions the following statements hold true[13,14].

1. Let $N(t)$ be the number of transitions of the process within an interval $(0, t)$. Then

$$\lim_{t\to\infty} \frac{N(t)}{t} = \frac{1}{\tau}$$

with probability 1.

2. Let $N_A(t)$ be the number of visits of the process to the states including to the set A within a time interval $(0, t)$. Then

$$\lim_{t\to\infty} \frac{N_A(t)}{t} = \frac{1}{\tau} \sum_{j\in A} \pi_j$$

with probability 1.

3. The mean time between two consecutive visits of the process to a state j equals τ/π_j.

4. If $S_A(t)$ denotes the sojourn time of the process in a state set A within a time interval $(0, t)$ then

$$\lim_{t\to\infty} \frac{S_A(t)}{t} = \sum_{j\in A} \frac{\tau_j \pi_j}{\tau}$$

with probability 1.

5. $\lim\limits_{t\to\infty} \dfrac{1}{t} \sum\limits_{j\in A} \int\limits_0^t p_j(x)\, dx = \sum\limits_{j\in A} \dfrac{\tau_j \pi_j}{\tau}.$

Assume that the following additional condition is satisfied. There is no positive δ for which all the increasing points of the function

$$\int_0^t Q_{ij}(t-x)\, dQ_{ji}(x), \quad t \geq 0,$$

are included to the set $\{0, \delta, 2\delta, 3\delta, \dots\}$ whatever a pair (i, j) is. (The continuity of function $Q_{ij}(t)$ is sufficient for this condition to be satisfied.) Then the following relations are true.

6. $\lim\limits_{t\to\infty} p_j(t) = \dfrac{\tau_j \pi_j}{\tau}, \quad \lim\limits_{t\to\infty} \sum\limits_{j\in A} p_j(t) = \sum\limits_{j\in A} \dfrac{\tau_j \pi_j}{\tau}.$

7. $H_j(t+\Delta)-H_j(t) \underset{t\to\infty}{\longrightarrow} \dfrac{\Delta\pi_j}{\tau}$, $\sum\limits_{j\in A}(H_j(t+\Delta)-H_j(t)) \underset{t\to\infty}{\longrightarrow} \Delta\sum\limits_{j\in A}\dfrac{\pi_j}{\tau}$.

8. Let $g_i(t)$ be integrated with a bounded $\sup\limits_j g_j(t)$ for any finite t. Define a new process $\gamma(t) = g_j(t-t_n)$ where $j = \nu(t)$ and t_n is the time of the last transition of the process within an interval $(0, t)$ if any. If there were no transitions in the interval then set $\gamma(t) = g_i^{(0)}$, where $i = \nu(0)$, $g_i^{(0)}$ are integrated functions having a common upper bound. Then we have

$$\mathbf{E}\,\gamma(t) \underset{t\to\infty}{\longrightarrow} \frac{1}{\tau}\sum_i \pi_j \int_0^\infty g_j(u)\,(1-Q_j(u))\,du.$$

In particular,

$$p_j(t, T) \underset{t\to\infty}{\longrightarrow} \frac{1}{\tau}\pi_j \int_T^\infty (1-Q_j(u))\,du.$$

9. Let $\gamma_1(t), \ldots, \gamma_m(t)$ be SMP independent with their initial states being known. Let each of them satisfy all the stated conditions. Let also A be an m-dimensional set; $N_A(t)$ be a number of visits of the m-dimensional process to the set A within an interval $(0, t)$. Then

$$\lim_{t\to\infty}\frac{N_A(t)}{t} = \sum_{k=1}^m \frac{1}{\tau^{(k)}}\sum{}' \frac{\pi_{i_1}^{(1)}\ldots\pi_{i_m}^{(m)}\,\tau_{i_1}^{(1)}\ldots\tau_{i_m}^{(m)}\,p_{i_kj}^{(k)}}{\tau^{(1)}\ldots\tau^{(m)}}$$

with probability 1 where $p_{ij}^{(k)}$, $\pi_j^{(k)}$ are one-step transition probabilities and ergodic probabilities of the embedded Markov chain of the process $\gamma_k(t)$; $\tau_i^{(k)}$, $\tau^{(k)}$ are variables similar to τ_i, τ and defined for $\nu_k(t)$ instead of $\nu(t)$. The symbol $\sum'$ means the sum in the set of indices $i_1, \ldots, i_m, j$ for which $(i_1, \ldots, i_m) \notin A$, $(i_1, \ldots, i_{k-1}, j, i_{k+1}, \ldots, i_m) \in A$.

10. Let $N_A^*(t)$ be the number of transitions of $(\nu_1(t), \ldots, \nu_m(t))$ to the states included in a set A of states. Then

$$\lim_{t\to\infty}\frac{N_A(t)}{t} = \sum_{k=1}^m \frac{1}{\tau^{(k)}}\sum{}'' \frac{\pi_{i_1}^{(1)}\ldots\pi_{i_m}^{(m)}\,\tau_{i_1}^{(1)}\ldots\tau_{i_m}^{(m)}\,p_{i_kj}^{(k)}}{\tau^{(1)}\ldots\tau^{(m)}}$$

where $\sum''$ denotes the sum over all the $(i_1, \ldots, i_m, j)$ for which $(i_1, \ldots$ $\ldots, i_{k-1}, j, i_{k+1}, \ldots, i_m) \in A$.

11. Let $S_A(t)$ be the sojourn time in a set A of states within an interval $(0, t)$ for the m-dimensional process $(\nu_1(t), \ldots, \nu_m(t))$. Then

$$\lim_{t\to\infty}\frac{S_A(t)}{t} = \sum_{(j_1,\ldots,j_m)\in A} \frac{\pi_{j_1}^{(1)}\ldots\pi_{j_m}^{(m)}\,\tau_{j_1}^{(1)}\ldots\tau_{j_m}^{(m)}}{\tau^{(1)}\ldots\tau^{(m)}}$$

with probability 1.

4.7 Applications of Semi-Markov Processes

SMPs are widely used, especially in queueing theory and reliability theory[15–17]. If all the operations being executed in a system are distributed exponentially then a Markov chain can be constructed for the explanation of the behavior of such a system; see Section 4.4. If there are some operations distributed generally then multidimensional models are required. Though if at most one such "general" operation may be executed each time t, then an SMP can usually be constructed as a model of the system behavior. By the method of SMP and Kendall's method of embedded Markov chains a great many results were derived.

Example 4.4. Consider a one-channel queueing system. A Poisson stationary flow $\{t_n\}$ of customers with a parameter λ falls into it. The customers are served in the order of their arrival. The sequence $\{\eta_n\}$ of service times of the customers is a sequence of independent random variables,

$$B(t) = \mathbf{P}\{\eta_n \le t\}, \quad \beta(s) = \int_0^\infty e^{-st}\, dB(t), \quad \tau = \mathbf{E}\,\eta_n, \quad \rho = \lambda\tau < 1.$$

Denote by $\xi(t)$ the queue length at a time t, and set $\xi_n = \xi(s_n + 0)$ where s_n is the time of the departure of the nth customer. Define a random process $\zeta(t) = \xi_n$, for $s_n \le t \le s_{n+1}$, $\zeta(t) = 0$ for $0 \le t < s_1$. The $\eta(t)$ is an SMP, and the $\{\xi_n\}$ is its embedded Markov chain. The existence and the form of the ergodic distribution $\{\pi_j\}$ of this Markov chain have been derived in Chapter 3 (see Sections 3.3 and 3.6). The generating function $\pi(z)$ of this distribution has the form

$$\pi(z) = \frac{(1-\rho)\,(1-z)\,\beta(\lambda(1-z))}{\beta(\lambda(1-z)) - z}.$$

Set $\pi_t(z) = \mathbf{E}\, z^{\xi(t)}$. Statement 8 of Section 4.6 implies the following equality:

$$\pi_t(z) \underset{t\to\infty}{\longrightarrow} \frac{1}{T} \sum_{j=0}^{\infty} \pi_j \int_0^\infty g_j(t,z)\,(1 - Q_j(t))\, dt$$

where $Q_j(t)$ is the distribution function of the time of the staying of the process $\zeta(t)$ in a state j, $g_j(t,z)$ is the conditional expectation of $\xi(s_n + t)$ provided $\xi_n = j$ and $s_{n+1} - s_n > t$.

If $j > 0$ then $Q_j(t) = B(t)$ and $g_j(t,z) = \mathbf{E}\, z^{j+\gamma_t}$ where γ_t is a Poisson random variable with the parameter λt. Hence $g_j(t,z)\,(1- -Q_j(t)) = z_j\, e^{-\lambda t\,(1-z)}\,(1 - B(t)),\ j > 0$. If $\xi_n = 0$ then the following two cases are the only possibilities:

(i) no customers came within the interval of the length t;

(ii) the first customer has arrived at time x, $s_n < x < s_n + t$, and γ_{t-x} customers have arrived in the interval $(x,\ s_n + t)$. Hence

$$g_0(t,z)\,(1 - Q_0(t)) = e^{-\lambda t} + \lambda z \int_0^t e^{-\lambda x - \lambda(1-z)(t-x)}\,(1 - B(t-x))\,dx.$$

Therefore

$$\sum_{j=0}^{\infty} \pi_j \int_0^{\infty} g_j(t,z)\,(1 - Q_j(t))\,dt =$$

$$= \pi_0 \int_0^{\infty} \left\{ e^{-\lambda t} + \lambda z \int_0^t e^{-\lambda x - \lambda(1-z)(t-x)}\,(1 - B(t-x))\,dx \right\} dt +$$

$$+ \sum_{j=1}^{\infty} \pi_j\, z^j \int_0^{\infty} e^{-\lambda t(1-z)}\,(1 - B(t))\,dt.$$

Some elementary transformations lead to the equality

$$\lim_{t\to\infty} \pi_t(z) = \pi(z).$$

This property was stated by Chinčin.

4.8 Linewise Markov Process

A homogeneous Markov process $\xi(t)$ is said to be **a linewise Markov process**[18] if the following conditions are satisfied.

1. The phase space $X = X_0 \cup X_1 \times \mathbb{R}^+$ where X_0 and X_1 are finite or countable sets. Elements of X_0 are designated by small letters: $i,\ j,\ \ldots$, whereas elements of $X_1 \times\ \mathbb{R}^+$ are designated by pairs: $(i,\ x)$ and so on. The X_0 and X_1 are disjoint sets.

2. If $\xi(t) = i \in X_0$ then the following transitions can occur within an interval dt: the transition to a state $j \in X_0$ with a probability $\lambda_{ij}\,dt$,

$j \in X_0$; the transition to a state $(j, 0)$ with a probability $\lambda_{ij}\, dt$, $j \in X_1$. It is assumed that the exit rates

$$\lambda_i = \sum_{j \in X_0} \lambda_{ij} + \sum_{j \in X_1} \lambda_{ij}$$

are finite.

3. If $\xi(t) = (i, x)$ then the following transitions can occur within an interval $(t, t + dt)$:

(i) $\xi(t + dt) = (i, x + dt)$ with a probability $1 - \lambda_i(x)\, dt$,
(ii) $\xi(t + dt) = j$ with a probability $\lambda_{ij}(x)\, dt$, $i \in X_0$,
(iii) $\xi(t + dt) = (j, 0)$ with a probability $\lambda_{ij}(x)\, dt$, $j \in X_1$,
(iv) $\xi(t + dt) = (j, x + dt)$ with a probability $\mu_{ij}(x)\, dt$, $i \neq j \in X_1$.
It is assumed that

$$\lambda_i(x) = \sum_{j \in X_0} \lambda_{ij}(x) + \sum_{j \in X_1} \lambda_{ij}(x) + \sum_{i \neq j \in X_1} \mu_{ij}(x)$$

is finite for each $i \in X_1$.

Denote by $p_i(t)$ the probability of a state i at a time $t \geq 0$. Denote also by $F_i(t, x)$ the probability of the set of states $\{i\} \times [0, x]$ at a time t. If $\lambda_i(x)$ are uniformly bounded in i and x and the initial distribution is described by an absolutely continuous probability measure then the following system of equation holds true

$$\frac{\partial F_j(t, x)}{\partial t} + \frac{\partial F_j(t, x)}{\partial x} = \int_0^x \lambda_j(t, u)\, dF_j(t, u) + \sum_{i \neq j} \lambda_{ij}(t, x)\, dF_i(t, x) + \left. + \frac{\partial F_j(t, u)}{\partial u} \right|_{u = 0}, \quad j \in X_1,$$

$$p_j'(t) = -\lambda_j\, p_j(t) + \sum_{i \neq j,\, i \in X_0} \lambda_{ij}\, p_i(t) + \sum_{i \in X_1} \int_0^\infty \lambda_{ij}(x)\, dF_i(t, x), \quad j \in X_0,$$

with boundary condition

$$\left. \frac{\partial F_j(t, u)}{\partial u} \right|_{u = 0} = \sum_{i \in X_0} \lambda_{ij}\, p_i(t) + \sum_{i \in X_1} \int_0^\infty \mu_{ij}(x)\, dF_i(t, x), \quad j \in X_1,$$

and appropriate initial conditions.

4.9 Shot Noise Process

Let $\eta(t)$, $t \in \mathbb{R}$, be a one-dimensional or multi-dimensional random process and $\eta_n(t)$ be its independent realizations. Let also a stationary Poisson flow $\{t_n\}$ of homogeneous events in $\mathbb{R}$ with the parameter λ be defined, independently of $\{\eta_n(t)\}$. Consider the random process

$$\xi(t) = \sum_n \eta_n(t - t_n)$$

where the convergence of the series means the pointwise convergence in probability. Such a random process is called **a shot noise process**. A random process $\eta_n(t - t_n)$ can be considered as an impulse with a shift t_n along $\mathbb{R}$. Therefore, a shot noise process is the sum (superposition) of purely random impulses. In most models $\eta_n(t) = 0$ for $t < 0$; in such cases t_n is the starting point of the nth impulse. But such a property is not essential for deriving the most results of the theory of shot noise processes. In an applied example impulses have the form $\eta_n(t) = \eta_n(0)\, e^{-\alpha t^2}$, $t \in \mathbb{R}$, where $\eta_n(0)$ is a random variable. In this case t_n is the center of the impulse, though it is convenient to call t_n the initial time of the nth impulse even in such two-sided cases.

Example 4.5. Let us conceive an electric capacity on which a flow of charged particles falls. Each particle carries some charge. An exponential discharge also takes place. In such a case $\eta(t) = 0$ for $t < 0$, $\eta(t) = \eta(0)\, e^{-\alpha t}$ for $t \geq 0$.

Example 4.6. Assume that a Poisson flow of airplanes enters into the zone serviced by the airport radar. For simplicity suppose that the time of staying within the zone is constant. Denote by $N(t)$ the number of airplanes present in the zone at a time t. Then $N(t)$ is a shot noise process, an impulse having the form $\eta(t) = 1$ if $0 \leq t \leq \Delta, otherwise$ $\eta(t) = 0$.

Assume that the following two conditions are satisfied:

(i) $\int_{-\infty}^{\infty} \mathbf{E}\,|\eta(t)|\, dt < \infty$,

(ii) $\int_{-\infty}^{\infty} \mathbf{E}\,|\eta^2(t)|\, dt < \infty$

for each of the components of a shot noise process. Then a shot noise process $\xi(t)$ is a stationary process both in a strict and broad sense. We have

$$\mathbf{E}\,\xi(t) = \lambda \int_{-\infty}^{\infty} \mathbf{E}\,\eta(t)\, dt.$$

For a one-dimensional process

$$\mathbf{Var}\,\xi(t) = \lambda \int_{-\infty}^{\infty} \mathbf{E}\,\eta^2(t)\,dt.$$

The covariance matrix $R(\tau) = (R_{ij}(\tau))$ of a multi-dimensional shot noise process $\eta(t) = (\eta_1(t), \ldots, \eta_m(t))$ has the following expression:

$$R_{ij}(\tau) = \mathbf{cov}(\xi_i(t), \xi_j(t+\tau)) = \lambda \int_{-\infty}^{\infty} \mathbf{E}\,\eta_i(s)\,\eta_j(s+\tau)\,ds.$$

Let $\xi(t)$ be a one-dimensional shot noise process. The joint distribution of $\xi(t_1), \ldots, \xi(t_n)$ is defined by the characteristic function

$$\mathbf{E}\exp\left(i\sum_{k=1}^{n} z_k\xi(t_k)\right) = \exp\left\{-\lambda\int_{-\infty}^{\infty}\left(1-\mathbf{E}\exp\left(i\sum_{k=1}^{n} z_k\eta(t_k-x)\right)\right)dx\right\}.$$

A similar formula is valid also for a multi-dimensional process; the products $z_k\,\xi(t_k)$ and $z_k\,\eta(t_k+x)$ should be changed by scalar products of appropriate vectors.

If $\varphi(t)$ is a continuous real-valued function then

$$\mathbf{E}\exp\left(i\int_a^b \varphi(t)\xi(t)dt\right) = \exp\left\{-\lambda\int_{-\infty}^{\infty}\left(1-\mathbf{E}\exp\left(i\int_a^b \varphi(t)\eta(t+x)\right)\right)dx\right\}.$$

Many real processes are characterized by small impulses and large λ. The following theorem justifies the Gaussian approximation for such cases.

Theorem 4.1. *Let the conditions 1 and 2 be satisfied. Consider a family of shot noise processes* $\{\xi_\lambda(t)\}$ *differing only in a value of the parameter* λ. *If* $\lambda \to 0$ *then the centered and normed processes*

$$\zeta_\lambda(t) = \frac{1}{\sqrt{\lambda}}(\xi_\lambda(t) - \mathbf{E}\,\xi_\lambda(t))$$

converge weakly to a stationary Gaussian process with a zero mean and the covariance matrix $R(\tau) = (R_{ij}(\tau))$ *of the form*

$$R_{ij}(\tau) = \int_{-\infty}^{\infty} \mathbf{E}\,\eta_i(s)\,\eta_j(s+\tau)\,ds.$$

A case of short-time impulses or rapidly damping ones is also typical for the applications. In such a situation a white-noise approximation is usually applied so that, for example, in the one-dimensional case under some limiting conditions the centered and normed integral of the process

$$X_\lambda(t) = \frac{1}{b\sqrt{\lambda}} \int_0^t (\xi_\lambda(x) - \mathbf{E}\,\xi_\lambda(x))\,dx$$

with

$$b^2 = \int_{-\infty}^{\infty} R(\tau)\,d\tau$$

converges in the weak sense to the standard Wiener process as $\lambda \to \infty$.

4.10 References

1. Chung K. L., *Markov Chains: With Stationary Transition Probabilities,* Springer, Berlin, 1967.
2. Cox, D. R., Miller, H. D., *The Theory of Stochastic Processes,* Chapman and Hall, London, 1972.
3. Feller, W., *An Introduction to Probability Theory and Its Applications,* 1, Wiley, New York, 1968.
4. Feller, W., *An Introduction to Probability Theory and Its Applications,* 2, Wiley, New York, 1971.
5. Gihman, I. I., Skorohod, A. V., *Theory of Stochastic Processes,* 1, Springer, Berlin, 1974.
6. Gihman, I. I., Skorohod, A. V., *Theory of Stochastic Processes,* 2, Springer, Berlin, 1975.
7. Gihman, I. I., Skorohod, A. V., *Theory of Stochastic Processes,* 3, Springer, Berlin, 1979.
8. Karlin, S., *A First Course in Stochastic Processes,* Acad. Press, New York, 1968.
9. Karlin, S., Taylor, H. M., *A Second Course in Stochastic Processes,* Acad. Press, New York, 1981.
10. Belyayev, Y. K., New results and generalizations of crossing problems, in *Stationary Random Processes,* Cramér, H., Leadbetter, M. R., Mir, Moscow, 341, 1969 (in Russian).
11. Haight, F. A., *Applied Probability,* Plenum Press, New York, 1981.
12. Grimmett, G., Stizzaker, D., *Probability and Random Processes,* Clarendon Press, Oxford, 1982.

13. Korolyuk, V. S., Turbin, A. F., *Semi-Markov Processes and Their Applications,* Naukova Dumka, Kiev, 1976 (in Russian).
14. Silvestrov, D. S., *Semi-Markov Processes with a Discrete Phase Space,* Kiev Univ. Publ., Kiev, 1971 (in Russian).
15. Barzilovich, E. Y., Kashtanov, B. A., *Some Mathematical Questions in the Theory of Complex Systems Maintenance,* Sov. Radio, Moscow, 1971 (in Russian).
16. Kovalenko, I. N., Moskatov, G. K., Barzilovich, E. Y., *Semi-Markov Models in Designing Problems of Controlling Systems in Flying Devices,* Mashinostroenye, Moscow, 1973 (in Russian).
17. Kredenzer, B. P., *Reliability Forecasting in Systems with Time Redundancy,* Naukova Dumka, Kiev, 1978 (in Russian).
18. Belyayev, Y. K., Linewise Markov processes and their application in the reliability theory, in *Fourth All-Union Confer. on Probab. Theory and Math. Statist.,* Lith. Publ., Vilnius, 309, 1962 (in Russian).

Chapter 5

Random Processes with Independent Increments

5.1 Multidimensional Brownian Motion

Let $\xi(t) = (\xi_1(t), \ldots, \xi_m(t))$ be an m-dimensional random process with independent increments defined over $\mathbb{R}^+$ and possessing the following properties: $\xi(0) = 0$ with the probability 1 and for each $h > 0$ and $t \geq 0$ the increment $\xi(t+h) - \xi(t)$ is an m-dimensional Gaussian random variable with the mean $a\,h = (a_1 h, \ldots, a_m h)$ and the covariance function $R_h = (R_{ij}\,h)$ where $R_1 = (R_{ij})$ is a nonnegatively defined symmetric matrix. The characteristic function of $\xi(t+h) - \xi(t)$ has the form

$$\psi_h(\lambda_1, \ldots, \lambda_m) = \mathbf{E}\, e^{i(\lambda_1 \Delta\xi_1 + \cdots + \lambda_m \Delta\xi_m)} = e^{i(\lambda,\, ah) - \frac{1}{2}\lambda^T R_h \lambda} =$$

$$= e^{i(\lambda,\, a)h - \frac{1}{2}\lambda^T R_1 \lambda h},$$

where

$$(\lambda,\, a) = \sum_{j=1}^{m} \lambda_j\, a_j, \quad \lambda = \begin{pmatrix} \lambda_1 \\ \vdots \\ \lambda_m \end{pmatrix}, \quad \lambda^T = (\lambda_1, \ldots, \lambda_m).$$

The defined process possesses the property that for any real $\lambda_1, \ldots, \lambda_m$ the linear combination of its components $\eta_\lambda(t) = \lambda_1\,\xi_1(t) + \cdots + \lambda_m\,\xi_m(t)$ is a one-dimensional process of brownian motion (**Wiener process**). Moreover $\xi(t)$ can be expressed via m independent Wiener processes. To do this choose such an orthogonal transformation C of $\mathbb{R}^m$ that $R_1 = C^T D C$

where D is a diagonal matrix. If the rank of R equals r, then, without a loss of the generality,

$$D = \begin{pmatrix} \sigma_1^2 & & & & & \\ & \ddots & & & & \\ & & \sigma_r^2 & & & \\ & & & 0 & & \\ & & & & \ddots & \\ & & & & & 0 \end{pmatrix}.$$

Consider r independent **standard Wiener processes** $w_1(t), \ldots, w_r(t)$ so that $\mathbf{E}\, w_k(t) = 0$, $\mathbf{E}\, w_k^2(t) = t$, $t \geq 0$; $\mathbf{E}\, w_k(t)\, w_j(t) = 0$ for $k \neq j$. Then

$$\xi(t) = at + C^T \begin{pmatrix} \sigma_1\, w_1(t) \\ \vdots \\ \sigma_r\, w_r(t) \\ 0 \\ \vdots \\ 0 \end{pmatrix}$$

where $\xi(t)$ is written as a column vector.

A general m-dimensional Wiener process[1–7] is defined as a linear transformation of a vector composed by independent standard Wiener processes, i.e.,

$$\Delta\xi(t) = \xi(t+h) - \xi(t) = ah + C^T \begin{pmatrix} \sigma_1\, \Delta w_1(t) \\ \vdots \\ \sigma_r\, \Delta w_r(t) \\ 0 \\ \vdots \\ 0 \end{pmatrix}$$

where $\Delta w_i(t) = w_i(t+h) - w_i(t)$.

A separable m-dimensional Wiener process possesses a continuous path function with the probability 1. Such a process can be approximated by a random broken line $\xi_h(t)$ being built in such a way. Generate $\xi(nh)$, $n \in \mathbb{N}$, via increments $\xi(nh) - \xi((n-1)h)$ and use the linear interpolation in intervals $((n-1)h,\, nh)$. It is possible to define $\xi_h(t)$ with $h = 1, 2^{-1}, 2^{-2}, \ldots$ on the same probability space in such a way that

$$\max_{0 \leq t \leq 1} |\xi(t) - \xi_h(t)| \longrightarrow 0$$

with probability 1 as $h = 2^{-n}$, $n \to \infty$.

Consider a standard separable Wiener process $w(t)$. Let a be a number, $a \neq 0$, and τ_a be the first time for the process to reach the level a. Consider the reflected process $\eta(t)$ defined as follows: $\eta(t) = \xi(t)$ if $t < \tau_a$; $\eta(t) = 2a - \xi(t)$ if $t \geq \tau_a$. Then the $\eta(t)$ is a standard Wiener process.

For many applications the distribution of the maximum of the process in a time interval is of interest. The following equality holds true:

$$\mathbf{P}\{\max_{0<t<T} w(t) > a\} = \sqrt{\frac{2}{\pi T}} \int_a^\infty \exp\{-\frac{x^2}{2T}\}\, dx.$$

5.2 Convergence of Sums of Infinitely Small Random Variables to the Process of Brownian Motion

One of the most valuable achievements of the probability theory is the Central Limit Theorem. While dealing with it, it is convenient to use the notation $\xi^0 = \xi - \mathbf{E}\xi$ for any random variable with a finite mean. For example if $\{\xi_k\}$ are independent random variables with a common distribution possessing a finite variance $\sigma^2 > 0$ then the distribution of the centered and normed sum

$$\frac{1}{\sigma\sqrt{n}}\,(\xi_1^0 + \cdots + \xi_n^0)$$

converges in the weak sense to the standard normal distribution as $n \to \infty$. Introduce a small parameter $\varepsilon > 0$ and consider a random process

$$\xi_\varepsilon(t) = \frac{\sqrt{\varepsilon}}{\sigma}\,(\xi_1^0 + \cdots + \xi_{[nt]}^0)$$

where $[x]$ is the notation for the integer part of a number x. Then under the stated condition $\{\xi(t),\ t \geq 0\}$ converges in the weak sense to the standard Wiener process.

Let $\xi_1, \xi_2, \ldots$ be a sequence of independent m-dimensional random variables with an identical distribution and a covariance matrix R. Then the random process

$$\xi_\varepsilon^0(t) = \sqrt{\varepsilon}\,(\xi_1^0 + \cdots + \xi_{[t/\varepsilon]}^0)$$

converges in the weak sense to an m-dimensional Wiener process $\eta(t)$ with the zero mean and the covariance matrix $R_1 = (\mathbf{cov}\,(\eta_i(1),\ \eta_j(1)))$.

More general theorems[2,8] concern the triangle scheme so that the distributions of random variables $\{\xi_k\}$ depend on k and/or ε. For one-dimensional variables $\{\xi_k\}$ the following theorem holds true.

Theorem 5.1. *Let the two following conditions be satisfied for a sequence of independent random variables* $\{\xi_k\}$*:*

(i) $\mathbf{Var}\,(\xi_1 + \cdots + \xi_{n_t}) \sim \gamma_\varepsilon^2 t,\ 0 < t \le T,$ *where* γ_ε *is an arbitrary positive function of* ε*,*

(ii) if $F_{\varepsilon k}(x)$ *is the distribution function of the* ξ_k^0 *then Lindeberg condition holds:*

$$\frac{1}{\gamma_\varepsilon^2}\sum_{k=0}^{n_T}\int\limits_{|x|\ge\delta\gamma_\varepsilon} x^2\,dF_{\varepsilon k}(x) \longrightarrow 0, \quad \delta > 0.$$

Then the random process

$$\xi_\varepsilon(t) = \frac{1}{\gamma_\varepsilon}\,(\xi_1^0 + \cdots + \xi_{n_t}), \quad 0 \le t \le T$$

converges in the weak sense to the standard Wiener process as $\varepsilon \to 0$.

For m*-dimensional independent random variables* $\xi_k = (\xi_{k1}, \ldots, \xi_{km})$ *the three conditions should be satisfied:*

(i) $\mathbf{Var}\,(\xi_{i1} + \cdots + \xi_{in_t}) \sim \gamma_{i\varepsilon}^2 t,\quad 0 < t \le T,\ 1 \le i \le m,\ \gamma_{i\varepsilon} > 0,$

(ii) $\mathbf{cov}\,(\xi_{i1} + \cdots + \xi_{in_t},\ \xi_{j1} + \cdots + \xi_{jn_t}) \sim \gamma_{i\varepsilon}\gamma_{j\varepsilon} r_{ij}\, t,\quad 0 < t \le T,$

(iii) If $F_{\varepsilon ik}(x)$ *is the distribution function of* ξ_{ik}^0*, then*

$$\frac{1}{\gamma_{i\varepsilon}^2}\sum_{k=0}^{n_T}\int\limits_{|x|\ge\delta\gamma_{i\varepsilon}} x^2\,dF_{\varepsilon ik}(x) \underset{\varepsilon\to 0}{\longrightarrow} 0, \quad \delta > 0,\ 1 \le i \le m.$$

Then the random process $\xi_\varepsilon(t) = (\xi_{1\varepsilon}(t), \ldots, \xi_{m\varepsilon}(t)),\ 0 \le t \le T,$ *with*

$$\xi_{i\varepsilon}(t) = \frac{1}{\gamma_{i\varepsilon}}\,(\xi_{i1}^0 + \cdots + \xi_{in_t}^0), \quad 1 \le i \le m,$$

converges in the weak sense to the m*-dimensional Wiener process* $\eta(t) = (\eta_1(t), \ldots, \eta_m(t)),\ 0 \le t \le T,$ *with parameters* $\mathbf{E}\eta(t) = 0,\ R_{ii} = \mathbf{Var}\,\eta_i(1) = 1,\ R_{ij} = \mathbf{cov}\,(\eta_i(1), \eta_j(1)) = r_{ij},\ i \ne j,$ *as* $\varepsilon \to 0$.

5.3 Characterization of a General Process with Independent Increments

Let $\{\xi(t),\ t \geq 0\}$ be an m-dimensional stochastically continuous separable random process with independent increments[2–6,9,10]. Such a process has no second kind discontinuities with the probability 1. Hence any finite interval can contain only finite discontinuities of a path function of a length exceeding ε. Let $A \subset \mathbb{R}^m$ be any Borel set apart from the origin. Set $\xi_A(t)$ the sum of all the jumps of the process $\xi(t)$ with values belonging to the set A, in a time interval $(0, t)$. Set also $\eta_A(t) = \xi(t) - \xi_A(t)$. Then $\eta_A(t)$ is a stochastically continuous separable process with independent increments, possessing no jumps from the set A. Denote by $\{t_n^A\}$ the set of times when the process $\xi(t)$ has jumps from set A. Then $\eta_A(t)$ is a non-stationary Poisson flow of homogeneous events (i.e., a memoryless simple point process). Denote by $\Pi(t, A)$ the mean number of its events in a time interval $(0, t)$. If $A_1, \ldots, A_N$ are disjoint Borel sets then the corresponding flows $\{t_n^{A_1}\}, \ldots, \{t_n^{A_N}\}$ are independent. The set function $\Pi(t, A)$ is a time-dependent measure which is finite with probability 1 for a set A apart from the origin. Choose a decreasing sequence $\{\varepsilon_n\}$ which tends to zero, and consider the sets

$$A_1 = \{|x| \geq \varepsilon_1\}, \quad A_2 = \{\varepsilon_2 \leq |x| < \varepsilon_1\}, \quad A_3 = \{\varepsilon_3 \leq |x| < \varepsilon_2\}, \quad \ldots$$

where $|x| = (x_1^2 + \cdots + x_m^2)^{1/2}$ for $x = (x_1, \ldots, x_m)$. The subtraction of the sum of processes $\xi_{A_1}(t), \ldots, \xi_{A_n}(t)$ from the process $\xi(t)$ gives a process with independent increments

$$\xi_n(t) = \xi(t) - \sum_{j=1}^{n} \xi_{A_j}(t)$$

having no jumps of a length exceeding ε_n. This process is independent of the $\{\xi_{A_j}(t),\ 1 \leq j \leq n\}$.

In some cases (usually in applied problems) the procedure leads to a process $\xi_0(t)$ with a continuous path function such that

$$\xi_n(t) \xrightarrow[n\to\infty]{w} \xi_0(t).$$

But in general the entire rate of jumps may be infinite. In such a case centered processes $\xi_{A_j}^0(t) = \xi_{A_j}(t) - \mathbf{E}\xi_{A_j}(t)$ are required. In a general case the sequence of random processes

$$\eta_n(t) = \xi(t) - \xi_{A_1}(t) - \sum_{j=1}^{n} \xi_{A_j}^0(t)$$

converges in the weak sense to a random process $\eta(t)$ with continuous path functions as $n \to \infty$.

Therefore the process $\xi(t)$ can be decomposed as follows:

$$\xi(t) = \xi(0) + \eta(t) + \xi_{A_1}(t) + \sum_{j=1}^{\infty} \xi^0_{A_j}(t)$$

where all the right-hand side terms are independent processes with independent increments, the $\eta(t)$ being a process with continuous path functions. The process $\eta^0(t) = \eta(t) - \mathbf{E}\,\eta(t)$ is called **the diffusion component** of the process $\xi(t)$.

Set, for the definiteness, $\varepsilon_1 = 1$. The processes $\xi_A(t)$, $\xi^0_A(t)$ satisfy the following additivity property: for disjoint sets $A_1, \ldots, A_n$ the equations

$$\xi_{A_1 \cup \ldots \cup A_n}(t) = \sum_{j=1}^{n} \xi_{A_j}(t), \qquad \xi^0_{A_1 \cup \ldots \cup A_n}(t) = \sum_{j=1}^{n} \xi^0_{A_j}(t)$$

hold true.

The application of a partition approach enables obtainment, as the result of a limiting procedure, of the following expression for the characteristic function $\varphi_t(\lambda)$ of a random variable $\xi(t)$:

$$\beta_t(\lambda) = \exp\left\{ \int\limits_{|x|>1} (e^{i(\lambda, x)} - 1)\,\Pi(t, dx) + \right.$$

$$\left. + \int\limits_{|x|\le 1} (e^{i(\lambda, x)} - 1 - i(\lambda, x))\,\Pi(t, dx) \right\}.$$

The random process $\eta(t)$ is Gaussian. Its characteristic function has the form

$$\alpha_t(\lambda) = \exp\left\{ i(\lambda, a(t)) - \frac{1}{2}(A(t)\lambda, \lambda) \right\}$$

where $a(t)$ is a continuous m-dimensional function; $\{A(t)\}$ is a family of nonnegatively defined functions. Moreover the matrices $A(t_2) - A(t_1)$ are also nonnegatively defined for $t_1 < t_2$. The λ is a column vector; $(\ ,\ )$ denotes a scalar product.

Let $\psi_0(\lambda)$ be the characteristic function of the random variable $\xi(0)$. Then the characteristic function of $\xi(t)$ has the form

$$\varphi_t(\lambda) = \psi_0(\lambda)\,\alpha_t(\lambda)\,\beta_t(\lambda), \quad t \ge 1.$$

The characteristic function of an increment of the process $\xi(\tau)-\xi(t)$ $(0 \le t < \tau)$ has the form

$$\frac{\alpha_\tau(\lambda)\,\beta_\tau(\lambda)}{\alpha_t(\lambda)\,\beta_t(\lambda)} = \exp\left\{ i(\lambda,\, a(\tau)-a(t)) - \frac{1}{2}((A(\tau)-A(t))\,\lambda,\,\lambda) + \right.$$

$$+ \int\limits_{|x|>1} (e^{i(\lambda,\,x)} - 1)\,(\Pi(\tau,\,dx) - \Pi(t,\,dx)) +$$

$$\left. + \int\limits_{|x|\le 1} (e^{i(\lambda,\,x)} - 1 - i\,(\lambda,\,x))\,(\Pi(\tau,\,dx) - \Pi(t,\,dx)) \right\}.$$

For a time-homogeneous process $\xi(t)$ we have

$$a(t) = a\,t, \quad A(t) = A(1)\,t, \quad \Pi(t,\,A) = \Pi(1,\,A)\,t,$$

hence $\varphi_t(\lambda) = \varphi_1^t(\lambda)$. In any case the random variable $\xi(t)-\xi(0)$ has an infinitely divisible distribution.

An alternative expression for the characteristic function of $\xi(t)$ satisfying the same conditions as the above stated ones has the form

$$\varphi_t(\lambda) = \psi_0(\lambda)\,\exp\left\{ i\,(\lambda,\,a_1(t)) - \frac{1}{2}(A(t)\,\lambda,\,\lambda) + \right.$$

$$\left. + \int\limits_{-\infty}^{\infty} \left(e^{i\,(\lambda,\,x)} - 1 - \frac{i\,(\lambda,\,x)}{1+|x|^2} \right) \frac{1+|x|^2}{|x|^2}\, G(t,\,dx) \right\}$$

where

$$a_1(t) = a(t) - \int\limits_{|x|\le 1} \frac{x\,|x|^2}{1+|x|^2}\,\Pi(t,\,dx) + \int\limits_{|x|>1} \frac{x}{1+|x|^2}\,\Pi(t,\,dx),$$

$$G(t,\,A) = \int\limits_{A} \frac{|x|^2}{1+|x|^2}\,\Pi(t,\,dx).$$

The general form of the characteristic function for processes with independent increments is due to Lévy (1934); for a recent approach see[2].

Example 5.1. A dam is characterized by the level of filling $\xi(t)$ at a time $t \ge 0$. This level increases at random times t_n by random values η_n. It is assumed that $\{t_n\}$ is a stationary Poisson point process with the parameter λ and $\{\eta_n\}$ are independent random variables with a characteristic function

$\psi(t)$. Besides, in the intervals between increases the level of water decreases with a constant rate γ. Assume for the simplicity that there is no restriction on the level either from above or from below. We have

$$\varphi_t(z) = \exp\{-i\gamma t z + \lambda t\,(\psi(z) - 1)\}$$

for the characteristic function of $\xi(t),\ t \geq 0$.

Example 5.2. Consider a flow of customers of m distinguishable kinds. In a time interval of a length dt a batch of customers can arrive with the probability $\lambda\, dt$. The arriving batch (if any) is characterized by a random vector $(\eta_1, \ldots, \eta_m)$ where η_k is the number of customers of the kth kind. Denote by $\psi(z) = \psi(z_1, \ldots, z_m)$ the characteristic function of $(\eta_1, \ldots, \eta_m)$. Then the following expression can be derived for the characteristic function $\varphi_t(z)$ of the random process $\xi(t) = (\xi_1(t), \ldots, \xi(t)_m)$ where $\xi_k(t)$ is the entire number of k kind customers arriving within a time interval $(0,\, t)$:

$$\varphi_t(z) = \mathbf{E} \exp\{i\,(z,\, \xi(t))\} = \exp\{-\lambda t\,(1 - \psi(z))\}.$$

5.4 Properties of Path Functions

In many areas random processes with stepwise path functions are being applied. In this case a path function for each finite T the interval $(0,\, T)$ contains a finite number of intervals $(t_0,\, t_1),\, (t_1,\, t_2),\, \ldots,\, (t_n,\, T)$, in each the path function is invariable.

A stochastically continuous separable random process $\xi(t)$ with independent increments possesses stepwise path functions iff its probabilistic structure can be described in the following way[3–5].

Consider a nonstationary Poisson process $\nu(t)$ with a leading function $\Lambda(t)$. Each time the Poisson process has a jump the process $\xi(t)$ has a jump too. If a single jump of the Poisson process occurs at a time t then the increment of $\xi(t)$ at the same time is a random variable $\eta(t)$ with a distribution function $H_t(x)$. In the case of a multiple jump so that $\nu(t + 0) - \nu(t - 0) = m$ the increment of $\xi(t)$ has the form $\eta_t^{(1)} + \cdots + \eta_t^{(m)}$ where $\{\eta_t^{(k)}\}$ are independent random variables with the same distribution $H_t(x)$. The random variables $\{\eta_t^{(k)}\}$ are independent over all the pairs $(t,\, k)$.

Let $\varphi_t(z)$ be the characteristic function of a random variable $\xi(t)$. The following expression holds true

$$\varphi_t(z) = \psi_0(z) \exp\left\{ \int_X (e^{i(z,x)} - 1)\, \Pi(t,\, dx) \right\}$$

where $\Pi(t, A)$ is a finite measure in A nondecreasing in t; $\psi_0(z)$ is the characteristic function of the initial value $\xi(0)$. We have

$$\pi(t, A) = \int_0^t \left\{ \int_A dH_\tau(x) \right\} d\Lambda(\tau),$$

$$\Lambda(t) = \Pi(t, X), \qquad dH_t(x) = \lim_{h\to 0} \frac{\Pi(t+h, dx) - \Pi(t, dx)}{\Lambda(t+h) - \Lambda(t)}.$$

For a path function to be nondecreasing with probability 1 the following condition is necessary and sufficient:

$$\varphi_t(z) = \psi_0(z) \exp\left\{ iz\gamma(t) + \int_0^\infty (e^{izx} - 1)\, \Pi(t, dx) \right\}$$

where $\varphi(t)$ is a nondecreasing function and $\Pi(t, A)$ is a measure for which

$$\Pi(\tau, A) - \Pi(t, A) \geq 0 \quad \text{if} \quad t < \tau \quad \text{and} \quad \int_x^1 x\, \Pi(t, dx) < \infty.$$

The functional

$$\operatorname*{var}_{a<t<b} x(t) = \sup_{n;\, a=t_0<t_1<\cdots<t_n=b} \sum_{k=0}^{n-1} |x(t_{k+1}) - x(t_k)|$$

is called **the variation** of a function $x(t)$ on a closed interval $[a, b]$. If $x(t)$ is a position of a particle at a time t then the defined functional equals the length of its path in the interval $[a, b]$. In particular, for an absolutely continuous function $x(t)$ the equality

$$\operatorname*{var}_{a<t<b} x(t) = \int_a^b |x'(t)|\, dt$$

holds. A path function of the Wiener process has an infinite variation on an arbitrary time interval $[t, t+\varepsilon]$ if $\varepsilon > 0$.

If a process with independent increments has a nonzero diffusion component then the variation of its path function is infinite with probability 1 on a time interval of a positive length. The "expression nonzero diffusion component" means that

$$\sum_{i=1}^{m} (\sigma_i^2(\tau) - \sigma_i^2(t)) > 0$$

for a time interval being considered, the $\sigma_i^2(t)$ denoting the ith diagonal element of the matrix $A(t)$.

Let $\xi(t)$ be a stochastically continuous separable process with independent increments. If $A(\tau) = A(t)$ and

$$\int_{|x|\le 1} |x| \, (\Pi(\tau, dx) - \Pi(t, dx)) < \infty$$

then $\xi(t)$ has a finite variation on the interval $[t, \tau]$ with the probability 1. If this condition is satisfied then the variation ζ of a path function on the interval $[t, \tau]$ is a random variable with the characteristic function

$$\varphi_\zeta(z) = \exp\left\{ iz \operatorname*{var}_{t\le u\le \tau} a(u) + \int_{|x|\le 1} (e^{iz|x|} - 1)\,(\Pi(\tau, dx) - \Pi(t, dx)) \right\}.$$

Consider a time-homogeneous stochastically continuous separable stochastic process $\xi(t)$ with independent increments. A function $g(t)$ is called **an upper function** of the process if

$$\overline{\lim_{t\to\infty}} \frac{\xi(t)}{g(t)} < 1$$

holds true with probability 1. A function $g(t)$ is called **a lower function** of the process if

$$\overline{\lim_{t\to\infty}} \frac{\xi(t)}{g(t)} > 1$$

with probability 1.

Consider a standard Wiener process $w(t)$. For each $\varepsilon > 0$ the function $(1+\varepsilon)\,(2t \ln\ln t)^{1/2}$ is its lower function. This statement is called **the law of the iterated logarithm**. A function $g(t)$, $t > 0$, is called **a function of regular increase** if functions $k_1(\lambda)$ and $k_2(\lambda)$ exist such that

$$k_1(\lambda)\, g(t) \le g(\lambda t) \le k_2(\lambda)\, g(t), \quad \lambda > 0,\ t > 0,$$

and

$$k_2(\lambda) \underset{\lambda\to 0}{\longrightarrow} 1, \quad k_1(\lambda) \underset{\lambda\to\infty}{\longrightarrow} \infty.$$

Let $\xi(t)$ be a symmetrically distributed time-homogeneous separable, stochastically continuous process with independent increments. If moreover

$$\int_0^\infty \frac{1}{t} \mathbf{P}\{\xi(t) > g(t)\}\, dt < \infty$$

and $g(t)$ is a function of regular increase then for a positive ε the function $(1+\varepsilon)\,g(t)$ is an upper function of $\xi(t)$.

Let $\xi(t)$ be a symmetrically distributed time-homogeneous process with independent increments and $g(t)$ be a function of regular increase for which the relation

$$\sum_{k=1}^{\infty} \mathbf{P}\{\xi(a^k) > g(a^k)\} = \infty$$

is satisfied for every $a > 1$. Then for any $\varepsilon > 0$ the function $(1-\varepsilon)\,g(t)$ is a lower function of the process.

5.5 Convergence of Sums of Independent Random Variables to a Process with Independent Increments

Assume that a series of n independent one-dimensional random variables $\xi_{n1}, \ldots, \xi_{nn}$ with distribution functions $F_{n1}(x), \ldots, F_{nn}(x)$ is defined for each $n = 1, 2, \ldots$. Set $s_{nk} = \xi_{n1} + \cdots + \xi_{nk}$, $1 \le k \le n$; $s_{n0} = 0$ and define a random process $\xi_n(t) = s_{n[nt]}$, $1 \le t \le n$. The question arises in the conditions for the weak convergence of the sequence of random processes to a stationary Poisson process $\nu(t)$ with a parameter λ in interval $(0, T)$.

It is assumed that all the $\{\xi_{nk}\}$ are infinitely small, i.e.,

$$\max_{1\le k\le n} \{\xi_{nk} \,\overline{\in}\, u_\varepsilon\} \underset{n\to\infty}{\longrightarrow} 0$$

where $u_\varepsilon = (-\varepsilon, \varepsilon)$.

Denote $R_\varepsilon = \mathbb{R} \setminus ((-\varepsilon, \varepsilon) \cup (1-\varepsilon, 1+\varepsilon))$ and assume the following conditions to be satisfied.

1. $\displaystyle\sum_{k=1}^{n} \mathbf{P}\{\xi_{nk} \in R_\varepsilon\} \underset{n\to\infty}{\longrightarrow} 0.$

2. $\displaystyle\sum_{k=1}^{[nt]} \mathbf{P}\{\xi_{nk} - 1 \in u_\varepsilon\} \underset{n\to\infty}{\longrightarrow} \lambda t, \quad \varepsilon < \varepsilon_0\ (\varepsilon_0 > 0).$

3. $\displaystyle\sum_{k=1}^{n} \int_{|x|<\varepsilon} x\, dF_{nk}(x) \underset{n\to\infty}{\longrightarrow} 0.$

4. $\displaystyle\sum_{k=1}^{n} \left\{ \int_{|x|<\varepsilon} x^2\, dF_{nk}(x) - \left(\int_{|x|<\varepsilon} x\, dF_{nk}(x) \right)^2 \right\} \underset{n\to\infty}{\longrightarrow} 0.$

The conditions 1-4 are necessary and sufficient for the desired convergence $\xi_n(t) \xrightarrow{w} \eta(t)$, $0 < t < T$.

The conditions for the convergence to a general process with independent increments are also known[6,7,9].

Let a series $(\xi_{n1}, \ldots, \xi_{nn})$ of independent one-dimensional random variables be defined for each $n \geq 1$. Assume that $F_n(x) = \mathbf{P}\{\xi_{nk} \leq x\}$, $1 \leq k \leq n$. Set

$$\gamma_n = n \int_{-\infty}^{\infty} \frac{x}{1+x^2}\, dF_n(x), \qquad G_n(x) = n \int_{-\infty}^{x} \frac{u}{1+u^2}\, n\, dF_n(u).$$

If $\gamma_n \underset{n\to\infty}{\longrightarrow} \gamma$, $G_n(x) \underset{n\to\infty}{\overset{w}{\longrightarrow}} G(x)$, where $G(x)$ is a nondecreasing bounded function, then the sequence of random processes $\xi_n(t) = \xi_{n1} + \cdots + \xi_{n\,[nt]}$ converges in the weak sense to a process $\xi(t)$ with independent increments, with the characteristic function of the form

$$\varphi_t(z) = \mathbf{E}e^{iz\xi(t)} =$$

$$= \exp\left\{t\left(iz\gamma + \int_{-\infty}^{\infty}\left(e^{izu} - 1 - \frac{izu}{1+u^2}\right)\frac{1+u^2}{u^2}\, dG(u)\right)\right\}.$$

Assume that a path function $\xi(t)$ belongs to the space D with probability 1; see Section 1.8. For any functional f continuous in Skorohod J-topology the distribution of $f(\xi_n(\cdot))$ converges to the distribution of $f(\xi(\cdot))$ in the weak sense as $n \to \infty$. In particular, the probability of the event $\{a \leq \xi_{n1} + \cdots + \xi_{nk} \leq b$ for every $k : 1 \leq k \leq n\}$ tends to $\mathbf{P}\{a \leq \xi(t) \leq b\}$ as $n \to \infty$ for almost all a, b.

If $g(z)$ is a continuous function then the distribution of the functional $\int_0^1 g(\xi_n(t))\, dt$ converges to the distribution of the functional $\int_0^1 g(\xi(t))\, dt$ as $n \to \infty$.

5.6 Distribution of a Functional of the Process

Let $\xi(t)$ be a time-homogeneous separable m-dimensional process with independent increments, $\xi(0) = 0$, and let $f(x)$ be a continuous bounded function. Define an additive functional

$$\varphi_x(t) = \int_0^t f(x + \xi(s))\,ds, \quad t \geq 0.$$

The joint distribution of such a process and the process $\xi(t)$ itself can be investigated via the integral transform

$$v(\lambda, x, t) = \mathbf{E}\, e^{-\lambda \varphi_x(t)}\, \Phi(x + \xi(t))$$

where $\Phi(x)$ is a continuous bounded function. The integral transform satisfies the following integral equation:

$$v(\lambda, x, t) = \int \Phi(y)\, dF_i(y - x) - \lambda \int_0^t \int_{-\infty}^{\infty} v(\lambda, y, t - s)\, f(y)\, dF_s(y - x)\, ds$$

where $F_t(x) = \mathbf{P}\{\xi(t) < x\}$. The stated equation has the unique solution for a sufficiently small $\lambda > 0$.

Let $\xi(t)$ be a time-homogeneous one-dimensional process with independent increments deprived of a diffusion component. Assume also that the process shift is negative and the rate λ of jumps is finite. In the stated conditions the characteristic function of the process has the form

$$\varphi_t(z) = \exp\left\{ t \left(-ia + \int_0^{\infty} (e^{izx} - 1)\, dB(x) \right) \right\}$$

where $a > 0$ (so that $\xi'(t) = -a$ in intervals between jumps), λ is the parameter of the point process of jumps of the process and $B(x)$ is the distribution function of the value of a jump; assume also that $B(0) = 0$.

For applications (queueing theory, storage theory etc.) the process $\gamma(t)$ defined as $\xi(t)$ being stopped at the origin up to the next jump is of wide use. Such a process can be defined as a solution of the stochastic differential equations $d\gamma(t) = -a\, E(\gamma(t))\, dt + a\,\xi(t)$, where $E(x) = 0$ if $x \leq 0$, $E(x) = 1$ if $x > 0$, satisfying an initial condition. In the queueing theory the case $a = 1$ is of the most interest. In such a case $\gamma(t)$ can be treated as **the virtual waiting time** in a one-channel queueing system with a Poisson input with the parameter λ, the $B(x)$ being the distribution function of a service time. The virtual waiting time $\gamma(t)$ is treated in such a way. If a customer arrives at random at a time t then its waiting time equals $\gamma(t)$.

Set $F(t,x) = \mathbf{P}\{\gamma(t) \le x\}$. The following **Takács integro-differential equation** holds true:

$$\frac{\partial F(t,x)}{\partial t} - \frac{\partial F(t,x)}{\partial x} = \lambda F(t,x) + \lambda \int_0^x B(x-y)\, dF(t,y), \quad x \ge 0,\ t > 0.$$

For the solution of the Takács equation see[3–6]. If

$$\tau = \int_0^\infty x\, dB(x), \quad \rho = \lambda\tau < \infty$$

then $F(t,x) \underset{t\to\infty}{\longrightarrow} F(x)$ and the Laplace-Stieltjes transform $\varphi(s) = \int_0^\infty e^{-sx}\, dF(x)$ is defined by **Chinčin formula**

$$\varphi(s) = \frac{1-\rho}{1-\lambda(1-\beta(s))/s}$$

where

$$\beta(s) = \int_0^\infty e^{-sx}\, dB(x).$$

In particular, the steady-state mean waiting time equals $\frac{\lambda(\tau^2+\sigma^2)}{2(1-\rho)}$ where σ^2 equals the variance of a service time.

5.7 References

1. Lévy, P., *Processus Stochastiques et Mouvement Brownien,* Revue et Augmentée, Paris, 1965.
2. Gihman, I. I., Skorohod, A. V., *Introduction to the Theory of Random Processes,* Saunders, Philadelphia, 1969.
3. Gihman, I. I., Skorohod, A. V., *Theory of Stochastic Processes,* 1, Springer, Berlin, 1974.
4. Gihman, I. I., Skorohod, A. V., *Theory of Stochastic Processes,* 2, Springer, Berlin, 1975.
5. Gihman, I. I., Skorohod, A. V., *Theory of Stochastic Processes,* 3, Springer, Berlin, 1979.
6. Skorohod, A. V., *Random Processes with Independent Increments,* Kluwer Acad. Publ., Dordrecht, 1991.

7. Grimmett, G., Stizzaker, D., *Probability and Random Processes,* Clarendon Press, Oxford, 1982.
8. Billingsley, P., *Convergence of Probability Measures,* Wiley, New York, 1968.
9. Gnedenko, B. V., Kolmogorov, A. N., *Limit Distributions for Sums of Independent Random Variables,* Addison-Wesley, Cambridge, 1954.
10. Loève, M., *Probability Theory,* Van Nostrand, New York, 1963.

Chapter 6

Processes Associated with a Poisson Process

6.1 Some Properties of Point Processes. Probability Generating Functional

The notion of a point process is one of the fundamental notions forming the basis of the theory of random processes[1–7]. Using point processes it is possible to study different classes of processes with jump-like trajectories (such as Poisson and associated processes, semi-Markov processes et al.) from the same point of view. It should be noted that these formulated definitions concerning point processes can be reformulated in terms of flows of homogeneous events (cf. Chapter 7).

Let $(\Omega, \mathfrak{F}, \mathbf{P})$ be some probability space and with every $\omega \in \Omega$ one can associate a countable sequence of real numbers $\{t_i\}$ without limit points. This sequence is called **a point process**. Let $N(A)$ be a number of points of the sequence $\{t_i\}$ having hit into a Borel set $A \in \mathfrak{B}$ where $\mathfrak{B}$ is a σ-algebra of Borel sets from $\mathbb{R} = (-\infty, +\infty)$. So $N(\cdot)$ is a nonnegative integer-valued and countably additive function defined on Borel sets. Such a function is called **a counting measure**. Since $\{t_i\}$ is a sequence without limit points, $N(\cdot)$ is **a Borel measure**, i.e., it is a nonnegative finite measure for any bounded Borel set from $\mathbb{R}$. Let M be a set of integer-valued Borel measures and $\mathfrak{M}$ be a corresponding σ-algebra of subsets generated by the following **cylindrical sets**

$$B = \{N(\cdot) : N(A_1) = r_1, \ldots, N(A_k) = r_k\} \tag{6.1}$$

with $k \geq 1$, $r_1 \geq 0, \ldots, r_k \geq 0$, $A_1, \ldots, A_n \in \mathfrak{B}$, $A_i \cap A_j = \emptyset$, $i \neq j$.

With every $\omega \in \Omega$ one can associate a collection $\{N(A),\ A \in \mathfrak{B}\}$ which is said to be **a counting process**. Since there exists a one-to-one correspondence between Ω and M, the distribution $\mathbf{P}(\cdot)$ generates a corresponding distribution $\mathbf{Q}(\cdot)$ on $\mathfrak{M}$.

A point process is denoted by $N(\cdot)$ where $N(A),\ A \in \mathfrak{B}$, is the number of process points belonging to A.

Introduce shift operators T_t and Θ_t, $t \in \mathbb{R}$, acting on the sets from $\mathfrak{B}$ and $\mathfrak{M}$ respectively:

$$T_t A = \{x \in \mathbb{R} : x + t \in A\}, \quad A \in \mathfrak{B},$$

$$\Theta_t B = \{N(\cdot) : N(T_t A_1) = r_1, \ldots, N(T_t A_k) = r_k\}$$

where $B \in \mathfrak{M}$ is defined by (6.1).

A point process is called a **stationary** one, if the equality

$$\mathbf{P}\{N(T_t A_1) = r_1, \ldots, N(T_t A_k) = r_k\} = \mathbf{P}\{N(A_1) = r_1, \ldots, N(A_k) = r_k\}$$

holds for any integer numbers $k \geq 1$, $r_1 \geq 0, \ldots, r_k \geq 0$ and for any $A_1, \ldots, A_k \in \mathfrak{B}$, $t \in \mathbb{R}$. In other words the equality $\mathbf{Q}\{\Theta_t B\} = \mathbf{Q}\{B\}$ must be satisfied for any $t \in \mathbb{R}$ and for any $\mathfrak{B} \in \mathfrak{M}$ defined by (6.1). If $N(\cdot)$ is a stationary point process, then $N(-\infty, +\infty)$ is equal either to zero or to infinity with the probability 1.

A set $B \in \mathfrak{M}$ is called **invariant** with respect to a shift operator Θ_t, if for any $t \in \mathbb{R}$ the sets $\Theta_t B$ and B coincide up to the set of a zero measure. A stationary point process is called an **ergodic** one, if the sets invariant with respect to a shift operator can be of the probability either 0 or 1. An equality

$$\lim_{t\to\infty} \frac{1}{t} \int_0^t \mathbf{Q}\{A \cap T_{-\tau} B\}\, d\tau = \mathbf{Q}\{A\}\,\mathbf{Q}\{B\}, \quad A,\ B \in \mathfrak{M},$$

is the necessary and sufficient condition for the ergodicity of a point process.

A stationary point process is said to be

(i) **strongly intermixing** with the intensity $\alpha(\tau) \underset{\tau\to\infty}{\longrightarrow} 0$, if

$$|\mathbf{Q}\{A \cap T_{-\tau} B\} - \mathbf{Q}\{A\}\,\mathbf{Q}\{B\}| \leq \alpha(\tau), \quad \tau \geq 0,\ A,\ B \in \mathfrak{M},$$

(ii) **weakly intermixing** with the intensity $\alpha(\tau) \underset{\tau\to\infty}{\longrightarrow} 0$, if

$$\left| \frac{1}{\tau} \int_0^\tau \mathbf{Q}\{A \cap T_{-t} B\}\, dt - \mathbf{Q}\{A\}\,\mathbf{Q}\{B\} \right| \leq \alpha(\tau), \quad \tau \geq 0,\ A,\ B \in \mathfrak{M},$$

(iii) φ-**intermixing** with the intensity $\varphi(\tau) \underset{\tau\to\infty}{\longrightarrow} 0$, if

$$|\mathbf{Q}\{A \cap T_{-\tau}B\} - \mathbf{Q}\{A\}\, Q(B)| \le \varphi(\tau)\, \mathbf{Q}\{B\}, \quad \tau \ge 0, \; A, B \in \mathfrak{M},$$

(iv) γ-**intermixing** with the intensity $\gamma(\tau) \underset{\tau\to\infty}{\longrightarrow} 0$, if

$$|\mathbf{Q}\{A \cap T_{-\tau}B\} - \mathbf{Q}\{A\}\, \mathbf{Q}\{B\}| \le \gamma(\tau)\, \mathbf{Q}\{A\}\, \mathbf{Q}\{B\}, \quad \tau \ge 0, \; A, B \in \mathfrak{M}.$$

The ergodicity and different kinds of the intermixing are related as follows: γ-intermixing $\Longrightarrow$ φ-intermixing $\Longrightarrow$ strong intermixing $\Longrightarrow$ weak intermixing $\Longrightarrow$ ergodicity.

By **the superposition** of point processes $N_1(\cdot), \ldots, N_n(\cdot)$ we mean a process consisting of all the points of these processes. An integral over a point process is defined by the relation

$$\int_{-\infty}^{+\infty} f(t)\, dN(t) = \sum_i f(t_i)$$

where $f(\cdot)$ belongs to a class of functions ensuring the sum in the right-hand side be absolute converging with the probability 1 (for example, $f(\cdot)$ is a bounded function vanishing outside of some bounded interval).

A moment structure of a point process is defined by **moment measures** $M_k(\cdot)$ defined on a k-multiple Cartesian product $\mathfrak{B} \times \cdots \times \mathfrak{B}$. The first and second moment measures are defined by relations

$$M_1(A) = \mathbf{E}\, N(A), \quad M_2(A \times B) = \mathbf{E}\{N(A)\, N(B)\}, \quad A, B \in \mathfrak{B}.$$

In the case of a stationary point process we have $M_1(A) = m\,|A|$ where $|A|$ is a Lebesgue measure of the set A and $m = \mathbf{E}\, N[0,1]$ is the intensity of the process $N(\cdot)$. However **factorial moment measures** $M_{(k)}(\cdot)$, $k \ge 1$, defined on a k-multiple Cartesian product $\mathfrak{B} \times \cdots \times \mathfrak{B}$ are more suitable for practical purposes[4]. The first two factorial moment measures are defined by relations:

$$M_{(1)}(A) = \mathbf{E}N(A), \; M_{(2)}(A \times B) = \mathbf{E}\{N(A)N(B)\} - \mathbf{E}N(A \cap B), \; A, B \in \mathfrak{B}.$$

Two main reasons stipulate the preferable use of factorial moment measures: (i) these measures can be relatively easily defined in terms of the distribution $\mathbf{P}(\cdot)$, (ii) under some hypotheses on $\mathbf{P}(\cdot)$ the measures $M_k(\cdot)$ are absolutely continuous with respect to the Lebesgue measure on $\mathfrak{B} \times \cdots \cdots \times \mathfrak{B}$, i.e., there exists **a factorial moment density** $\mu_{(k)}(x_1, \ldots, x_k)$ with the following meaning: $\mu_{(k)}(x_1, \ldots, x_k)\, h_1 \ldots h_k + o(h_1 + \cdots + h_k)$

is the probability that each of intervals $(x_1,\, x_1 + h_1), \ldots, (x_k,\, x_k + h_k)$ contain at least one point of the process. Such densities can be efficiently used in the reliability analysis of complex systems[8].

Laplace-Stieltjes transform and generating functions are the main tools used for the investigation of sequences of continuous and discrete random variables. **A probability generating functional** plays the same role with respect to point processes. It is defined by the relation

$$G[z] = \mathbf{E} \exp \left\{ \int_{-\infty}^{+\infty} \ln z(t)\, dN(t) \right\} = \mathbf{E} \left\{ \prod_i z(t_i) \right\}$$

where a function $z(\cdot)$ is taken from a special class of functions and $\{t_i\}$ is a sequence of points of the process.

Later on we assume the function $z(\cdot)$ belongs to one of the following classes:

(a) $z(\cdot) \in V$, if $z(t)$ is a measurable function, $0 \le z(t) \le 1,\ t \in \mathbb{R}$, and there exist finite real numbers a, b such that $z(t) = 1$ for $t \overline{\in} (a,\, b)$,

(b) $z(\cdot) \in L(N)$, if $z(t)$ is a measurable function, $0 \le z(t) \le 1,\ t \in \mathbb{R}$, and

$$\int_{-\infty}^{+\infty} |\ln z(t)|\, M_1(dt) < \infty$$

where $M_1(A) = \mathbf{E}\, N(A),\ A \in \mathfrak{B}$.

Example 6.1. A probability generating functional for the stationary Poisson process is defined by the formula

$$G[z] = \exp \left\{ -\lambda \int_{-\infty}^{+\infty} (1 - z(t))\, dt \right\}.$$

Example 6.2. A probability generating functional for the nonstationary Poisson process with the intensity measure $\Lambda(\cdot)$ (see Section 6.5) is defined by the relation

$$G[z] = \exp \left\{ - \int_{-\infty}^{+\infty} (1 - z(t))\, \Lambda(dt) \right\}.$$

Properties of the probability generating functional:

(i) $0 \le G[z] \le 1$,

(ii) if for any $t \in \mathbb{R}$ $\quad z_1(t) \le z_2(t)$, then $G[z_1] \le G[z_2]$,

(iii) if $1+\lambda z \in V$ or $1+\lambda z \in L(N)$ and all factorial moment measures $M_{(k)}(A_1,\ldots,A_k)$, $k \geq 1$, $A_1,\ldots,A_k \in \mathfrak{B}$, are finite, then

$$G[1+\lambda z] = 1 + \sum_{k=1}^{\infty} \frac{\lambda^k}{k!} \int_{-\infty}^{+\infty} \ldots \int_{-\infty}^{+\infty} z(t_1)\ldots z(t_k)\, M_{(k)}(dt_1,\ldots,dt_k).$$

6.2 Cluster Processes

Cluster processes[5,9–22] belong to those few classes of processes that not only find wide practical applications but also establish a good theory. So, these processes were used as the model for a spatial distribution of galaxies[9,11], for the investigation of infection spreading in ecological problems[10], for studying traffic flows[13], as the model for the analysis of computer failure patterns[14] and for the description of the occurrence and development of earthquakes[18].

Later on we consider only one-dimensional cluster processes. Let $N_c(\cdot)$ be a point process and $\{N_i(\cdot),\ i = \ldots,-2,-1,0,1,2,\ldots\}$ be mutually independent and independent of $N_c(\cdot)$ point processes with the same distribution. Assume that every point t_i of the process $N_c(\cdot)$ generates a process $\Theta_{-t_i}N_i(\cdot)$, which is called **a process of cluster members** or simply **a cluster**. Here Θ_t is a shift operator introduced in Section 6.1; t_i is **a cluster center**.

A cluster process $N(\cdot)$ is the superposition of all clusters (depending on the considered problem, the superposition can include or not the process $N_c(\cdot)$ called **a process of cluster centers**). If $N_c(\cdot)$ is a Poisson process, then $N(\cdot)$ is called **a Poisson cluster process**.

Let $G_1[z]$ be a generating functional of the process $N_c(\cdot)$ and $G_2[z/t]$ be a generating functional of a cluster provided that its center is located at a point t. The independence between clusters implies that a generating functional $G[z]$ of the process $N(\cdot)$ equals $\prod_i G_2[z/t_i]$ provided that all the points $\{t_i\}$ of the process $N_i(\cdot)$ are fixed. Consequently $G[z] =$ $= G_1[G_2[z/t]]$.

Examples of cluster processes.

1. **A nonordinary Poisson process**. All the members of the cluster are situated at the center of this cluster. Let $P(x)$, $P(1) = 1$, be a generating function for the number of members in one cluster and $N_c(\cdot)$ be a Poisson process defined by the intensity measure $\Lambda(\cdot)$ (see Section 6.5).

Then

$$G[z] = \exp\left\{ \int_{-\infty}^{+\infty} (1 - P(z(t)))\, \Lambda(dt) \right\}.$$

2. **A queueing system** $G|GI|\infty$. Consider a queueing system with an infinite number of servers. Let an input flow of customers be a process $N_c(\cdot)$ of cluster centers defined by a generating functional $G_1[z]$. Service times are mutually independent random variables with the same distribution function $F(x)$. Each cluster consists of only one member displaced to the right from its center at a distance η with the distribution function $F(x)$. As a cluster process we have an output flow of demands with a generating functional

$$G[z] = G_1 \left[- \int_{0}^{+\infty} z(t+x)\, dF(x) \right].$$

3. **A Neyman-Scott process**[9,11,18,19]. A process of cluster centers is a Poisson one with the intensity λ. Let $P(x)$, $P(1) = 1$, be a generating function for the number of members in one cluster. The displacements of cluster members from its center are independent, identically distributed random variables with the distribution function $F(x)$. Then a generating functional for such cluster process is defined by the formula

$$G[z] = \exp\left\{ -\lambda \int_{-\infty}^{+\infty} \left(1 - P\left(- \int_{-\infty}^{+\infty} z(t+x)\, dF(x) \right) \right) dt \right\}.$$

4. **A model for the earthquake description**[18]. Assume that a sequence of earthquakes occurs according to a Poisson process with the intensity λ (the process $N_c(\cdot)$). Each earthquake generates a series of secondary shocks, being a nonhomogeneous Poisson process with the intensity $\mu(t)$, $t \geq 0$, and $m = \int_0^{\infty} \mu(t)\, dt < \infty$. Then

$$G[z] = \exp\left\{ -\lambda \int_{-\infty}^{+\infty} \left(1 - \exp\left\{ - \int_{0}^{+\infty} \mu(x)\, (1 - z(t+x))\, dx \right\} \right) dt \right\}.$$

This process is equivalent to the Neyman-Scott process having the following properties: (i) the number of members in one cluster has a Poisson distribution with the mean m, (ii) the displacement of every cluster member from its center is an independent random variable with the density $\frac{\mu(t)}{m}$, $t \geq 0$.

5. **A Bartlett-Lewis process**[14,15,17,20] (or **a branching Poisson process**). A process $N_c(\cdot)$ is a Poisson one with the intensity λ. Every point of this process generates a cluster disposed on the right of its center. If a cluster center is located at the point t, then the cluster members are disposed at points $t+Y_1,\ t+Y_1+Y_2,\ \ldots,\ t+Y_1+\cdots+Y_\nu$ where ν is an integer-valued positive random variable and $Y_1,\ Y_2,\ \ldots$ is a sequence of independent nonnegative identically distributed random variables. If $N(\cdot)$ is a nonhomogeneous Poisson process with the intensity $\lambda(t)$, then we have **a nonhomogeneous branching Poisson process**[16,21]. A generating functional of a branching Poisson process (both homogeneous and nonhomogeneous) can't be found in the explicit form (without taking into account a representation in the form of multiple integrals). If $N_c(\cdot)$ is a renewal process, then we have **a branching renewal process**.

A cluster process is called **regular**, if any finite interval contains only a finite number of process points with the probability 1.

Properties of cluster processes.

1. A cluster process having a finite number of members in one cluster with the probability 1 is regular iff for any bounded interval I

$$\int_{-\infty}^{\infty} \rho_I(t)\, dN_c(t) < \infty \quad \text{with the probability 1.} \tag{6.2}$$

Here $\rho_I(t) = \mathbf{P}\{N_i(I) > 0 \mid \text{a cluster center is located at } t\}$. In particular, a relation (6.2) is true, if for any bounded interval I

$$\int_{-\infty}^{\infty} \rho_I(t)\, dM_c(dt) < \infty$$

where $M_c(A) = \mathbf{E}\, N_c(A),\ A \in \mathfrak{B}$. There is an example[12] of a Neyman-Scott process having with the probability 1 an infinite number of points in any bounded interval I.

2. A stationary Poisson cluster process is regular iff for some bounded interval I

$$\int_{-\infty}^{\infty} \rho_I(t)\, dt < \infty.$$

3. A Bartlett-Lewis process is regular iff $\mathbf{E}\,\nu < \infty$.

4. Let $N(\cdot)$ be a Neyman-Scott process such that $1 - F(s) \sim s^{-\alpha} L(s)$ for some $\alpha \geq 0$ as $s \to \infty$ where $L(s) \geq 0$ is a slowly changing function[23]. Then a relation

$$\mathbf{E}\max\{\nu^\beta,\ \nu^\beta L(\nu^\beta)\} < \infty$$

is a sufficient condition for the regularity of a stationary Neyman-Scott process. Here ν is the number of members in one cluster, $\beta = (1+\alpha)^{-1}$.

5. If $N(\cdot)$ is a regular stationary cluster process and $N_c(\cdot)$ is a strongly intermixing process, then $N(\cdot)$ is also a strongly intermixing process.

6. Let $N(t)$ be a number of points in $[0, t)$ of a Bartlett-Lewis process and ν be a number of members in one cluster. If $\mathbf{E}\,\nu^2 < \infty$, then

$$Z(t) = \frac{N(t) - \mathbf{E}\,N(t)}{(\lambda t \mathbf{E}\,\nu^2)^{1/2}}$$

has an asymptotic normal distribution with a zero mean and a unit variance as $t \to \infty$. Let $\mathbf{E}\,\nu^2 < \infty$. A random variable

$$Z^*(t) = \frac{N(t) - \lambda t \mathbf{E}\,\nu}{(\lambda t \mathbf{E}\,\nu^2)^{1/2}}$$

has an asymptotic normal distribution with a zero mean and a unit variance as $t \to \infty$ iff

$$\frac{1}{\sqrt{t}} \sum_{j=1}^{\infty} \int_0^t \mathbf{P}\{\nu \geq j;\ Y_1 + \cdots + Y_j > u)\,du \underset{t\to\infty}{\longrightarrow} 0$$

where $Y_1, Y_2, \ldots$ is a sequence of independent identically distributed random variables used in the definition of a Bartlett-Lewis process.

A multidimensional Neyman-Scott cluster process is used as a model for the description of the spatial distribution of galaxies[9]. A hypothesis that a number of galaxies in a given volume has a Poisson distribution can't be confirmed by the existing statistical data. As a rule, galaxies are disposed as clusters. Therefore a multidimensional Neyman-Scott process is apparently the most suitable model. This model is defined by the following assumptions.

1. Let $\gamma(\alpha)$ be a number of cluster centers in a domain α having a volume V. Then $\gamma(\alpha)$ is a random variable with the distribution depending only on V (but not on the shape and disposal of the domain V). Denote by $G_\gamma(t/V)$ a generating function for $\gamma(\alpha)$.

2. If $\alpha_1, \alpha_2, \ldots$ are disjoint domains, then $\gamma(\alpha_j)$, $j = 1, 2, \ldots$, are mutually independent random variables.

3. Let α be an arbitrary domain with a positive volume V containing exactly $n > 1$ cluster centers $c_1, \ldots, c_n$ and let β be an arbitrary part of α. If $a_1 < a_2 < \cdots < a_m$ are arbitrary m numbers ($m < n$) chosen from $1, 2, \ldots, n$, then the probability that a domain β contains exactly m cluster centers $c_{a_1}, \ldots, c_{a_m}$ does not depend on the set $a_1, \ldots, a_m$ and is equal to

the conditional probability that $\gamma(\beta) = m$ provided $\gamma(\alpha) = n$, divided by the number of possible selections of m identical objects from n objects, i.e.,

$$\mathbf{P}\{\gamma(\beta) = m,\ c_{a_1} \in \beta,\ \ldots,\ c_{a_m} \in \beta \,|\, \gamma(\alpha) = n\} =$$

$$= \frac{m!\,(n-m)!}{n!}\,\mathbf{P}\{\gamma(\beta) = m \,|\, \gamma(\alpha) = n\}.$$

4. With every cluster center one can associate a random variable ν, which means a number of galaxies belonging to one cluster. This random variable has the same distribution for any cluster. Its generating function is $G_\nu(t)$.

5. Numbers $\nu_1, \nu_2, \ldots$ of galaxies in different clusters are mutually independent random variables.

6. Assume that a cluster center is located at a point $(u,\ v,\ w)$. Then a position of every galaxy belonging to this cluster is determined by a random variable with a probability density $f(\eta)$ depending solely on the distance $\eta = ((x-u)^2 + (y-v)^2 + (z-v)^2)^{1/2}$ between the cluster center $(u,\ v,\ w)$ and the galaxy position $(x,\ y,\ z)$. The function $f(\eta)$ is assumed to be continuous on η and is the same for different clusters.

7. Coordinates $(x_m,\ y_m,\ z_m)$, $m = 1, 2, \ldots$, of galaxies (independent of whether they belong to one cluster or to different clusters) are mutually independent random vectors.

8. The motions of cluster centers and galaxies in clusters are neglected.

For the described cluster process a joint generation function $G_{n_1,n_2}(t_1, t_2)$ was found for the numbers n_1, n_2 of galaxies visible on the perfect photo in two different domains ω_1 and ω_2.

6.3 Secondary Processes

Assume the following objects are defined: $N(\cdot)$ is a point process, $\{t_i\}$ is its sample path; $\{\chi_i\}$ is a sequence of independent random variables taking values in a measurable space $(X,\ \mathfrak{U})$ and independent of a point process $N(\cdot)$; $f(u,\ v,\ x)$, $u,\ v \in \mathbb{R}$, $x \in X$, is some function defined on $\mathbb{R} \times \mathbb{R} \times X$ taking values in $\mathbb{R}$ and $\mathfrak{B} \times \mathfrak{B} \times \mathfrak{U}$-measurable.

A random process

$$\eta(t) = \sum_{i:t_i \le t} f(t,\ t_i,\ \chi_i), \quad t \in \mathbb{R},$$

is called **a filtered point process**. In particular, if $N(\cdot)$ is a Poisson process, then $\eta(t)$ is a **filtered Poisson point process**.

Random process $N(\cdot)$ and $f(t, t_i, \chi_i)$ are called **primary** and **secondary** processes respectively. In the case of a cluster process, $N_c(\cdot)$ is a primary process and $N_i(\cdot)$ is a secondary process.

A shot noise process is of great practical importance. It can be defined as a special case of the process $\eta(t)$:

$$\xi(t) = \sum_{i:t_i \le t} f(t - t_i, \chi_i), \quad t \in \mathbb{R}.$$

It is supposed that $f(t, \chi_i) = 0$ for $t < 0$ and $\{\chi_i\}$ are independent and identically distributed random variables. A function $f(t - t_i, \chi_i)$ is called **a shot noise**. Shot noise processes were introduced by studying the change of the anode current in thermoelectronic diodes[24]. These processes can also be used for the investigation of noise in electronic devices[25], in acoustics[26], in optics[27], in astrophysics[28] and in other fields.

The main properties of shot noise processes were described in Section 4.9. One more property is the following one. Let $N(\cdot)$ be a non-homogeneous Poisson process with the intensity $\lambda(t)$, $t \ge 0$ (in particular, $\lambda(t) \equiv \lambda$). Let $f(t, \chi) = \chi e^{-\alpha t}$, $t \ge 0$, $\alpha \ge 0$ and $\chi \ge 0$ with the probability 1 (this is an important case in the theory of counters). Then a process $\xi(t)$ can be investigated by analytical methods. Denote

$$\zeta(t, z) = \sum_{k: t-z<t_k \le t} f(t - t_k, \chi_k),$$

$$F(z, x, t) = \mathbf{P}\{\zeta(t, z) \le x\}, \quad H(x) = \mathbf{P}\{\chi \le x\}.$$

Then the following integro-differential equation

$$\frac{\partial F(z, x, t)}{\partial z} = \lambda(t - z) \left\{ \int_0^x H((x - y)e^{\alpha z})\, d_y F(z, y, t) - F(z, x, t) \right\}$$

holds. This equation has a unique solution $F(z, x, t)$ satisfying an initial condition $F(0, x, t) \equiv 1$, $x \ge 0$, $t \ge 0$.

If $f(t, \chi)$ is an arbitrary function, then $\xi(t)$ is not a Markov process. If $\lambda(t) \equiv \lambda$ and

$$\int_0^\infty \int_0^\infty |f(z, x)|\, dH(x)\, dz < \infty,$$

then with the probability 1 there exist limits $\xi(t) = \lim_{z\to\infty} \zeta(t, z)$ and $F(x) = \lim_{z\to\infty} F(z, x, t)$.

Shot noise processes were investigated in detail with a renewal process[29] and a filtered point process[30] as $N(\cdot)$ (in particular, conditions were formulated providing the convergence of the filtered Poisson process to a Gaussian one).

Example 6.3. By hitting the cathode of the electronic amplifier, the photons knock out electrons[31]. The secondary emission implies the rapid increase of the flow of electrons between the cathodes. Let $p_0(k)$ be the probability that hitting the cathode particle knocks out k electrons, and r be a number of reinforcing plates in the electronic amplifier. Let $p(k)$ be the probability that a hitting on the plate electron knocks out k secondary electrons. We suppose that different electrons generate mutually independent secondary flows of electrons. Denote

$$f_0(s) = \sum_{k=0}^{\infty} p_0(k)\, s^k, \quad f(s) = \sum_{k=0}^{\infty} p(k)\, s^k.$$

Then a generating function for the number of electrons hitting the nth reinforcing plate is defined by the following recurrence relation:

$$g_n(s) = f(g_{n-1}(s)), \quad n = 1, 2, \ldots, r, \quad g_0(s) = f_0(s).$$

This relation enables finding a generation function for the number of electrons hitting the anode. If the number of electrons reaching the anode equals k, then a condenser in the anode circuit obtains a voltage of $x = ke/C$ volt where $e = 1.6 \cdot 10^{-19}$ coulomb is a charge of the electron. Since k takes discrete values $0, 1, 2, \ldots$ and e is a very small value, we can assume that x is a continuous random variable. A condenser with the initial charge of χ volt discharging with a time across a resistance R, i.e., a charge exponentially decreases according to the function $\chi e^{-t/(RC)}$. Here χ is a random variable that is equal to a total charge transmitted on a condenser by all electrons reaching the anode. This random variable has a distribution function $H(u)$. In the considered example

$$f(u, \chi) = \begin{cases} \chi e^{-\alpha u} & \text{if } u \geq 0 \\ 0 & \text{if } u < 0 \end{cases} \quad \text{with } \alpha = (RC)^{-1}.$$

6.4 Self-Exciting and Mutually Exciting Point Processes

A self-exciting point process[30,32–34] can be considered as the generalization of a nonhomogeneous Poisson process; its intensity depends on the previous process's behavior. Such processes are used to describe the functioning of different electron tubes or detectors of radiation. They also have a wide application for the describing "contagious" processes (such as grippe or measles) where every event of the process increases the probability of the appearance in the nearest future of further events. Moreover, a multidimensional case (mutually exciting processes) describes the situation, when several point processes can interact with each other (so it can be taken into account that a movement of ill people from one region to another promotes the rising and spreading of contagious diseases).

Let $Q = \{t_j,\ j = 0, \pm 1, \pm 2, \ldots\}$ be an ordinary point process without any fixed atoms on a real line defined on some probability space $(\Omega, \mathfrak{F}, \mathbf{P})$. Let $N(A)$ be a counting measure defined for any Borel set $A \in \mathfrak{B}$ and equal to the number of points of the sequence Q hitting into A. Denote by $\{\mathcal{H}_t,\ t \in \mathbb{R}\}$ a flow of σ-algebras ($\mathcal{H}_t \in \mathfrak{F}$, $\mathcal{H}_{t_1} \subset \mathcal{H}_{t_2}$ for $t_1 < t_2$) such that a random variable $N(A)$ is measurable with respect to $\mathcal{H}_t$ for any $A \in \mathfrak{B}$, $A \subset (-\infty, t)$. $\mathcal{H}_t$ determines the prior history of the sequence Q to the moment t. A function

$$\Lambda(t\,|\,\mathcal{H}_t) = \lim_{\delta \to 0} \frac{1}{\delta} \mathbf{P}\{N(t, t+\delta) > 0\,|\,\mathcal{H}_t\}, \quad t \in \mathbb{R}, \tag{6.3}$$

is called **a total intensity function** of the point process. It is assumed that $\Lambda(t\,|\,\mathcal{H}_t)$ is finite with the probability 1 for any $t \in \mathbb{R}$. The following lemma holds.

Lemma. *By fixed σ-algebra $\mathcal{H}_0$ there exists at most one ordinary point process Q^+ defined for $t \geq 0$ and satisfying the relation (6.3) with a given intensity function $\Lambda(\cdot\,|\,\cdot)$.*

An ordinary point process $N(\cdot)$ with the function $\Lambda(t\,|\,\mathcal{H}_t)$ is called **a self-exciting process.**

A class of self-exciting processes is a proper subclass of all ordinary point processes. If the intensity $\Lambda(t\,|\,\mathcal{H}_t)$ depends only on the few last values $t_i < t$, then we deal with a process having a bounded memory.

A self-exciting process $N(t), t \geq 0$, has **a bounded m-memory**, if its intensity $\Lambda(t\,|\,\mathcal{H}_t)$ depends only on t, $N(t) = N(0, t]$ and $t_{N(t)-m+1}$, $t_{N(t)-m+2}, \ldots, t_{N(t)}$. If $m = 0$, then $\Lambda(t\,|\,\mathcal{H}_t)$ depends only on t, $N(t)$ and we deal with a Markov process.

If $\Lambda(t\,|\,\mathcal{H}_t) = h(t - t_{N(t)})$ for some function $h(t) \geq 0$, $t \geq 0$, then a self-exciting process is a renewal process.

Consider **a self-exciting Hawkes process** with a function $\Lambda(t\,|\,\mathcal{H}_t)$ satisfying the relation

$$\Lambda(t\,|\,\mathcal{H}_t) = \nu + \int_{-\infty}^{t} \gamma(t-u)\, dN(u) \tag{6.4}$$

where $\nu > 0$, $\gamma(u) \geq 0$, $u \geq 0$, and $m = \int_0^\infty \gamma(u)\,du < 1$ ($\gamma(u)$ is called **a reaction function**).

Point processes $N_1(k), \ldots, N_k(t)$ are called **mutually exciting Hawkes processes** if the following relations are satisfied:

$$\mathbf{P}\{N_i(t, t+\Delta) = 1 \,|\, \mathbf{N}(s),\, s \leq t\} = \Lambda_i(t)\,\Delta + o(\Delta),$$

$$\mathbf{P}\{N_i(t, t+\Delta) > 1 \,|\, \mathbf{N}(s),\, s \leq t\} = o(\Delta), \quad i = 1, \ldots, k,$$

with

$$\Lambda(t) = \mathbf{v} + \int_{-\infty}^{t} \gamma(t-u)\,d\mathbf{N}(u)$$

where $\gamma(t) = (\gamma_{ij}(t))$ is a matrix, $\gamma_{ij}(t) \geq 0$, $i, j = 1, \ldots, k$, and $\mathbf{N}(t)$, $\Lambda(t)$, $\mathbf{v}$ are vectors $\mathbf{N}(t) = (N_1(t), \ldots, N_k(t))^T$, $\Lambda(t) = (\Lambda_1(t), \ldots, \Lambda_k(t))^T$, $\mathbf{v} = (v_1, \ldots, v_k)^T$, $v_i \geq 0$, $i = 1, \ldots, k$, $t \geq 0$ (the symbol "T" denotes the operation of a transposition).

The next theorem states that under some hypotheses with a self-exciting process one can associate a Poisson cluster process.

Theorem 6.1. *If $\nu > 0$ and $m = \int_0^\infty \gamma(u)\,du < 1$, $\gamma(u) \geq 0$, $u \geq 0$, then there exists a unique stationary ordinary point process $N(\cdot)$ with a finite intensity $\nu/(1-m)$ and with a function $\Lambda(t \,|\, \mathcal{H}_t)$ satisfying the equality (6.4).*

Such a process can be represented as a Poisson cluster process having the following structure. A process of cluster centers is a Poisson one with the intensity ν. Every point generates a secondary nonhomogeneous Poisson process with the intensity $\gamma(t)$. At the same time every point of the secondary process generates a new nonhomogeneous Poisson process with the intensity $\gamma(t)$ and so on. A cluster process is defined as the superposition of all these processes (including the process of cluster centers). Hence a generating functional of a self-exciting Hawkes process is determined by the formula

$$G[z(\cdot)] = \exp\left\{ -\nu \int_{-\infty}^{+\infty} (1 - F[z_t(\cdot)])\,dt \right\}$$

where $z_t(\cdot) = z(t + \cdot)$ and $F[z(\cdot)]$ is a generating functional of a cluster with a center at zero (a center belongs to the cluster). This functional $F[z(\cdot)]$

satisfies the equation

$$F[z(\cdot)] = z(0)\exp\left\{-\int_0^{+\infty}(1 - F[z_t(\cdot)])\,\gamma(t)\,dt\right\}.$$

If we set $z(x) = y$ for any $x \geq 0$, then $F[z(\cdot)] = \pi(y)$ is a generating function for the number of members in one cluster satisfying the equation

$$\pi(y) = y\exp\{m\,(\pi(y) - 1)\}.$$

Let $N(0, t]$ denote a number of points at $(0, t]$ of the stationary or nonstationary (i.e., starting at time 0) self-exciting Hawkes process.

Theorem 6.2. *If* $\int_0^\infty u\gamma(u)\,du < \infty$, *then*

$$\mathbf{P}\left\{\frac{N(0,t] - \nu t/(1-m)}{(\nu t/(1-m)^3)^{1/2}} \leq y\right\} \underset{t\to\infty}{\longrightarrow} \frac{1}{\sqrt{2\pi}}\int_{-\infty}^{y} e^{-u^2/2}\,du.$$

Example 6.4. Consider a detector with the input nonhomogeneous Poisson flow of particles having the intensity $\mu(t)$. Assume that some particle arrives at time t. Then all particles coming in the interval $(t, t+\tau)$ can't be discovered by the detector. Here τ is a random variable called a dead time. We assume that particles coming in $(t, t+\tau)$ don't prolong the dead time. Let $N(t)$, $t \geq 0$, be a number of detected particles at $(0, t]$. Then $\{N(t),\, t \geq 0\}$ is a self-exciting process with the intensity

$$\Lambda(t\,|\,t_1, \ldots, t_{N(t)}) = \begin{cases} 0 & \text{if } \; t_{N(t)} \leq t < t_{N(t)} + \tau \\ \mu(t) & \text{if } \; t_{N(t)} + \tau \leq t. \end{cases}$$

6.5 Doubly Stochastic Poisson Process (Cox Process)

A doubly stochastic Poisson process (or a Cox process) is a popular model used in nuclear medicine, in optics, in electrophysiology and in other fields[19,30,35–40].

Let $(\Omega, \mathfrak{F}, \mathbf{P})$ be some probability space, $\mathfrak{B}$ be a σ-algebra of Borel sets from $\mathbb{R}$, M be a set of integer-valued countably additive Borel measures, $\mathfrak{M}$ be a corresponding σ-algebra, generated by the sets of the form (6.1) and $\mathbf{Q}(\cdot)$ be a distribution on $\mathfrak{M}$ generated by a corresponding distribution $\mathbf{P}(\cdot)$

on $\mathfrak{F}$. Let moreover $N(\cdot)$ be a counting measure and $\Lambda(\cdot) = \mathbf{E}N(\cdot) \in W$ where $(W, \mathfrak{U})$ is a measurable space of countably additive Borel measures with a σ-algebra $\mathfrak{U}$ of subsets from W generated by cylindrical sets of the form

$$C = \{\Lambda(\cdot) : \Lambda(A_i) = x_i, \ i = 1, \ldots, k\},$$

$$A_i \in \mathfrak{B}, \ A_i \cap A_j = \emptyset, \ i \neq j, \ x_i \geq 0, \ i, j = 1, \ldots, k, \ k \geq 1.$$

A point process $N(\cdot)$ is called **a Poisson process with the intensity measure** $\Lambda(\cdot) \in W$ if for any bounded set $B \in \mathfrak{B}$ the relation

$$\mathbf{P}\{N(B) = k\} = \frac{(\Lambda(B))^k}{k!} e^{-\Lambda(B)}, \quad k = 0, 1, 2, \ldots, \tag{6.5}$$

holds and $N(B_1), \ldots, N(B_n)$ are independent random variables for any $n \geq 1$, $B_i \in \mathfrak{B}$, $B_i \cap B_j = \emptyset$, $i \neq j$, $i, j = 1, \ldots, n$. In particular, $\Lambda(B) = \lambda |B|$ for a stationary Poisson process with the intensity $\lambda < \infty$ where $|B|$ denotes a Lebesgue measure of the set B.

For any measure $\Lambda(\cdot) \in W$ there exists a unique measure $\mathbf{Q}_\Lambda(\cdot)$ defined on $\mathfrak{M}$ and being a distribution of a point process $N(\cdot)$ for which the relation (6.5) is satisfied[41].

Let $\pi(\cdot)$ be some probability measure defined on $(W, \mathfrak{U})$. A point process $N(\cdot)$ with a distribution $\int_W \mathbf{Q}_\Lambda(\cdot)\, \pi(d\Lambda)$ is called **a doubly stochastic Poisson process (a Cox process)**. In other words a doubly stochastic Poisson process has a random intensity measure $\Lambda(\cdot) \in W$ with the distribution $\pi(\cdot)$.

If $N(\cdot)$ is a doubly stochastic Poisson process and a random measure $\Lambda(\cdot)$ is distributed according to $\pi(\cdot)$, then

$$\mathbf{P}\{N(B) = k\} = \int_W \frac{(\Lambda(B))^k}{k!} e^{-\Lambda(B)}\, \pi(d\Lambda) = \mathbf{E}\left\{\frac{(\Lambda(B))^k}{k!} e^{-\Lambda(B)}\right\}.$$

Example 6.5. Let $N(\cdot)$ be a doubly stochastic Poisson process defined for $t \geq 0$ and with $\Lambda(t) = \Lambda(0, t] = \xi\, G(t)$, $t \geq 0$, where ξ is a nonnegative random variable having a gamma-distribution with the density

$$f(t) = \begin{cases} \frac{\alpha^{\beta+1}}{\Gamma(\beta+1)}\, t^\beta e^{-\alpha t} & \text{if } t \geq 0 \\ 0 & \text{if } t < 0, \end{cases}$$

$\Gamma(\beta + 1) = \int_0^{+\infty} t^\beta e^{-t}\, dt$, $\beta > -1$, $\alpha > 0$, and $G(t)$ is a nonnegative monotone increasing function, $G(0) = 0$. A point process $N(t) = N(0, t]$, $t \geq 0$, with the distribution

$$\mathbf{P}\{N(t)=n\} = \frac{\Gamma(n+\beta+1)}{\Gamma(n+1)\,\Gamma(\beta+1)}\left(\frac{\alpha}{\alpha+G(t)}\right)^{\beta+1}\left(\frac{G(t)}{\alpha+G(t)}\right)^{n}, \quad n \ge 0,$$

is called **a nonhomogeneous Polya process**[30].

Stated results below associate doubly stochastic processes with other classes of processes.

Let $\Lambda(x) = \Lambda(0, x]$. Then $\{\Lambda(x),\ x \ge 0\}$ is a nondecreasing continuous on the right random process with $\Lambda(0) \ge 0$. A process $\Lambda^{-1}(x) = \sup\{y : \Lambda(y) \le x\}$ is called an inverse one to $\Lambda(\cdot)$. The case $\mathbf{P}\{\Lambda^{-1}(x) = \infty\} > 0$ for some $x > 0$ is not excluded.

Let $F_0(x)$ and $F(x)$, $F(0) < 1$, be distribution functions defining a general renewal process, i.e., $F_0(x)$ is the distribution function of the time to the first renewal and $F(x)$ is the distribution function of the times between successive renewals. Let

$$f^*(s) = \int_0^{+\infty} e^{-s\,x}\, dF(x), \qquad f_0^*(s) = \int_0^{+\infty} e^{-s\,x}\, dF_0(x).$$

Theorem 6.3.

1. A doubly stochastic Poisson process defined by a random measure $\Lambda(\cdot)$ is a renewal process iff $\Lambda^{-1}(x)$ is a process with stationary and independent increments.

2. A renewal process is a doubly stochastic Poisson process iff

$$f^*(s) = \frac{1}{1-\ln g^*(s)}, \qquad f_0^*(s) = g_0^*(s)\, f^*(s)$$

where $g^*(s) = \int_0^{+\infty} e^{-s\,x}\, dG(x)$ for some infinitely divisible distribution function $G(x)$, $G(0) < 1$, and $g_0^*(s) = \int_0^{+\infty} e^{-s\,x}\, dG_0(x)$ for some distribution function $G_0(x)$.

3. These two representations are connected by the relations.

$$\mathbf{E}\, e^{-s\Lambda^{-1}(0)} = g_0^*(s), \qquad \mathbf{E}\, e^{-s\,(\Lambda^{-1}(1)-\Lambda^{-1}(0))} = g^*(s).$$

Suppose

$$\Lambda(x) = \int_0^{x} \lambda(u)\, du$$

where $\{\lambda(u),\ u \geq 0\}$ is a random process not identically equal to zero. Let

$$F_0(x) = \frac{1}{\mu} \int\limits_0^x (1 - F(x))\,dx, \qquad \mu = \int\limits_0^{+\infty} (1 - F(x))\,dx,$$

i.e., a renewal process is a stationary one.

Theorem 6.4. *A stationary renewal process defined by distribution function $F(x)$ is a doubly stochastic Poisson process iff there exist $\lambda > 0$ and a nondecreasing function $K(z)$, $z > 0$, $\int\limits_0^{+\infty} z\,dK(z) < \infty$, such that for any $s \geq 0$*

$$f^*(s) = \lambda \left(\lambda + s + \int\limits_0^{+\infty} (1 - e^{-sz})\,dK(z) \right)^{-1}. \tag{6.6}$$

A corresponding process $\lambda(t)$ can take only two values λ and 0. Let condition (6.6) be satisfied and $K = K(+\infty) - K(0) < \infty$. If $K = 0$, then we have a Poisson process with the intensity λ. If $K > 0$, then $G(z) = K^{-1}[K(z) - K(0)]$ is a distribution function and in this case a process $\lambda(t)$ takes values λ and 0 alternately in the intervals with mutually independent random lengths: $\lambda(t) = \lambda$ in intervals having exponentially distributed lengths with the mean K^{-1} and $\lambda(t) = 0$ in intervals with lengths having a distribution function $G(x)$.

A linear stochastic Poisson process is a doubly stochastic Poisson process having as the intensity $\lambda(t)$ a linear process

$$\lambda(t) = \int\limits_{-\infty}^{+\infty} f(t-u)\,dX(u) \tag{6.7}$$

where $X(t)$ is a random process with stationary independent and nonnegative increments, $\mathbf{E}X^2(t) < \infty$ for any $t \in \mathbb{R}$ and $f \in B_2$ (a function $f \in B_n$, $n \geq 1$, if $|f|^p$ is an integrable function for $p = 1, \ldots, n$). An integral in (6.7) exists in the sense of the mean square convergence.

Since $X(t)$ is a process with stationary and independent increments, the following relation

$$\tau\,\varphi(\theta) = \ln \mathbf{E} \exp\{-\theta\,(X(t+\tau) - X(t))\} = -\tau \left\{ \gamma\theta + \int\limits_0^{+\infty} (1 - e^{-\theta x}) \frac{dK(x)}{x} \right\}$$

holds where $\gamma \geq 0$ is a constant and $K(x)$ is a bounded nondecreasing function, $K(0) = 0$. A generating functional of the linear stochastic process is the following one:

$$G[z] = \exp\left\{ \int_{-\infty}^{+\infty} \varphi\left(\int_{-\infty}^{+\infty} f(t)\,(1 - z(t+u))\,dt \right) du \right\}$$

where $z(\cdot) \in V$ (see Section 6.1).

Theorem 6.5. *A linear stochastic Poisson process is a Neyman-Scott process* (see Section 6.2).

Theorem 6.6. *Let $N(\cdot)$ be a stationary point process and $f(\cdot)$ be a nonnegative integrable function. Then a doubly stochastic Poisson process with the intensity $\lambda(t) = \int_{-\infty}^{+\infty} f(t-u)\,dN(u)$ is equivalent to a cluster process with $N_c(\cdot) = N(\cdot)$ and with $N_i(\cdot)$ being a nonstationary Poisson process with the intensity $f(x)$.*

If $\lambda = \int_{-\infty}^{+\infty} f(x)\,dx$ and $g(x) = \lambda^{-1} f(x)$, then this cluster process is a Neyman-Scott process with the number of members in one cluster having a Poisson distribution with a parameter λ. Every member of the cluster is shifted from its center on a random length with a distribution function $F(x) = \int_{-\infty}^{x} g(u)\,du$.

Theorem 6.7. *Let $N(\cdot)$ be a doubly stochastic Poisson process with a random intensity $\lambda(t)$, $\mathbf{E}\,\lambda(t) < \infty$ for any $t \in \mathbb{R}$. Then $N(\cdot)$ is a self-exciting process with the intensity*

$$\Lambda(t\,|\,\mathcal{H}_t) = \mathbf{E}\{\lambda(t)\,|\,\mathcal{H}_t\}, \quad t \in \mathbb{R},$$

where $\{\mathcal{H}_t,\ t \in \mathbb{R}\}$ is a flow of σ-algebras (see Section 6.4).

So for a nonhomogeneous Polya process

$$\Lambda(t\,|\,\mathcal{H}_t) = (N(t) + \beta + 1)\,(\alpha + G(t))^{-1}, \quad t \geq 0.$$

Theorem 6.8. *A renewal process with interrenewal times having gamma-distribution with a density*

$$\frac{\rho\,(\rho x)^{\alpha-1}\,e^{-\rho x}}{\Gamma(\alpha)}, \quad \alpha > 0, \ \ \rho > 0, \ \ x > 0, \ \ \Gamma(\alpha) = \int_0^\infty u^{\alpha-1}\,e^{-u}\,du,$$

is a doubly stochastic Poisson process iff $\alpha \in (0, 1]$.

Example 6.6. A problem of air contamination is of great importance in ecology. Suppose it is necessary to evaluate a distribution of the time from the entrance of an aerosol particle into the atmosphere to the moment of its removal. The intensity of the particle's removal essentially depends on the weather. In the example we suppose this intensity depends only on two states of the weather: either it rains or not. Let λ_r and λ_d be the intensities of the particle's removal during rainy and dry periods respectively. Then $\lambda(t)$ is a random process defined by

$$\lambda(t) = \begin{cases} \lambda_r & \text{if it rains at time } t \\ \lambda_d & \text{otherwise.} \end{cases}$$

Suppose $\lambda(t)$ is a homogeneous Markov chain with a continuous time having transition intensities

$$q_r = \lim_{h\to 0} \frac{1}{h} \mathbf{P}\{\lambda(h) = \lambda_d \,|\, \lambda(0) = \lambda_r\}, \quad q_d = \lim_{h\to 0} \frac{1}{h} \mathbf{P}\{\lambda(h) = \lambda_r \,|\, \lambda(0) = \lambda_d\}$$

and an initial distribution

$$p_r = \mathbf{P}\{\lambda(0) = \lambda_r\}, \quad p_d = \mathbf{P}\{\lambda(0) = \lambda_d\}, \quad p_r + p_d = 1.$$

Let a particle enter the atmosphere at time $t = 0$ and T be the time of its removal. Then

$$G(t) = \mathbf{P}\{T > t\} = \mathbf{E}\exp\left\{-\int_0^t \lambda(u)\,du\right\}.$$

This example can be generalized in the case when $\lambda(t)$ depends on K states of the weather, i.e., $\lambda(t)$ is a Markov chain with a continuous time and with K states.

6.6 Bivariate Poisson Process

A bivariate point process $N(\cdot)$ is a couple of not necessarily independent point processes $N_1(\cdot)$ and $N_2(\cdot)$. It is supposed that $N_1(\cdot)$ and $N_2(\cdot)$ are defined on a common probability space. Similar to a one-dimensional case, the investigation of the process $N(\cdot) = (N_1(\cdot), N_2(\cdot))$ is based on such convenient tools as **two-dimensional generating functionals**

$$G[z_1, z_2] = \mathbf{E}\exp\left\{\int_{-\infty}^{+\infty} \ln z_1(t)\,dN_1(t) + \int_{-\infty}^{+\infty} \ln z_2(t)\,dN_2(t)\right\}$$

where $z_1(\cdot)$ and $z_2(\cdot)$ belong to a special class of functions (for example, $z_1,\, z_2 \in V$ - see Section 6.1).

A generating functional uniquely defines and is defined by joint distributions of $N_1(\cdot)$ and $N_2(\cdot)$. Let $G_i[z]$, $i = 1, 2$, be generating functionals for processes $N_i(\cdot)$, $i = 1, 2$, and $\mathbf{1}$ be a function identically equal to unit. Then the following properties are fulfilled:

(i) $G_1[z_1] = G[z_1,\, \mathbf{1}]$, $G_2[z_2] = G[\mathbf{1},\, z_2]$,

(ii) if $N_1(\cdot)$ and $N_2(\cdot)$ are independent processes, then

$$G[z_1,\, z_2] \;=\; G_1[z_1]\, G_2[z_2], \quad z_i \in V,\; i = 1, 2,$$

(iii) if $N_1(\cdot)$ and $N_2(\cdot)$ are independent processes, then $G_1[z]\, G_2[z]$ is a generating functional of the superposition $N_1(\cdot) + N_2(\cdot)$.

Let $\Lambda(\cdot)$ be a nonnegative countably additive Borel measure, i.e., $\Lambda(\cdot) \in$ $\in W$ (see Section 6.5). A generating functional of a Poisson process with the intensity measure $\Lambda(\cdot)$ is equal to

$$\exp\left\{\int\limits_{-\infty}^{+\infty} (z(t) - 1)\, \Lambda(dt)\right\}, \quad z \in V.$$

A bivariate Poisson process[42,43] is a two-dimensional point process $N(\cdot) = (N_1(\cdot),\, N_2(\cdot))$ possessing the following two properties:

(i) marginal processes $N_1(\cdot)$ and $N_2(\cdot)$ are Poisson processes with intensity measures $\Lambda_1(\cdot)$ and $\Lambda_2(\cdot)$,

(ii) there exists a functional $\Phi[z]$, $1 - z \in V$, such that for $z_1, z_2 \in V$ the equality

$$G[z_1,\, z_2] \;=\; G[z_1,\, \mathbf{1}]\; G[\mathbf{1},\, z_2]\; \Phi[(1 - z_1)\,(1 - z_2)]$$

holds.

Let $\Lambda_1(B) = \Lambda_2(B) = \Lambda(B)$, $B \in \mathfrak{B}$. It holds the following characterizing theorem.

Theorem 6.9. *Marginal processes $N_1(\cdot)$ and $N_2(\cdot)$ of the bivariate Poisson process can be represented as*

$$N_1(\cdot) \;=\; M_1(\cdot) \,+\, M_2(\cdot), \quad N_2(\cdot) \;=\; M_3(\cdot) \,+\, M_2(\cdot) \tag{6.8}$$

where $M_1(\cdot)$, $M_2(\cdot)$ and $M_3(\cdot)$ are mutually independent doubly stochastic Poisson processes defined by random measures $\nu(\cdot)$, $\mu(\cdot)$ and $\nu(\cdot)$ respectively (with the probability 1 $\mu(\cdot) \le \Lambda(\cdot)$ and $\nu(\cdot) = \Lambda(\cdot) - \mu(\cdot)$). Moreover a generating functional is defined by the relation

$$\mathbf{E}\exp\left\{\int\limits_{-\infty}^{+\infty} (z_1(t) - 1)\, \nu(dt) \,+\, \int\limits_{-\infty}^{+\infty} (z_1(t)\, z_2(t) - 1)\, \mu(dt) +\right.$$

$$+ \int_{-\infty}^{+\infty} (z_2(t) - 1)\, \nu(dt) \Bigg\} = \exp \Bigg\{ \int_{-\infty}^{+\infty} (z_1(t) - 1)\, \Lambda(dt) +$$

$$+ \int_{-\infty}^{+\infty} (z_2(t) - 1)\, \Lambda(dt) \Bigg\} \; \mathbf{E} \exp \Bigg\{ \int_{-\infty}^{+\infty} (z_1(t) - 1)\, (z_2(t) - 1)\, \mu(dt) \Bigg\}.$$

The main properties[42,43] of bivariate Poisson processes follow from the representation (6.8).

6.7 Gauss-Poisson Process

A class of Gauss-Poisson processes is a generalization of Poisson processes. It is a two-parameter collection of point processes. These parameters are the first two factorial moment measures (see Section 6.1). We suppose the process is defined on a real line (generalizations on the more complex spaces are obvious). Consider a functional

$$G[z] = \exp \Bigg\{ m \int_{-\infty}^{+\infty} (z(t) - 1)\, Q(dt) +$$

$$+ \frac{1}{2} m^2 \int_{-\infty}^{+\infty} \int_{-\infty}^{+\infty} (z(t) - 1)\, (z(u) - 1)\, \rho(t, u)\, Q_2(dt \times du) \Bigg\}, \quad z \in V, \qquad (6.9)$$

where $Q(\cdot)$ and $Q_2(\cdot)$ are probability distributions on $\mathfrak{B}$ and $\mathfrak{B} \times \mathfrak{B}$, $m > 0$ and $\rho(\cdot, \cdot)$ is a bounded function on $\mathbb{R} \times \mathbb{R}$, $\rho(t, u) = \rho(u, t)$, $Q_2(A_1 \times \times A_2) = Q_2(A_2 \times A_1)$, $t, u \in \mathbb{R}$, $A_1, A_2 \in \mathfrak{B}$.

Under some hypotheses[44,45] relation (6.9) defines a generating functional of a point process. In particular, if $\rho(x_1, x_2) \equiv 0$, then $G[z]$ is a generating functional of a nonhomogeneous Poisson process. Therefore a functional (6.9) defines a nontrivial generalization of a class of Poisson processes.

A multidimensional normal distribution is defined by its first and second order moments. Moreover its generating function

$$g(z_1, \ldots, z_n) = \exp \left\{ \sum_{i=1}^{n} z_i m_i + \frac{1}{2} \sum_{i,j=1}^{n} z_i z_j \sigma_{ij} \right\}$$

has a form similar to (6.9). If $G[z]$ is a generating functional of a point process, then this process is uniquely defined by the first two factorial moment measures (similarly to a Gaussian process). At the same time this process is invariant with respect to the operation of superposition (analogous to a Poisson process). Therefore it is natural to call a process with a generating functional (6.9) as **a Gauss-Poisson process**.

Consider a more general functional than (6.9):

$$G[z] = \exp\left\{\int_{-\infty}^{+\infty} (z(t)-1)\,\lambda(dt) + \right.$$

$$\left. + \frac{1}{2}\int_{-\infty}^{+\infty}\int_{-\infty}^{+\infty} (z(t)-1)\,(z(u)-1)\,H(dt \times du)\right\} \tag{6.10}$$

where $z \in V$ and $\lambda(\cdot)$, $H(\cdot,\cdot)$ are Borel measures on $\mathfrak{B}$ and $\mathfrak{B} \times \mathfrak{B}$.

Theorem 6.10. *Relation (6.10) defines a generating functional of a point process iff for any bounded sets A, $B \in \mathfrak{B}$*

$$H_s(A \times B) \le \min(\lambda(A), \lambda(B)) \tag{6.11}$$

where

$$H_s(A \times B) = \frac{1}{2}(H(A \times B) + H(B \times A)).$$

In particular, it follows from (6.11) that $H_s(A \times \mathbb{R}) < \infty$, $H_s(\mathbb{R} \times B) < \infty$.

A point process $N(\cdot)$ is called **stationary of the nth order**, if for any $k_i \ge 0$, $A_i \in \mathfrak{B}$, $i = 1, \dots, n$, $t \in \mathbb{R}$ the relation

$$\mathbf{P}\{N(A_i + t) = k_i,\ i = 1, \dots, n\} = \mathbf{P}\{N(A_i) = k_i,\ i = 1, \dots, n\} \tag{6.12}$$

holds where $A + t = \{x + t : x \in A\}$. If relation (6.12) holds for any $n \ge 1$, then we have **a stationary point process**.

Properties of a Gauss-Poisson process.

Theorem 6.11. *If*

$$\lambda(A+t) = \lambda(A), \quad H((A+t) \times (A+t)) = H(A \times A)$$

for any $t \in \mathbb{R}$ and any $A \in \mathfrak{B}$, then a Gauss-Poisson process is stationary of the first order.

Theorem 6.12. *If a Gauss-Poisson process is stationary of the first order, then it is a stationary process.*

This property is intrinsic also to a Poisson process. A Gauss-Poisson process is not necessarily an ordinary one. Let $H_t = H([0,t] \times [0,t])$.

Theorem 6.13. *If*

$$\lim_{t\to 0} t^{-1} H_t = 0,$$

then a stationary Gauss-Poisson process is an ordinary one.

If a Gauss-Poisson process possesses a memoryless property, then $H(A\times \times B) = 0$ for $A \cap B = \emptyset$. Therefore

$$G[z] = \exp\left\{ \int_{-\infty}^{+\infty} (z(t) - 1)\,\lambda(dt) + \int_{-\infty}^{+\infty} (z(t) - 1)^2\, H_1(dt) \right\}$$

where $H_1(dt) = H(dt \times dt)$.

Theorem 6.14. *A stationary Gauss-Poisson process is a strongly intermixing and hence an ergodic process.*

Theorem 6.15. *Gauss-Poisson processes are invariant with respect to the following operations:*

(i) by the superposition of independent Gauss-Poisson processes with parameters $(\lambda_i,\ H_i)$, $i = 1, \ldots, n$, we obtain a Gauss-Poisson process with parameters $(\sum_{i=1}^{n} \lambda_i,\ \sum_{i=1}^{n} H_i)$ satisfying the relation (6.11);

(ii) by a shift operation: let $\{t_i\}$ be points of a Gauss-Poisson process $N(\cdot)$ and $t_i + \eta_i$ be corresponding points of the shifted process $N_0(\cdot)$ where $\{\eta_i\}$ is a sequence of mutually independent and identically distributed random variables with a distribution function $F(x)$; a generating functional of the process $N_0(\cdot)$ is equal to

$$G_0[z] = G\left[\int_{-\infty}^{+\infty} z(t+x)\,F(dx) \right], \quad z \in V,$$

where $G[z]$ is a generating functional of the process $N(\cdot)$. A shifted Gauss-Poisson process is defined by parameters

$$\lambda_0(A) = \int_{-\infty}^{+\infty} \lambda(A - x)\,F(dx), \quad A \in \mathfrak{B},$$

$$H_0(A \times B) = \int_{-\infty}^{+\infty}\int_{-\infty}^{+\infty} H((A-x) \times (B-y))\,F(dx)\,F(dy), \quad A,\ B \in \mathfrak{B};$$

(iii) by the operation of thinning: every point of the process $N(\cdot)$ is deleted with the probability $1-q$ and remains with the probability q. Then

$$G_q[z] = G[1-q+q\,z] = \exp\left\{ q \int_{-\infty}^{+\infty} (z(t)-1)\,\lambda(dt) + \right.$$

$$\left. + \frac{1}{2} q^2 \int_{-\infty}^{+\infty} \int_{-\infty}^{+\infty} (z(t)-1)\,(z(u)-1)\,H(dt \times du) \right\},$$

i.e.,

$$\lambda_q(A) = q\,\lambda(A), \quad H_q(A \times B) = q^2\,H(A \times B), \quad A, B \in \mathfrak{B}.$$

Theorem 6.16. *A stationary Gauss-Poisson process can be represented as a Poisson cluster process of the following structure: a process of cluster centers has an intensity λ, a cluster has no points with the probability p_0, has only one point located at its center with the probability $1-p_0$ and has two points with the probability p_0, one of which is located at the center of a cluster and the other is shifted to the right from the center on a random distance with some distribution function $W(x)$.*

Theorem 6.17. *Solely a Poisson process is simultaneously a renewal process and a Gauss-Poisson process.*

Some properties of Gauss-Poisson processes are inherent in processes of statistical mechanics. Gauss-Poisson processes are used for the construction of stochastic models in population genetics, in ecology, in other biological sciences and also in various problems of queueing theory and operation research.

6.8 References

1. Wold, H., Sur les processus stationnaires ponctueles, *Colloq. Int.*, 13, 75, 1948.
2. Bartlett, M. S., Processus stochastiques ponctueles, *Ann. Inst. Henri Poincarè,* 14, 35, 1954.
3. Bartlett, M. S., *An Introduction to Stochastic Processes: with Special Reference to Methods and Applications,* Cambridge Univ. Press, Cambridge, 1978.
4. Moyal, J. E., The general theory of stochastic population processes, *Acta Math.,* 108, 1, 1962.

5. Vere-Jones, D., Some applications of probability generating functionals to the study of input-output streams, *J. Roy. Statist. Soc.*, B, 30, 321, 1968.
6. Westcott, M., The probability generating functionals, *J. Austral. Math. Soc.*, 14, 448, 1972.
7. Laslett, G. M., Mixing of cluster point processes, *J. Appl. Probab.*, 15, 715, 1978.
8. Kovalenko, I. N., *Studies in Analysis of Reliability of Complex Systems*, Naukova Dumka, Kiev, 1975 (in Russian).
9. Neyman, J., Scott, E. L., A theory of the spatial distribution of galaxies, *Astrophys. J.*, 116, 144, 1952.
10. Thompson, H. R., Spatial point processes with applications to ecology, *Biometrica*, 42, 102, 1955.
11. Neyman, J., Scott, E. L., A statistical approach to problems of cosmology, *J. Roy. Statist. Soc.*, B, 20, 1, 1958.
12. Neyman, J., On finiteness of the process of clustering, *Sankhya*, A, 25, 69, 1963.
13. Bartlett, M. S., The spectral analysis of point processes, *J. Roy. Statist. Soc.*, B, 25, 264, 1963.
14. Lewis, P. A. W., A branching Poisson process model for the analysis of computer failure patterns, *J. Roy. Statist. Soc.*, B, 26, 398, 1964.
15. Lewis, P. A. W., Implications of a failure model for the use and maintenance of computers, *J. Appl. Probab.*, 1, 347, 1964.
16. Lewis, P. A. W., Nonhomogeneous branching Poisson process, *J. Roy. Statist. Soc.*, B, 29, 343, 1967.
17. Lewis, P. A. W., Asymptotic properties and equilibrium conditions for branching Poisson processes, *J. Appl. Probab.*, 6, 355, 1969.
18. Vere-Jones, D., Stochastic models for earthquake occurrence, *J. Roy. Statist. Soc.*, B, 32, 1, 1970.
19. Westcott, M., On existence and mixing results for cluster point processes, *J. Roy. Statist. Soc.*, B, 33, 290, 1971.
20. Westcott, M., Results in the asymptotic and equilibrium theory of Poisson cluster processes, *J. Appl. Probab.*, 10, 807, 1973.
21. Westcott, M., A note on the nonhomogeneous Poisson cluster process, *J. Appl. Probab.*, 14, 396, 1977.
22. Daley, D. J., Vere-Jones, D., The extended probability generating functional, with application to mixing properties of cluster point processes, *Math. Nachr.*, 131, 311, 1987.
23. Feller, W., *An Introduction to Probability Theory and Its Applications*, 2, Wiley, New York, 1971.
24. Schottky, W., Über spontane Stromschwankungen in verschiedenen Elektrizitätsleitern, *Ann. Phys.*, 57, 541, 1918.

25. Conference *on Physical Aspects of Noise in Electronic Devices,* Peregrinus, Stevenage, Hertfordshire, 1968.
26. Kuno, A., Ikegaya, K., A statistical investigation of acoustic power radiated by a flow of random point sources, *J. Acoust. Soc. Jap.,* 29, 662, 1973.
27. Picinbono, B., Bendjaballah, C., Pouget, J., Photoelectron shot noise, *J. Math. Phys.,* 11, 2166, 1970.
28. Srinivasan, S. K., Iyer, K. S. S., Random processes associated with random points on a line, *Zastos. Mat.,* 8, 221, 1966.
29. Takács, L., On secondary processes generated by recurrent processes, *Acta Math. Acad. Sci. Hung.,* 7, 17, 1956.
30. Snyder, D. L., *Random Point Processes,* Wiley, New York, 1975.
31. Takács, L., On secondary processes generated by a Poisson process and their application in physics, *Acta Math. Acad. Sci. Hung.,* 5, 203, 1954.
32. Hawkes, A. G., Spectra of self-exciting and mutually exciting point processes, *Biometrica,* 58, 83, 1971.
33. Hawkes, A. G., Point spectra of some mutually exciting point processes, *J. Roy. Statist. Soc.,* B, 33, 438, 1971.
34. Hawkes, A. G., Oakes, D., A cluster process representation of a self-exciting process, *J. Appl. Probab.,* 11, 493, 1974.
35. Cox, D. R., Some statistical methods related with series of events, *J. Roy. Statist. Soc.,* B, 17, 129, 1955.
36. Smith, W. L., Regenerative stochastic processes, *Proc. Roy. Soc.,* A, 232, 6, 1955.
37. Kingman, J. F. C., On doubly stochastic Poisson processes, *Proc. Cambridge Phil. Soc.,* 60, 923, 1964.
38. Westcott, M., Identifiability in linear processes, *Z. Wahrscheinlichkeitstheorie und verw. Geb.,* 16, 39, 1970.
39. Grandell, J., Doubly stochastic Poisson processes, *Lect. Notes Math.,* 529, 1, 1976.
40. Yannaros, N., On Cox processes and gamma renewal processes, *J. Appl. Probab.,* 25, 423, 1988.
41. Kerstan, J., Matthes, K., Mecke, J., *Unbegrenzt teilbare Punktprozesse,* Akademie-Verlag, Berlin, 1974.
42. Griffiths, R. C., Milne, R. K., A class of bivariate Poisson processes, *J. Multivar. Anal.,* 8, 380, 1978.
43. Sclove, S. L., The occurrence of fingerprint characteristics as a two-dimensional Poisson process, *Commun. Statist.,* A9, 675, 1980.
44. Newman, D. S., A new family of point processes which are characterized by their second moment properties, *J. Appl. Probab.,* 7, 338, 1970.
45. Milne, R. K., Westcott, M., Further results for Gauss-Poisson processes, *Adv. Appl. Probab.,* 4, 151, 1972.

Chapter 7

Random Flows of Events

7.1 Main Definitions

A random flow of events is one of the main mathematical models used for the description of different phenomena of nature. Stated below definitions[1–3] for random flows can be reformulated in terms of point processes in general state spaces (cf. Sections 2.5 and 6.1).

Let $(X, \mathfrak{U})$ be a measurable space. A set of random variables $\{N(A), A \in \mathfrak{U}\}$ defined on the probability space $(\Omega, \mathfrak{F}, \mathbf{P})$ is called **a random flow** $N(\cdot)$, if the following conditions are satisfied:

(i) $N(\cdot)$ is a countable additive function of sets $A \in \mathfrak{U}$,

(ii) $N(A)$ can take only integer values 0, 1, 2, ... for any $A \in \mathfrak{U}$.

In other words, a random flow $N(\cdot)$ is a nonnegative integer-valued random function defined on $\mathfrak{U}$. An event "$N(A) = k$" means that a set $A \in \mathfrak{U}$ contains k events of the flow (for example, k customers arrived in a queueing system or it was k failures at a given time interval A).

A set X must be chosen depending on the problem being investigated. So in the renewal theory $X = [0, \infty)$ and the moments of the event's appearance in a random flow are uniquely determined by a random process having nondecreasing step path functions with integer jumps. If we study one-dimensional point processes (see Section 6.1), then $X = (-\infty, +\infty)$. But in real problems the set X has a more complex structure. For example, by studying the flow of earthquakes beside the moment of an earthquake occurrence it is important to know the coordinates of its epicenter. So as X we can choose a set of points (φ, θ, r, t) where φ and θ are the latitude and the longitude, r is the depth of the earthquake center and t is the moment of its occurrence.

Just as in the theory of point processes a generating functional is an important tool for the investigation of random flows. A notion of a stochastic integral $\int_X \ln z(t)\, N(dt)$ can be introduced for a wide range of functions (see Section 6.1). Then **a generating functional** is defined as

$$G[z] = \mathbf{E} \exp\left\{ \int_X \ln z(t)\, N(dt) \right\}.$$

A random flow is defined by multidimensional distributions

$$\mathbf{P}\{N(A_i) = k_i,\ i = 1, \ldots, m\},$$

$$m \geq 1,\ \ k_i = 0, 1, 2, \ldots,\ \ A_i \in \mathfrak{U},\ \ i = 1, \ldots, m.$$

A Poisson flow is of special importance for solving real engineering problems. Let $\Lambda(A)$ be a nonnegative countable additive function of sets $A \in \mathfrak{U}$. If for any $m \geq 1$ and for any disjoint sets $A_i \in \mathfrak{U}$, $i = 1, \ldots, m$,

$$\mathbf{P}\{N(A_i) = k_i,\ i = 1, \ldots, m\} = \prod_{i=1}^{m} \frac{(\Lambda(A_i))^{k_i}}{k_i!} e^{-\Lambda(A_i)},$$

$$k_i \geq 0,\ \ i = 1, \ldots, m,$$

then a random flow $N(\cdot)$ is called **a Poisson flow**. Its main property is **the memoryless**, i.e., for any disjoint sets $\{A_i\}$ the numbers of events hitting these sets are mutually independent random variables.

Assume a shift operator T_t is introduced in the space X and let $T_t A$ be a shifted set A. A random flow $N(\cdot)$ is called a **stationary** one (with respect to this shift operator), if for any $m \geq 1$, $k_i \geq 0$, $A_i \in \mathfrak{U}$, $i = 1, \ldots, m$, and for any t

$$\mathbf{P}\{N(T_t A_i) = k_i,\ i = 1, \ldots, m\} = \mathbf{P}\{N(A_i) = k_i,\ i = 1, \ldots, m\}.$$

Let $X = \mathbb{R}$ and let $N(\cdot)$ be a stationary random flow. A mean number of events arising in the time interval of the unit length is called **an intensity** μ of the stationary flow. For stationary flows a limit always exists

$$\lambda = \lim_{t \to 0} \frac{1}{t} \mathbf{P}\{N(0, t) > 0\},$$

which is called **a parameter** of a random flow (a case $\lambda = \infty$ is not excluded). In general $\lambda \leq \mu$, but under some weak hypotheses the equality $\lambda = \mu$ holds (see Section 7.3). A random stationary flow defined on $\mathbb{R}$ is called an **ordinary** one, if

$$\lim_{t \to 0} \frac{1}{t} \mathbf{P}\{N(0, t) \geq 2\} = 0.$$

Let $\alpha(\cdot)$ be some measure defined on sets of $\mathfrak{U}$ (for example, a space or a volume can be taken as such measure for $X = \mathbb{R}^2$ or $X = \mathbb{R}^3$). A random flow is called an **ordinary flow** (with respect to this measure $\alpha(\cdot)$), if for any $A \in \mathfrak{U}$

$$\lim_{\alpha(A)\to 0} \frac{1}{\alpha(A)} \mathbf{P}\{N(A) \geq 2\} = 0.$$

In terms of introduced notions a stationary Poisson flow (i.e., Poisson flow with $\lambda(A) = \mu\,\alpha(A)$; if $X = \mathbb{R}^n$, then $\alpha(\cdot)$ is a Lebesgue measure) is uniquely determined by three properties: it is a stationary, ordinary and memoryless flow. A flow having these three properties is called **a simplest flow**.

Later on we suppose $X = [0, \infty)$ and $\mathfrak{U}$ is a σ-algebra of Borel sets of X. In many engineering problems the physical conditions of the event's appearance form a natural basis for the hypothesis that a considered flow is ordinary and memoryless. But at the same time a flow can't be considered as stationary. Such flow is called **a simplest nonstationary flow**. The probability of the appearance of $k \geq 0$ events at a time interval $(t,\, t+\tau)$ depends not only on τ, but also on t. Denote this probability as $p_k(t,\, t+\tau)$. If for any bounded interval $(a,\, b)$ the equality

$$\lim_{\tau\to 0} \frac{1}{\tau} \sum_{k=2}^{\infty} p_k(t,\, t+\tau) = 0$$

holds uniformly on $t \in (a,\, b)$, then a nonstationary flow is an ordinary one.

Assume that for any $t \geq 0$ there exists a limit

$$\lim_{\tau\to 0} \frac{1}{\tau} \left(1 - p_0(t,\, t+\tau)\right) = \lambda(t)$$

that is called **an instantaneous parameter value**. Then

$$p_k(t,\, t+\tau) = \frac{1}{k!} \left(\Lambda(t,\, t+\tau)\right)^k e^{-\Lambda(t,\, t+\tau)}, \quad k = 0, 1, 2, \ldots,$$

where $\Lambda(t,\, t+\tau)$ is a mean number of events appearing in the interval $(t,\, t+\tau)$:

$$\Lambda(t,\, t+\tau) = \int_t^{t+\tau} \lambda(z)\, dz.$$

The mean intensity of the flow in $(t,\, t+\tau)$ is

$$\frac{1}{\tau}\, \mu(t,\, t+\tau) = \frac{1}{\tau} \sum_{k-0}^{\infty} k\, p_k(t,\, t+\tau) = \frac{1}{\tau}\, \Lambda(t,\, t+\tau).$$

The instantaneous intensity $\mu(t)$ of the flow is defined by the formula

$$\mu(t) = \lim_{\tau \to 0} \frac{1}{\tau} \mu(t, t+\tau) = \lim_{\tau \to 0} \frac{1}{\tau} \Lambda(t, t+\tau) = \lambda(t).$$

Hence the instantaneous intensity of the ordinary memoryless flow coincides with the instantaneous parameter value. Moreover the following important property holds: if it is known that an interval $(t, t+\tau)$ contains n events of the ordinary memoryless flow, then moments of the events occurrence are independent random variables with the same distribution function $\Lambda(t, x)/\Lambda(t, t+\tau),\ t < x \le t+\tau$.

7.2 Random Memoryless Flows

Let $\Lambda(t),\ t \ge 0$, be a mean number (finite or infinite) of flow events appearing in a half-interval $[0, t)$. This nonnegative and nondecreasing function $\Lambda(t)$ is called **a leading function**. A random flow is called a **finite** one, if $\Lambda(t) < \infty$ for any $t \ge 0$.

If $\Lambda(t)$ is a nonnegative finite nondecreasing function, then a formula

$$p_k(\alpha, \beta) = \exp\{-(\Lambda(\beta)-\Lambda(\alpha))\} \frac{(\Lambda(\beta) - \Lambda(\alpha))^k}{k!}, \quad 0 \le \alpha < \beta,\ k \ge 0, \quad (7.1)$$

defines some memoryless flow with a leading function $\Lambda(t)$. However such flows form only a special case of all finite memoryless flows.

A moment t is called a **regular** or **singular** point of a finite flow depending on what relations $p_0(t) = 1$ or $p_0(t) < 1$ are satisfied where

$$p_k(t) = \lim_{h \to 0} p_k(t-h, t+h), \quad k \ge 0,\ t > 0. \quad (7.2)$$

A finite flow is called a **regular** one, if it has no singular points, i.e., $p_0(t) = 1$ for any $t \ge 0$.

The following simple, but useful, properties are satisfied.

1. If $0 \le \alpha < \beta$, then $\sum\limits_{k=0}^{\infty} p_k(\alpha, \beta) = 1$.

2. A limit on the right-hand side of formula (7.2) exists for any $k \ge 0$ and $t > 0$.

3. All continuity points of the function $\Lambda(t)$ are regular points of the flow and all discontinuity points of $\Lambda(t)$ are singular points of the flow.

4. A finite flow can have at most the countable set of singular points.

5. The continuity of a leading function is a necessary and sufficient condition for the regularity of the flow.

6. If a flow is a regular one, then $p_0(t-h,\, t+h) \underset{h\to 0}{\longrightarrow} 1$ uniformly on $t \in [\alpha,\, \beta]$, $0 < \alpha < \beta$.

7. The inequality $p_0(\alpha,\, \beta) > 0$ holds for any regular memoryless flow and for any $\alpha,\, \beta,\ 0 \le \alpha < \beta$.

8. The generating function

$$\Phi(x;\, \alpha,\, \beta) = \sum_{k=0}^{\infty} p_k(\alpha,\, \beta)\, x^k, \quad 0 \le \alpha < \beta < \infty, \ x \ge 0,$$

of any regular memoryless flow can be represented in the form

$$\Phi(x;\, \alpha,\, \beta) = \exp\left\{ \sum_{k=0}^{\infty} (\varphi_k(\beta) - \varphi_k(\alpha))\, x^k \right\} \tag{7.3}$$

where functions $\varphi_k(t),\ \varphi_k(0) = 0,\ k \ge 0$, are

(i) continuous for all values of $t \ge 0$,

(ii) monotone on t for any $k \ge 0$ (nondecreasing for $k > 0$ and nonincreasing for $k = 0$),

(iii) series $\sum\limits_{k=0}^{\infty} k\, \varphi_k(t)$ converges for any $t \ge 0$,

(iv) $\sum\limits_{k=0}^{\infty} \varphi_k(t) = 0$ for any $t \ge 0$.

Theorem 7.1. *Every regular memoryless flow has a generating function $\Phi(x;\, \alpha,\, \beta)$ of the form (7.3) with functions $\{\varphi_k(t)\}$ satisfying conditions (i) - (iv); vice versa every such function $\Phi(x;\, \alpha,\, \beta)$ is a generating function of some regular memoryless flow.*

A finite flow $N(\cdot)$ is called a **singular** one, if three properties are satisfied:

(i) events of the flow $N(\cdot)$ can appear only at some defined in advance moments forming at most the countable set of points $t_1,\, t_2,\, \ldots$,

(ii) numbers of events appearing at different moments $\{t_i\}$ are mutually independent random variables,

(iii) if $q_k^{(i)}$ is the probability that at moment t_i there appear k events $(k \ge 0)$ of the flow $N(\cdot)$, then the relation

$$\sum_{i:\, 0 \le t_i \le t} \ \sum_{k=0}^{\infty} k\, q_k^{(i)} < \infty$$

holds for any $t \ge 0$

The property (ii) expresses the memoryless and the property (iii) defines a finiteness of the flow. A singular flow is uniquely defined by two sequences $\{t_j\}$ and $\{q_k^{(i)}\}$.

Theorem 7.2. *Any finite memoryless flow can be represented as the superposition of two independent flows of the same type, one of which is regular and the other is singular. The moments of the events' appearance in a singular flow coincide with singular points of the original flow and corresponding probabilities* $\{q_k^{(i)}\}$ *are defined by the relation*

$$q_k^{(i)} = \lim_{h\to 0} p_k(t_i - h,\, t_i + h), \quad k \geq 0,\ i = 1, 2, \ldots .$$

Let $p_k(\alpha, \beta)$, $0 \leq \alpha < \beta < \infty$, $k \geq 0$, be the probabilities defined in Section 7.1 and let $\Lambda(t)$, $t \geq 0$, be a leading function of the flow, i.e., $\Lambda(t)$ is a mean number of events appearing in $[0, t)$. Suppose the relation (7.1) is satisfied. If $\Lambda(t) = \lambda t$, $t \geq 0$, for some $\lambda > 0$, then we have a stationary Poisson (or simplest) flow of events. In general, $\Lambda(t)$ is a nonnegative finite and nondecreasing function of t, $\Lambda(t) = 0$ for $t \leq 0$. Denote

$$\psi_k(\alpha, \beta) = \sum_{i=k}^{\infty} p_i(\alpha, \beta), \quad k \geq 0.$$

A random flow (in general, nonstationary) is called an **ordinary** one, if for any $t > 0$ and $\varepsilon > 0$ there exist such $\delta > 0$ that an inequality

$$\psi_2(\alpha, \beta) \leq \varepsilon\, \psi_1(\alpha, \beta)$$

holds for all α, β such that $0 \leq \alpha < \beta < \alpha + \delta < t$.

Theorem 7.3. *An equality (7.1) holds for a regular memoryless flow iff it is an ordinary flow.*

A random flow is called **a singular Poisson flow**, if it is a singular one and $\{q_k^{(i)}\}$ are determined by the equalities

$$q_k^{(i)} = \frac{\lambda_i^k}{k!}\, e^{-\lambda_i}, \quad \lambda_i > 0,\ i = 1, 2, \ldots,\ k \geq 0.$$

A generating function

$$\Gamma(x;\, \alpha, \beta) = \sum_{k=0}^{\infty} p_k(\alpha, \beta)\, x^k$$

of a singular Poisson flow can be represented in the form

$$\Gamma(x;\, \alpha, \beta) = \exp\left\{ (x-1) \sum_{i:\, \alpha \leq t_i < \beta} \lambda_i \right\}.$$

It is assumed that series

$$C(t) = \sum_{i:\,\alpha \le t_i < t} \lambda_i$$

converges for any $t \ge 0$. Then

$$\Gamma(x;\,\alpha,\,\beta) = \exp\{(C(\beta) - C(\alpha))\,(x-1)\}.$$

Theorem 7.4. *A memoryless flow satisfies (7.1) iff it can be represented as a superposition of two mutually independent flows, one of which is a regular one and the other is a singular Poisson flow with* $C(t) < \infty$ *for any* $t \ge 0$.

Such decomposition is defined directly by the leading function $\Lambda(t)$ of the original flow: as the moments $\{t_i\}$ we take the discontinuity points of $\Lambda(t)$, $\lambda_i = \Lambda(t_i + 0) - \Lambda(t_i - 0)$. As the leading functions of singular and regular components we take respectively $C(t) = \sum_{i:\,t_i<t} \lambda_i$ and $\Lambda^*(t) = \Lambda(t) - C(t)$.

The property of the flow to be a Poisson one is preserved (remains invariant) under the following transformations.

1. If $N_1(\cdot), \ldots, N_k(\cdot)$ are mutually independent Poisson flows with leading functions $\Lambda_1(t), \ldots, \Lambda_k(t)$, then a superposition of flows $N(\cdot) = \sum_{i=1}^{k} N_i(\cdot)$ is a Poisson one with a leading function $\Lambda(t) = \sum_{i=1}^{k} \Lambda_i(t)$.

2. Let $\{t_i\}$ be points of a stationary flow $N(\cdot)$ with an intensity λ and $\{t_i + \eta_i\}$ be points of a shifted flow $N_0(\cdot)$ where $\{\eta_i\}$ is a sequence of mutually independent identically distributed random variables with distribution function $F(x)$. Then $N_0(\cdot)$ is a stationary Poisson flow with an intensity λ.

3. If every point of a Poisson flow with an intensity λ is deleted with the probability q and remains with the probability $1 - q$, then a flow of remaining points is a Poisson one with the intensity $\lambda\,(1-q)$.

4. Consider a single-server queueing system with an input Poisson flow of the intensity λ. Assume the service times are mutually independent identically distributed random variables with the distribution function $1 - e^{-\mu t}$, $\lambda < \mu$, $t > 0$. Then in a steady-state the sequence of service terminations is a Poisson flow with the intensity λ.

7.3 Stationary Flows

Stationary flows have the following properties[2–4].

1. There exists a parameter λ (finite or infinite) for any stationary flow. If μ is an intensity of the stationary flow, then the inequality $\mu \geq \lambda$ holds.

2. Among all stationary memoryless flows the equality $\mu = \lambda$ holds only for the simplest flow. For all other flows $\mu > \lambda$.

3. If a stationary flow has a finite intensity μ, then the equality $\mu = \lambda$ implies the ordinariness of the flow.

4. **Korolyuk Theorem.** *The ordinariness of the stationary flow implies the equality $\mu = \lambda$ (the case $\mu = \lambda = \infty$ is not excluded).*

5. In a stationary memoryless flow the moments of the events' occurrence form a simplest flow. At each such moment k events occur with the probability $a_k \geq 0$, $\sum\limits_{k=1}^{\infty} a_k = 1$. A generating function for the number of events of a stationary memoryless flow occurring in $[0, t)$ is defined by the relation

$$F(t, x) = e^{\lambda t (\Phi(x) - 1)}$$

where

$$\Phi(x) = \sum_{i=1}^{\infty} a_i x^i, \quad 0 \leq x \leq 1.$$

In order to investigate stationary flows Palm (1943) has introduced a function $\varphi_0(t)$ defined as the conditional probability of "no" events of the flow in the interval $(t_0, t_0 + t)$ provided that an event has occurred at the moment t_0. For the most interesting practical cases this conditional event has a zero probability. Hence the function $\varphi_0(t)$ can't be considered as strictly defined. Chinčin (1963) gave a more precise definition. He introduced an infinite sequence of functions, $\varphi_k(t)$, $t \geq 0$, $k \geq 0$, analogous to a Palm function (these functions are known as Palm-Chinčin functions).

Let $N(\cdot)$ be a stationary random flow. **A Palm-Chinčin function** of the kth order is defined as the limit

$$\varphi_k(t) = \lim_{\tau \to 0} \frac{\mathbf{P}\{N(0, \tau) > 0, \ N(\tau, \tau + t) = k\}}{\mathbf{P}\{N(0, \tau) > 0\}},$$

i.e., $\varphi_k(t)$ is a conditional probability of the occurrence of k events in the half-interval $(0, t]$ provided that at least one of the events has occurred at the initial moment $t = 0$.

Palm-Chinčin functions exist for any stationary flow with a finite parameter λ and satisfy equations

$$p_0(t) = 1 - \lambda \int_0^t \varphi_0(z)\,dz, \quad p_k(t) = \lambda \int_0^t (\varphi_{k-1}(z) - \varphi_k(z))\,dz, \ k \geq 1,$$

where $p_k(t) = \mathbf{P}\{N(0, t) = k\}, \ k \geq 0$.

7.4 Flows with Bounded Aftereffect

Let $t_1, t_2, \ldots$ be moments of the events' occurrence, $t_{k+1} \geq t_k, \ k \geq 1$. Let $z_k = t_k - t_{k-1}, \ k \geq 1, \ t_0 = 0$. If the distribution of a random vector $(z_1, \ldots, z_n)$ is given for any $n \geq 1$, then a random flow of events will be defined as well.

If random variables $z_1, z_2, \ldots$ are mutually independent, then a corresponding flow is called **a flow with a bounded aftereffect**[5]. Such flow is uniquely defined by a set of functions $A_k(t) = \mathbf{P}\{z_k < t\}, \ k \geq 1$.

If $A_2(t) = A_3(t) = \cdots = A(t)$, then a flow with a bounded aftereffect is called **a recurrent flow with a delay**. In the case of $A_1(t) = A(t)$ we have **a recurrent flow**. Assume that a sequence of moments, where events can occur, forms a recurrent flow with a delay defined by functions $A_1(t)$ and $A(t)$, $A_1(+0) < 1$, $A(+0) < 1$. Assume also that at each such moment k events ($k \geq 1$) can occur with the probability $a_k \geq 0, \ \sum_{k=1}^{\infty} a_k = 1$, independently of the number of events that occurred earlier. So defined flow is called **a quasirecurrent flow with a delay**. If $a_1 = 1$, then we have a recurrent flow with a delay.

Theorem 7.5. *A random flow is a stationary flow with a bounded aftereffect and a finite positive intensity iff it is a quasirecurrent flow with a delay defined by the functions*

$$A_1(t) = \frac{1}{\tau} \int_0^t (1 - A(u))\,du, \quad A(t), \quad a_n = 1$$

where n is some positive integer number and $\tau = \int_0^{\infty} (1 - A(u))\,du < \infty$.

The following theorem describes a structure of a stationary ordinary flow with a bounded aftereffect and a finite positive intensity (**a Palm flow**).

Theorem 7.6. *A random flow is a Palm flow iff it is a recurrent flow with a delay defined by functions*

$$A_1(t) = \frac{1}{\tau}\int_0^t (1-A(u))\,du \text{ and } A(t),\ A(+0)=0,\ \tau=\int_0^\infty (1-A(u))\,du < \infty.$$

For stationary ordinary flows the memoryless implies the boundedness of the aftereffect. If the condition of ordinariness is not satisfied, then this statement is false.

Limit theorems for the superposition of flows and for thinned flows (see Sections 7.5 and 7.6) are the basis to suppose a random flow in many applied problems is the simplest one.

Till now unbounded flows of events, i.e., flows consisting of unbounded number of events were considered. A Bernoulli flow[5] is an example of a bounded flow. Suppose there are n independent sources and each of them can produce only one event in the interval $(0,\,T)$, $T>0$. It is also assumed that each event can appear in the interval of the length Δ with the probability $\frac{\Delta}{T}$. A superposition of all these one-event flows is called **a Bernoulli flow**. Let $P_k(t)$ be the probability that k events of a Bernoulli flow appear in $(0,\,t) \subseteq (0,\,T)$. Then

$$P_k(t) = C_n^k\left(\frac{t}{T}\right)^k\left(1-\frac{t}{T}\right)^{n-k}, \quad k=0,\,1,\,\ldots,\,n.$$

If $t_1,\ldots,t_n$ are successive moments of the events' occurrence and $z_k = t_k - t_{k-1}$, $k=1,\ldots,n$, $t_0=0$, then

$$\mathbf{P}\{z_k > t\} = \left(1-\frac{t}{T}\right)^n, \quad k=0,\,1,\,\ldots,\,n.$$

7.5 Superposition of Point Processes

Assume that point processes $\{X_i(t)\}$ are defined over the same probability space; $X_i(t)$ denoting the counting process equals the number of events in a time interval $[0,\,t)$. Then a point process $X(t)=\sum_i X_i(t)$ is called **the superposition** of point processes $\{X_i(t)\}$.

In reality the case of the superposition of a large number of point processes of small intensities is typical.

Some conditions[2,3,6,7] are established which guarantee the closeness of the superposition to a Poisson point process. For point processes defined via their event times, the conditions are as follows.

Assume that for each n we have n independent point processes $\{t_{nkj}\}$, $1 \le k \le n$, where t_{nkj} equals the time of the jth event of the kth point process. Denote $F_{nk}(t) = \mathbf{P}\{t_{nk1} < t\}$, $G_{nk}(t) = \mathbf{P}\{t_{nk2} < t\}$. The condition for individual point processes to be **infinitely small** is formulated in such a way:

$$\max_{1 \le k \le n} F_{nk}(t) \underset{n\to\infty}{\longrightarrow} 0$$

for any fixed t. Assume that this condition is satisfied and moreover

$$\sum_{k=1}^{n} F_{nk}(t) \underset{n\to\infty}{\longrightarrow} \Lambda(t)$$

where $\Lambda(t)$ ia a nondecreasing function, and also

$$\sum_{k=1}^{n} G_{nk}(t) \underset{n\to\infty}{\longrightarrow} 0$$

for any fixed t. Then $X_n(t) \overset{w}{\underset{n\to\infty}{\longrightarrow}} X(t)$ where $X(t)$ is a Poisson process with the leading function $\Lambda(t)$. These conditions are necessary and sufficient for the superposition of infinitely small point processes to converge to a Poisson process.

A consequence (Chinčin-Ososkov Theorem[2,6]). *Let $\{t_{nkj}\}$ be independent Palm point processes with the same distribution function $F_{n1}(t)$ of the moment of the first event. If $n\,F_{n1}(t) \underset{n\to\infty}{\longrightarrow} \lambda t$ for any fixed $t > 0$, then the superposition converges to a stationary Poisson point process with the parameter λ.*

In what follows we formulate conditions for the convergence of a superposition of infinitely small point processes to a memoryless point process.

Let $X(t)$ be a process with independent increments, $X(0) = 0$, and

$$\varphi_t(z) = \mathbf{E}\, e^{izX(t)} = \exp\left\{ \sum_{r=1}^{\infty} \Pi_r(t)\,(e^{izr} - 1) \right\},$$

i.e., $X(t)$ is a superposition of independent Poisson processes, rth of which is characterized by the leading function $\Pi_r(t)$ and can have only jumps of the size r. Assume also that

$$\sum_r \Pi_r(t) = \Pi(t) < \infty.$$

We shall call the superposition of point processes defined in such a way to be characterized by the sequence $\{\Pi_r(t)\}$.

Consider a family of independent point processes $\{t_{n1j}\}, \ldots, \{t_{nnj}\}$ for each $n \geq 1$. Denote

$$F_{nk}(t) = \mathbf{P}\{t_{nk1} < t\},$$

$$F^*_{nkr}(t) = \mathbf{P}\{t_{nk1} = \ldots = t_{nkr} < t, \;\; t_{nkr} < t_{nkr+1}\},$$

$$G_{nk}(t) = \mathbf{P}\left\{\bigcup_{r=2}^{\infty} \{t_{nk1} < t_{nkr} < t\}\right\}$$

and assume the following conditions to be satisfied:

(i) $\max\limits_{1 \leq k \leq n} F_{nk}(t) \longrightarrow 0$ (the condition for the point processes to be infinitely small).

(ii) $\sum\limits_{k=1}^{n} F_{nk}(t) \underset{n \to \infty}{\longrightarrow} \Pi(t), \quad t > 0.$

(iii) $\sum\limits_{k=1}^{n} F^*_{nkr}(t) \underset{n \to \infty}{\longrightarrow} \Pi_r(t), \quad t > 0.$

(iv) $\sum\limits_{k=1}^{n} G_{nk}(t) \underset{n \to \infty}{\longrightarrow} 0.$

Then the superposition of point processes $\{t_{n1j}\}, \ldots, \{t_{nnj}\}$ weakly converges as $n \to \infty$ to a finite memoryless point process characterized by the sequence $\{\Pi_r(t)\}$.

Let $a_0 < a_1 < \cdots < a_m$ be fixed numbers, η_i be the number of events of the point process $X_n(t)$ in a half-interval $[a_{i-1}, a_i)$, $\mathbf{x} = (x_1, \ldots, x_m)$, $F_n(\mathbf{x})$ be the distribution function of the random variable $(\eta_1, \ldots, \eta_m)$, and $P(y, \lambda)$ be the distribution function of a Poisson random variable with a parameter λ. Denote

$$\Lambda_n(t) = \sum_{k=1}^{n} F_{nk}(t), \quad \lambda_i = \Lambda_n(a_i) - \Lambda_n(a_{i-1}), \;\; 1 \leq i \leq m.$$

Then according to **Grigelionis Theorem**[7] we have

$$|F_n(\mathbf{x}) - \prod_{i=1}^{m} P(x_i, \lambda_i)| \leq 2m \sum_{i=1}^{m} A_n(a_{i-1}, a_i) + 2(m+1)\, B_n(a_0, a_m)$$

where $A_n(a_{i-1}, a_i)$ means the sum of squared probabilities of one event of the point process $\{x_{nkj}\}$ in the half-interval $[a_{i-1}, a_i)$ (the sum is taken over k, $1 \leq k \leq n$); $B_n(a_0, a_m)$ means the sum of the probabilities of at least two events of point processes $\{x_{nkj}\}$ in a half-interval

$[a_0, a_m)$ $(1 \le k \le n)$. Consider a class of possible limiting point processes for superpositions of recurrent point processes.

Assume that for each $n \ge 1$ we have a superposition of independent recurrent point processes characterized by the distribution functions $F_{nk0}(x)$ of the moment of the first event and $F_{nk}(x)$ of the interval between successive events. We suppose the summands to be infinitely small point processes and moreover assume that the mean total number of all events in $(0, t)$ is bounded. Then, if a limiting flow (in a weak sense convergence) exists, it can have only the following structure. Consider a Poisson flow $\{t_k^{(0)}\}$ with a leading function Λ. With every event of such a flow one can associate a finite or infinite sequence of events $\{t_k^{(j)}\}$ defined as follows. If $t_k^{(0)} = t$, then $t_k^{(j)} - t_k^{(j-1)}$ are independent random variables with distribution function $F_t(x)$ with $F_t(\infty) \le 1$. Then a limiting flow is a superposition of flows $\{t_k^{(j)} < \infty\}$. These sequences are independent for different k given a fixed $t_k^{(0)}$.

7.6 Limit Theorems for Thinned Flows

For studying different systems, a statistical structure of a flow of failures is of a special interest. The main difficulty is that the interfailure times are, as a rule, dependent random variables. This difficulty can be often overcome by using asymptotic methods. In many cases under some weak hypotheses the thinned flows can be approximated by a Poisson flow. A general setting of a problem can be formulated in the following way.

Assume we have a family of operators L_ε depending on some parameter $\varepsilon > 0$ and transforming a flow $N(\cdot)$ into a flow $N_\varepsilon(\cdot)$: $N_\varepsilon(A) = L_\varepsilon[N(A)]$, $A \in \mathfrak{B}$ where $\mathfrak{B}$ is a σ-algebra of Borel sets from $\mathbb{R}$. We consider a flow $N(\cdot)$ as an input flow and a flow $N_\varepsilon(\cdot)$ as an output flow. Different queueing or redundant systems can be used as an operator L_ε. If for any A with $0 < |A| < \infty$ ($|A|$ is a Lebesgue measure of $A \in \mathfrak{B}$) the relation

$$\mathbf{P}\{N_\varepsilon(A) > 0\} \underset{\varepsilon\to 0}{\longrightarrow} 0$$

holds then the output flow is called **a thinned flow**. For specific operator L_ε we formulate some theorems[1,3,8–11] giving sufficient conditions ensuring the convergence

$$N_\varepsilon\left(\frac{A}{k(\varepsilon)}\right) \underset{\varepsilon\to 0}{\overset{w}{\longrightarrow}} N_0(A), \quad A \in \mathfrak{B},$$

where $N_0(\cdot)$ is some flow, $k(\varepsilon) \underset{\varepsilon\to 0}{\longrightarrow} 0$ and "$\overset{w}{\longrightarrow}$" means a weak convergence (see Section 1.9).

Assume the operator L_ε acts on the input flow in the following way. Let the events of the input flow $N(\cdot)$ occur at moments $\{t_i\}$, $t_i \leq t_{i+1}$. Then the events of the output flow $N_\varepsilon(\cdot)$ can occur only at moments $\{t_i\}$; at each of these moments an event occurs with the probability ε and does not occur with the probability $1-\varepsilon$. As an example of such operator L_ε a detection of defect details can be mentioned: a flow of undetected defect details is an output flow $N_\varepsilon(\cdot)$.

Theorem 7.7. *If there exists a constant $\lambda > 0$ such that for any $\delta > 0$ the relation*

$$\lim_{|A|\to\infty} \mathbf{P}\left\{\left|\frac{N(A)}{|A|} - \lambda\right| > \delta\right\} = 0$$

holds uniformly on all intervals A of a finite length, then

$$N_\varepsilon\left(\frac{A}{\varepsilon}\right) = L_\varepsilon\left[N\left(\frac{A}{\varepsilon}\right)\right] \xrightarrow[\varepsilon\to 0]{w} N_0(A)$$

where $N_0(\cdot)$ is a Poisson flow with the intensity λ.

Theorem 7.8. *If for some nonnegative random variable ξ with a distribution function $G(x)$ and for any $\delta > 0$ a relation*

$$\lim_{|a|\to\infty} \mathbf{P}\left\{\left|\frac{N(A)}{|A|} - \xi\right| > \delta\right\} = 0$$

holds uniformly on all intervals of a finite length, then

$$N_\varepsilon\left(\frac{A}{\varepsilon}\right) = L_\varepsilon\left[N\left(\frac{A}{\varepsilon}\right)\right] \xrightarrow[\varepsilon\to 0]{w} N_0(A)$$

where $N_0(A)$ is a Poisson process with a random intensity ξ having a distribution function $G(x)$.

Consider a recurrent input flow defined by the distribution function $F(x)$. Suppose the same operator L_ε acts on it. Then the output flow is also a recurrent one defined by a distribution function $F_\varepsilon(x)$. Denote

$$\varphi(s) = \int_0^\infty e^{-sx}\, dF(x), \qquad \varphi_\varepsilon(s) = \int_0^\infty e^{-sx}\, dF_\varepsilon(x).$$

These functions satisfy the relation

$$\varphi_\varepsilon(s) = \frac{\varepsilon\,\varphi(s)}{1-(1-\varepsilon)\,\varphi(s)}.$$

The following problem was stated and solved[3]: it is necessary to describe a set of all possible limits of functions $\varphi_\varepsilon(\delta s)$ as $\delta \to 0$ where $0 < \varepsilon = = \varepsilon(\delta) \underset{\delta\to 0}{\longrightarrow} 0$. Denote

$$\varphi_0(s) = \lim_{\delta\to 0} \frac{\varepsilon(\delta)\,\varphi(\delta\, s)}{1-(1-\varepsilon(\delta))\,\varphi(\delta\, s)}.$$

Theorem 7.9. *The only possible form of the function $\varphi(s)$ is either*

$$\varphi_0(s) = \frac{1}{1+c\,s^\beta}, \qquad \operatorname{Re} s \geq 0,\ \ 0<\beta\leq 1, \tag{7.4}$$

or

$$\varphi_0(s) \equiv 1. \tag{7.5}$$

The function $\varphi_0(s)$ has the form (7.4) iff the equality

$$\varphi(s) = 1 - a(s)\,s^\beta + o(a(s)\,s^\beta), \quad s\to 0,\ \ s>0, \tag{7.6}$$

holds where $a(s) > 0$ is a slowly changing function as $s \to 0$. In this case

$$\varepsilon(\delta) = \frac{1}{c}\,a(\delta)\,\delta^\beta(1+o(1)), \quad \delta\to 0.$$

A function $\varphi_0(s)$ has the form (7.5) iff the equality

$$1-\varphi(\delta\, s) = o(\varepsilon(\delta)), \quad \delta\to 0,$$

holds.

A distribution with a density function

$$p(x) = \begin{cases} 0 & \text{if } x<1 \\ \frac{\beta}{x^{\beta+1}} & \text{if } x\geq 1, \end{cases}$$

where $0 < \beta \leq 1$, is an example of function $F(x)$ satisfying a representation (7.6). For such $F(x)$ we have

$$\varphi(s) = 1 - c\,s^\beta + o(s), \quad s\to 0.$$

This considered problem can be generalized by "the scheme of series". Assuming that the distribution function $F(x)$ depends on a small parameter $\delta > 0$ a wider class of limit distributions can be obtained[10,11]. This problem can be formulated as follows.

Assume a system's behavior is described by some regenerative process $\chi(t)$ with regeneration moments $0 = t_0 < t_1 < t_2 < \cdots$, $\xi_n = t_n - t_{n-1}$, $n \geq 1$. On each such regeneration period $(t_n, t_{n+1}]$ it can cause an event

A independently of n and of the process behavior outside this period. The entrance of the process in a given set of states or crossing of a given level are typical examples of such event A. It is necessary to evaluate the distribution of the first moment τ of the event A occurrence. Denote: $t_n + \eta_n$ is the moment of the event A occurrence provided that this event has occurred in $(t_n, t_{n+1}]$, χ_n is an indicator of the event A_n,

$$\zeta_n = \begin{cases} \xi_n & \text{if } \chi_n = 0 \\ \eta_n & \text{if } \chi_n = 1 \end{cases}, \quad q = \mathbf{P}\{A_n\}, \quad \overline{q} = \sup_{z>0} \frac{\mathbf{E}\,(1 - e^{-z\zeta_n})\,\chi}{\mathbf{E}\,(1 - e^{-z\xi_n})},$$

$$q_0 = \max(q, \overline{q}).$$

Theorem 7.10. *If the distribution of a random vector (ξ_n, η_n, ζ_n) changes such that $q > 0$, $q_0 \to 0$ and $\gamma\tau$ converges to a nonsingular random variable for some normalizing multiplier γ, then*

$$\mathbf{E}\,e^{-\gamma z \tau} \longrightarrow \frac{1}{1 + \omega(z)} \tag{7.7}$$

where

$$\omega(z) = \int_0^\infty \frac{1}{x}\,(1 - e^{-zx})\,dP(x)$$

and $P(x)$ is a monotone nondecreasing function such that

$$P(0) = 0, \qquad \int_1^\infty \frac{1}{x}\,dP(x) < \infty.$$

The relation (7.7) is satisfied iff

$$\int_0^x \frac{t}{q}\,dF\left(\frac{t}{\gamma}\right) \longrightarrow P(x)$$

for any $x > 0$.

Corollary 1. *If $T = \mathbf{E}\,\zeta_n < \infty$, then*

$$\mathbf{P}\left\{\frac{q}{T}\tau > x\right\} \longrightarrow e^{-x}, \quad x > 0, \tag{7.8}$$

holds iff

$$\int_x^\infty \frac{t}{q}\,dF\left(\frac{tT}{q}\right) \longrightarrow 0$$

for any $x > 0$.

Corollary 2. *If for some $p \in (1, 2]$*

$$q \left[\frac{\mathbf{E}\,\zeta_n^p}{(\mathbf{E}\,\zeta_n)^p} \right]^{\frac{1}{p-1}} \longrightarrow 0,$$

then the relation (7.8) is true.

Consider a case[9] when a condition $\mathbf{E}\,\zeta_n < \infty$ is not necessarily satisfied. Let $H(x)$ denote a conditional distribution function of ξ_n provided that an event A has not occurred in $(t_n, t_{n+1}]$.

Theorem 7.11. *Assume that $H(x)$ depends on a small parameter $q > 0 : H(x) = H_q(x)$ and suppose there exist functions $\rho = \rho(q) = o(q)$, $T = T(q)$ and $a = a(q)$ such that*

$$\frac{Tq}{a} = o(1), \quad q \to 0, \quad \int_0^T x\,dH(x) = a, \quad 1 - H(T) = \rho.$$

Then

$$\mathbf{P}\left\{ \frac{q\tau}{a} > x \right\} \longrightarrow e^{-x}$$

uniformly on $x \geq 0$.

7.7 Marked Point Processes. Main Definitions

Not only the moments of events' occurrence, but also a characterization of these events is of great practical interest. So, particles recorded by a counter have different speeds, and customers in a queueing system have different service times, priorities and so on. Real practical problems have initiated the introduction and investigation of marked point processes. Every event of such process is characterized not only by the moment of its occurrence, but also by some mark k giving a quality description of this event.

Let $(\Omega, \mathfrak{F}, \mathbf{Q})$ be some probability space, $(K, \mathfrak{K})$ be a measurable space, called **the space of marks**, $\mathbb{R} = (-\infty, +\infty)$ and $\mathfrak{B}$ be a σ-algebra of Borel sets from $\mathbb{R}$. Suppose with every $\omega \in \Omega$ one can associate a countable sequence $\Phi = \{(x_i, k_i), x_i \in \mathbb{R}, k_i \in K\}$. We suppose also that $x_n \to \infty$ with probability 1 as $n \to \infty$, and $x_i \neq x_j$ for $i \neq j$. A sequence Φ is called **a random marked point process.**

Let M_K be a set of all marked point sequences and $\mathfrak{M}_{\mathfrak{K}}$ be a minimal σ-algebra of subsets from M_K such that a mapping $\Phi \longrightarrow N_\Phi(A \times L)$ is a measurable one for any bounded $A \in \mathfrak{B}$ and $L \in \mathfrak{K}$ where $N_\Phi(A \times L)$ is the number of points $\{(x_i, k_i)\}$ of the sequence Φ with $x_i \in A$ and $k_i \in L$. In this case a distribution $\mathbf{Q}\{\cdot\}$ on $\mathfrak{F}$ generates a corresponding distribution $\mathbf{P}\{\cdot\}$ on $\mathfrak{M}_{\mathfrak{K}}$.

Let T_t, $t \in \mathbb{R}$, be a shift operator on M_K, i.e., $(x, k) \in T_t\Phi$ iff $(x+t, k) \in \in \Phi$. The operator T_t is a one-to-one measurable mapping of $(M_K, \mathfrak{M}_{\mathfrak{K}})$ into itself.

Random marked point process and a distribution $\mathbf{P}\{\cdot\}$ are called **stationary**, if an equality

$$\mathbf{P}\{N_{T_t\Phi}(A_1 \times L_1) = r_1, \ldots, N_{T_t\Phi}(A_k \times L_k) = r_k\} =$$

$$= \mathbf{P}\{N_\Phi(A_1 \times L_1) = r_1, \ldots, N_\Phi(A_k \times L_k) = r_k\}$$

holds for any integer $k \geq 1$, $r_1 \geq 0, \ldots, r_k \geq 0$ and $A_1, \ldots, A_k \in \mathfrak{B}$, $L_1, \ldots, L_k \in \mathfrak{K}$, $t \in \mathbb{R}$.

For any stationary distribution $\mathbf{P}\{\cdot\}$ on $\mathfrak{M}_{\mathfrak{K}}$ and for any $L \in \mathfrak{B}$ it holds with probability 1 either $N_\Phi(\mathbb{R} \times L) = 0$ or $N_\Phi((-\infty, 0) \times L) = = N_\Phi((0, +\infty) \times L) = \infty$. An equality

$$\nu_P(B \times L) = \int_{M_K} N_\Phi(B \times L)\, \mathbf{P}\{d\Phi\}, \quad B \in \mathfrak{B}, \ L \in \mathfrak{K},$$

defines on $\mathfrak{B} \times \mathfrak{K}$ the so-called intensity measure of the distribution $\mathbf{P}\{\cdot\}$.

The following statements are true.

1. The equality

$$\int_{M_K} \sum_{a \in \Phi} f(a)\, \mathbf{P}\{d\Phi\} = \int_{R \times K} f(a)\, \nu_P(da)$$

holds for any distribution $\mathbf{P}\{\cdot\}$ on $\mathfrak{M}_{\mathfrak{K}}$ and for any nonnegative measurable real function $f(\cdot)$ defined on $(\mathbb{R} \times K, \mathfrak{B} \times \mathfrak{K})$.

2. If $\mathbf{P}\{\cdot\}$ is a stationary distribution, then a measure $\nu_P(\cdot)$ and a Lebesgue measure $\mu(\cdot)$ satisfy the relation

$$\nu_P(B \times L) = \lambda_P(L)\, \mu(B), \quad B \in \mathfrak{B}, \ L \in \mathfrak{K},$$

where

$$\lambda_P(L) = \int_{M_K} N_\Phi([0, 1) \times L)\, \mathbf{P}\{d\Phi\}$$

is **the intensity** of L with respect to $\mathbf{P}(\cdot)$.

3. For any stationary distribution $\mathbf{P}\{\cdot\}$ on $\mathfrak{M}_{\mathfrak{K}}$ with a finite intensity $\lambda_P(K)$ and any $L \in \mathfrak{K}$ the relation

$$\frac{1}{t} N_\Phi([0,\, t) \times L) \underset{t\to\infty}{\longrightarrow} \sigma_L(\Phi) \text{ with the probability 1}$$

holds where $\sigma_L(\Phi)$ is a random variable invariant with respect to a shift operator T_x (i.e., distributions of random variables $\sigma_L(T_x\Phi)$ and $\sigma_L(\Phi)$ coincide for any $x \in \mathbb{R}$). Moreover

$$\int_{M_K} \sigma_L(\Phi)\,\mathbf{P}\{d\Phi\} = \lambda_P(L). \tag{7.9}$$

If $\mathbf{P}\{\emptyset\} = 0$, then $\sigma_K(\cdot)$ is positive almost everywhere. Under such a hypothesis the relative frequency of the appearance of events with marks from L equals $\sigma_L(\Phi)/\sigma_K(\Phi)$ for almost all Φ. The probability that an at random chosen point from Φ has a mark from L is defined by formula

$$\chi_P(L) = \int_{M_K} \sigma_L(\Phi)\,(\sigma_K(\Phi))^{-1}\,\mathbf{P}\{d\Phi\}.$$

A function $\chi_P(\cdot)$ is a distribution on $\mathfrak{K}$. It follows from (7.9) that a measure $\chi_P(\cdot)$ coincides with $\lambda_P(\cdot)\,(\lambda_P(K))^{-1}$, if $\sigma_K(\Phi) = \lambda_P(K)$ fast for all Φ. This condition is satisfied, if $\mathbf{P}\{\cdot\}$ is **an ergodic distribution**, i.e., if there is not a set $S \in \mathfrak{M}_{\mathfrak{K}}$ which is invariant with respect to a shift operator T_x, $x \in \mathbb{R}$, and at the same time $0 < \mathbf{P}\{S\} < 1$ (see Section 6.1).

7.8 Palm Distribution

Let

$$\begin{aligned}
{}^L M_K &= \{\Phi : \Phi \in M_K,\ N_\Phi((-\infty,\, 0) \times L) = N_\Phi((0,\, +\infty) \times L) = \infty\},\\
M_K^L &= \{\Phi : \Phi \in {}^L M_K,\ N_\Phi(\{0\} \times L) \neq 0\},\\
{}^L\mathfrak{M}_{\mathfrak{K}} &= \{A : A \in \mathfrak{M}_{\mathfrak{K}},\ A \subset {}^L M_K\}, \quad \mathfrak{M}_{\mathfrak{K}}^L = \{A : A \in \mathfrak{M}_{\mathfrak{K}},\ A \subset M_K^L\}
\end{aligned}$$

for $L \in \mathfrak{K}$. Assume now that we have chosen a fixed $L \in \mathfrak{K}$ and a stationary distribution $\mathbf{P}\{\cdot\}$ on $\mathfrak{M}_{\mathfrak{K}}$ such that $\mathbf{P}\{N_\Phi(\mathbb{R} \times L) = 0\} = 0$. Then it follows from Section 7.7 that $\mathbf{P}\{{}^L M_K\} = 1$ and $\lambda_P(L) > 0$. With every original sequence Φ we associate a new sequence

$$\Phi^* = \{(x,\, T_x\Phi) : N_\Phi(\{x\} \times L) \neq 0\}$$

with M_K^L as a space of marks. Hence we have a measurable one-to-one mapping of $({}^L M_K,\ {}^L\mathfrak{M}_{\mathfrak{K}})$ into $(M_{M_K^L},\ \mathfrak{M}_{\mathfrak{M}_{\mathfrak{K}}^L})$. Moreover a distribution $\mathbf{P}\{\cdot\}$ generates a corresponding distribution $\mathbf{P}^*\{\cdot\}$ on $\mathfrak{M}_{\mathfrak{M}_{\mathfrak{K}}^L}$, which is a stationary or ergodic one, if the distribution $\mathbf{P}\{\cdot\}$ is of the same type. It is obvious that $\lambda_{P^*}(M_K^L) = \lambda_P(L) > 0$.

If we in addition suppose $\lambda_P(L) < \infty$, then

$$\mathbf{P}_L\{\cdot\} = (\lambda_P(L))^{-1}\lambda_{P^*}(\cdot)$$

is a probability distribution on $\mathfrak{M}_{\mathfrak{K}}^L$, which is called **a Palm distribution**[12–14] for $\mathbf{P}\{\cdot\}$ with respect to L.

If $S \in \mathfrak{M}_{\mathfrak{K}}^L$, then $\mathbf{P}_L\{\cdot\}$ can be interpreted as the probability that $T_x\Phi \in S$ for some "chosen at random" point $(x,\, k) \in \Phi,\ k \in L$. If $\mathbf{P}\{\cdot\}$ is an ergodic distribution, then the relation

$$\frac{N_\Phi((0,\, t] \times L;\ S)}{N_\Phi((0,\, t] \times L)} \underset{t\to\infty}{\longrightarrow} \mathbf{P}_L\{S\}$$

holds for almost all $\Phi \in {}^L M_K$ where $N_\Phi((0,\, t] \times L;\ S)$ is the number of points $(x,\, k) \in \Phi$ such that $x \in [0,\, t),\ k \in L,\ T_x\Phi \in S$.

Let $\Phi \in {}^L M_K$. Renumber points $(x,\, k),\ x \in \mathbb{R},\ k \in L$, of the sequence Φ in the increasing order of their first coordinates. Denote the corresponding values by $x_i^L(\Phi)$ where

$$x_0^L(\Phi) = \min\{x : x > 0,\ N_\Phi(\{x\} \times L) \neq 0\}.$$

These values $x_i^L(\Phi)$, $i = 0, \pm 1, \pm 2, \dots,$ are measurable functions of Φ. Distributions $\mathbf{P}_L\{\cdot\}$ and $\mathbf{P}\{\cdot\}$ satisfy the following equality:

$$\mathbf{P}_L\{S\} = (\lambda_P(L))^{-1} \int\limits_{{}^L M_K} \left(\frac{1}{t} \sum_{i=0}^{n_\Phi(t)-1} k_S(T_{x_i^L(\Phi)}\Phi) \right) \mathbf{P}\{d\Phi\}$$

for all $t > 0$ and $S \in \mathfrak{M}_{\mathfrak{K}}^L$ where $n_\Phi(t) = N_\Phi((0,\, t] \times L)$ and $k_S(\cdot)$ is an indicator function. Let

$$x_L(\Phi) = x_0^L(\Phi),\ \ y_L(\Phi) = x_{-1}^L(\Phi),\ \ \Phi \in {}^L M_K,\ \ \Theta_L\Phi = T_{x_L(\Phi)}\Phi,$$

$$\Theta_L^{-1}\Phi = T_{x_{-2}^L(\Phi)}\Phi,\ \ \Phi \in M_K^L.$$

If $\Phi \in M_K^L$, then $y_L(\Phi) = 0$. A Palm distribution is not invariant with respect to all T_t. Nevertheless the following statement holds.

Theorem 7.12. *If* $\mathbf{P}\{\cdot\}$ *is a stationary distribution on* $\mathfrak{M}_{\mathfrak{K}}$, $L \in \mathfrak{K}$, $\mathbf{P}\{N_\Phi(\mathbb{R} \times L) = 0\} = 0$ *and* $\lambda_P(L) < \infty$, *then*

$$\mathbf{P}_L\{\Theta_L(S)\} = \mathbf{P}_L\{S\}$$

for all $S \in \mathfrak{M}^L_{\mathfrak{K}}$.

Distribution $\mathbf{P}\{\cdot\}$ is uniquely determined by a Palm distribution $\mathbf{P}_L\{\cdot\}$.

Theorem 7.13. *Under the hypotheses of Theorem 7.12 the relation*

$$\mathbf{P}\{S\} = \lambda_P(L) \int_{M^L_K} \left(\int_0^{x_L(\psi)} k_S(T_\tau \psi)\, d\tau \right) \mathbf{P}_L\{d\psi\} =$$

$$= \lambda_P(L) \int_0^\infty \mathbf{P}_L\{x_L(\psi) > \tau,\ T_\tau \psi \in S\}\, d\tau$$

holds for all $S \in \mathfrak{M}_{\mathfrak{K}}$.

Theorem 7.14. *Under the hypotheses of Theorem 7.12 the following relations hold:*

1. $$(\lambda_P(L))^{-1} = \int_{M^L_K} x_L(\psi)\, \mathbf{P}_L\{d\psi\}.$$

2. $$\mathbf{P}\{N_\Phi((0,t] \times L) = 1\} = \lambda_P(L)\, t + o(t), \quad t \to +0,$$
$$\mathbf{P}\{N_\Phi((0,t] \times L) > 1\} = o(t), \quad t \to +0.$$

3. $$\mathbf{P}\{-y_L(\Phi) > u,\ x_L(\Phi) > v\} = \lambda_P(L) \int_{u+v}^\infty (1 - F(x))\, dx,$$

$$\mathbf{P}\{x_L(\Phi) - y_L(\Phi) > u\} = \lambda_P(L) \int_u^\infty x\, dF(x),$$

$$\mathbf{P}\{-y_L(\Phi) > u\} = \mathbf{P}\{x_L(\Phi) > u\} = \lambda_P(L) \int_u^\infty (1 - F(x))\, dx,$$

for any $u, v \geq 0$ *where*

$$F(x) = \mathbf{P}\{x_L(\psi) < x\}, \quad \lambda_P(L) = \int_0^\infty x\, dF(x).$$

4. $\mathbf{P}_L\{S\} = \lim_{t\to 0} \mathbf{P}\{T_{x_L(\Phi)}\Phi \in S \,|\, x_L(\Phi) < t\}$ *for any* $S \in \mathfrak{M}^L_{\mathfrak{K}}$.

5. *A distribution* $\mathbf{P}\{\cdot\}$ *is ergodic with respect to* T_t *iff* $\mathbf{P}_L\{\cdot\}$ *is ergodic with respect to* Θ_L.

The next theorem implies that properties of a Palm distribution given in Theorems 7.12 and 7.14 (the first relation) are its characteristic properties.

Theorem 7.15. *Let* L *be a nonempty set from* $\mathfrak{K}$ *and* $\mathbf{Q}\{\cdot\}$ *be a distribution on* $\mathfrak{M}^L_{\mathfrak{K}}$. *Then there exists a stationary distribution* $\mathbf{P}\{\cdot\}$ *on* $\mathfrak{M}_{\mathfrak{K}}$ *satisfying conditions of Theorem 7.12 and such that* $\mathbf{P}_L\{\cdot\} = \mathbf{Q}\{\cdot\}$ *iff the distribution* $\mathbf{Q}\{\cdot\}$ *is invariant with respect to* Θ_L *and*

$$\Delta = \int_{M^L_K} x_L(\psi)\, \mathbf{Q}\{d\psi\} < \infty.$$

A random marked point process ψ with a distribution $\mathbf{Q}\{\cdot\}$ is called a **synchronous** one with respect to L, if it possesses the following properties:

(i) $\psi \in M^L_K$,

(ii) $\{(\alpha^L_i(\psi),\, k_i(\psi)),\ i = 0, \pm1, \pm2, \dots\}$ is a stationary in a strict sense sequence where $\alpha^L_i(\psi) = x^L_{i+1}(\psi) - x^L_i(\psi)$,

(iii) $\Delta = \mathbf{E}\alpha^L_1(\psi) < \infty$ and $\alpha^L_1(\psi) > 0$ with the probability 1.

Theorem 7.15 implies that a random marked point process ψ is a synchronous one, if $\mathbf{Q}\{\cdot\}$ is a Palm distribution $\mathbf{P}_L\{\cdot\}$ corresponding to some stationary distribution $\mathbf{P}\{\cdot\}$ of the random marked point process Φ.

7.9 Processes with Embedded Marked Point Processes

With every random marked point process Φ with a space of marks K one can associate a random process $\eta(t)$, $t \in \mathbb{R}$, taking values in K and with piecewise constant trajectories: $\eta(t) = k_j$ if $x_j(\Phi) < t < x_{j+1}(\Phi)$. If $L \in \mathfrak{K}$ is fixed and $\lambda_P(L) > 0$, then a marked point process Φ (and therefore a process $\eta(t)$, $t \in \mathbb{R}$) can be represented as a marked point process $\Phi_L = \{(x^L_i(\Phi),\, \Phi^L_i),\ i = 0, \pm1, \pm2, \dots\}$ where Φ^L_i is a random counting measure (see Section 6.1) defining a process Φ in $(x^L_i,\, x^L_{i+1}]$. The same process $\eta(t)$, $t \in \mathbb{R}$, has different representations depending on the choice of $L \in \mathfrak{K}$. This process $\eta(t)$ is an example of the process with an embedded marked point process Φ.

Let $\{\xi_i(t),\ 0 < t \le \tau_i,\ i = 0, \pm1, \pm2, \dots\}$ be a family of terminated random processes taking values in some measurable space $(E, \mathfrak{L})$ and defined on the same probability space. Let

$$\Phi = \{(x_i, k_i), \ i = 0, \pm 1, \pm 2, \dots\} \tag{7.10}$$

where $k_i = \xi_i(t)$, $0 < t \le \tau_i = x_{i+1} - x_i$, i.e., random processes $\{\xi_i(t)\}$, are used as marks.

If Φ is a random marked point process of the form (7.10), then a random process $\eta(t)$, $t \in \mathbb{R}$, such that $\eta(t) = k_i = \xi_i(t)$, $x_i < t \le x_{i+1}$, is called **a process with an embedded marked point process** Φ.

Points $x_i(\Phi)$ and marks $k_i(\Phi)$, $i = 0, \pm 1, \pm 2, \dots$, are called **basis points** and **cycles** of the process $\eta(t)$. If the sequence of basis points is fixed, then there exists a one-to-one correspondence between a random process and an embedded marked point process. If $x_0(\Phi) = 0$ and $\{\tau_i\}$, $\{\xi_i(t)\}$ are sequences of independent and identically distributed random variables and processes, then $\eta(t)$ is a regenerative process.

A process $\eta(t)$, $t \in \mathbb{R}$, is called a **stationary** or a **synchronous** one, if the corresponding process Φ is of the same type.

7.10 Principle of Intensity Conservation

Let $\eta(t)$, $t \in \mathbb{R}$, be a stationary Markov process with an embedded stationary random marked point process Φ. With every Φ of the form (7.10) one can associate a stationary random marked point process

$$\psi = \{(x_i, (\xi_{i-1}(\tau_{i-1}), \xi_i(+0))), \ i = 0, \pm 1, \pm 2, \dots\}$$

where $\tau_{i-1} = x_i - x_{i-1}$. The mark x_i describes a state of the process $\eta(t)$ directly before and after the basis point x_i.

Transition probabilities

$$p_h(y, Y) = \mathbf{P}\{\eta(u+h) \in Y \mid \eta(u) = y, \ N_\psi([u, u+h] \times E^2) = 0\},$$

$$u \in \mathbb{R}, \ h > 0, \ y \in E, \ Y \in \mathfrak{L},$$

describe a behavior of the process $\eta(t)$ in intervals between basis points. Here $N_\psi([a, b] \times A \times B)$, $-\infty < a < b < +\infty$, $A, B \in \mathbb{R}$, denotes the number of points of the sequence ψ hitting $[a, b]$ and with marks from $A \times B$. Since processes $\eta(t)$ and ψ are generated by a stationary random marked point process Φ, the transition probabilities $p_h(y, Y)$ don't depend on $u \in \mathbb{R}$, i.e., they are homogeneous in time.

Let $\lambda(\cdot)$ be an intensity of the stationary marked point process ψ, $\lambda = \lambda(E \times E)$. A point $u \in \mathbb{R}$ has a property (C) with respect to a set L

and a mapping $f : \mathbb{R} \to E$, if for any open interval U containing u there exist $t_1, t_2 \in U$ such that $f(t_1) \in Y$ and $f(t_2) \in \overline{Y}$, where $\overline{Y} = E \setminus Y$.

Such point $u \in \mathbb{R}$ possessing a property (C) with respect to Y and f can be interpreted as the point of crossing a boundary of the domain Y by a mapping f. Introduce σ-algebras $\mathfrak{L}_1$ and $\mathfrak{L}_2$ such that $\mathfrak{L}_2 \subset \mathfrak{L}_1 \subset \mathfrak{L}$.

Let $\mathfrak{L}_1$ be a σ-algebra containing all $Y \in \mathfrak{L}$ satisfying two conditions:

(i) almost all trajectories $\eta_0(t)$ of the process $\eta(t)$ possess the following property: if x is a basis point of the trajectory $\eta_0(t)$, then for any $\varepsilon > 0$ there exist $t_1 \in (x - \varepsilon, x)$ and $t_2 \in (x, x + \varepsilon)$ such that

$$\begin{aligned} &\eta_0(t_1) \in Y \;\; (\text{or } \overline{Y}) \quad \text{if } \eta_0(x-0) \in Y \;\; (\text{or } \overline{Y}), \\ &\eta_0(t_2) \in Y \;\; (\text{or } \overline{Y}) \quad \text{if } \eta_0(x+0) \in Y \;\; (\text{or } \overline{Y}); \end{aligned}$$

(ii) a sequence of points $u \in \mathbb{R}$ possessing a property (C) with respect to Y and to trajectories of the process $\eta(t)$ is a stationary point process $\psi_Y^{(C)}$ with a finite intensity.

Let $\mathfrak{L}_2$ be a σ-algebra containing all $Y \in \mathfrak{L}_1$ satisfying an additional condition:

(iii) for any $\varepsilon > 0$ there exists a set $D \in \mathfrak{L}_1$ such that

$$\lim_{h \to +0} p_h(y, Y) = \chi_Y(y)$$

uniformly on $y \in D$, $\lambda(D \times E) > \lambda - \varepsilon$. Here $\chi_Y(y)$ is an indicator function:

$$\chi_Y(y) = \begin{cases} 1 & \text{if } y \in Y \\ 0 & \text{if } y \in \overline{Y}. \end{cases}$$

The point $u \in \mathbb{R}$ is called **the moment of entrance (or exit)** of the mapping $f : \mathbb{R} \to E$ in (or out of) a set Y, if there exists $\varepsilon > 0$ such that

$$f(u-t) \in \overline{Y}, \;\; f(u+t) \in Y, \qquad (f(u-t) \in Y, \;\; f(u+t) \in \overline{Y})$$

for any $t \in (0, \varepsilon)$. It is obvious that entrance and exit moments satisfy the property (C) with respect to Y and f.

It follows from conditions (i) and (ii) that a basis point x satisfies a property (C) with respect to $Y \in \mathfrak{L}_1$ and a trajectory $\eta_0(t)$, if $\eta_0(x-0) \in \overline{Y}$, $\eta_0(x+0) \in Y$ or $\eta_0(x-0) \in Y$, $\eta_0(x+0) \in \overline{Y}$. For $Y \in \mathfrak{L}_1$ this condition means that x is the moment of entrance in Y (or a moment of exit out of Y) of the trajectory $\eta_0(t)$. Consequently entrance moments (and exit moments) being basis points form a stationary point process ψ_Y^1 (or ψ_Y^0) with a finite intensity λ_Y^1 (or λ_Y^0):

$$\lambda_Y^1 = \lambda(\overline{Y} \times Y), \quad \lambda_Y^0 = \lambda(Y \times \overline{Y}).$$

Theorem 7.16. *The relation*

$$\mathcal{A}_{\mathcal{P}}(Y) = \lambda(Y \times \overline{Y}) - \lambda(\overline{Y} \times Y)$$

holds for any $Y \in \mathfrak{L}_2$ *where*

$$\mathcal{A}_{\mathcal{P}}(Y) = \lim_{h \to +0} \frac{1}{h} \left(\int_E p_h(y, Y)\, \mathcal{P}\{dy\} - \mathcal{P}\{Y\} \right)$$

and $\mathcal{P}\{\cdot\}$ *is a one-dimensional distribution of the process* $\eta(t)$, $t \in \mathbb{R}$, *at the arbitrary moment.*

Corollaries.

1. Let $Y : (0, \infty) \longrightarrow \mathfrak{L}$ be a mapping such that $Y(u) \in \mathfrak{L}_2$ for almost all $u \geq 0$ and there exists an integral $\int_0^\infty \mathcal{A}_{\mathcal{P}}(Y(u))\, du$. Then

$$\int_0^\infty \mathcal{A}_{\mathcal{P}}(Y(u))\, du = \int_0^\infty (\lambda(Y(u) \times \overline{Y}(u)) - \lambda(\overline{Y}(u) \times Y(u)))\, du.$$

2. The equality

$$\mathcal{A}_{\mathcal{P}}(Y) = \lambda(\mathcal{P}^-\{Y\} - \mathcal{P}^+\{Y\})$$

holds for any $Y \in \mathfrak{L}_2$ where

$$\mathcal{P}^-\{\cdot\} = \frac{1}{\Lambda}\lambda((\cdot) \times E), \quad \mathcal{P}^+\{\cdot\} = \frac{1}{\Lambda}\lambda(E \times (\cdot)), \quad \Lambda = \lambda(E \times E),$$

are distributions of the process $\eta(t)$ directly before and after a chosen at random basis point.

3. The equality

$$\psi_Y^{(C)} = \psi_Y^1 + \psi_Y^0 + \varphi_Y^1 + \varphi_Y^0$$

holds for any $Y \in \mathfrak{L}_1$ where $\psi_Y^{(C)}$, ψ_Y^1 and ψ_Y^0 are introduced in the above processes and φ_Y^1, φ_Y^0 are point processes generated by the moments of entrance in Y or exit out of Y, which are not basis points. Let processes φ_Y^1 and φ_Y^0 be stationary. Denote their finite intensities by α_Y^1 and α_Y^0. Then we obtain **the principle of intensity conservation**[14]**:** the equalities

$$\alpha_Y^1 + \lambda_Y^1 = \alpha_Y^0 + \lambda_Y^0, \quad \mathcal{A}_{\mathcal{P}}\{Y\} = \alpha_Y^1 - \alpha_Y^0$$

hold for any $Y \in \mathfrak{L}_2$ where λ_Y^1 and λ_Y^0 are the intensities of stationary random marked point processes ψ_Y^1 and ψ_Y^0.

4. If $\eta(0)$ and x_1 are independent, then

$$\mathbf{P}\{Y\} = \mathcal{P}^-\{Y\} = \frac{1}{\Lambda}\lambda(Y \times E)$$

for any $Y \in \mathfrak{L}_1$.

Since $\eta(t)$, $t \in \mathbb{R}$, is a stationary process, the independence of $\eta(0)$ and x_1 means that the process value at the arbitrary point and a time to the next basis point are independent random variables. For example, this condition is satisfied, if the sequence of basis points form a Poisson process and a random value $\eta(0)$ is independent of the Poisson process behavior in a half-interval $[0, \infty)$.

7.11 References

1. Belyayev, Y. K., Random flows and a renewal theory, in *Renewal Theory,* Cox, D., Smith, W. L., Sov. Radio, Moscow, 238, 1967 (in Russian).
2. Chinčin, A. J., *Mathematical Methods in the Theory of Queueing,* Griffin, London, 1969.
3. Gnedenko, B. V., Kovalenko, I. N., *Introduction to Queueing Theory,* Birkhäuser, Boston, 1989.
4. Palm, C., Intensitätsschwankungen im Fernsprechverkehr, *Ericksson Techn.,* 44, 1, 1943.
5. Klimov, G. P., *Bedienungsprozesse,* Birkhäuser, Basel, 1979.
6. Ososkov, G. A., A limit theorem for flows of homogeneous events, *Teor. Veroyatn. Primen.,* 1, 274, 1956 (in Russian).
7. Grigelionis, B. I., On the convergence of sums of stepwise random processes to a Poisson one, *Teor. Veroyatn. Primen.,* 8, 189, 1963 (in Russian).
8. Belyayev, Y. K., Limit theorems for thinned flows, *Teor. Veroyatn. Primen.,* 8, 175, 1963 (in Russian).
9. Kovalenko, I. N., *Analysis of Rare Events in Evaluation of Systems Efficiency and Reliability,* Sov. Radio, Moscow, 1980 (in Russian).
10. Solovyev, A. D., Standby with rapid renewal, *Engin. Cybernetics,* 8, 49, 1970.
11. Solovyev, A. D., Asymptotic behavior of the time of first occurrence of a rare event, *Engin. Cybernetics,* 9, 1038, 1971.
12. König, D., Matthes, K., Nawrotzki, K., *Verallgemeinerungen der Erlangschen und Engsetschen Formeln,* Akademie-Verlag, Berlin, 1967.
13. König, D., Matthes, K., Nawrotzki, K., Unempfindlichkeitseigenschaften von Bedienungsprozessen, in *Einführung in die Bedienungstheorie,* Gnedenko, B. V., Kovalenko, I. N., Akademie-Verlag, Berlin, 359, 1971.

14. Franken, P., König, D., Arndt, U., Schmidt, V., *Queues and Point Processes,* Akademie-Verlag, Berlin, 1981.

Chapter 8

Classes of Constructively Defined Random Processes

8.1 Chains with Complete Connections

Chains with complete connections[1–6] were introduced as one possible generalization of usual Markov chains. Denote:

$(\Omega, \mathfrak{F}, \mathbf{P})$ is a probability space,

$(X, \mathfrak{U})$ is a measurable space,

$X^{(n)} = X \times \cdots \times X$ is a n-multiple Cartesian product of the set X and $\mathfrak{U}^{(n)}$ is a corresponding σ-algebra, $n = 1, 2, \ldots,$

$M = M(X, \mathfrak{U})$ is the set of all probability measures defined on $\mathfrak{U}$ and $\mathfrak{M}$ is a σ-algebra of subsets from M generated by **cylindrical sets**

$$C = \{\mu(\cdot) : \mu(A_i) = x_i,\ i = 1, \ldots, k\},$$

$A_i \in \mathfrak{U},\ A_i \cap A_j = \varnothing,\ i \neq j,\ x_i \geq 0,\ i, j = 1, \ldots, k,\ x_1 + \cdots + x_k \leq 1,\ k \geq 1,$

$\{T_x^{(n)},\ x \in X,\ n = 1, 2, \ldots\}$ is the set of mappings of M into M,

$T^{(n)}_{x^{(n)}} = T^{(n)}_{x_1, \ldots, x_n} = T^{(n)}_{x_n} \ldots T^{(1)}_{x_1}$ where $x^{(n)} = (x_1, \ldots, x_n) \in X^{(n)}$.

A sequence of random variables $\{\xi_n,\ n = 1, 2, \ldots\}$ defined on Ω and taking values in X forms **a chain with complete connections**, if conditional probabilities

$$p_{x_1, \ldots, x_n}(A) = \mathbf{P}\{\xi_{n+1} \in A \,|\, \xi_j = x_j,\ 1 \leq j \leq n\},\quad A \in \mathfrak{U},$$

satisfy the equalities

$$p_{x_1,\dots,x_{n+1}}(\cdot) = T^{(n+1)}_{x_{n+1}} p_{x_1,\dots,x_n}(\cdot), \quad p_{x_1}(\cdot) = T^{(1)}_{x_1} p(\cdot)$$

for any $n = 1, 2, \dots$ and $(x_1, \dots, x_{n+1}) \in X^{n+1}$ where $p(A) = \mathbf{P}\{\xi_1 \in A\}$, $A \in \mathfrak{U}$. This definition implies that

$$p_{x_1,\dots,x_n}(\cdot) = T^{(n+1)}_{x_1,\dots,x_n} p(\cdot)$$

for any $x^{(n)} = (x_1, \dots, x_n) \in X^{(n)}$. In other words a sequence $\{\xi_n\}$ forms a chain with complete connections, if the conditional probability of $\xi_{n+1} \in A$ provided that $\xi_1 = x_1, \dots, \xi_n = x_n$ depends not only on x_n, but also on corresponding conditional distribution at time n.

An algorithm for the successive construction of $\xi_1, \xi_2, \dots$ can be formulated as follows. According to the distribution $p(\cdot)$ we construct a random variable ξ_1. If $\xi_1 = x_1$, then we construct ξ_2 according to the distribution $Q_2(\cdot) = T^{(1)}_{x_1} p(\cdot)$. Assume ξ_n is determined according to the distribution $Q_n(\cdot)$ and its constructed value equals x_n. Then a random variable ξ_{n+1} has a distribution $Q_{n+1}(\cdot) = T^{(n)}_{x_n} Q_n(\cdot)$ being used to construct a next sampling value x_{n+1}.

If the operator $T^{(n)}_x$, $x \in X$, does not depend on n, then we have **a homogeneous chain with complete connections**.

If for some $P(x; \cdot) \in M$, $x \in X$, the relation $T^{(n)}_x p(\cdot) = P(x; \cdot)$ holds for any $n = 1, 2, \dots$, $x \in X$ and $p(\cdot) \in M$, then $\{\xi_n, n = 1, 2, \dots\}$ is a simple homogeneous Markov chain.

The chains with complete connections considered above are called **simple**. If the conditional distribution of ξ_{n+1} depends on several preceding observations and on several preceding conditional distributions, then we have **a complex chain with complete connections**.

If X is a finite set, i.e., $X = \{x_1, \dots, x_{m+1}\}$, then as M it is sufficient to take

$$M = \left\{ (p_1, \dots, p_m) : \ p_j \geq 0, \ j = 1, \dots, m, \ \sum_{j=1}^{m} p_j \leq 1 \right\}$$

and as $\{T^{(n)}_x, x \in X\}$ we take vector functions

$$\{\varphi^{(n)}_i, \ i = 1, \dots, m\} : \ \Delta \longrightarrow \Delta',$$

where

$$\Delta = (p_1, \dots, p_m), \quad \Delta' = (p'_1, \dots, p'_m), \quad \Delta, \Delta' \in M,$$

i.e.,

$$p'_j = \varphi_{ij}^{(n)}(p_1, \ldots, p_m), \quad 1 \le i \le m.$$

In other words $\varphi_{ij}^{(n)}(p_1, \ldots, p_m)$ is a conditional probability that $\xi_{n+1} = j$ provided that $\xi_n = i$ and we know the conditional distribution at time n defined by vector $\Delta = (p_1, \ldots, p_m)$. If functions $\{\varphi_{ij}^{(n)}(\Delta)\}$ don't depend on Δ, then we have a usual Markov chain with a finite set of states.

If functions $\{\varphi_{ij}^{(n)}(\Delta)\}$ can be presented in the form

$$\varphi_{ij}^{(n)}(p_1, \ldots, p_m) = a_{0ij}^{(n)} + \sum_{k=1}^{m} a_{kij}^{(n)} p_k,$$

then we have a **linear** chain with complete connections.

Theorem 8.1 (the existence theorem). *For a given measurable space $(X, \mathfrak{U})$, probability measure $p(\cdot) \in M$ and a set of mappings $\{T_x^{(n)}, x \in X, n = 1, 2, \ldots\}$ of M into M there exists a probability space $(\Omega, \mathfrak{F}, \mathbf{P})$ and a chain with complete connections defined on this probability space taking values in a set of states X and with initial distribution $p(\cdot)$ and transition mappings $\{T_x^{(n)}, x \in X, n = 1, 2, \ldots\}$.*

Processes with complete connections can be introduced analogously[2]. Chains with complete connections are used in the theory of adaptive systems and for the investigation of random learning models[2,7].

Simple and complex homogeneous chains with complete connections can be investigated from the same point of view using random systems with complete connections[8,9].

Let $(\Omega, \mathfrak{F}, \mathbf{P})$ be a probability space, $(X, \mathfrak{U})$ and $(W, \mathfrak{R})$ be some measurable spaces, $u(\cdot, x)$ be a mapping of W into W for any $x \in X$, $u(\cdot, x^{(n)}) = u(u(\cdot, x^{(n-1)}), x_n)$, $x^{(n)} = (x_1, \ldots, x_n)$, and let the condition of measurability be satisfied:

$$\{(c, x^n) : \ u(c, x^{(n)}) \in V\} \in \mathfrak{R} \times \mathfrak{U}^{(n)} \quad \text{for any } V \in \mathfrak{R}, \ n = 1, 2, \ldots .$$

Moreover suppose some real function $P(\cdot, \cdot)$ is defined on $W \times \mathfrak{U}$: $P(c, \cdot)$ is a probability measure on $\mathfrak{U}$ for any $c \in W$ and $P(\cdot, A)$ is a $\mathfrak{R}$-measurable function for any $A \in \mathfrak{U}$.

A sequence of random variables $\{\xi_n, n = 1, 2, \ldots\}$ defined on Ω and taking values in X forms **a homogeneous random system with complete connections**, if there exist $c \in W$ and functions $u(\cdot, \cdot)$, $P(\cdot, \cdot)$ satisfying the above stated conditions and such that the equality

$$\mathbf{P}\{\xi_{n+1} \in A \,|\, \xi_j = x_j, \ j = 1, \ldots, n\} = P(u(c, x^{(n)}), A)$$

holds for any $n = 1, 2, \ldots,\ x^{(n)} \in X^{(n)}$ and $A \in \mathfrak{U}$.

A statement analogous to Theorem 8.1 holds also for random systems with complete connections.

Introduce I_φ processes associated with such systems. Let $K(\xi_j,\ j \in \Lambda)$, $\Lambda \subset \{1, 2, \ldots\}$, denote a σ-algebra generated by random variables $\xi_j,\ j \in \Lambda$.

A sequence of random variables $\{\xi_n,\ n = 1, 2, \ldots\}$ defined on Ω and taking values in X is called **an I_φ process**, if there exists a real function $\varphi(\cdot, \cdot, \cdot)$ defined on $\{1, 2, \ldots\} \times \{1, 2, \ldots\} \times \{1, 2, \ldots\}$, satisfying the inequalities

$$0 \le \varphi(k+1,\, l,\, n) \le 1, \quad \varphi(k,\, l,\, n+1) \le \varphi(k+1,\, l,\, n), \quad k, l, n \in \{1, 2, \ldots\},$$

and such that the inequality

$$|\mathbf{P}\{(\xi_{k+n}, \ldots, \xi_{k+n+l-1}) \in A^{(l)} \,|\, K(\xi_j,\ 1 \le j \le k)\} - \\ - \mathbf{P}\{(\xi_{k+n}, \ldots, \xi_{k+n+l-1}) \in A^{(l)}\}| \le \varphi(k,\, l,\, n)$$

holds with the probability 1 for any $k,\, l,\, n \in \{1, 2, \ldots\}$, $A^{(l)} \in \mathfrak{U}^{(l)}$.

Theorem 8.2. *A homogeneous random system with complete connections is an I_φ process for any $c \in W$.*

8.2 Processes Associated with Semi-Markov Process

1. **(J, X)-processes**[10–14]. Let $\{(J_n, X_n),\ n = 0, 1, 2, \ldots\}$ be a two-dimensional random process defined on the probability space $(\Omega, \mathfrak{F}, \mathbf{P})$ and taking values in the state space $I_r \times \mathbb{R}$ where $I_r = \{1, 2, \ldots, r\}$. Let moreover $\{Q_{jk}(x),\ j, k = 1, 2, \ldots, r\}$ be a set of nondecreasing functions defined on $\mathbb{R}$ and satisfying conditions $Q_{jk}(-\infty) = 0$ and $\sum\limits_{k=1}^{\infty} Q_{jk}(\infty) = 1$, $j = 1, \ldots, r$.

A process $\{(J_n, X_n),\ n = 0, 1, 2, \ldots\}$ is called **a (J, X)-process**, if $X_0 = 0$ with the probability 1,

$$\mathbf{P}\{J_0 = k\} = p_k, \quad k = 1, \ldots, r, \quad \sum_{k=1}^{\infty} p_k = 1,$$

and for any $(k, x) \in I_r \times \mathbb{R}$ with the probability 1

$$\mathbf{P}\{J_n = k,\ X_n \le x \,|\, (J_0, X_0), \ldots, (J_{n-1}, X_{n-1})\} = Q_{J_{n-1}k}(x).$$

If $Q_{jk}(0) = 0,\ j, k = 1, \ldots, r,$ then a sequence $\{(J_n, X_n)\}$ defines a usual semi-Markov process with a finite number of states.

Let $\{R_n(j, k, t),\ j, k \in I_r,\ t \in \mathbb{R},\ n = 1, 2, \ldots\}$ be some set of random variables. Under some hypotheses[10] concerning this set and a stochastic matrix $\{Q_{jk}(\infty)\}_{j,k=1}^{r}$ there exists $\mu \in \mathbb{R}$ such that

$$\frac{1}{\sqrt{n}} \left(\sum_{k=1}^{n} Y_k - n\,\mu \right)$$

converges in distribution as $n \to \infty$ to a normal distributed random variable. Here $Y_n = R_n(J_{n-1}, J_n, X_n),\ n \geq 1$. Moreover, if $Y_1 + \cdots + Y_n \to \infty$ with the probability 1, then under the same hypotheses

$$\frac{1}{\sqrt{t}} \left(N(t) - \frac{t}{\mu} \right)$$

converges in distribution as $t \to \infty$ to a normal distributed random variable where

$$N(t) \; = \; \max\{n \geq 1 : \; Y_1 + \cdots + Y_n < t\}, \quad t \geq 0.$$

A statement analogous to a Renewal Theorem for renewal processes holds also for (J, X)-processes.

Theorem 8.3. *Suppose the following conditions are satisfied:*

(i) there exists a positive integer number m such that a matrix P^m has only positive components where $P = (p_{jk})$ is a matrix with components $p_{jk} = Q_{jk}(\infty),\ j, k = 1, \ldots, r,$

(ii) for any $j, k = 1, \ldots, r$ a conditional distribution of X_n provided that $J_{n-1} = j$ and $J_n = k$ is a nonlatticed one with a finite second moment.

Then

$$\lim_{x \to \infty} \sum_{n=1}^{\infty} \mathbf{P}\{x < X_1 + \cdots + X_n \leq x + h\} \; = \; \begin{cases} h/\mu & \text{if } \mu > 0 \\ 0 & \text{if } \mu < 0 \end{cases}$$

for any fixed $h > 0$ where μ is a constant such that

$$\frac{1}{n}(X_1 + \cdots + X_n) \underset{n \to \infty}{\longrightarrow} \mu$$

in probability.

2. **Semi-Markov process of multiplication**[15–18]. A Semi-Markov process of multiplication is a natural generalization of a Markov process of multiplication[19].

A real-valued random process $\xi(t)$, $t \geq 0$, is called **a semi-Markov process of multiplication**, if the following conditions are satisfied:

(i) $\xi(0) = x > 0$ for some constant x,

(ii) trajectories of $\xi(t)$ are continuous on the right and $d\xi(t)/dt = -1$ everywhere with the exception of points

$$t_i = \sum_{k=1}^{i} \tau_k, \quad i \geq 1, \; t_0 = 0,$$

where the relations

$$\xi(t_n) = \gamma_n \, \xi(t_n - 0), \quad n = 1, 2, \dots ,$$

are satisfied. Here $\{\tau_i\}_{i=1}^{\infty}$ and $\{\gamma_i\}_{i=1}^{\infty}$ are two independent sequences of positive independent and identically distributed random variables. If $\mathbf{P}\{\tau_i > x\} = e^{-\alpha x}$, then $\xi(t)$ is called **a Markov process of multiplication**.

When studying semi-Markov processes of multiplication the main attention is concentrated on the following characteristics:

(i) the probability of degeneration

$$f(x) = \mathbf{P}\{\zeta(x) < \infty\}$$

where

$$\zeta(x) = \inf\{t : \, \xi(t) \leq 0 \,|\, \xi_0 = x\}$$

is a first moment of degeneration, and

(ii) the probability of degeneration to the moment of $(n+1)$th jump:

$$f_n(x) = \mathbf{P}\{\xi(t_{n+1}) \leq 0 \,|\, \xi(0) = x\}, \quad n = 0, 1, \dots .$$

The definition of a semi-Markov process of multiplication implies that

$$f(x) = \lim_{n \to \infty} f_n(x) = \mathbf{P}\{\xi \geq x\}, \quad f_n(x) = \mathbf{P}\{\xi_n \geq x\}$$

where

$$\xi_n = \sum_{i=1}^{n+1} \frac{\tau_i}{\gamma_0 \, \gamma_1 \, \cdots \, \gamma_{i-1}}, \quad \xi = \sum_{i=1}^{\infty} \frac{\tau_i}{\gamma_0 \, \gamma_1 \, \cdots \, \gamma_{i-1}}, \quad \gamma_0 = 1.$$

Theorem 8.4. *If* $\mathbf{E}|\ln \gamma_1| < \infty$, *then* $f(x) \not\equiv 1$, $x > 0$, *iff*

$$\mathbf{E} \ln \gamma_1 > 0, \quad \mathbf{E} \ln (1 + \tau_1) < \infty.$$

Theorem 8.5. *If*

$$\mathbf{E}\ln\gamma_1=-1,\quad \mathbf{Var}(\ln\gamma_1)=1,\quad \mathbf{E}\ln(1+\tau_1)<\infty,$$

then

$$\lim_{x\to\infty}\mathbf{P}\left\{\frac{\zeta(x)-\ln x}{\sqrt{\ln x}}<y\right\}=\Phi(y)=\frac{1}{\sqrt{2\pi}}\int_{-\infty}^{y}\exp\left\{-\frac{t^2}{2}\right\}dt,\quad y\in\mathbb{R}.$$

Theorem 8.6. *If*

$$\mathbf{E}\ln\gamma_1=0,\quad \mathbf{Var}(\ln\gamma_1)=1,\quad \lim_{x\to\infty}\mathbf{P}\{\tau_1>x\}\ln^2 x=0,$$

then

$$\lim_{x\to\infty}\mathbf{P}\left\{\frac{\zeta(x)}{\ln^2 x}<y\right\}=2\left(1-\Phi\left(\frac{1}{\sqrt{y}}\right)\right)$$

for any $y>0$.

Described semi-Markov process of multiplication can be interpreted in the following way[19]. Consider an enterprise with the input flow $\{t_n\}$ of orders. The size of the order is proportional to the means of an enterprise at the time of the order's arrival. Between orders the means of the enterprise are spent with a constant speed. If the initial capital equals $x>0$, then $\zeta(x)$ is the ruin time of the enterprise and $1-f(x)$ is the probability that the enterprise does not ruin during its functioning.

3. **Complex randomized semi-Markov process**[20–23]. These processes generalize usual semi-Markov processes in the case of certain dependence on the past of what is inherent in real systems. These processes are supposed to be continuous on the right.

A usual semi-Markov process $\xi(t)$ with a state space $\{0,1,2,\ldots\}$ is defined by the initial distribution $p^0=\{p_i^0,\ i=0,1,2,\ldots\}$, $p_i^0=\mathbf{P}\{\xi(0)=$ $=i\}$, and by a semi-Markov matrix of transition probabilities $P(x)=$ $=(P_{ij}(x))_{i,j=0}^{\infty}$, $P_{ij}(x)=\mathbf{P}\{\xi_n=j,\ \zeta_n\le x\,|\,\xi_{n-1}=i\}$ where $\zeta_n=\kappa_n-$ $-\kappa_{n-1}$, κ_n is the nth transition moment of the process $\xi(t)$, and $\{\xi_n=$ $=\xi(\kappa_n)\}$ is a Markov chain embedded in a process $\xi(t)$.

Let $\nu(t)=\max\{n:\ \kappa_n\le t\}$ and $\zeta(t)=\kappa_{\nu(t)}-\kappa_{\nu(t)-1}=\zeta_{\nu(t)}$ be a time between two last transitions of the process $\xi(t)$ in $[0,t]$. Denote:

$$Y(t)=(\xi(t),\zeta(t))=(\xi_{\nu(t)},\zeta_{\nu(t)})=Y_{\nu(t)},$$

$Y_n=Y(\kappa_n)=(\xi_n,\zeta_n)$ is a Markov chain embedded in $Y(t)$; $\eta(t)=$ $=t-\kappa_{\nu(t)}$ is the time from the last transition moment κ_n of the process $\xi(\cdot)$ to the moment t; at the transition moment $\eta(\kappa_n)=0$.

A process $\xi(t) \in \{0,1,2,\ldots\}$ is called **r-complex semi-Markov process** ($r > 0$), if it is defined by $(r+1)$-dimensional matrix of transition probabilities

$$f(x\,|\,y) = \{\, f_{ij}(x\,|\,y),\ \ i = (i_1,\ldots,i_r),\ \ i_m, j \in \{0,1,2,\ldots\},$$
$$m = 1,\ldots,r,\ \ x \geq 0,\ \ y = (y_1,\ldots,y_r)\,\}$$

and by initial distribution $f_i^0 = \{f_i^0(y,\,x)\}$,

$$f_i^0(y,\,x) = \mathbf{P}\{\,Y_{\nu(0)-m} = (i_m,\,y_m),\ m = 0,1,\ldots,r-1;\ \eta(0) < x\,\}$$

of $(2r+1)$-dimensional process

$$X(t) = \{\,Y_{\nu(t)-m},\ \ m = 0,1,\ldots,r-1;\ \ \eta(t)\,\}.$$

This process takes into account the dependence of the current state on r preceding states and on the sojourn times in these states. If $r = 1$, then we have **a one-complex semi-Markov process**. A complex semi-Markov process $\xi(t)$ is called **randomized**, if

$$f_{ij}(x\,|\,y) = \sum_{\alpha\in A} c_i^\alpha(y)\, f_{ij}^\alpha(x),\ \ i = (i_1,\,\ldots,\,i_m),\ y = (y_1,\,\ldots,\,y_m),$$

where α is a parameter of a randomization taking values in a discrete space A, $c_i^\alpha(y) \geq 0$, $\sum_{\alpha\in A} c_i^\alpha(y) = 1$, and $f_i^\alpha(x) = \sum_j f_{ij}^\alpha(x)$ is a distribution function for any i, α, $f_i^\alpha(\infty) = 1$.

For described processes the distribution of a sojourn time in a given subset of states[20,22] and some ergodic properties[21] were studied.

4. **Class of processes close to a semi-Markov process.** It is well known[24] that classes of Markov and semi-Markov processes are not invariant with respect to the operation of enlargement, i.e., if $\nu(t)$ is a Markov process, then an enlarged process $\nu^*(t) = f(\nu(t))$ is not generally speaking a Markov process. Therefore it was expedient to introduce and investigate a class of processes invariant with respect to the operation of enlargement. This is a class of processes close to a semi-Markov process[25].

Let $\nu(t)$, $t \geq 0$, be a random process with a finite or countable set of states X having probability 1 continuous on the right trajectories with only a finite number of jumps in any finite interval. A probability description of $\nu(t)$ is given by multidimensional distributions of random vectors $(\nu_0;\, t_1,\, \nu_1,\, \ldots,\, t_n,\, \nu_n)$, $n \geq 0$, where t_n, $n \geq 1$, are ordered moments $(t_i \leq t_{i+1})$ of process jumps, $t_0 = 0$, $\nu_n = \nu(t_n)$, $n \geq 0$. These finite-dimensional distributions can be defined by the initial distribution

$p_i = \mathbf{P}\{\nu(0) = i\}$, $i \in X$, and by functions $p_{ij}(\nu_t)$, $F_{ij}(x;\, \nu_t)$, $i, j \in X$, $t \geq 0$, depending on the whole trajectory ν_t of the process in $[0,\, t]$, $\nu_t = \{\nu(\tau),\ 0 \leq \tau \leq t\}$. Namely let t be one of the moments, when the process $\nu(t)$ changes its state. Then $p_{ij}(\nu_t)$ is the probability that the next state is equal to j provided that $\nu(t) = i$ and the trajectory ν_t is known; $F_{ij}(x;\, \nu_t)$ is the distribution function of the time $\xi_{ij}(\nu_t)$ from t to the next jump of the process provided that $\nu(t) = i$, the trajectory ν_t is known and j is the next state of the process.

A random process defined by p_i, $p_{ij}(\nu_t)$ and $F_{ij}(x;\, \nu_t)$, $i, j \in X$, $t \geq 0$, is called **close to a semi-Markov process** if there exist:

(i) nonnegative values p_{ij}, $i, j \in X$, do not necessarily equal unit in the sum on j,

(ii) nonnegative functions $g_{ij}(x)$ and $f_{ij}(x)$ $g_{ij}(x) \leq f_{ij}(x)$, $i, j \in X$, $x \geq 0$,

(iii) nonnegative values $d_{ij} < 1$, $\Delta_{ij} < 1$, $c_{ij} < \infty$, $\tau_{ij} < \infty$, $i, j \in X$, such that with the probability 1

$$\left|\frac{p_{ij}(\nu_t)}{p_{ij}} - 1\right| \leq d_{ij}, \quad \left|\frac{\mathbf{E}\,\xi_{ij}(\nu_t)}{\tau_{ij}} - 1\right| \leq \Delta_{ij}, \quad \mathbf{E}\,\xi_{ij}^2(\nu_t) \leq c_{ij},$$

$$g_{ij}(x) \leq F_{ij}(x;\, \nu_t) \leq f_{ij}(x), \quad x > 0, \; i, j \in X.$$

If the original process is close to a semi-Markov one, then an enlarged process is of the same type. Relations[25] were obtained to evaluate characteristics of the enlarged process by corresponding characteristics of the original process.

8.3 Some Generalizations of Regenerative Processes

1. **Strongly regenerative processes**[26]. A separable regular real-valued random process $\xi(t)$, $t \geq 0$, is called **a strongly regenerative process**, if

(i) there exists a state x_0 (without loss of generality $x_0 = 0$) such that

$$\mathbf{P}\{\xi(u + t_j) \leq x_j,\ j = 1, \ldots, n \,|\, \xi(t),\ t < u,\ \xi(u) = 0\} =$$

$$= \mathbf{P}\{\xi(t_j) \leq x_j,\ j = 1, \ldots, n \,|\, \xi(0) = 0\}$$

for any $n \geq 1$, $u \geq 0$, $0 \leq t_1 < t_2 < \cdots < t_n$, $x_j \in \mathbb{R}$, and

(ii) $\mathbf{P}\{\tau \geq x \,|\, \xi(0) = 0\} = e^{-\lambda x}$, $\lambda > 0$, where $\tau = \inf\{t :\ \xi(t) \neq 0\}$.

In other words a strongly regenerative process is a regenerative process with a regeneration period consisting of two independent parts, one of which has an exponential distribution.

Different properties of the process were investigated

$$\tau(z) = \inf\left\{t: \int_0^t \varphi(\xi(u))\,du \geq z\right\}, \quad z \geq 0,$$

where $\xi(t),\ t \geq 0$, is a strongly regenerative process and $\varphi(x),\ x \in \mathbb{R}$, is a measurable bounded function such that $\varphi(0) > 0$ and $\varphi(x) \leq 0$ for $x \neq 0$.

2. **General renewal processes**[27]. Let $\{\xi_n,\ n = 1, 2, \ldots\}$ be a sequence of independent random variables defined on a probability space $(\Omega,\ \mathfrak{F},\ \mathbf{P})$. Let $S_0 = 0$ and $S_n = S_{n-1} + \xi_n,\ n \geq 1$. **A general renewal process** $\{N_\alpha(t),\ t > 0\}$ for $\alpha \in [0,\ 1)$ is defined by the relation

$$N_\alpha(t) = \inf\{n \geq 1 :\ S_n > t\,n^\alpha\}, \quad t > 0.$$

Assume that

$$0 < \mu_n = \mathbf{E}X_n \to \mu \in (0,\ \infty) \text{ as } n \to \infty \tag{8.1}$$

and

$$\sigma_n^2 = \mathbf{E}X_n^2 - \mu_n^2 \to \sigma^2 \in (0,\ \infty) \text{ as } n \to \infty. \tag{8.2}$$

Theorem 8.7. *If conditions (8.1), (8.2) hold and*

(i) $\{\xi_n,\ n = 1, 2, \ldots\}$ *obeys the Strong Law of Large Numbers, i.e.,* $n^{-1} S_n = \mu + o(1)$ *almost sure as* $n \to \infty$, *then*

$$t^{-p/(1-\alpha)}\,(N_\alpha(t))^p \to \mu^{-p/(1-\alpha)} + o(1)$$

almost as surely as $t \to \infty,\ p > 0$.

Theorem 8.8. *If conditions (8.1), (8.2) and (i) in Theorem 8.7 hold, and in addition*

(ii) $\{\xi_n,\ n = 1, 2, \ldots\}$ *satisfies a General Limit Property, i.e.,*

$$\sum_{j=1}^n (\xi_j - \mu_j) \Big/ \left(\sum_{j=1}^n \mathbf{E}\xi_j^2\right)^{1/2} \longrightarrow N(0,\ 1)$$

in law as $n \to \infty$, *then*

$$\frac{\mu\,(1-\alpha)\,(N_\alpha(t) - (t/\mu)^{1/(1-\alpha)})}{\sigma\,(t/\mu)^{1/(2(1-\alpha))}} \longrightarrow N(0,\ 1)$$

in law as $n \to \infty$ where $N(0, 1)$ is a normal random variable with a zero mean and a unit variance.

A m-general renewal process[28] forms a special case of a general renewal process. Let $\{\xi_n, \ n = 1, 2, \ldots\}$ be a sequence of independent nonnegative random variables with distribution functions $F_1(x), \ F_2(x), \ \ldots$ and m be a given nonnegative integer. A process $\{\xi_n, \ n = 1, 2, \ldots\}$ is called **a m-general renewal process**, if $F_{m+1}(x) = F_{m+i}(x), \ F_i(0) = 0$, for any $i = 1, 2, \ldots$. If $m = 0$, then we have a simple renewal process.

A process $\{\xi_n, \ n = 1, 2, \ldots\}$ is called **latticed**, if there exists d being the maximal positive number such that the relation

$$\mathbf{P}\{\xi_n \in \{\pm kd, \ k = 1, 2, \ldots\}\} = 1, \quad n = 1, 2, \ldots,$$

holds. Otherwise $\{\xi_n\}$ is a **continuous** process.

Theorem 8.9. *If $\{\xi_n, \ n = 1, 2, \ldots\}$ is a continuous m-general renewal process and $\mu_{i1} = \mathbf{E}\xi_i < \infty, \ i = 1, \ldots, m+1, \ \mu_{m+12} = \mathbf{E}\xi_{m+1}^2 < \infty$, then*

$$H(t) = \frac{t}{\mu_{m+11}} + \frac{\mu_{m+12}}{2\,\mu_{m+11}^2} - 1 + \sum_{i=1}^{m}\left(1 - \frac{\mu_{i1}}{\mu_{m+11}}\right) + o(1)$$

as $t \to \infty$ where $H(t) = \mathbf{E}\max\{k : \ \xi_1 + \cdots + \xi_k \le t\}$.

A similar theorem can be formulated for a latticed process $\{\xi_n, \ n = 1, 2, \ldots\}$.

3. **Regenerative alternating process**[29]. Let $0 = t_1' < t_1'' < t_2' < t_2'' < \cdots$ be an alternating renewal process, i.e., $\{Y_i' = t_i'' - t_i'\}_{i=1}^{\infty}$ and $\{Y_i'' = t_{i+1}' - t_i''\}_{i=1}^{\infty}$ are independent sequences consisting of independent and identically distributed in every sequence random variables. Let moreover $\{\xi_i'(t), \ 0 \le t < Y_i'\}_{i=1}^{\infty}$ and $\{\xi_i''(t), \ 0 \le t < Y_i''\}_{i=1}^{\infty}$ be independent sequences consisting of independent random processes taking values in the same measurable space $(X, \mathfrak{U})$. It is supposed that processes of the same sequence have the same finite-dimensional distributions.

A process

$$\zeta(t) = \begin{cases} \xi_i'(t - t_i') & \text{if } t_i' \le t < t_i'' \\ \xi_i''(t - t_i'') & \text{if } t_i'' \le t < t_{i+1}', \quad i \ge 1, \end{cases}$$

is called **a regenerative alternating process**.

Processes of such a type arise by studying the failure flow in systems operating in two modes. So, suppose the transition moments from one mode to another are independent of a failure flow and form an alternating process. In view of external effects, the intensities of failure flows in every mode are random processes with characteristics depending on the mode. Then the intensity of the failure flow forms a regenerative alternating process.

For these processes sufficient conditions ensuring the existence of the limit were obtained

$$\lim_{t\to\infty} \mathbf{P}\{\zeta(t) \in B\}, \quad B \in \mathfrak{U},$$

and its value was evaluated[29].

4. **Regenerative processes with several types of regeneration points[30].** Let $\zeta(t),\ t \geq 0$, be a process defined by the relation

$$\zeta(t) = \xi_{\nu(t)}(t - S_{N(t)}), \quad t \geq 0,$$

where $\nu(t)$ is a semi-Markov process with a finite or countable space of states $\{1, 2, \ldots\}$; $N(t) = \sum_i N_i(t)$; $N_i(t)$ is the number of visits of the state i in $[0,\, t)$ by the process $\nu(t)$; $S_{N(t)}$ is the sum of random variables controlled by the embedded Markov chain, i.e., $S_{\nu(t)}$ is the last moment in $[0,\, t)$ when $\nu(t)$ changes its state; $\xi_i(t),\ i = 1, 2, \ldots$, are independent one-dimensional or multidimensional random processes with a space of states X. It is supposed that their probability characteristics depend only on i.

A process $\zeta(t),\ t \geq 0$, is called **a regenerative process with several types of regeneration points** or **a regenerative semi-Markov process.** Its probability characteristics at time t depend only on the state of a semi-Markov component $\nu(t)$ and are independent of the process behavior before the moment $S_{N(t)}$. This process can be defined constructively. So, let $\{(\xi_i(t),\ \tau_i),\ i = 1, 2, \ldots\}$ be a given set of terminating random processes where τ_i is a nonnegative random variable and $\xi_i(t)$ is a random process defined on the half-interval $0 \leq t < \tau_i$. If ν_n denotes the state of a semi-Markov component $\nu(t)$ directly after the nth transition, then $q_{ij}(u) = \mathbf{P}\{\nu_{n+1} = j \,|\, \nu_n = i,\ \tau_i = u\}$ is the transition probability from the state i into the state j provided that the sojourn time of $\nu(t)$ in i equals u. Then a regenerative semi-Markov process is constructed in the following way. We choose an initial state i according to a given distribution $\{\pi_i^{(0)},\ i = 1, 2, \ldots\}$. Then $\zeta(t) = \xi_i(t)$ for $0 \leq t < \tau_i$ where a random time τ_i has a given distribution function $B_i(t) = \mathbf{P}\{\tau_i \leq t\}$. We choose the next state j according to the distribution $\{q_{ij}(\tau_i),\ j = 1, 2, \ldots\}$. We set $\zeta(t) = \xi_j(t - \tau_i),\ \tau_i \leq t < \tau_i + \tau_j$, where a time τ_j has a distribution function $B_j(t)$. Then we repeat a described algorithm and find the next state of the process.

Described process is a generalization of semi-Markov, regenerative, line-wise and alternating regenerative processes. A process $\zeta(t)$ is sometimes called **a semiregenerative process**[31] (with $\nu(t)$ being a semi-Markov process with an arbitrary space of states).

5. **Regenerative multivariate point processes**[32]. Consider a point process. With every point one can associate some event. There are $q+1$ different types of events: $0, 1, \ldots, q$. Let $N_i(A)$, $0 \leq i \leq q$, denote a number of events of the type i that occurred in a Borel set A. A point process $N(\cdot)$ is called a **regenerative multivariate** one, if (i) events of the type 0 form a renewal process, and (ii) for any $u > 0$ a joint distribution of $\{N_0(t, t+u), \ldots, N_q(t, t+u)\}$ depends only on u and is independent of t and the process behavior before t provided that an event of type 0 occurred at moment t.

For this process a system of equations was obtained for generating functions of the number of events in the interval of a fixed length, statements similar to a Key Renewal Theorem and other asymptotic results[32].

8.4 Cumulative Processes

Cumulative processes arise in a natural way by studying regenerative processes[33,34].

Let $\xi_1, \xi_2, \ldots$ be a sequence of independent identically distributed random variables, $S_n = \xi_1 + \cdots + \xi_n$, $n \geq 1$. A process $w(t)$, $t \geq 0$, is called **a cumulative process**, if it satisfies the following conditions:

(C1) $w(S_1) - w(0)$, $w(S_2) - w(S_1)$, $w(S_3) - w(S_2)$, $\ldots$ is a sequence of mutually independent and identically distributed random variables;

(C2) $w(t)$ with the probability 1 has a bounded variation on any finite time interval;

(C3) a variation $\tilde{w}(t) = \int_0^t |dw(u)|$ of the process $w(t)$ satisfies a condition (C1).

A cumulative process of the form

$$w(t) = \begin{cases} \sum_{i=1}^{N(t)} w_i & \text{if } N(t) = 1, 2, \ldots \\ 0 & \text{if } N(t) = 0 \end{cases}$$

is a special case of described process. Here

$$N(t) = \max\{n : \xi_1 + \cdots + \xi_n \leq t\}$$

and $\{w_i\}$ is a sequence of independent and identically distributed random variables such that w_i, $i = 1, 2, \ldots$, are independent of ξ_j for $j \neq i$ (but

w_i and ξ_i can be dependent). If w_i and ξ_i are independent for $i = 1, 2, \ldots$ and $\{\xi_i\}$ have exponential distribution, then $w(t)$ is a generalized Poisson process.

Cumulative processes have wide applications. For example:

(i) if w_i is the cost of replacement by the ith failure of equipment, then $w(t)$ is the total cost of replacements at the time t,

(ii) if some device is subjected to blows occurring according to a renewal process and w_i is a wear caused by the ith blow, then $w(t)$ is a total wear of the device at time t,

(iii) in queueing systems described by regenerative processes the following cumulative processes are of great interest:

- a total busy time in the interval $(0, t)$,
- a total time spent by customers in a queue in $(0, t)$,
- a total time in $(0, t)$ spent by a system in a given set of states.

Without any loss of generality we can suppose $w(0) = 0$. Denote:

$$k_r = \mathbf{E}(w(\xi_1))^r, \quad \tilde{k}_r = \mathbf{E}(\tilde{w}(\xi_1))^r, \quad \mu_r = \mathbf{E}\xi_1^r, \quad r \geq 1,$$

$$\sigma_1^2 = \mu_2 - \mu_1^2, \quad \sigma_2^2 = k_2 - k_1^2, \quad \rho = \sigma_1^{-1}\sigma_2^{-1}\,\mathbf{E}(\xi_1 - \mu_1)(w(\xi_1) - k_1).$$

Some properties of cumulative processes.

1. If $\mu_1 < \infty$ and $\tilde{k}_1 < \infty$, then with the probability 1

$$\frac{w(t)}{t} \underset{t\to\infty}{\longrightarrow} \frac{k_1}{\mu_1} \quad \text{and} \quad \frac{\mathbf{E}\,w(t)}{t} \underset{t\to\infty}{\longrightarrow} \frac{k_1}{\mu_1}.$$

2. If $\mu_2 < \infty$ and $\tilde{k}_2 < \infty$, then

$$\mathbf{Var}(w(t)) \sim \frac{1}{\mu_1}\,t\left\{\sigma_2^2 - 2\rho\,\sigma_1\,\sigma_2\,\frac{k_1}{\mu_1} + \sigma_1^2\left(\frac{k_1}{\mu_1}\right)^2\right\} \quad \text{as } t \to \infty.$$

3. If $\sigma_2^2 < \infty$, $\mu_1 < \infty$ and $k_2 < \infty$, then for any $x \in \mathbb{R}$

$$\mathbf{P}\left\{w(t) - k_1\,N(t) \leq x\,\sigma_2\sqrt{\frac{t}{\mu_1}}\right\} \underset{t\to\infty}{\longrightarrow} \Phi(x) = \frac{1}{\sqrt{2\pi}}\int_{-\infty}^{x} e^{-u^2/2}\,du$$

where $N(t) = \max\{n : S_n \leq t\}$.

4. If $\sigma_2^2 < \infty$, $\mu_2 < \infty$ and $\tilde{k}_2 < \infty$, then for any $x \in \mathbb{R}$

$$\mathbf{P}\left\{w(t) - \frac{k_1\,t}{\mu_1} \leq x\,(\gamma\,t\,\mu_1^{-1})^{1/2}\right\} \underset{t\to\infty}{\longrightarrow} \Phi(x)$$

where $\gamma = \sigma_2^2 - 2\rho\sigma_1\sigma_2 k_1/\mu_1 + \sigma_1^2 k_1^2/\mu_1^2$.

5. A relation $w(t)/N(t) \to k_1$ holds with the probability 1 as $t \to \infty$.

Strongly cumulative processes[34] form a subclass of the above described cumulative processes. These processes are defined by (C1), (C2) and

(C3′) a process

$$V(t) = \begin{cases} w(t) - w(0) & \text{if } 0 \le t < S_1 \\ w(t) - w(S_{N(t)}) & \text{if } t \ge S_1 \end{cases}$$

is a regenerative process. The majority of known cumulative processes are strongly cumulative processes.

Theorem 8.10. *If* $\{w(t),\ t \ge 0\}$ *is a strongly cumulative renewal process such that* $\tilde{k}_1 < \infty$ *and* $F(x) = \mathbf{P}\{\xi_1 < x\}$ *is a nonlatticed distribution function, then for any* $h > 0$

$$\lim_{t\to\infty} \mathbf{E}(w(t+h) - w(t)) = \frac{h k_1}{\mu_1}.$$

An example of a Central Limit Theorem's application for cumulative processes. Suppose we have some particle with the initial energy t. In collisions with other particles it loses a part of its energy and generates new secondary particles. If after some collision the particle's energy sinks lower than a given level, then this particle can't generate any other particles in the future. It is assumed that a joint distribution of the number of arising secondary particles and the energy loss of the original particle in one collision is independent of the original particle's energy at this moment (this condition is not satisfied only when the energy of the original particle reaches some comparatively low level). It is necessary to find a distribution of the total number of secondary particles generated in all collisions. Mark on the energy axis successive values of the original particle's energy directly after collisions. Under some weak hypotheses (in particular supposing a great value of the initial energy t) we can consider the intervals between these "regeneration moments" as mutually independent and identically distributed random variables. A total number of particles generated by the original particle with the initial energy t forms a cumulative process $w(t)$. It follows from limit theorems that a distribution of $w(t)$ is close to a normal distribution for large values of t.

8.5 Theory of Counters

A theory of counters of the first and second kinds and corresponding theory of Geiger-Müller counters[35–38] form one of most fruitful fields of applications of a renewal theory.

Let $\{\xi_n,\ n \geq 1\}$ and $\{\tau_n,\ n \geq 1\}$ be independent renewal processes, $S_n = \xi_1 + \cdots + \xi_n$, $n \geq 1$, $S_0 = 0$. A renewal process $\{\xi_n^*,\ n \geq 1\}$ is said to be obtained of $\{\xi_n,\ n \geq 1\}$ by means of **a counter with a dead time of nonprolonged type (a counter of the first kind)**, if

$$\xi_k^* = S_{n_k} - S_{n_{k-1}}, \quad S_k^* = S_{n_k}, \quad k \geq 1, \quad S_{n_0} = 0,$$

where

$$n_1 = \min\{n \geq 1 : \ S_n > \tau_1\}, \quad n_k = \min\{n : \ S_n > S_{n_{k-1}} + \tau_k\}, \quad k \geq 2.$$

Thus a dead interval with a random length τ_1 starts at time $t = 0$. A counter can't record renewal points $\{\xi_n\}$ at its dead time. But it records the first renewal point ξ_{n_1} following the dead interval. A new dead interval of the length τ_2 starts at time ξ_{n_1}.

A renewal process $\{\xi_n^*,\ n \geq 0\}$ is said to be obtained of $\{\xi_n,\ n \geq 1\}$ by means of **a counter with a dead time of prolonged type (a counter of the second kind)**, if it is defined in the following way. If $S_n > S_m + \tau_{m+1}$ for all $0 \leq m < n$, $S_0 = 0$, then we record a renewal moment S_n. In this case $\{\xi_n^*\}$ are the times between successive recorded moments $\{S_n^*\}$, $S_0^* = 0$. In other words, with every renewal point S_n one can associate a dead interval of the random length τ_{n+1} starting at this point. We record only those renewal points that don't belong to any of these intersecting intervals.

Consider some generalizations of counters of the first and second kind.

1. Assume an input flow of particles forms a semi-Markov process $\{\xi(t),\ t \geq 0\}$. Then a flow of recorded particles forms also a semi-Markov process $\{\eta(t),\ t \geq 0\}$, where $\eta(t) = j$, if $\xi(t_i) = j$ at the last recording moment t_i.

2. Assume some flow of particles enters the counter of the first or second kind. Denote $S_0^* = 0$, S_1^*, $S_2^*, \ldots$ the successive recording moments of entering particles. At each such moment S_i^*, $i \geq 0$, there arises an impulse of a random length τ_i. It is supposed that τ_0 depends only on the type of the recorded particle and τ_i $(i \geq 1)$ depends only on types of particles recorded at moments S_j^*, $j = 0, 1, \ldots, i$, and on lengths $S_{j+1}^* - S_j^*$, τ_j, $j = 0, 1, \ldots, i-1$.

3. There exist different methods to define lengths of impulses generated by unrecorded particles in counters of a second kind. For example[39,40]:

(i) a distribution function of the impulse's length depends only on whether the particle is recorded or not,

(ii) lengths of impulses have identical distribution functions, but every impulse arises only with the probability p_r, $r \geq 0$, $p_0 = 0$, where r is the number of impulses present at the time of particle's arrival.

8.6 Cascade Processes

Cascade processes arise by studying particle partition[41–45]. Assume that at the initial moment $t = 0$ there is one particle with the energy X_0. Let T be its random lifetime. At this moment a particle divides into N other particles with the energies $X_1, \ldots, X_N$. Their behavior is similar to the original particle and is independent of other particles and of the previous history of the process. Random variables T, N, $X_1, \ldots, X_N$ are defined by distributions

$$G(t) = \mathbf{P}\{T \leq t\}, \quad q_j = \mathbf{P}\{N = j\}, \quad j = 1, 2, \ldots,$$

$$\Phi_j(x_1, \ldots, x_j \,|\, X_0) = \mathbf{P}\{X_1 \leq x_1, \ldots, X_j \leq x_j \,|\, X_0, \ N = j\}.$$

Let $N(x, t)$ be a number of particles at time t with the energy not less than x. A process $\{N(x, t),\ t \geq 0\}$ is called **a cascade process** or **a branching process with the energy**. Let

$$p_n(x, t \,|\, X_0) = \mathbf{P}\{N(x, t) = n \,|\, A(X_0)\},$$

$$p_n(x, t) = \mathbf{P}\{N(x, t) = n \,|\, A(1)\}.$$

An event $A(X_0)$ means that there was only one particle with the energy X_0 at the moment $t = 0$.

Since $N(0, t)$ is the number of particles at time t, the process $\{N(0, t),$ $t \geq 0\}$ is a branching process. If $q_2 = 1$, then we have **a binary cascade process**. Later on we assume $\mathbf{P}\{X_1 + \cdots + X_N \leq X_0\} = 1$, i.e., a new energy can't arise by partition of the particle.

A function $\Phi_j(x_1, \ldots, x_j)$ is called **homogeneous**, if for any $k > 0$

$$\Phi_j(kx_1, \ldots, kx_j \,|\, k X_0) = \Phi_j(x_1, \ldots, x_j \,|\, X_0).$$

If a binary process has a homogeneous function $\Phi_2(x_1, x_2 \,|\, X_0)$ being absolutely continuous with respect to its arguments and $G(t) = 1 - e^{-\lambda t}$, $t \geq 0$, $\lambda > 0$, then a process is called **a binary nucleon cascade process**.

For $p_n(x, t \,|\, X_0)$ an integral equation was obtained having a unique bounded solution being a probability function[42]. Moreover under some hypotheses

$$\frac{p_n(x,t)}{p_m(x,t)} \underset{t\to\infty}{\longrightarrow} 0$$

where $n > m,\ \ m \le [x^{-1}]$ and

$$\frac{p_n(x'',t)}{p_n(x',t)} \underset{t\to\infty}{\longrightarrow} 0$$

where $0 \le x' < x'' \le 1,\ \ 0 < n \le [(x')^{-1}]$ (here $[\cdot]$ denotes an entire part).

Theorem 8.11. *If for a binary cascade process* $X_0 = 1,\ \Phi(x_1, x_2) = \Phi_2(x_1, x_2 \mid 1)$ *is a symmetrical function and* $G(t) = 1 - e^{-\lambda t},\ t \ge 0,\ \lambda > 0,$ *then a ratio* $N(x_t, t)/N(0, t)$ *converges in probability as* $t \to \infty$ *to*

$$\Phi(z) = \frac{1}{\sqrt{2\pi}} \int_{-\infty}^{z} e^{-u^2/2}\, du$$

where

$$x_t = \exp\{-2\lambda\mu t - z\,(2\lambda t\,(\mu^2 + \sigma^2))^{1/2}\},$$
$$\mu = \mathbf{E}\,(-\ln X_1), \quad \sigma^2 = \mathbf{Var}\,(-\ln X_1).$$

A similar statement holds for the number of particles with given sizes obtained by the subdivisions[41].

8.7 Extremal Processes

Let $\xi_1, \xi_2, \dots$ be a sequence of independent identically distributed random variables, $M_n = \max(\xi_1, \dots, \xi_n),\ n \ge 1$. Under some well known necessary and sufficient conditions[46] there exist sequences $\{a_n\}$ and $\{b_n\}$ such that for any $t \in \mathbb{R}$ there exists a limit

$$\lim_{n\to\infty} \mathbf{P}\,\{b_n^{-1}\,(M_n - a_n) < t\}$$

equal to a nonsingular distribution function $F(x)$ having one of the following forms:

$$(1)\ \ F_1(t,\lambda) = \begin{cases} \exp\{-\lambda t^{-\alpha}\} & \text{if } t > 0 \\ 0 & \text{if } t \le 0, \end{cases}$$

$$(2)\ \ F_2(t,\lambda) = \begin{cases} \exp\{-\lambda(-t)^{\alpha}\} & \text{if } t \le 0 \\ 0 & \text{if } t > 0, \end{cases}$$

$$(3)\ \ F_3(t,\lambda) = \exp\{-\lambda\exp(-t)\}, \quad -\infty < t < +\infty.$$

There were introduced and studied[47,48] processes $\{\eta(t),\ t \ge 0\}$ being the "limits" as $n \to \infty$ of the processes

$$\eta_n(t) = (M_{[tn]+1} - a_n)/b_n, \quad 0 \le t < \infty.$$

There exist three limit processes corresponding to different $F_i(t, \lambda)$, $i = 1, 2, 3$. Extremal processes $\eta(t)$ are defined similarly to stable processes $\zeta(t)$ being the limits of the processes $\zeta_n(t) = (S_{[tn]} - c_n)/d_n$, $t \ge 0$, where $S_n = \zeta_1 + \cdots + \zeta_n$, $n = 1, 2, \ldots$. Here $\{\zeta_i\}$ is the sequence of independent identically distributed random variables and $\{c_n\}$, $\{d_n\}$ are sequences of constants such that $\mathbf{P}\{(S_n - c_n)/d_n < x\}$ converges to a nonsingular distribution function of a stable type.

A process $\eta_i(t)$, $t \ge 0$, $i = 1, 2, 3$, is called **an extremal process of the type** i, if for any $k \ge 1$, $0 \le t_1 < t_2 < \cdots < t_k$ the joint distribution of $\eta_i(t_1), \ldots, \eta_i(t_k)$ is the same as the joint distribution of

$$\eta_1^{(i)},\ \max(\eta_1^{(i)},\ \eta_2^{(i)}),\ \ldots,\ \max(\eta_1^{(i)},\ \eta_2^{(i)}, \ldots, \eta_k^{(i)})$$

where $\eta_1^{(i)}, \ldots, \eta_k^{(i)}$ are independent random variables with distribution functions $F_i(t, \lambda t_1)$, $F_i(t, \lambda(t_2 - t_1))$, $\ldots$, $F_i(t, \lambda(t_k - t_{k-1}))$. Denote

$$\eta_{n1}(t) = M_{[tn]+1}/n^{1/\alpha}, \quad \eta_{n2}(t) = M_{[tn]+1}\, n^{1/\alpha}, \quad \eta_{n3}(t) = M_{[tn]+1} - \ln n.$$

Theorem 8.12. *Processes $\eta_{ni}(t)$, $t \ge 0$, $i = 1, 2, 3$, converge as $n \to \infty$ to extremal processes $\eta_i(t)$, $t \ge 0$, $i = 1, 2, 3$, in the sense of convergence of finite-dimensional distributions.*

Theorem 8.13. *If η_s, $\zeta_1, \zeta_2, \ldots$, $\omega_1, \omega_2, \ldots$ are independent sequences of independent random variables (η_s has a distribution function $F_1(x, \lambda s)$, $\{\zeta_i\}$ are exponentially distributed with parameter 1 random variables, $\{\omega_i\}$ are uniformly distributed in [0, 1] random variables), then a process $\eta(t)$, $t \ge s$, is stochastically equivalent (in a broad sense) to a process $\tilde{\eta}_1(t)$:*

$$\tilde{\eta}_1(t) = \begin{cases} \eta_s & \text{if } t \in [s,\ s + \eta_s^\alpha \zeta_1) \equiv [s,\ s_1) \\ \eta_s/\omega_1^{1/\alpha} & \text{if } t \in [s_1,\ s_1 + \eta_s^\alpha \zeta_2/\omega_1) \equiv [s_1,\ s_2) \\ \eta_s/(\omega_1\omega_2)^{1/\alpha} & \text{if } t \in [s_2,\ s_2 + \eta_s^\alpha \zeta_3/(\omega_1\omega_2)) \equiv [s_2,\ s_3) \\ \cdots & \cdots\cdots \end{cases}$$

Analogous theorems hold for extremal processes of the types 2 and 3. A more general definition of the extremal process[47,48] can be introduced. A process $\eta(t)$, $t \ge 0$, is called **extremal**, if for any $k \ge 1$, $0 \le t_1 <$

$< t_2 < \cdots < t_k$ a joint distribution of $\eta(t_1), \ldots, \eta(t_k)$ is the same as a joint distribution of $\eta_1, \max(\eta_1, \eta_2), \ldots, \max(\eta_1, \eta_2, \ldots, \eta_k)$ where $\eta_1, \ldots, \eta_k$ are independent random variables and

$$\mathbf{P}\{\eta_i < x\} = \mathbf{P}\{\eta(t_i - t_{i-1}) < u\}, \quad i = 1, \ldots, k, \ t_0 = 0.$$

In other words, there exists a distribution function $F(x)$ such that for any $n \geq 1, \ x_1, \ldots, x_n \geq 0, \ 0 \leq t_1 < t_2 < \cdots < t_n < \infty$

$$\mathbf{P}\{\eta(t_1) < x_1, \ldots, \eta(t_n) < x_n\} = [F(\min\{x_1, \ldots, x_n\})]^{t_1} \times$$
$$\times \ [F(\min\{x_2, \ldots, x_n\})]^{t_2 - t_1} \ \ldots \ [F(x_n)]^{t_n - t_{n-1}}. \tag{8.3}$$

A process having such finite-dimensional distributions is called ***F*-extremal** or ***Q*-extremal** where $Q(x) = -\ln F(x)$. It follows from (8.3) that

$$\mathbf{P}\{\eta(t) < x\} = \exp\{-t\,Q(x)\}.$$

Extremal processes of types 1, 2, 3 are the special cases of Q-extremal processes with

$$\begin{aligned} Q_1(x) &= \lambda x^{-\alpha}, \quad x > 0, \ \alpha, \lambda > 0, \\ Q_2(x) &= \lambda(-x)^{\alpha}, \quad x < 0, \quad \alpha, \lambda > 0, \\ Q_3(x) &= -\lambda e^{-x}, \quad -\infty < x < +\infty, \ \lambda > 0. \end{aligned}$$

Some properties of extremal processes[47–54].

1. Let $\{\xi_n, \ n \geq 1\}$ be a sequence of independent exponentially distributed random variables with a parameter 1, $\eta(t), \ t \geq s$, be an extremal process and

$$R_y(u) = \begin{cases} 0 & \text{if } u \leq y \\ 1 - \frac{Q(u)}{Q(y)} & \text{if } u > y. \end{cases}$$

Let moreover $\tilde{\eta}(t), \ t \geq s > 0$, satisfy conditions: (i) $\tilde{\eta}(s)$ has the same distribution as $\eta(s)$, and (ii) if $\tilde{\eta}(s) = x$, then $\tilde{\eta}(t) = x$ during a time $\xi_1/Q(x)$ and then a process jumps on a new level $\gamma_1 > x$ (γ_1 is a random variable with a distribution function $R_x(u)$) where its sojourn time equals $\xi_2/Q(\gamma_1)$. Then a process jumps on a level $\gamma_2 > \gamma_1$ with a distribution function $R_{\gamma_1}(u)$ and so on. Then $\eta(t), \ t \geq s$, and $\tilde{\eta}(t), \ t \geq s$, are equivalent in a broad sense Markov processes.

2. Let $\eta(t)$ be an extremal process, $Q(x)$ be a continuous function and

$$c(t_1, t_2) = \begin{cases} 0 & \text{if } \eta(t_2) > \eta(t_1) \\ 1 & \text{otherwise}, \qquad t_1 < t_2. \end{cases}$$

Then:

(i) if $0 < t_1 < t_2 < \cdots < t_k$, $k \geq 2$, then $c(t_1, t_2), \ldots, c(t_{k-1}, t_k)$, are independent random variables,

(ii) for any $0 < s < u$ a process $\eta(t)$ has with probability 1 a finite number of jumps in (s, u) being a Poisson distributed random variable with a parameter $\ln(u/s)$,

(iii) let $0 < s < u$, $x < v$ and $J(x, v)$ be a number of jumps of $\eta(t)$ in the interval from s to the first moment z when $\eta(z) > v$ provided that $\eta(s) = x$. Then $J(x, v)$ has a Poisson distribution with a parameter $\ln(Q(x)/Q(v))$.

8.8 Piecewise Linear Markov Processes

Piecewise linear Markov processes were introduced in order to describe a wide class of queueing and redundant systems[55,56].

A piecewise linear process has a discrete component $\nu(t)$ defining a "qualitative" system's state and continuous components $\xi_1(t), \xi_2(t), \ldots$ characterizing the operations going on in the system at time t. A discrete component $\nu(t)$ takes values in a finite or countable set N. With every $\nu \in N$ one can associate a nonnegative integer number $|\nu|$ named a "rank" of the state ν. This rank equals the number of operations going on by the system's state ν.

A piecewise linear process is a homogeneous continuous on the right Markov process of the form

$$\zeta(t) = \{\nu(t);\ \xi_1(t), \ldots, \xi_{|\nu(t)|}(t)\}, \quad t \geq 0. \tag{8.4}$$

Here $\xi_j(t)$ denotes a residual work which must be done to complete the jth operation from $|\nu(t)|$ operations happening at time t. Evolution of the process $\zeta(t)$ is described in the following way.

1. There is given an initial distribution

$$\mathbf{P}\{\nu(0) = \nu;\ \xi_1(0) < x_1, \ldots, \xi_{|\nu|}(0) < x_{|\nu|}\}.$$

2. With every ν such that $|\nu| > 0$ one can associate a vector $\alpha_\nu = (\alpha_{\nu 1}, \ldots, \alpha_{\nu|\nu|})$ with nonnegative components where $\alpha_{\nu j}$ denotes the speed of the jth operation's fulfillment by the system's state ν, i.e., a variable $\xi_j(t)$ decreases with the speed $\alpha_{\nu j}$:

$$\frac{d\xi_j(t)}{dt} = -\alpha_{\nu j}.$$

3. If $\xi_j(t)$ vanishes (i.e., the jth operation was terminated at the moment t) and the system's state is equal to ν, then $\nu(t + 0) = \mu$ with

the probability $p_{\nu\mu}^{(j)}$ where $\{p_{\nu\mu}^{(j)}\}$ is the probability distribution on N. A vanished component is deleted from a vector (8.4), i.e., $|\mu| = |\nu| - 1$.

4. Along with the termination of operations spontaneous jumps of the process $\zeta(t)$ can occur; if $\nu(t) = \nu$, then in the interval $(t, t+h)$ the system can pass to the state μ, $|\mu| > |\nu|$, with the probability $\lambda_{\nu\mu} h + o(h)$ (it is assumed that $\sum_{\mu \in N} \lambda_{\nu\mu} < \infty$ for any $\nu \in N$). In this case to the variables $\xi_1(t), \ldots, \xi_{|\nu|}(t)$ we add from the right additional variables $\eta_1, \ldots, \eta_{|\mu|-|\nu|}$ being independent of the process behavior before the moment t. These new variables have a joint distribution function

$$H(x_1, \ldots, x_{|\mu|-|\nu|}) = \mathbf{P}\{\eta_1 < x_1, \ldots, \eta_{|\mu|-|\nu|} < x_{|\mu|-|\nu|}\}.$$

It is assumed that $\mathbf{P}\{\alpha\eta_i + \beta\eta_j = x\} = 0$ for any x, α, β, i, j, i.e., two or more additional variables can't vanish simultaneously. In particular this condition is satisfied when $\eta_1, \eta_2 \ldots$ have a multidimensional density.

All introduced values have a simple interpretation. Consider an example taken in the reliability theory.

Example 8.1. Let a system consist of three components that can fail. There are two devices for their repair. The size of the work for the repair of a failed component is a random variable with a distribution function $H(x)$. Failure-free operation times of components have a distribution function $1 - e^{-\lambda x}$, $\lambda > 0$. A repair speed of the failed component is proportional to the total number of failed components. Let $\nu(t)$ denote the number of failed components at time t. If $\nu(t) = \nu$, then

$$\alpha_{\nu i} = c\nu \quad \text{if} \quad i \le \nu, \ \nu = 1, 2;$$

$$\alpha_{\nu i} = \begin{cases} 3c & \text{if } i = 1, 2 \\ 0 & \text{if } i = 3 \end{cases} \qquad \text{for } \nu = 3;$$

$$\lambda_{\nu\mu} = \begin{cases} \lambda(3-\nu) & \text{if } \mu = \nu + 1 \\ 0 & \text{if } \mu \ne \nu + 1. \end{cases}$$

The ergodicity of a random process is one of the most important properties. The following theorem gives simple sufficient conditions ensuring the ergodicity of a piecewise linear Markov process.

Theorem 8.14. *If N is a finite set of states, all η_j have finite mean values and $\sum_{j=1}^{|\nu|} \lambda_{\nu j} > 0$ for any ν with $|\nu| > 0$, then there exists an ergodic distribution of the process $\zeta(t)$:*

$$F(\nu; x) = \lim_{t\to\infty} \mathbf{P}\{\nu(t) = \nu; \ \xi_j(t) < x_j, \ 1 \le j \le |\nu|\},$$

$$\sum_{\nu \in N} \lim_{x_1, \ldots, x_{|\nu|} \to \infty} F(\nu;\, x) = 1.$$

Enumerate some possible generalizations of a described piecewise linear Markov process[57,58].

1. Suppose a jth operation terminates at the moment t, i.e.,

$$\zeta(t-0) = (\nu;\ \xi_1,\ \ldots,\ \xi_{j-1},\ 0,\ \xi_{j+1},\ \ldots,\ \xi_{|\nu|})$$

and $\xi_k > 0,\ k \neq j$. Then $\nu(t) = \mu$ with the probability $p_{\nu\mu}^{(j)}$ where $|\mu| \geq |\nu| - 1$. If $|\mu| \geq |\nu|$, then we construct a random vector $\eta_{\nu\mu}^{(j)} = (\eta_1,\ \ldots,\ \eta_{|\mu|-|\nu|+1})$ with positive components and a multidimensional distribution function $H_{\nu\mu}^{(j)}(x_1,\ \ldots,\ x_{|\mu|-|\nu|+1})$. In this case

$$\zeta(t) = (\mu;\ \xi_1,\ \ldots,\ \xi_{j-1},\ \xi_{j+1},\ \ldots,\ \xi_{|\nu|},\ \eta_1,\ \eta_2,\ \ldots,\ \eta_{|\mu|-|\nu|+1}) = $$

$$= (\mu;\ \xi_1,\ \ldots,\ \xi_{|\mu|}),$$

which means that by the termination of jth operation there appears a demand for $m = |\mu| - |\nu| + 1$ additional operations with corresponding works $\eta_1, \ldots, \eta_m$.

2. If

$$\nu(t-0) = \nu,\ \ \xi_{j_1}(t-0) = \cdots = \xi_{j_s}(t-0) = 0,$$
$$(1 \leq s \leq |\nu|,\ \ \{j_1,\ \ldots,\ j_s\} \subset \{1,\ \ldots,\ |\nu|\}),$$

then with the probability $p(\nu,\, j_1,\, \ldots,\, j_s;\ \mu)$ the system goes into the state μ with $|\mu| \leq |\nu| - s$. According to a given distribution we set $l = |\nu| - s - |\mu|$ variables to be equal to zero and delete them from the representation (8.4).

3. If $\nu(t) = \nu$, then in the interval $(t,\, t+h)$ the system can pass to the state μ with the probability $\lambda_{\nu\mu}\, h + o(h)$. Then according to given distributions we add s new variables ($s \geq 0$) and set l existing variables ($l \geq 0$) to be equal to zero and delete them from the representation, i.e., $|\mu| = |\nu| + s - l$. New additional variables are positive random variables $\eta_{\nu\mu 1},\ \ldots,\ \eta_{\nu\mu s}$ with a multidimensional distribution function $H_{\nu\mu}(x_1,\ \ldots,\ x_s)$.

8.9 References

1. Onicescu, O., Mihoc, G., Sur les chaines de variables statistiques, *Bull. Sci. Math.*, 59, 174, 1935.
2. Ciucu, G., Theodorescu, R., *Procese cu Legături Complete,* Acad. RPR, Bucuresti, 1960.
3. Cohn, H., Le probleme limite central pour les systemes aléatoires à liaisons complétes, *C. r. Acad. Sci.*, 259, 3423, 1964.

4. Iosifescu, M., Theodorescu, R., On properties of chains with complete connections, *Ukr. Mat. Shurn.*, 16, 93, 1964 (in Russian).
5. Iosifescu, M., Processus aléatoires à liaisons complétes purement discontinus, *C. r. Acad. Sci.*, 266, 1159, 1968.
6. Oprişan, G., Popescu, G., The renewal equation for renewal processes with complete connections, *Stud. şi Cerc. Mat.*, 41, 501, 1989.
7. Sragovich, V. G., *Theory of Adaptive Systems*, Nauka, Moscow, 1976 (in Russian).
8. Iosifescu, M., Random systems with complete connections and with an arbitrary set of states, *Rev. Math. Pures et Appl.*, 8, 611, 1963 (in Russian).
9. Botez, M., Theodorescu, R., Sur la convergence des series de variables à leatoires formant in I_φ processus, *C. r. Acad. Sci.*, 260, 4404, 1965.
10. Pyke, R., Markov renewal process: definitions and preliminary properties, *Ann. Math. Statist.*, 32, 1231, 1961.
11. Hatori, H., A limit theorem on (J, X)-processes, *Kodai Math. Semin. Repts*, 18, 317, 1966.
12. Hatori, H., Mori, T., An improvement of a limit theorem on (J, X)-processes, *Kodai Math. Semin. Repts*, 18, 347, 1966.
13. Hatori, H., Mori, T., Oodaira, H., A renewal theorem on (J, X)-processes, *Kodai Math. Semin. Repts*, 19, 159, 1967.
14. Hatori, H., Mori, T., Oodaira, H., A remark concerning a renewal theorem on (J, X)-processes, *Kodai Math. Semin. Repts*, 19, 189, 1967.
15. Lev, G. S., Semi-Markov process of multiplication with a drift, *Teor. Veroyatn. Primen.*, 17, 160, 1972 (in Russian).
16. Lev, G. S., Asymptotic properties of probability of degeneration after time t for semi-Markov processes of multiplication, *Teor. Veroyatn. Primen.*, 20, 162, 1975 (in Russian).
17. Lev, G. S., Distribution of the absorption moment of a semi-Markov process of multiplication, *Teor. Probab. Appl.*, 24, 844, 1979.
18. Lev, G. S., Generalized process of multiplication, *Teor. Probab. Appl.*, 32, 686, 1987.
19. Gaver, D. P., An absorption probability problem, *J. Math. Anal. Appl.*, 9, 384, 1964.
20. Ezhov, I. I., Zakharin, A. M., A sojourn time of a complex randomized semi-Markov process in a given subset of states, *Dopovidi Akad. Nauk Ukr.SSR*, A, 345, 1978 (in Russian).
21. Zakharin, A. M., Ergodic theorem on one-complex randomized semi-Markov processes, *Ukr. Mat. Shurn.*, 30, 809, 1978 (in Russian).
22. Zakharin, A. M., Complex randomized semi-Markov processes, *Cybernetics*, 14, 872, 1978.

23. Zakharin, A. M., Some Markov models of systems with a partial aftereffect, *Cybernetics,* 16, 205, 1980.
24. Korolyuk, V. S., Turbin, A. F., *Semi-Markov Processes and Their Applications,* Naukova Dumka, Kiev, 1976 (in Russian).
25. Kovalenko, I. N., *Studies in Analysis of Reliability of Complex Systems,* Naukova Dumka, Kiev, 1975 (in Russian).
26. Borozdin, O. P., Ezhov, I. I., Strongly regenerative random processes, *Dopovidi Akad. Nauk Ukr.SSR,* A, 771, 1976 (in Russian).
27. Niculescu, S. P., On general renewal processes, *Stud. şi Cerc. Mat.,* 40, 507, 1988.
28. Niculescu, S. P., An asymptotic renewal theorem for m-general renewal processes, *Bull. Math. Soc. Sci. Math. Rep. Soc. Roum.,* 22, 55, 1978.
29. Yastrebenetsky, M. A., A class of regenerative random processes, *Engin. Cybernetics,* 7, 45, 1969.
30. Rykov, V. V., Yastrebenetsky, M. A., On regenerative processes with several types of regeneration moments, *Kibernetika,* No. 3, 82, 1971 (in Russian).
31. McDonald, D., On semi-Markov and semiregenerative processes, II, *Ann. Probab.,* 6, 995, 1978.
32. Berman, M., Regenerative multivariate point processes, *Adv. Appl. Probab.,* 10, 411, 1978.
33. Smith, W. L., Regenerative stochastic processes, *Proc. Roy. Soc.,* A, 232, 6, 1955.
34. Brown, M., Ross, S. M., Asymptotic properties of cumulative processes, *SIAM J. Appl. Math.,* 22, 93, 1972.
35. Feller, W., On probability problems in the theory of counters, *Courant Anniversary Vol.,* 105, 1948.
36. Goldansky, V. I., Kuzenko, A. V., Podgoretsky, M. I., *Statistics in Recording of Nuclear Particles,* Fizmatgiz, Moscow, 1959 (in Russian).
37. Albert, G. E., Nelson, L., Contribution to the statistical theory of counter data, *Ann. Math. Statist.,* 24, 9, 1953.
38. Smith, W. L., On renewal theory, counter problems and quasi-Poisson processes, *Proc. Cambr. Phil. Soc.,* 53, 175, 1957.
39. Ramakrishnan, A., Mathews, P. M., A stochastic problem relating to counters, *Phil. Mag.,* 44, 1122, 1953.
40. Takács, L., On a coincidence problem concerning particle counters, *Ann. Math. Statist.,* 32, 739, 1961.
41. Kolmogorov, A. N., On a logarithmically normal distribution of particles sizes by a subdivision, *Doklady Akad. Nauk SSSR,* 21, 99, 1941 (in Russian).

42. Ney, P. E., Ratio limit theorems for cascade processes, *Z. Wahrscheinlichkeitstheor. und verw. Geb.*, 3, 32, 1964.
43. Ney, P. E., The limit distribution of a binary cascade process, *J. Math. Anal. Appl.*, 10, 30, 1965.
44. Harris, T. E., *The Theory of Branching Processes,* Springer, Berlin, 1963.
45. Sevast'janov, B. A., *Verzweigungsprozesse,* Oldenburg, München, 1975.
46. Gnedenko, B. V., Sur la distribution limite du terme maximum d'une série aléatoire, *Ann. Math.*, 44, 423, 1943.
47. Dwass, M., Extremal processes, I, *Ann. Math. Statist.*, 35, 1718, 1964.
48. Dwass, M., Extremal processes, II, *Illinois Math.*, 10, 381, 1966.
49. Lamperti, J., On extreme order statistics, *Ann. Math. Statist.*, 35, 1726, 1964.
50. Resnick, S. I., Extremal processes and record value times, *J. Appl. Probab.*, 10, 864, 1973.
51. Resnick, S. I., Inverses of extremal processes, *Adv. Appl. Probab.*, 6, 392, 1974.
52. Ballerini, R., Resnick, S. I., Embedding sequences of successive maxima in extremal processes, with applications, *J. Appl. Probab.*, 24, 827, 1987.
53. Resnick, S. I., *Extreme Values, Regular Variation and Point Processes,* Springer, Berlin, 1987.
54. Dagsvik, J. K., Markov chains generated by maximizing components of multidimensional extremal processes, *Stochast. Processes and Appl.*, 28, 31, 1988.
55. Buslenko, N. P., Kalashnikov, V. V., Kovalenko, I. N., *Lectures in Theory of Complex Systems,* Sov. Radio, Moscow, 1973 (in Russian).
56. Gnedenko, B. V., Kovalenko, I. N., *Introduction to Queueing Theory,* Birkhäuser, Boston, 1989.
57. Gillert, H., Ein Beitrag zur Theorie der stückweise linearen Markovschen Prozesse, *Math. Nachr.*, 70, 99, 1975.
58. Kovalenko, I. N., On the boundedness of a mean sojourn time of a customer in a queueing system with a bounded queue, *Problemy Ustoichivosti Stokhasticheskikh Modelei, Trudy Seminara,* VNIISI, Moscow, 66, 1981 (in Russian).

Chapter 9

Some Special Classes of Processes

9.1 Stable Processes

Stable distributions stand out as natural generalizations of a normal distribution. Stable processes play the same role with respect to a process of Brownian motion.

Let ξ and η be one-dimensional random variables defined on some probability space $(\Omega,\ \mathfrak{F},\ \mathbf{P})$. Later on a notation $\xi \stackrel{d}{=} \eta$ means that distributions of ξ and η coincide.

Let $\xi,\ \xi_1,\ \xi_2,\ \dots$ be mutually independent random variables having the same distribution function $F(x)$ and defined on the probability space $(\Omega,\ \mathfrak{F},\ \mathbf{P})$, $S_n = \xi_1 + \cdots + \xi_n$, $n \geq 1$. A distribution function $F(x)$ is called **stable in a broad sense** if it is not concentrated at zero and for any $n \geq 1$ there exist constants $c_n > 0$ and γ_n such that

$$S_n \stackrel{d}{=} c_n \xi + \gamma_n. \tag{9.1}$$

A distribution $F(x)$ is called **stable in a strict sense** if it satisfies (9.1) with $\gamma_n = 0$.

Properties of stable distributions[1]:

(i) values $c_n = n^{1/\alpha}$ with $0 < \alpha \leq 2$ are the only possible normalizing constants where α is called **a characteristic index** of the distribution $F(x)$,

(ii) all stable in a broad sense distributions are continuous functions,

(iii) if the distribution function $F(x)$ is stable in a strict sense with a characteristic index α, then for any $s, t \geq 0$ the relation

$$s^{1/\alpha}\xi_1 + t^{1/\alpha}\xi_2 \stackrel{d}{=} (s+t)^{1/\alpha}\xi$$

holds,

(iv) for any distribution stable in a broad sense with a characteristic index $\alpha \neq 1$ there exists a constant b such that a distribution $F(x+b)$ is stable in a strict sense,

(v) if $\alpha = 1$, then corresponding centering constants in (9.1) have the form $\gamma_n = \gamma n \ln n$.

Later on we use the following equivalent definition of a stable distribution. A distribution $F(x)$ is called **stable** if a characteristic function

$$f(z) = \mathbf{E}\, e^{iz\xi}, \quad z \in \mathbb{R},$$

satisfies the relation

$$\ln f(z) = \begin{cases} i\gamma - \lambda |z|^\alpha \left\{1 - i\beta \operatorname{sign}(z) \operatorname{tg}\left(\frac{\pi\alpha}{2}\right)\right\} & \text{if } \alpha \neq 1 \\ i\gamma - \lambda \left\{|z| + i\beta \frac{2}{\pi} z \ln |z|\right\} & \text{if } \alpha = 1 \end{cases} \tag{9.2}$$

where $0 < \alpha \leq 2$, $|\beta| \leq 1$, $\lambda > 0$, $\gamma \in \mathbb{R}$.

A distribution with $|\beta| = 1$ is called **a completely asymmetric stable distribution**. If $\beta = +1$ ($\beta = -1$), then $F(x)$ is a limit distribution of the normed sums of positive (negative) independent and identically distributed random variables. A distribution with $\beta = 0$ is called **a symmetric stable distribution**. If $\alpha = 2$ and $\lambda = 0.5$, then $f(z)$ is a characteristic function of a standard normal distribution. It follows from (9.2) that $S_n \stackrel{d}{=} n^{1/\alpha}\xi_1$ for $\alpha \neq 1$ and $S_n \stackrel{d}{=} n\xi_1 + \frac{2}{\pi}\beta\lambda n \ln n$ for $\alpha = 1$.

A random process $\xi(t)$, $t \geq 0$, with stationary independent increments defined on the probability space $(\Omega, \mathfrak{F}, \mathbf{P})$ is called **a one-dimensional stable process**, if

$$\psi(s, z) = \mathbf{E}\, e^{iz(\xi(t+s)-\xi(t))} = e^{s \ln f(z)}, \quad t, s \geq 0, \quad z \in \mathbb{R},$$

where $\ln f(z)$ is defined by (9.2). It is supposed that $\xi(0) = 0$ and trajectories of the process are continuous on the right and at any point there exists a limit on the left. All definitions for the stable distributions automatically carry over stable processes with the replacement of "distribution" by "process".

Generally speaking the trajectories of a stable process are not continuous. Completely asymmetric stable processes have either only positive jumps (if $\beta = +1$) or only negative jumps (if $\beta = -1$). If $\beta = +1$ and

$0 < \alpha < 1$, then trajectories of the process are nondecreasing with the probability 1. If $1 \leq \alpha \leq 2$, then this property of monotonicity is absent.

A notion of a multidimensional stable process can be introduced by analogy with a one-dimensional case. **A stable distribution** in $\mathbb{R}^n$ has a n-dimensional characteristic function $\psi(z)$, $z \in \mathbb{R}^n$, of the form

$$\ln \psi(z) = i(a, z) - \lambda |z|^\alpha \int_{S_n} w_\alpha(z, \theta)\, \mu(d\theta) \tag{9.3}$$

where $0 < \alpha \leq 2$, $\lambda > 0$, a is a fixed point from $\mathbb{R}^n$, (α, z) is a scalar product,

$$w_\alpha(z, \theta) = \left(1 - i\,\mathrm{sign}(z, \theta)\,\mathrm{tg}\left(\frac{1}{2}\pi\alpha\right)\right) \left|\left(\frac{z}{|z|}, \theta\right)\right|^\alpha, \quad \alpha \neq 1,$$

$$w_\alpha(z, \theta) = \left|\left(\frac{z}{|z|}, \theta\right)\right| + \frac{2i}{\pi}\left(\frac{z}{|z|}, \theta\right) \ln|(z, \theta)|, \quad \alpha = 1,$$

and $\mu(\cdot)$ is a probability measure concentrated on the surface of the sphere $S_n \subset \mathbb{R}^n$ with a unit radius. It is assumed that $\mu(\cdot)$ is a nonsingular distribution, i.e., $\mu(\cdot)$ does not concentrate on $S_m \subset \mathbb{R}^m$ for some $m < n$.

Without any loss of generality we can assume $a = 0$. If $\alpha = 2$, then an imaginary part of $w_2(z, \theta)$ is absent and a stable distribution is a multidimensional normal distribution. If $\mu(\cdot)$ is a uniform distribution on S_n, then a stable distribution is called **symmetric** and $\ln \psi(z) = -\lambda |z|^\alpha$, $0 < \alpha \leq 2$.

A random process $\xi(t)$, $t \geq 0$, with stationary independent increments defined on the probability space $(\Omega, \mathfrak{F}, \mathbf{P})$ and taking values in $\mathbb{R}^n$ is called **a multidimensional stable process**, if

$$\psi(s, z) = \mathbf{E}\, e^{i(z, \xi(t+s) - \xi(t))} = e^{s \ln \psi(z)}, \quad z \in \mathbb{R}^n, \ t, s > 0,$$

where $\ln \psi(z)$ is defined by the relation (9.3).

There were investigated[2] properties of functionals

$$V_1(a) = \inf\{t : |\xi(t)| \geq a\}, \quad V_2(a) = \int_0^\infty I_a(\xi(t))\, dt,$$

$$V_3(a) = \sup_{0 \leq \tau \leq t} |\xi(\tau)|, \quad 0 \leq t < \infty,$$

of trajectories of a multidimensional stable process. Here $I_a(\cdot)$ is an indicator function of a closed sphere in $\mathbb{R}^n$ with a center at a point 0 and with a radius a, $0 < a < \infty$. For some other properties of stable processes see[3–6].

9.2 Cauchy Process

A Cauchy process is a special case of the general stable process. Let $\xi(t)$, $t \geq 0$, be a one-dimensional process with stationary independent increments without a Gaussian component, i.e., its characteristic function has the form[7]

$$\mathbf{E}\, e^{iz(\xi(t+s)-\xi(t))} = e^{s\varphi(z)}, \quad z \in \mathbb{R}, \quad t,\, s \geq 0, \tag{9.4}$$

where

$$\varphi(z) = iaz + \int_{-\infty}^{\infty} \left\{ e^{izy} - 1 - \frac{izy}{1+y^2} \right\} \nu(dy)$$

for some $a \in \mathbb{R}$ and with a measure $\nu(\cdot)$ satisfying conditions

$$\nu(\mathbb{R}) = \infty, \quad \int_{-\infty}^{\infty} \min(1, y^2)\, \nu(dy) < \infty, \quad \nu(\{0\}) = 0.$$

If a measure $\nu(\cdot)$ is chosen in the form

$$\nu(dy) = \begin{cases} C_1 y^{-2}\, dy & \text{if } y > 0 \\ C_2 y^{-2}\, dy & \text{if } y < 0 \end{cases}$$

with $C_1 \neq C_2$, $C_1, C_2 > 0$, then we have **a asymmetric Cauchy process**. Its characteristic function has the form (9.4) where

$$\varphi(z) = -|z| - i\frac{\beta}{2\pi} z \ln|z| \tag{9.5}$$

(without any loss of generality we can assume $a = 0$). Here $\beta = p - q$, $q = 1 - p$, $0 \leq p \leq 1$ (p is a function of C_1 and C_2; in particular $p = 0.5$ for $C_1 = C_2$). If $\beta = 0$, then we have **a symmetric Cauchy process**. If $|\beta| = 1$, then a process is called **a completely asymmetric Cauchy process**. It follows from (9.5) that a Cauchy process is a stable process with a characteristic index $\alpha = 1$.

Some properties of a Cauchy process[8,9].

1. Let

$$T(y) = \inf\{t > 0 : \ \xi(t) = y\}.$$

If $\beta \neq 0$, then the relations

$$\mathbf{P}\{T(y) < \infty \,|\, \xi(0) = x\} > 0, \quad \mathbf{P}\{T(x) = 0 \,|\, \xi(0) = x\} = 1$$

hold for any $x, y \in \mathbb{R}$. If $\beta = 0$, then for any $x, y \in \mathbb{R}$

$$\mathbf{P}\{T(y) < \infty \,|\, \xi(0) = x\} = 0.$$

2. For a completely asymmetric Cauchy process with $\beta = 1$ the relation

$$\lim_{t\to\infty} \frac{\xi(t)}{t \ln t} = 1$$

holds with the probability 1.

A multidimensional Cauchy process[10] is defined as a multidimensional stable process with a characteristic index $\alpha = 1$.

9.3 χ^2-Process, Bessel and Rayleigh Processes

I. χ^2-process. Let $\xi_1(t), \ldots, \xi_n(t)$, $t \in \mathbb{R}$, be independent separable Gaussian processes having the same mean values and covariance functions. Then a process

$$\xi(t) \;=\; \xi_1^2(t) + \cdots + \xi_n^2(t), \quad t \in \mathbb{R}, \tag{9.6}$$

is called **a χ^2-process** .

χ^2-process is of applied interest in a number of stochastic modelling situations (for example, for studying empirical data) and of considerable theoretical interest since it is one of the few processes for which the main results can be obtained in a simple closed form. A process $\xi(t)$ can be interpreted as a squared distance of a vector process $(\xi_1(t), \ldots, \xi_n(t))$ to the origin of coordinates.

It follows from (9.6) that trajectories of the process $\xi(t)$ are continuous and differentiable, if trajectories of processes $\xi_i(t)$, $i = 1, \ldots, n$, are of the same type. Assume that $\xi_1(t), \ldots, \xi_n(t)$ are stationary processes with $\mathbf{E}\,\xi_i(t) = 0$, $r(t) = \mathbf{E}\,\xi_i(s)\,\xi_i(s+t)$, $i = 1, \ldots, n$. Let $U_{nu}(0, T)$ be the number of upcrossings of a level $u > 0$ by the process $\xi(t)$ in the interval $(0, T)$.

Some properties of the stationary χ^2-process[11,12].

1. $$\mathbf{E}\,U_{nu}(0, 1) \;=\; (\lambda_2/\pi)^{1/2}\,(u/2)^{(n-1)/2}\,e^{-u/2}/\Gamma(n/2) \tag{9.7}$$

where

$$\lambda_2 \;=\; \mathbf{Var}\,(\xi_i'(0)) = -r''(0), \quad \Gamma\left(\frac{n}{2}\right) \;=\; \int_0^\infty u^{n/2-1}\,e^{-u}\,du.$$

If $\lambda_2 = \infty$, then a right-hand side of (9.7) equals ∞.

2. Let

$$r(t) = 1 - \lambda_2 \frac{t^2}{2!} + \lambda_4 \frac{t^4}{4!} + o(t^4) \quad \text{as } t \to +0 \tag{9.8}$$

where $\lambda_2 < \infty$, $\lambda_4 < \infty$, and let for some $\alpha > 0$

$$r(t) = O(t^{-\alpha}) \quad \text{as } t \to \infty. \tag{9.9}$$

Then for any $\tau > 0$ and for $T = \tau / \mathbf{E}\, U_{nu}(0,\, 1)$ the relation

$$\lim_{u \to \infty} \mathbf{P}\{U_{nu}(0,\, T) = k\} = \frac{\tau^k}{k!} e^{-\tau}, \quad k = 0,\, 1,\, 2,\, \dots,$$

holds.

3. If conditions (9.8), (9.9) are satisfied and

$$\mu = (\lambda_2/\pi)^{1/2} (u/2)^{(n-1)/2} e^{-u/2} / \Gamma(n/2), \quad T = e^{-z}/\mu,$$

then

$$\lim_{u \to \infty} \mathbf{P}\{\max_{0 \le t \le T} \xi(t) < u\} = \exp\{-e^{-z}\}.$$

Here u and T satisfy the equality

$$u - (n-1)\ln(u/2) = 2\{\ln T + z + \ln((\lambda_2/\pi)^{1/2}/\Gamma(n/2))\}$$

or

$$u = 2z + 2\ln T + (n-1)\ln\ln T + 2\ln((\lambda_2/\pi)^{1/2}/\Gamma(n/2)) + o(1)$$

as $T \to \infty$.

4. If conditions (9.8), (9.9) are satisfied, then

$$\lim_{T \to \infty} \mathbf{P}\{v_T \min_{0 \le t \le T} \xi(t) < z\} = 1 - \exp\{-z^{(n-1)/2}\}, \quad z \ge 0,$$

where

$$v_T = \frac{1}{2}\left(\frac{\sqrt{\lambda/\pi}}{\Gamma(n/2)}\, T\right)^{2(n-1)}.$$

II. Bessel process. Let $\xi(t)$, $t \ge 0$, be a one-dimensional diffusion process defined by a shift coefficient $a(t,\, x)$ and a diffusion coefficient $b(t,\, x)$, $t \in \mathbb{R}$. Under some hypotheses[7,13] on the transition probability of a diffusion process and on the coefficients $a(t,\, x)$, $b(t,\, x)$ the function $u(t,\, x)$ being the solution of the equation

$$-\frac{\partial u(t,\, x)}{\partial t} = a(t,\, x)\frac{\partial u(t,\, x)}{\partial x} + \frac{1}{2} b(t,\, x)\frac{\partial^2 u(t,\, x)}{\partial x^2}, \tag{9.10}$$

$$u(s,\, x) \;=\; \varphi(x), \quad x \in \mathbb{R}, \; t \in (0,\, s),$$

uniquely defines a transition probability of a diffusion process. Here $\varphi(\cdot)$ belongs to a class of functions being dense everywhere in the space of continuous functions. The relation (9.10) is called an inverse Kolmogorov equation. If $b(t,\, x) = 1$ and $a(t,\, x) = \frac{\alpha-1}{2x}$, then a one-dimensional diffusion process $\xi(t)$, $t \geq 0$, defined by a differential operator L_α on $[0,\, \infty)$

$$L_\alpha \;=\; \frac{1}{2}\left(\frac{d^2}{dx^2} + \frac{\alpha - 1}{x}\,\frac{d}{dx}\right)$$

is called **a Bessel process**[14–16] with index α.

For $\alpha \leq 0$ the point $x = 0$ is absorbing, for $0 < \alpha < 2$ this is a reflecting point and for $\alpha \geq 2$ the point $x = 0$ is inaccessible. If α equals a positive integer n, then a Bessel process $\xi(t)$ is a radial part of the n-dimensional process of a Brownian motion, i.e., if $B_1(t), \ldots, B_n(t)$ are n mutually independent one-dimensional processes of a Brownian motion, then

$$\xi(t) \;=\; \sqrt{B_1^2(t) + \cdots + B_n^2(t)}.$$

In particular, if $\xi(t)$ and $\eta(t)$ are independent Bessel processes with indexes n and m, then $\zeta(t) \;=\; \sqrt{\xi^2(t) + \eta^2(t)}$ is also a Bessel process with index $n + m$.

III. Rayleigh process. This process is a special case of a Bessel process. Let $B_1(t)$ and $B_2(t)$ be independent stationary one-dimensional processes of a Brownian motion with a zero mean and the same covariance function. Then a process

$$\xi(t) \;=\; \sqrt{B_1^2(t) + B_2^2(t)}$$

is called **a Rayleigh process**[17]. This process has **a Rayleigh distribution**, i.e., for any t

$$F(x) \;=\; \mathbf{P}\left\{\xi(t) \leq x\right\} \;=\; 1 - \exp\{-x^2/(2\sigma^2)\},$$

where σ^2 is a variance of the processes $B_1(t)$ and $B_2(t)$.

Assume that a covariance function of processes $B_1(t)$ and $B_2(t)$ satisfies the conditions providing the trajectories of these processes to be differentiable with the probability 1 and twice differentiable at zero. In different engineering problems the necessity often arises to evaluate the distribution of $\max\limits_{0 \leq t \leq T} \xi(t)$ for some $T > 0$. An asymptotic distribution of this maximum for large values of T can be found by means of the distribution of the crossing moment of the level x, which in turn can be determined by a joint distribution of $\xi(t)$ and $\xi'(t)$.

A joint density $f(x, y)$ of processes $\xi(t)$ and $\xi'(t)$ and a mean number $N(x)$ of upcrossings of the level x per unit time are defined by the relations

$$f(x, y) = x\,(2\pi m_2)^{-1/2}\,\sigma^{-2}\,\exp\left\{-\frac{1}{2}\,(x^2\,\sigma^{-2} + y^2\,m_2^{-1})\right\},$$

$$N(x) = \int_0^\infty y\,f(x, y)\,dy = \frac{x}{\sigma^2}\left(\frac{m_2}{2\pi}\right)^{\frac{1}{2}}\exp\left\{-\frac{x^2}{2\sigma^2}\right\},$$

where $m_2 = \int_0^\infty x^2\,dF(x)$.

9.4 Ornstein-Uhlenbeck Process

By investigating the Brownian motion, Einstein and Smoluchovsky in 1905 discovered: if we consider a coordinate $\xi(t)$ of a particle at time t as a random variable, then $\xi(t+s) - \xi(s)$, $s, t \geq 0$, has a Gaussian distribution with a zero mean and a variance αt (α is a positive constant defined by physical characteristics of a moving particle and a liquid). Moreover $\xi(t)$ is a process with independent increments. Wiener proved that trajectories of the process $\xi(t)$ can be chosen continuous with the probability 1. At the same time they are nowhere differentiable and have unbounded variation in any finite time interval. Ornstein and Uhlenbeck introduced in 1930 a process without these demerits. A new distribution of $\xi(t+s) - \xi(s)$, $s, t \geq 0$, is Gaussian with a zero mean and a variance $\frac{\alpha}{\beta}\,(e^{-\beta t} - 1 + \beta t)$, which is approximately equal to αt for large values of t and to $\alpha\,\beta\,t^2/2$ for small values of t. Here $\beta > 0$ is a second constant defined by physical considerations.

Let $U(t)$ denote a derivative of the process $\xi(t)$. All distributions of $\xi(t)$ can be obtained by corresponding distributions of $U(t)$. Denote

$$p(s, x;\ s+t, y) = \mathbf{P}\,\{U(s+t) < y\,|\,U(s) = x\}.$$

This distribution is independent of s and can be calculated by the formula

$$p(s, x;\ s+t, y) = \frac{1}{\sqrt{2\pi}}\int_{-\infty}^{y} \exp\{(v - x e^{-\beta t})/(\sigma^2\,(1 - e^{-2\beta t}))\}\,dv \quad (9.11)$$

with $\sigma^2 = \frac{1}{2}\alpha\beta$. **An Ornstein-Uhlenbeck process** is a stationary Markov process $U(t)$ possessing two properties:

(i) $U(t)$ is a normal distributed random variable with a zero mean and a variance σ^2 for any t,

(ii) a process $U(t)$ has a transition function (9.11).

A covariance function of the Ornstein-Uhlenbeck process is defined by the relation

$$\mathbf{E}\, U(s)\, U(s+t) \;=\; \sigma^2\, e^{-\beta t}, \quad t \geq 0.$$

A joint distribution of $U(t_1)$ and $U(t_2)$, $t_1 < t_2$, is a two-dimensional normal distribution with zero means, identical variances σ^2 and a coefficient of correlation $e^{-\beta(t_2-t_1)}$.

Theorem 9.1. *Let $\{U(T),\ t \in \mathbb{R}\}$ be a one-parameter family of random variables defining a random process with the following properties:*

(i) it is a stationary Markov process,

(ii) for any $s, t,\ s \neq t$, random variables $U(s)$ and $U(t)$ have two-dimensional normal distribution with $m = \mathbf{E}\, U(t)$ and $\sigma_2 = \mathbf{E}\,(U(t) - m)^2$.

Then a described process can be one of two following types:

(A) for any $n \geq 1$ and $t_1 < \cdots < t_n$ random variables $U(t_1), \ldots, U(t_n)$ are independent and have a normal distribution with a mean m and a variance σ^2,

(B) $\{U(t),\ t \in \mathbb{R}\}$ is a Gaussian process with the covariance function

$$\mathbf{E}\,\{(U(t) - m)\,(U(s) - m)\} \;=\; \sigma^2\, e^{-\beta|t-s|}$$

where $\beta > 0$ is some constant (Ornstein-Uhlenbeck process).

Without any loss of generality we can suppose $m = 0$. If follows from the relation

$$\mathbf{E}\,(U(s+t) - U(s))^2 \;=\; 2\sigma^2\,(1 - e^{-\beta|t|}) \;\sim\; 2\sigma^2\,\beta\,|t| \quad \text{as } t \to 0$$

that $U(s+t) - U(s)$ is proportional to $|t|^{\frac{1}{2}}$ and therefore $U'(s)$ do not exist, i.e., a moving particle can't have finite accelerations.

Theorem 9.2. *Trajectories of the Ornstein-Uhlenbeck process are continuous with the probability 1. Moreover*

$$\overline{\lim_{t\to 0}}\, \frac{U(t) - U(0)}{\left(4\sigma^2\beta t \ln\ln \frac{1}{t}\right)^{1/2}} \;=\; 1, \quad \overline{\lim_{t\to\infty}}\, \frac{U(t)}{(2\sigma^2 \ln t)^{1/2}} \;=\; 1$$

with the probability 1.

Let

$$\xi(t) - \xi(0) \;=\; \int_0^t U(s)\, ds.$$

Then $\xi(t)-\xi(0)$ and $U(t)$ have a two-dimensional normal distribution with zero means, variances $2\sigma^2(e^{-\beta t}-1+\beta t)/\beta^2$ and σ^2 respectively and with a coefficient of correlation

$$\frac{1-e^{-\beta t}}{\sqrt{2}(e^{-\beta t}-1+\beta t)^{1/2}}.$$

Let

$$B(t) = \beta(\xi(t)-\xi(0)) + U(t) - U(0).$$

Then for any t a random variable $B(t)$ has a normal distribution with a zero mean and a variance $2\sigma^2\beta|t|$. Moreover a distribution of $B(s+t)-B(s)$ does not depend on s; it is a normal distribution with a zero mean and a variance $2\sigma^2\beta|t|$. Furthermore

$$\mathbf{E}\{(B(t_2)-B(t_1))(B(s_2)-B(s_1))\} = 0$$

for any $s_1 < s_2 \le t_1 < t_2$.

Properties of the Ornstein-Uhlenbeck process[18–24] (for $\sigma=\beta=1$).

1. If

$$E_a = \sup\{t: \sup_{0\le s\le t} U(s) < a\}, \quad a>0,$$

then for some $\delta>0$

$$\mathbf{P}\{E_a > t \,|\, U(0)=0\} = \lambda e^{\gamma(a)t} + o(e^{(\gamma(a)-\delta)t})$$

as $t\to\infty$. Here $\gamma(a)$ is a solution of the equation

$$f(a,x)=0$$

belonging to the interval $(-1,0)$ where

$$f(a,x) = \int_0^\infty (t-a)\,t^x\,e^{at-t^2/2}\,dt.$$

Furthermore $\lambda=-\theta/\gamma(a)$ where

$$\theta = \lim_{x\to\gamma(a)+0} \frac{x-\gamma(a)}{f(a,x)} \int_0^\infty t^{x+1}\,e^{-t^2/2}\,dt.$$

An analogous statement holds for

$$E_{a,-b} = \sup\{t: -b < \inf_{0\le s\le t} U(s) \le \sup_{0\le s\le t} U(s) < a\}, \quad a,b>0.$$

2. For any fixed $t > 0$

$$\mathbf{P}\{\sup_{0\le s\le t}|U(s)| > c\} \sim \frac{2}{\sqrt{2\pi}}\,t\,c\,e^{-c^2/2} \quad \text{as } c \to \infty.$$

A distribution of $U(t)$ provided given $U(0) = z$ is normal with a mean

$$\mathbf{E}\{U(t)\,|\,U(0) = z\} = z\,e^{-\beta t}, \quad \beta \ge 0,$$

and a variance

$$\mathbf{Var}\{U(t)\,|\,U(0) = z\} = \frac{\sigma_1^2}{2\beta}(1 - e^{-2\beta t})$$

where $\sigma_1^2 = 2\beta\sigma^2$. Coefficients $-\beta$ and σ_1^2 are called infinitesimal mean and variance respectively:

$$\frac{\partial}{\partial t}\mathbf{E}\{U(t)\,|\,U(0) = z\}_{t=0} = -\beta\,z, \qquad \frac{\partial}{\partial t}\mathbf{Var}\{U(t)\,|\,U(0) = z\}_{t=0} = \sigma_1^2.$$

The Ornstein-Uhlenbeck process is a diffusion process with a shift coefficient $-\beta\,z$ and a diffusion coefficient σ_1^2.

Under some general hypotheses and due to its properties Ornstein-Uhlenbeck process can be represented as a limit of the sequence of Markov processes. Thus, if $\{z_N(t)\}$ is a sequence of one-dimensional or multidimensional birth and death processes, then under some natural hypotheses a process

$$w_N(t) = \frac{1}{\sqrt{N}}\,(z_N(t) - a\,N)$$

converges in some metric to the Ornstein-Uhlenbeck process (one-dimensional or multidimensional) as $N \to \infty$. Here a is some constant.

A stationary Markov process $U(t)$ is called **a n-dimensional Ornstein-Uhlenbeck process** if $U(t)$ has a n-dimensional normal distribution with zero means and a covariance matrix C and the distribution of $U(t+s)$ provided given $U(s) = z$ is normal with a vector of mean values

$$\mathbf{E}\{U(t)\,|\,U(0) = z\} = (\exp(B\,t))^T\,z, \quad t \ge 0,$$

and a covariance matrix

$$\mathbf{E}\{\{U(t) - (\exp(B\,t))^T z\}\,\{U(t) - (\exp(B\,t))^T z\}^T\,|\,U(0) = z\} =$$

$$= C - (\exp(B\,t))^T\,C\exp(B\,t), \quad t \ge 0,$$

where a superscript "T" denotes the operation of transposition, B is some matrix; if

$$B = \begin{pmatrix} b_{11} & \dots & b_{1n} \\ \vdots & & \vdots \\ b_{n1} & \dots & b_{nn} \end{pmatrix},$$

then

$$\exp(B\,t) = \begin{pmatrix} \exp(b_{11}\,t) & \dots & \exp(b_{1n}\,t) \\ \vdots & & \vdots \\ \exp(b_{n1}\,t) & \dots & \exp(b_{nn}\,t) \end{pmatrix}.$$

9.5 Subadditive Processes

Let $\mathbb{N}^+$ be a set of nonnegative integer numbers and $\{\xi(s,\,t),\ s < t,\ s, t \in \mathbb{N}^+\}$ be a set of real random variables defined on the same probability space $(\Omega,\,\mathfrak{F},\,\mathbf{P})$ and satisfying the following conditions.

S_1. For any $s < t < u$

$$\xi(s,\,u) \le \xi(s,\,t) + \xi(t,\,u)$$

with the probability 1.

S_2. Distribution of $\xi(s,\,t)$ depends only on $t - s$.

S_3. For any $u \in \mathbb{N}^+$ a random process $\eta(t) = \xi(t\,u,\ (t+1)\,u),\ t \in \mathbb{N}^+$, is stationary.

S_4. $E\,|\xi(0,\,t)| < \infty$ for any $t \in \mathbb{N}^+$ and there exists a constant $A > 0$ such that

$$g(t) = E\,\xi(0,\,t) \ge -A\,t$$

for any $t \ge 1$.

A process satisfying these conditions is called **a subadditive process**[25–29]. Conditions S_1 and S_2 imply that

$$E\,\xi(s,\,t) = g(t-s) \text{ and } g(u-s) \le g(t-s) + g(u-t), \quad s < t < u.$$

According to the theory of subadditive functions

$$\lim_{t\to\infty} \frac{g(t)}{t} = \gamma$$

where

$$\gamma = \inf_{t\ge 1} \frac{g(t)}{t} < \infty.$$

A condition S_2 is insufficiently strong for the construction of the ergodic theory for subadditive processes. Therefore instead of S_2 a stronger condition was used[28].

S_2': joint distributions of the process $\xi(s+1,\ t+1)$ coincide with joint distributions of $\xi(s,\ t)$ for $s,\ t \in \mathbb{N}^+$.

There exist examples of processes satisfying S_2 and not satisfying S_2'.

Theorem 9.3. *If $\xi(s,\ t)$ is a subadditive process satisfying S_1, S_2', S_3 and S_4, then with the probability 1 and in mean there exists a finite limit*

$$\zeta = \lim_{t\to\infty} \frac{\xi(0,\ t)}{t} \tag{9.12}$$

with $\mathbf{E}\zeta = \gamma$.

Let $\mathbb{R}_+ = [0,\ \infty)$. Then $\xi(s,\ t)$ is **a subadditive process** with continuous parameters, if it satisfies conditions $S_1\ (s,\ t \in \mathbb{R}_+)$, $S_3\ (u,\ t \in \mathbb{R}_+)$, $S_4\ (t > 0)$ and S_2' (for any $\tau > 0$ joint distributions of the process $\xi(s+\tau,\ t+\tau)$ coincide with corresponding distributions of the process $\xi(s,\ t),\ s,\ t \in \mathbb{R}_+$). There exists a finite limit

$$\gamma = \lim_{t\to\infty} \frac{g(t)}{t} = \inf_{t>0} \frac{g(t)}{t}, \quad g(t) = \mathbf{E}\xi(0,\ t).$$

Let $\xi(s,\ t)$ be **an additive process** satisfying conditions S_1 (with the equality), S_2', S_3 and S_4. Then $\xi(0,\ t),\ t > 0$, is a process with stationary increments. Let $\eta(t),\ t > 0$, be an arbitrary process with stationary increments and finite mean value. Then $\xi(s,\ t) = \eta(t) - \eta(s),\ s < t$, satisfies S_1 (with the equality), S_2', S_3 and S_4. This remark combined with Theorem 9.4 implies that the statement of Theorem 9.3 is, generally speaking, false for subadditive processes with continuous parameters.

Theorem 9.4. *For any positive monotone increasing function $\Gamma(t)$, $t \geq 0$, there exists a process $\{\eta(t),\ t \geq 0\}$ with stationary increments and a finite mean value such that its trajectories have derivatives of any order and the relation*

$$\mathbf{P}\{|\eta(t)| \leq \Gamma(t) \ \ \textit{for any large value}\ \ t\} = 0$$

is satisfied.

Suppose

$$\mathbf{E} \sup_{s,t\in I,\, s<t} |\xi(s,\ t)| < \infty \tag{9.13}$$

for some bounded interval $I \subset [0,\ \infty)$.

Theorem 9.5. *If $\xi(s,\ t)$ is a separable subadditive process with continuous parameters satisfying (9.13), then a limit in (9.12) exists with the probability 1 and in the mean,* $\mathbf{E}\zeta = \gamma$.

9.6 Random Fields with Independent Increments

A random field is a set of real random variables $\{\xi(t_1, \ldots, t_n),\ t_1, \ldots, t_n \in \mathbb{R}\}$ defined on the same probability space $(\Omega, \mathfrak{F}, \mathbf{P})$, i.e., it is a random function of several real variables. Random fields with independent increments form a special but important class of general random fields[30–35]. Denote:

$$\mathbb{R}_+ = [0, \infty), \quad t = (t_1, \ldots, t_n), \quad a = (a_1, \ldots, a_n),$$

$$\begin{aligned} & b = (b_1, \ldots, b_n), \quad a_i \le b_i, \ i = 1, \ldots, n, \\ & C[a, b] = \{t : a_i \le t_i \le b_i, \ i = 1, \ldots, n\}, \qquad (9.14) \\ & \Delta^{(i)}_{[a_i, b_i]} F(t) = F(t_1, \ldots, t_{i-1}, b_i, t_{i+1}, \ldots, t_n) - \\ & \qquad - F(t_1, \ldots, t_{i-1}, a_i, t_{i+1}, \ldots, t_n), \\ & \Delta_{C[a, b]} (F(t)) = \Delta^{(1)}_{[a_1, b_1]} \cdots \Delta^{(n)}_{[a_n, b_n]} F(t) \end{aligned}$$

is a mixed difference. If $n = 2$, then

$$\Delta_{C[a, b]} (F(t)) = F(b_1, b_2) - F(b_1, a_2) - F(a_1, b_2) + F(a_1, a_2).$$

A random field $\xi(t)$, $t \in \mathbb{R}^n_+$, is called **a random field with independent increments**, if the following conditions are satisfied:

(i) if $t_i = 0$ for some $i \in \{1, \ldots, n\}$, then $\xi(t) = 0$ with the probability 1,

(ii) for any integer number $m \ge 2$ and for any disjoint m-dimensional rectangles $C_1, \ldots, C_m$ of the form (9.14) random variables $\Delta_{C_1}(\xi(t)), \ldots, \Delta_{C_m}(\xi(t))$ are independent.

If furthermore a distribution of a random variable $\Delta_{C[a, b]}(\xi(t))$ depends only on the measure of the rectangle $C[a, b]$ (i.e., on $(b_1 - a_1) \cdots (b_n - a_n)$), then $\xi(t)$ is called **a random field with independent homogeneous increments.**

If $\{\eta_{ij}\}_{i,j=0}^{\infty}$ is a sequence of independent random variables, then

$$\xi(t, s) = \sum_{i<t} \sum_{j<s} \eta_{ij}, \quad t \ge 0, \ s \ge 0,$$

is a random field with independent increments. In the case of identically distributed random variables $\{\eta_{ij}\}_{i,j=0}^{\infty}$ we have a random field with homogeneous independent increments.

Properties of random fields with independent increments depending on two parameters (**biadditive processes**).

A function $\varphi(t,\ s)$, $t > 0$, $s > 0$, is called **a locally upper function** of a random field $\xi(t,\ s)$, if

$$\mathbf{P}\{\overline{\lim_{\substack{t\downarrow 0\\ s\downarrow 0}}}\frac{\xi(t,\ s)}{\varphi(t,\ s)} \le 1\} = 1$$

and a **locally lower function**, if

$$\mathbf{P}\{\overline{\lim_{\substack{t\downarrow 0\\ s\downarrow 0}}}\frac{\xi(t,\ s)}{\varphi(t,\ s)} \ge 1\} = 1.$$

Theorem 9.6. *Let $\xi(t,\ s)$, $t \ge 0$, $s \ge 0$ be a stochastically continuous field with independent homogeneous increments such that*

$$\mathbf{E}\,\xi(t,\ s) = 0, \quad \sigma^2(t,\ s) = \mathbf{E}\,\xi^2(t,\ s), \quad \varphi(t,\ s) = \psi(t\,s)$$

where $\psi(u)$, $u \ge 0$, satisfies conditions:

(i) in the neighborhood of zero it is a continuous nonnegative increasing function,

(ii) $\lim_{v\downarrow 1}\sup_{u>0}|\psi(u\,v)/\psi(u) - 1| = 0,$

(iii) $\sigma(u)/\psi(u) \underset{u\to 0}{\longrightarrow} 0.$

If

$$I_1 = \int_0^{t_0}\int_0^{s_0}(t\,s)^{-1}\mathbf{P}\{\xi(t,\ s) > \varphi(t,\ s)\}\,dt\,ds < \infty$$

for some small values of $t_0 > 0$ and $s_0 > 0$, then $\varphi(t,\ s)$ is a locally upper function of the field $\xi(t,\ s)$.

Theorem 9.7. *Let $\xi(t,\ s)$ be the same field as in Theorem 9.6 and $\varphi(t,\ s) = \psi(t\,s)$ where $\psi(u)$, $u \ge 0$, satisfies conditions (i), (ii) of Theorem 9.6 and moreover for any $\varepsilon > 0$ there exists $\alpha_\varepsilon > 0$ such that*

$$\mathbf{P}\{|\xi(t,\ s)| < \varepsilon\,\varphi(t,\ s)\} \ge \alpha_\varepsilon.$$

If $I_1 = \infty$, then $\varphi(t,\ s)$ is a locally lower function of the field $\xi(t,\ s)$.

Corollary 1. *Assume that $\xi(t,\ s)$ and $\varphi(t,\ s)$ satisfy all the conditions of Theorems 9.6 and 9.7. Then $\varphi(t,\ s)$ is an upper (lower) function of the field $\xi(t,\ s)$, if the integral I_1 is finite (infinite).*

Wiener field is one of the most important examples of a random field with independent increments. A separable random field with independent homogeneous increments is called **a Wiener field**, if for any n-dimensional rectangle $C[a, b]$ a random variable $\Delta_{C[a,\, b]}(\xi(t))$ has a normal distribution with a zero mean and a variance $\prod_{i=1}^{n}(b_i - a_i)$. Let $W(t, s)$, $t \geq 0$, $s \geq 0$, be a two-dimensional Wiener field.

Corollary 2. *Let $q(u)$, $u \geq 0$, be a continuous and nonnegative function in the neighborhood of zero such that $u\,q(u)$ increases, $q(u) \uparrow \infty$ as $u \to 0$ and*

$$\lim_{v \downarrow 1} \sup_{u>0} \left| \frac{q(u\,v)}{q(u)} - 1 \right| = 1.$$

Then a function $\varphi(t, s) = \sqrt{t\,s}\, q(t\,s)$ is a locally upper (lower) function of the field $W(t, s)$, if for some small values $t_0 > 0$ and $s_0 > 0$ the integral

$$I_2 = \int_0^{s_0}\int_0^{t_0} (t\,s\,q(t\,s))^{-1} \exp\left\{ -\frac{1}{2}\, q^2(t, s) \right\} dt\, ds$$

is finite (infinite).

In particular, $I_2 < \infty$ for $q_1(ts) = (1+\varepsilon)\sqrt{4 \ln|\ln(ts)|}$ and $I_2 = \infty$ for $q_2(ts) = (1-\varepsilon)\sqrt{4ts \ln|\ln(ts)|}$ where $\varepsilon \in (0, 1)$.

Corollary 3 (a local law of iterated logarithm). *The relation*

$$\mathbf{P}\left\{ \lim_{\substack{s\downarrow 0 \\ t \downarrow 0}} \frac{|W(t, s)|}{\sqrt{4t\,s \ln|\ln(t\,s)|}} = 1 \right\} = 1$$

holds.

A random field with independent increments $\xi(t, s)$, $\mathbf{E}\,\xi(t, s) = 0$, $t \geq 0$, $s \geq 0$, is called **a sum of independent jumps** to the moment t, if there exists a countable (with respect to y) set of independent random variables $\{\alpha_y(s),\ s \geq 0\}$ with $\mathbf{E}\,\alpha_y(s) = 0$ for any y, s such that

$$\xi(t, s) = \sum_{y \leq t} \alpha_y(s).$$

A random field with independent increments $\xi(t, s)$, $\mathbf{E}\,\xi(t, s) = 0$, $t \geq 0$, $s \geq 0$, is called **a sum of independent jumps** to (t, s), if there exists a countable set of independent random variables $\{\beta(x, y)\}$ with $\mathbf{E}\,\beta(x, y) = 0$ for any x, y such that

$$\xi(t, s) = \sum_{x \leq t}\sum_{y \leq s} \beta(x, y).$$

Theorem 9.8. *A random field $\xi(t, s)$ with independent increments can be represented in the form*

$$\xi(t, s) = f(t, s) + \xi_1(t, s) + \xi_2(t, s) + \xi_3(t, s) + \xi_4(t, s)$$

where $f(t, s)$ is a determinate function, $\xi_1(t, s)$ is a sum of independent jumps to (t, s), a field $\xi_2(t, s)$ is continuous in probability with respect to s for any $t \geq 0$ and $\xi_2(t, s)$ is a sum of independent jumps to the moment t for any $s \geq 0$, a field $\xi_3(t, s)$ is continuous in probability with respect to t for any $s \geq 0$ and $\xi_3(t, s)$ is a sum of independent jumps to the moment s for any $t \geq 0$, and a field $\xi_4(t, s)$ is continuous in probability with respect to $t \geq 0$ and $s \geq 0$.

9.7 Periodic Random Processes

Periodic processes often arise in applications where the described phenomenon has a periodic nature (for example in radio engineering devices).

Let $\xi(t)$, $t \in \mathbb{R}$, be a separable real process defined on the probability space $(\Omega, \mathfrak{F}, \mathbf{P})$. Denote

$$F_{t_1,\dots,t_n}(x_1, \dots, x_n) = \mathbf{P}\{\xi(t_1) < x_1, \dots, \xi(t_n) < x_n\}, \quad n \geq 1, \ t_i, x_i \in \mathbb{R}.$$

A random process $\xi(t)$ is called **periodic** with a period T, if for any $n \geq 1$, $t_1, \dots, t_n, x_1, \dots, x_n \in \mathbb{R}$ a function $F_{t_1+\tau,\dots,t_n+\tau}(x_1, \dots, x_n)$ is periodic with respect to τ with a period T. For example, if $\eta(t)$ is a stationary process, then $\eta(t)\cos t$ and $\eta(t) + \cos t$ are periodic processes with a period 2π.

Properties of periodic processes[36,37].

1. A probability $\mathbf{P}\{\xi(t+\tau) \in A, \ a \leq t \leq b\}$ is a periodic function of τ.

2. If there exists finite $m(t) = \mathbf{E}\xi(t)$, then $m(t)$ is a periodic function with a period T. If there exists finite

$$m_{k_1,\dots,k_n}(t_1, \dots, t_n) = \mathbf{E}\{(\xi(t_1))^{k_1} \cdots (\xi(t_n))^{k_n}\},$$

then $m_{k_1,\dots,k_n}(t_1 + \tau, \dots, t_n + \tau)$ is a periodic function with a period T.

3. Let $\varphi(t, x)$ be a periodic with respect to t function with a period T and a Borel function with respect to x and let $\eta(t)$ be a stationary process. Then $\varphi(t, \eta(t))$ is a periodic process with a period T.

Let $\xi_1(t), \dots, \xi_n(t)$ be real random processes defined on the same probability space $(\Omega, \mathfrak{F}, \mathbf{P})$. Periodic processes $\xi_1(t), \dots, \xi_n(t)$ are called **periodically connected**, if for any integer number $m_1 > 0, \dots, m_n > 0$ and for any $t^j_{k_j}$, $x^j_{k_j} \in \mathbb{R}$, $k_j = 1, \ldots, m_j$, $j = 1, \ldots, n$, a function

$$\mathbf{P}\{\xi_j(t^j_{k_j} + \tau) < x^j_{k_j},\ k_j = 1, \ldots, m_j,\ j = 1, \ldots, n\}$$

is periodic with respect to τ with a period T.

It follows from this definition that periodic processes $\xi_1(t), \ldots, \xi_n(t)$ have the same period T. If $\xi_1(t), \ldots, \xi_n(t)$ are periodically connected processes, then $\xi_1(t) + \cdots + \xi_n(t)$ and $\xi_1(t) \cdots \xi_n(t)$ are periodic processes periodically connected with any of $\xi_i(t)$, $i = 1, \ldots, n$. Moreover, if a function $\varphi(t, x_1, \ldots, x_n)$ is periodic with respect to t with a period T and is Borel with respect to $x_1, \ldots, x_n$, then $\varphi(t, \xi_1(t), \ldots, \xi_n(t))$ is a periodic process periodically connected with $\xi_i(t)$, $i = 1, \ldots, n$.

Let $\xi(t)$, $t \in \mathbb{R}$, be a complex-valued random process defined on the probability space $(\Omega, \mathfrak{F}, \mathbf{P})$ and having a zero mean and a finite variance. A random process $\xi(t)$ is called **periodically correlated** with a period T, if its covariance function $B(t, s) = \mathbf{E}\,\xi(t)\,\overline{\xi(s)}$ exists, continuous and for any t, s satisfies the relation

$$B(t, s) = B(t + T, s + T).$$

The above given definitions suppose the periodicity of some characteristics of the process. This is not always valid even for processes with periodic trajectories. So, if α is uniformly distributed in $[-1, 1]$ random variable, then a process $\xi(t) = \cos(\alpha t)$ is not periodic or periodically correlated, what follows from the relations

$$\mathbf{E}\,\xi(t) = \frac{\sin t}{t}, \quad \mathbf{E}\{(\xi(s) - \mathbf{E}\,\xi(s))\,(\xi(t) - \mathbf{E}\,\xi(t))\} =$$

$$= \frac{1}{2}\left(\frac{\sin(t+s)}{t+s} + \frac{\sin(t-s)}{t-s}\right) - \frac{\sin t \cdot \sin s}{t\,s}.$$

A random process $\xi(t)$, $t \in \mathbb{R}$, defined on the probability space $(\Omega, \mathfrak{F}, \mathbf{P})$ is called **strictly periodic (in a strict sense)**, if there exists a positive random variable α that is measurable with respect to $(\Omega, \mathfrak{F}, \mathbf{P})$, takes values in some finite interval and such that for any $n \geq 1$, $t_1, \ldots, t_n \in \mathbb{R}$ a joint distribution of random variables $\xi(t_1 + k\,\alpha), \ldots, \xi(t_n + k\,\alpha)$ is independent of an integer k. This random variable α is called a period of a strictly periodic process. If a process is periodic, then it is also a strictly periodic process. In this case α is a determinate value. An inverse statement is false. However, if $\xi(t)$ is a strictly periodic process with a period α, then $\eta(t) = \xi(\alpha t/T)$ for some fixed $T > 0$ is a periodic process with a period T. Strictly periodic processes are often the solutions of differential equations describing the oscillations with random amplitudes and frequencies.

Therefore these processes are widely used in radio engineering and in the theory of automatic control.

A random process $\xi(t)$ is called **strictly periodic in a broad sense**, if for any t, t_1, $t_2 \in \mathbb{R}$ and for any $k = \pm 1, \pm 2, \ldots$ the equalities

$$\mathbf{E}\,\xi(t) = \mathbf{E}\,\xi(t + k\,\alpha), \quad \mathbf{E}\,\{\xi(t_1 + k\,\alpha)\,\xi(t_2 + k\,\alpha)\} = \mathbf{E}\,\{\xi(t_1)\,\xi(t_2)\}$$

hold where α is a positive random variable taking values in some finite interval. If for any $t \in \mathbb{R}$ there exist $\mathbf{E}\,\xi(t)$ and $\mathbf{E}\,\xi^2(t)$, then a process strictly periodic in a narrow sense is strictly periodic in a broad sense. Generally speaking such a process is not periodically correlated.

9.8 Random Processes Used by Description of Complex Systems

Describe some models of general processes that are used in the theory of queueing systems and networks, branching and migration processes, in the analysis of stochastic dynamic systems with random perturbations, random movements and have other different applications.

1. A system consists of m components which can fail and be repaired[38]. Let $F_j(x)$ be a distribution function of a random variable ξ_j $(1 \le j \le m)$ which is said to be a reliability resource of the jth component. Denote by $\xi_j(t)$ the residual reliability resource of the jth component at time t. Vanishing of $\xi_j(t)$ is the criterion for the failure of the jth component. In this case we instantaneously replace a failed component with a new one with a reliability resource independent of the past.

Let $\{(\eta_z^{(j)}(t,\omega),\ \tau_z^{(j)}(\omega)),\ 0 \le t < \tau_z^{(j)}(\omega)\}$, $1 \le j \le m$, be a set of terminating random processes taking values in some space Z with $0 \notin Z$; a subscript $z \in Z \cup \{0\}$. A repair of failed components can be described in the following way.

Suppose all components are in operating states at time t. Then $z(t) = 0$. If jth component fails at some moment t_0 and $z(t_0 - 0) = 0$, then there arises a failure impulse with the lifetime $\tau_0^{(j)}(\omega)$ and we set

$$z(t) = \begin{cases} \eta_0^{(j)}(t - t_0,\, \omega) & \text{if } 0 \le t - t_0 < \tau_0^{(j)}(\omega) \\ 0 & \text{if } t - t_0 \ge \tau_0^{(j)}(\omega) \end{cases}$$

provided that there are no other failures. Here $\omega = \omega_{t_0}$ where $\{\omega_t\}$ is a set independent for different t identically distributed random variables.

Assume that jth component fails at time s and $z(s-0) = z'$. Then we set

$$z(t) = \begin{cases} \eta_{z'}^{(j)}(t-s,\, \omega_s) & \text{if } 0 \le t-s < \tau_{z'}^{(j)}(\omega_s) \\ 0 & \text{if } t-s \ge \tau_{z'}^{(j)}(\omega_s) \end{cases}$$

up to the next failure. Thus an existing impulse changes its structure at any failure moment.

The rate of the jth component's wear at time t is $-\xi_j'(t)$. It is assumed that

$$-\xi_j'(t) = \alpha_j(z(t);\, \xi_1(t),\, \ldots,\, \xi_m(t))$$

where

$$0 < \alpha_0 \le \alpha_j(0;\, \xi_1,\, \ldots,\, \xi_m) \le \alpha_1, \quad |\alpha_j(z;\, \xi_1,\, \ldots,\, \xi_m)| \le C, \quad z \ne 0,$$

for any values of arguments (thus by the existing failure impulse a negative rate of a wear is not excluded). It is assumed also that this system of differential equations satisfies with the probability 1 the conditions providing the existence and uniqueness of its solution for all $t > 0$ and under given initial conditions. So it is defined a random process

$$\zeta(t) = (z(t);\, \xi_1(t),\, \ldots,\, \xi_m(t)).$$

This is a non-Markov process possessing the property: if $z(t) = 0$, then the whole dependence on the past is concentrated in values of $\xi_j(t)$, $1 \le j \le$ $\le m$. A method was developed enabling the construction of an embedded Markov chain for such processes[38,39].

2. Let $(\Omega,\, \mathfrak{U},\, \mathbf{P})$ and $(\Theta,\, \mathfrak{B},\, \mathbf{Q})$ be probability spaces, X and Z be measurable spaces. Introduce[40,41] a set of terminating random processes $\{(\xi(t,\, \theta,\, x,\, z,\, \omega),\ \tau(\theta,\, x,\, z,\, \omega)),\ 0 \le t < \tau(\theta,\, x,\, z,\, \omega)\}$, $x \in X$, $z \in Z$, $\theta \in \Theta$, defined on $(\Omega,\, \mathfrak{U},\, \mathbf{P})$, and a function $\lambda(A,\, t,\, z,\, \omega)$, $\omega \in \Omega$, $z \in Z$, $t \ge 0$, A belonging to a σ-algebra of subsets of X. It is assumed that $\lambda(A,\, t,\, z,\, \omega)$ is σ-additive with respect to the first argument by fixed t, z, ω and is measurable with respect to t, z, ω by fixed A.

Let 0 be a given fixed state. As usual we construct a random process $z(t) = z(t,\, \omega)$ for every fixed ω. Later on we assume that $\{\theta_t,\ t \ge 0\}$ is a random process with independent values from Θ and θ_t has a distribution $Q(\cdot)$.

Let $z(0) = 0$. If $z(t) = 0$, then a failure of the type $x \in A$ can occur in the interval $(t,\, t+dt)$ with the probability $\lambda(A,\, t,\, 0,\, \omega)\, dt$. If such event has occurred at the moment t', then $z(t) = \xi(t-t',\, \theta_{t'},\, x,\, 0,\, \omega)$ for $t \in [t',\, t'+\tau(\theta_{t'},\, x,\, 0,\, \omega))$ provided that there were no other failures in this half-interval. In this case $z(t) = 0$ at the end of the half-interval and the situation is repeated.

If a failure of the type $y \in A$ occurs at some moment s of this half-interval (the probability of such event at the interval $(s,\, s + ds)$ equals $\lambda(A,\, s,\, z(s),\, \omega)\, ds$), then a previously defined trajectory $z(t)$, $t > s$, must be replaced by $z(t) = \xi(t - s,\, \theta_s,\, y,\, z(s),\, \omega)$ for $t \in [s,\, s + \tau(\theta_s,\, y,\, z(s),\, \omega))$. In turn this new trajectory can be changed at the next failure moment.

3. The operation of a wide range of stochastic systems can be described in terms of random processes with trajectories characterized by spontaneous switchings being random functionals of the previous process behavior. A special class of random processes with discrete interference of a chance named switching processes was introduced[42–45].

These processes can be described as two-component processes $(x(t),\, \zeta(t))$, $t \geq 0$, such that there exists a sequence of moments $t_1 < t_2 < \cdots$ and on each half-interval $[t_k,\, t_{k+1})$ $x(t) = x(t_k)$ and the behavior of the process $\zeta(t)$ depends on the value $(x(t_k),\, \zeta(t_k))$ only. The moments t_k are switching moments and $x(t)$ is the discrete switching component.

Switching processes are the natural generalization of such well-known classes of processes as Markov processes homogeneous on the second component, processes with independent increments and Markov or semi-Markov switchings, piecewise Markov aggregates, Markov processes with Markov and semi-Markov switchings (random evolutions). Such processes as constructively defined are very suitable for asymptotic investigation of stochastic systems with "rare" and "fast" switchings.

Let

$$F_k = \{(\zeta_k(t,\, x,\, \alpha),\, \tau_k(x,\, \alpha),\, \beta_k(x,\, \alpha)),\ t \geq 0,\, x \in X,\, \alpha \in \mathbb{R}^r\},\quad k \geq 0,$$

be jointly independent on index k parametric families where $(X,\, \mathfrak{U})$ is a measurable space, $\zeta_k(t,\, x,\, \alpha)$ for each fixed $k,\, x,\, \alpha$ is a random process with trajectories belonging to the Skorohod space D^r_∞ (a space of continuous on the right functions without discontinuities of the second kind taking values in $\mathbb{R}^r$) and $\tau_k(x,\, \alpha)$, $\beta_k(x,\, \alpha)$ are possibly dependent on $\zeta_k(\cdot,\, x,\, a)$ random variables, $\tau_k(\cdot) \geq 0$, $\beta_k(\cdot) \in X$. It is supposed that vectors from $\mathbb{R}^r$ are column vectors and introduced variables are measurable on the pair $(x,\, a)$ with respect to a σ-algebra $\mathfrak{U} \times \mathfrak{B}$ where $\mathfrak{B}$ is a σ-algebra of Borel sets from $\mathbb{R}^r$. Let also $(x_0,\, S_0)$ be independent of F_k, $k \geq 0$, initial vector in $X \times \mathbb{R}^r$. Introduce the following recurrent sequences:

$$t_0 = 0,\quad t_{k+1} = t_k + \tau_k(x_k,\, S_k),\quad S_{k+1} = S_k + \xi_k(x_k,\, S_k),$$

$$x_{k+1} - \beta_k(x_k,\, S_k),\quad k \geq 0,$$

where

$$\xi_k(x, \alpha) = \zeta_k(\tau_k(x, \alpha), x, \alpha).$$

Then a two-component process $(x(t), \zeta(t))$, $t \geq 0$, is called **a switching process**, where

$$\zeta(t) = S_k + \zeta_k(t - t_k, x_k, S_k), \quad x(t) = x_k \text{ as } t_k \leq t < t_{k+1}, \ t \geq 0.$$

A process

$$S(t) = S_k \text{ as } t_k \leq t < t_{k+1}, \ t \geq 0,$$

is called **a recurrent process of a semi-Markov type**.

The limit theorems concerning the convergence of one switching process to another (in the class of switching processes) were proved in the case of "rare" switchings and these results allow construction of a theory of the asymptotic consolidation (enlargement) of the states of stochastic systems. This approach is realized for homogeneous and nonhomogeneous in time Markov and semi-Markov processes with nonergodic (asymptotically splitting on different classes) finite and arbitrary state space of a hierarchic structure[42,43]. Applications for Markov and non-Markov queueing systems in random environments are studied and asymptotic behavior of the flows of rare events and asymptotic enlargement in the case of fast service are investigated[45]. There were also proved theorems[44] on diffusion approximation enabling the investigation of transition phenomena for service processes in overloading queueing systems and networks of a Markov and semi-Markov type, state-dependent branching processes with asymptotically large number of particles in random environment and dynamic systems with fast random switchings.

9.9 References

1. Feller, W., *An Introduction to Probability Theory and Its Applications,* 2, Wiley, New York, 1971.
2. Taylor, S. J., Sample path properties of a transient stable process, *J. Math. Mech.,* 16, 1229, 1967.
3. Mijnheer, J. L., Properties of the sample functions of the completely asymmetric stable process, *Z. Wahrscheinlichkeitstheor. und verw. Geb.,* 27, 153, 1973.
4. Kasahara, Yu., Makoto, M., Weighted sums of i.i.d. random variables attracted to integrals of stable processes, *Probab. Theory and Related Fields,* 78, 75, 1988.
5. Nolan, J. P., Continuity of symmetric stable processes, *J. Multivar. Analysis,* 29, 84, 1989.

6. Samorodnitsky, G., Szulga, J., An asymptotic evaluation of the tail of a multiple symmetric α-stable integral, *Ann. Probab.*, 17, 1503, 1989.
7. Gihman, I. I., Skorohod, A. V., *Introduction to the Theory of Random Processes,* Saunders, Philadelphia, 1969.
8. Port, S. C., Stone, C. J., The asymmetric Cauchy processes on the line, *Ann. Math. Statist.*, 40, 137, 1969.
9. Millar, P. W., Some remarks on asymmetric processes, *Ann. Math. Statist.*, 43, 597, 1972.
10. Hawkes, J., Measure function properties of the asymmetric Cauchy process, *Mathematika (Gr. Brit.),* 17, 68, 1970.
11. Sharpe, K., Some properties of the crossings process generated by a stationary χ^2-process, *Adv. Appl. Probab.*, 10, 373, 1978.
12. Aronowich, M., Adler, R. J., Extrema and level crossings of χ^2-processes, *Adv. Appl. Probab.*, 18, 901, 1986.
13. DeBlassie, R. D., Stopping times of Bessel processes, *Ann. Probab.*, 15, 1044, 1987.
14. Molchanov, S. A., Ostrovsky, E., Symmetric stable processes as the traces of singular diffusion processes, *Teor. Veroyatn. Primen.*, 14, 127, 1969 (in Russian).
15. Shiga, T., Watanabe, S., Bessel diffusions as a one-parameter family of diffusion processes, *Z. Wahrscheinlichkeitstheor. und verw. Geb.*, 27, 37, 1973.
16. Penrose, M. D., The minimum of a large number of Bessel processes, *J. London Math. Soc.*, 38, 566, 1988.
17. Hasofer, A. M., On the derivative and the upcrossing of the Rayleigh process, *Austral. J. Statist.*, 12, 150, 1970.
18. Doob, J. L., The Brownian movement and stochastic equations, *Ann. Math.*, 43, 351, 1942.
19. McNeil, D. R., Schach, S., Central limit analogue for Markov population processes, *J. Roy. Statist. Soc.*, B, 35, 1, 1973.
20. Beekman, J. A., Asymptotic distributions for Ornstein-Uhlenbeck process, *J. Appl. Probab.*, 12, 107, 1975.
21. Dirkse, J. P., An absorption probability for the Ornstein-Uhlenbeck processes, *J. Appl. Probab.*, 12, 595, 1975.
22. Nobile, A. G., Ricciardi, L. M., Sacerdote, L., Exponential trends of Ornstein-Uhlenbeck first-passage-time densities, *J. Appl. Probab.*, 22, 360, 1985.
23. Ricciardi, L. M., Sato, S., First-passage-time density and moments of the Ornstein-Uhlenbeck process, *J. Appl. Probab.*, 25, 43, 1988.
24. Gsörgo, M., Lin, Z., On moduli of continuity for Gaussian and χ^2-processes generated by Ornstein-Uhlenbeck processes, *Math. Repts Acad. Sci. Can.*, 10, 203, 1988.

25. Hammersley, J. M., Welsh, J. A. D., First-passage percolation, subadditive processes, stochastic networks and generalized renewal theory, in *Bernoulli-Bayes-Laplace Anniversary Volume,* Neyman, J. and Le Cam, L. M., Eds., Springer, Berlin, 1965.
26. Kingman, J. F. C., Continuous time Markov processes, *Proc. London Math. Soc.,* 13, 539, 1963.
27. Kingman, J. F. C., The ergodic theory of subadditive stochastic processes, *J. Roy. Statist. Soc.,* B, 30, 499, 1968.
28. Kingman, J. F. C., Subadditive ergodic theory, *Ann. Probab.,* 1, 883, 1973.
29. Liggett, T. M., An improved subadditive ergodic theorem, *Ann. Probab.,* 13, 1279, 1985.
30. Chenzov, N. N., Wiener random fields of several parameters, *Doklady Akad. Nauk SSSR,* 106, 607, 1956 (in Russian).
31. Hudson, W. N., A decomposition theorem for biadditive processes, *Pacif. J. Math.,* 42, 333, 1972.
32. Hudson, W. N., Continuity of sample functions of biadditive processes, *Pacif. J. Math.,* 42, 343, 1972.
33. Zinchenko, N. M., On a ψ-variation of a Wiener random field, in *Theory of Random Processes,* Inst. of Math., Kiev, 1974 (in Russian).
34. Zinchenko, N. M., On the exit probability of a Wiener random field beyond some surface, *Teor. Veroyatn. Mat. Statist.,* 13, 62, 1975 (in Russian).
35. Zinchenko, N. M., A local increase of random fields with independent increments, *Teor. Probab. Appl.,* 24, 184, 1979.
36. Dorogovtsev, A. Y., Remarks on differential equations perturbed by periodic random processes, *Ukr. Mat. Shurn.,* 14, 119, 1962 (in Russian).
37. Gladyshev, E. G., Periodic and fast periodic correlated random processes with continuous time, *Teor. Veroyatn. Primen.,* 8, 184, 1963 (in Russian).
38. Kovalenko, I. N., Remarks on the construction of an embedded renewal process for essentially multidimensional queueing processes, *Teoriya Massovogo Obsluzhivaniya, Trudy Seminara,* VNIISI, Moscow, 95, 1981 (in Russian).
39. Kovalenko, I. N., Kuznetsov, N. Yu., Renewal process and rare events limit theorems for essentially multidimensional queueing processes, *Math. Operationsforsch. und Statist.,* Ser. Optimization, 12, 211, 1981.
40. Kovalenko, I. N., Limit theorems for reliability theory, *Cybernetics,* 13, 902, 1977.

41. Kovalenko, I. N., Analysis of Rare Events in Evaluation of Systems Efficiency and Reliability, Sov. Radio, Moscow, 1980 (in Russian).
42. Anisimov, V. V., Switching processes, *Cybernetics,* 13, 590, 1977.
43. Anisimov, V. V. Limit theorems for processes with semi-Markov switchings and their applications, *Random Operators and Stochast. Equations,* 2, 333, 1994.
44. Anisimov, V. V., Switching processes: averaging principle, diffusion approximation and applications, *Acta Appl. Math.,* 40, 95, 1995.
45. Anisimov, V. V., Sztrik, J., Asymptotic analysis of some complex renewable systems operating in random environments, *Europ. J. Oper. Res.,* 41, 162, 1989.

Chapter 10

Stability of Random Processes

10.1 Stability of Complex Systems

Stability of a system operation is one of the most important problems arising from its design. From the stability we understand the continuous dependence of some systems' characteristics on its parameters. The solution of this problem makes it possible to evaluate whether the system's real functioning corresponds to the forecasted one (simple approximate models are usually used by such evaluations).

There exist many different definitions for the notion of stability. A specific type of stability is used depending on the considered problem and on the system destination. Moreover a system can be stable with respect to some perturbations and at the same time not stable with respect to other perturbations in the sense of the same definition.

Some definitions of the stability of complex systems[1–3].

1. If a queueing system with waiting is studied, then a notion of **Loynes stability** is usually used. Let $W_n(t)$ be a distribution function of the nth customer waiting time. A queueing system is said to be:

(i) **stable functioning**, if the sequence $\{W_n(t)\}$ converges in a weak sense as $n \to \infty$ to a nonsingular distribution function $W(t)$,

(ii) **substable functioning**, if the sequence $\{W_n(t)\}$ satisfies the inequality $W_n(t) \geq W(t)$ for any n and t where $W(t)$ is some nonsingular distribution function.

The system stability implies its substability, which means that the sequence $\{w_n\}$ of waiting times is bounded in probability. This is the probability analog of a Lagrange stability of the solution of a system of differential equations. In the case of stability in addition to the boundedness in probability the convergence of distribution functions is required, i.e., the existence of the steady-state regime.

As the system parameters subjected to perturbations one can take $W_0(t)$, characteristics of the input flow, of the service process and characteristics describing the system structure. Let α be a system parameter taking values in some set A and let $Q \subset A$. Then a queueing system **operates stable with respect to** Q, if one of the mentioned above properties holds for any $\alpha \in Q$.

2. Assume a system operation depends on some parameter α taking values out of some set A. Let $z(t, \alpha)$, $t \in I$, for any fixed α be a random process taking values in a measurable space $(Z, \mathfrak{U})$ and describing a system behavior in the interval I. Let $Q \in \mathfrak{U}$, $Q_0 \in A$, $T \in I$, $\varepsilon > 0$. A process $z(t, \alpha)$ is said to be **stable** with respect to (ε, T, Q_0, Q), if

$$\inf_{\alpha \in Q_0} \mathbf{P}\{z(t, \alpha) \in Q, \ t \in T\} \geq 1 - \varepsilon.$$

A set Q_0 defines restrictions to a perturbation and Q is a set of admissible values of a process defined on some set T.

The so-defined notion of the stability plays an important role in the systems reliability analysis. Thus a set Q can be interpreted as a set of operating system states and a set $Z \setminus Q$ is considered as system failure states. Then $\mathbf{P}\{z(t, \alpha) \in Q, \ t \in T\}$ is the probability of a failure-free system operation at moments out of a fixed set T. Such a definition of system stability makes it possible to evaluate its reliability.

3. Let $z(t, \alpha)$, $t \geq 0$, $\alpha \in A$, be a random process taking values in a measurable space $(Z, \mathfrak{U})$. Assume that bounded and closed subsets $Q_\gamma \in \mathfrak{U}$ were chosen forming a collection $V = \{Q_\gamma\}$.

A process $z(t, \alpha)$ is said to be **stable**, if for any $\varepsilon \in (0, 1)$ and $\alpha \in S_0 \subset \subset A$ (S_0 is a given subset of A) there exist a set $Q_\gamma \in V$ and a random time $\tau(\alpha)$ such that $\mathbf{P}\{\tau(\alpha) < \infty\} \geq 1 - \varepsilon$ and $z(t, \alpha) \in Q_\gamma$ for all $t > \tau(\alpha)$.

This definition is destined for the investigation of the stability of systems operating in several regimes defined by sets Q_γ. System stability means that with the probability no less than a given value $1 - \varepsilon$ the system enters some set Q_γ and remains there.

10.2 Boundedness of Random Processes

A boundedness of a random process is one of the most important of its qualitative properties[4]. In the case of random processes this notion is a stochastic analog to a Lagrange stability of the solution of a system of differential equations.

Let $z(t)$, $t \geq 0$, be a random process taking values in a measurable space $(Z, \mathfrak{U})$ with an introduced metric $\rho(x, y)$, $x, y \in Z$. Choose some fixed $b \in Z$ and set $\|x\| = \rho(x, b)$.

Assume that a considered process is **regular**, i.e.,

$$\mathbf{P}\{\theta < \infty \,|\, z(0) = z\} = 0$$

for any initial state z. Here $\theta = \lim_{m\to\infty} \theta_m$ where θ_m is a first exit moment of the process $z(t)$, $t \geq 0$, out of the set $\{x : \|x\| < m\}$. A sufficient condition ensuring the regularity of $z(t)$ is the following one:

$$\lim_{c\to\infty} \mathbf{P}\{\theta_c < t \,|\, z(0) = z\} = 0$$

for any initial state z and any fixed $t > 0$.

A random process $z(t)$, $t \geq 0$, with an initial state z is said to be **bounded in probability**, if

$$\lim_{c\to\infty} \inf_{t\geq 0} \mathbf{P}\{\|z(t)\| < c \,|\, z(0) = z\} = 1.$$

A random process $z(t)$, $t \geq 0$, is said to be **bounded in mean along the time**, if

$$\lim_{c\to\infty} \inf_{t\geq 0} \frac{1}{t} \int_0^t \mathbf{P}\{\|z(s)\| < c \,|\, z(0) = z\}\, ds = 1.$$

A regularity of the process is the necessary condition for its boundedness in the sense of the above definition. The boundedness of the process in probability implies its mean boundedness along the time, but not necessarily vice versa. If there exists a nonsingular final distribution

$$P_z(B) = \lim_{t\to\infty} \mathbf{P}\{z(t) \in B \,|\, z(0) = z\}, \quad z \in Z,\ B \in \mathfrak{U}, \quad P_z(Z) = 1,$$

then a process is bounded in probability for an initial state z. If a process is bounded in probability, then a final distribution does not necessarily exist. As such an example we can refer to a finite irreducible periodic Markov chain.

10.3 Stability of Markov Chains

Let $Z = \mathbb{R}^m$ be a space of real-valued m-dimensional vectors, i.e., every state z is a vector $z = (z_1, \ldots, z_m)$ where $\{z_i\}$ are real numbers, and $\mathfrak{B}$ is a σ-algebra of Borel sets from Z. Let $z(n)$, $n = 0, 1, \ldots,$ be a homogeneous Markov chain taking values in Z and with a transition function $P(z; \cdot)$. Define an operator A acting on the function $V(z)$, $z \in Z$, as follows:

$$AV(z) = \int\limits_Z V(y)\, P(z; dy) - V(z). \tag{10.1}$$

A value $AV(z)$ is the mean one-step increment of the function $V(z(\cdot))$ by given initial state z. Let D_A denote a set of measurable functions $V(\cdot)$ with a finite right hand side of the relation (10.1) for any $z \in Z$. In particular all bounded functions $V(\cdot)$ belong to the set D_A.

Let $\{\xi(n),\ n \geq 0\}$ be a sequence of independent identically distributed random variables taking values in a measurable space $(\Xi, \mathfrak{U})$, and $F(z, \xi)$ be a function defined on $Z \times \Xi$ and taking values in Z. Then a relation

$$z(n+1) = F(z(n), \xi(n)), \quad n \geq 0, \tag{10.2}$$

defines a homogeneous Markov chain with a transition probability

$$P(z, B) = \mathbf{P}\{F(z, \xi(n)) \in B\}, \quad B \in \mathfrak{B}.$$

Parallel with $\{z(n),\ n \geq 0\}$ consider a Markov chain $\{z^*(n),\ n \geq 0\}$ defined by the relation

$$z^*(n+1) = F(z^*(n), \xi^*(n)), \quad n \geq 0. \tag{10.3}$$

It is assumed that $(z(0), z^*(0))$, $(\xi(0), \xi^*(0))$, $(\xi(1), \xi^*(1))$ ... is a sequence of independent couples, but random variables in a couple can be dependent. Let ρ_Z and ρ_Ξ be metrics introduced in spaces Z and Ξ. Consider a process $x(n) = (z(n), z^*(n))$ taking values in $X = Z \times Z$. It follows from (10.2) and (10.3) that $\{x(n),\ n \geq 0\}$ is a homogeneous Markov chain. Let A_X be a generating operator acting on the function $W(x)$, $x \in X$, i.e.,

$$A_X W(x) = \int\limits_{Z\times Z} W(y)\, \mathbf{P}\{x(n+1) \in (y, y+dy) \mid x(n) = x\} - W(x).$$

Consider two definitions of the process $\{z(n),\ n \geq 0\}$ stability, one of which takes into account joint definition of the processes $\{z(n),\ n \geq 0\}$

and $\{z^*(n),\ n \geq 0\}$, and the other is based on marginal characteristics of these processes.

Process $\{z(n), n \geq 0\}$ is said to be **stable in mean along the time,** if for any $\varepsilon > 0$ there exist constants $\delta_1 > 0$ and $\delta_2 > 0$ such that the inequality

$$\overline{\lim_{k\to\infty}} \frac{1}{k} \sum_{n=0}^{k-1} \mathbf{P}\{\rho_Z(z(n),\ z^*(n)) \geq \varepsilon\} < \varepsilon$$

holds provided that

$$\mathbf{E}\,\rho_Z(z(n),\ z^*(n)) < \delta_1, \quad \mathbf{E}\,\rho_\Xi(\xi(k),\ \xi^*(k)) < \delta_2.$$

Let S denote a set of sequences $(z(0),\ \xi(1),\ \xi(2),\ \ldots)$ with final distributions

$$Q(B) = \lim_{n\to\infty} \mathbf{P}\{z(n) \in B\}, \quad Q^*(B) = \lim_{n\to\infty} \mathbf{P}\{z^*(n) \in B\}, \quad B \in \mathfrak{B},$$

being independent of the initial states $z(0)$ and $z^*(0)$. Let μ_1 be some distance between random variables $z(0)$ and $z^*(0)$ defined by marginal distributions of these variables and μ_2 be an analogous distance between $\xi(n)$ and $\xi^*(n)$.

A process $\{z(n),\ n \geq 0\}$ is said to be **stable in a limit (weakly)**, if for any $\varepsilon > 0$ there exists a constant $\delta > 0$ such that the inequality

$$\pi(Q,\ Q^*) < \varepsilon$$

holds provided that

$$(z(0),\ \xi(0),\ \ldots) \in S, \quad (z^*(0),\ \xi^*(0),\ \ldots) \in S$$

and

$$\mu_1(z(0),\ z^*(0)) < \delta, \quad \mu_2(\xi(k),\ \xi^*(k)) < \delta.$$

Here π denotes **a Lévy-Prokhorov distance**:

$$\pi(Q,\ Q^*) = \inf\{\varepsilon_1 > 0 :\ Q(B) \leq Q^*(B^{\varepsilon_1}) + \varepsilon_1,\ Q^*(B) \leq Q(B^{\varepsilon_1}) + \varepsilon_1 \text{ for any } B \in \mathfrak{B}\}$$

where

$$B^{\varepsilon_1} = \{z \in Z :\ \rho_Z(z,\ B) < \varepsilon_1\}$$

is an ε_1-neighborhood of the set B and

$$\rho_Z(z,\ B) = \inf_{y\in B} \rho_Z(z,\ y), \quad z \in Z,\ B \in \mathfrak{B}.$$

This definition requires the weak convergence of final distributions of $\{z^*(n), n \geq 0\}$ by diminishing perturbations measured by distances μ_1 and μ_2. The method of test functions[4] is one of the most widely used methods applied to obtain useful qualitative and quantitative stability estimates. It is a generalization of a direct Ljapunov method. Consider some results obtained by this method.

Theorem 10.1. *Assume that* $(z(0), \xi(0), \xi(1), \dots) \in S$, $(z^*(0), \xi^*(0), \xi^*(1), \dots) \in S$ *and a process* $\{z(n), n \geq 0\}$ *is stable in mean along the time for any fixed initial state* $(z(0), z^*(0))$. *Then* $\pi(Q, Q^*) \to 0$ *as* $\mathbf{E}\,\rho_{\Xi}(\xi(k), \xi^*(k)) \to 0$.

Suppose the nonperturbed process is bounded in mean along the time for any $z(0) \in Z$, i.e.,

$$\inf_{n>0} \frac{1}{n} \sum_{j=0}^{n-1} \mathbf{P}\{\|z(j)\| < c\} \underset{c\to\infty}{\to} 1. \tag{10.4}$$

Denote by $\mu(\alpha, \alpha^*)$ some "deviation measure" of the sequence $\alpha^* = (z^*(0), \xi^*(0), \xi^*(1), \dots)$ from $\alpha = (z(0), \xi(0), \xi(1), \dots)$. For example

$$\mu(\alpha, \alpha^*) = \max\{\mathbf{E}\,\rho_Z(z(0), z^*(0)),\ \mathbf{E}\,\rho_{\Xi}(\xi(k), \xi^*(k))\},$$

$$\mu(\alpha, \alpha^*) = \max\{\mu_1(z(0), z^*(0)),\ \mu_2(\xi(k), \xi^*(k))\}.$$

Let further $\beta_1(y_1, \alpha, \alpha^*)$ and $\beta_2(y_1, y_2, \alpha, \alpha^*)$, $y_1, y_2 \in (0, \infty)$, denote nonnegative functions satisfying the condition: for any fixed $y_1, y_2 \in (0, \infty)$

$$\beta_1(y_1, \alpha, \alpha^*) \to 0, \quad \beta_2(y_1, y_2, \alpha, \alpha^*) \to 0$$

as $\mu(\alpha, \alpha^*) \to 0$.

Theorem 10.2. *Assume that a process* $\{z(n), n \geq 0\}$ *satisfies a condition (10.4) and there exists a nonnegative function* $W(x)$, $x \in X$, *such that for any* $K > 0$ *and* $\delta > 0$ *the following inequalities hold:*

(i) $\displaystyle\sup_{\|z\|<K,\, z^*\in Z} A_X W(z, z^*) \leq \beta_1(K, \alpha, \alpha^*)$,

(ii) $\displaystyle\sup_{x\in X} A_X W(x) \leq a < \infty$,

(iii) $\displaystyle\sup_{\rho_Z(z, z^*)\geq\delta,\, \|z\|<K} A_X W(x) \leq -c(\delta, K) + \beta_2(\delta, K, \alpha, \alpha^*)$

where $a > 0$ *and* $c(\delta, K) > 0$ *are some constants. Then for any* $\varepsilon > 0$ *the equality*

$$\overline{\lim_{k\to\infty}} \frac{1}{k} \sum_{j=0}^{k-1} \mathbf{P}\{\rho_Z(z(j), z^*(j)) \geq \varepsilon\} = 0$$

holds as $\mu(\alpha, \alpha^*) \to 0$.

Theorem 10.3. *If the hypotheses of Theorem 10.2 hold for* $\mu(\alpha, \alpha^*) = \mathbf{E}\,\rho_\Xi(\xi(k), \xi^*(k))$, *then a process* $\{z(n), n \geq 0\}$ *is stable in mean along the time for any fixed initial state* $x(0) = (z, z^*)$ *of the process* $\{x(n), n \geq 0\}$.

Let $\mathfrak{R}$ be some set of random variables and $\mu(\xi, \eta)$ be a distance between them defined generally speaking in terms of their joint distribution. Denote by $\mathcal{H}(F, G)$ a set of couples (ξ, η), $\xi, \eta \in \mathfrak{R}$, with different joint distributions and fixed marginal distributions $F(B) = \mathbf{P}\{\xi \in B\}$, $G(B) = \mathbf{P}\{\eta \in B\}$. A distance

$$\mu_0(F, G) = \inf_{(\xi, \eta)\in\mathcal{H}(F, G)} \mu(\xi, \eta)$$

depends only on marginal distributions of ξ and η and is called **a minimal distance** with respect to μ in a set $\mathfrak{R}$.

Theorem 10.4. *If the hypotheses of Theorem 10.3 hold and for all "perturbations" there exist final distributions* Q *and* Q^*, *then a process* $\{z(n), n \geq 0\}$ *is stable in a limit (weakly) for any initial states* $z(0)$, $z^*(0)$ *and for a distance* $\mu_2(\xi(k), \xi^*(k))$ *being minimal with respect to* $\mathbf{E}\,\rho_\Xi(\xi(k), \xi^*(k))$.

Up to now we have considered a stability either in mean along the time or in a limit case. Such notions are very useful and can give important information on the system behavior for a long-time functioning system. At the same time there often arise practical problems requiring the study of system stability that is uniform in time.

Let $h(z(n), z^*(n))$ be some distance between random variables $z(n)$ and $z^*(n)$ defined by their joint distribution and let

$$\mu(\alpha, \alpha^*) = \max\{\mu_1(z(0), z^*(0)),\ \mu_2(\xi(k), \xi^*(k))\}.$$

Note that stability and corresponding quantitative estimates depend on chosen distances h and μ, which in turn to a great extent depend on the problem under consideration and on the used method.

A process $\{z(n), n \geq 0\}$ is said to be **(h, μ)-stable** at the point α, if for any $\varepsilon > 0$ there exist $\delta = \delta(\varepsilon) > 0$ such that

$$\sup_n h(z(n), z^*(n)) < \varepsilon \tag{10.5}$$

provided that $\mu(\alpha, \alpha^*) < \delta$. This number $\delta = \delta(\varepsilon)$ is called **a stability module**.

The next definition takes into account a different functional nature of the initial state $(z(0), z^*(0))$ and a controlling sequence $\{\xi(k), \xi^*(k)\}$. A process $\{z(n), n \geq 0\}$ is said to be (h, μ_1, μ_2)**-stable** at the point α, if for any $\varepsilon > 0$ there exist $\delta_1 = \delta_1(\varepsilon) > 0$ and $\delta_2 = \delta_2(\varepsilon) > 0$ such that a relation (10.5) holds provided that $\mu_1(z(n), z^*(n)) < \delta_1$ and $\mu_2(\xi(k), \xi^*(k)) < \delta_2$. For example, if we take **a Ki-Fan distance**

$$\kappa(z(n), z^*(n)) = \inf\{\varepsilon > 0 : \mathbf{P}\{\rho_Z(z(n), z^*(n)) \geq \varepsilon\} < \varepsilon\} \tag{10.6}$$

as $h(z(n), z^*(n))$, then

$$\sup_n \kappa(z(n), z^*(n)) < \varepsilon$$

provided that

$$\sup_n \mathbf{P}\{\rho_Z(z(n), z^*(n)) \geq \varepsilon\} < \varepsilon. \tag{10.7}$$

Therefore conditions sufficient for (10.7) ensure the stability in the sense of a metric (10.6).

Denote by $S(V_0, R, L)$, $0 < R \leq L$, $V_0(z) \geq 0$, $z \in Z$, the following class of functions:

$$S(V_0, R, L) = \{W(z, z^*) : R\rho_Z(z, z^*) \leq W(z, z^*) \leq$$

$$\leq V_0(z) + L\rho_Z(z, z^*), \quad z, z^* \in Z\}.$$

Theorem 10.5. *Let there exist a function $V_0(z) \geq 0$, $z \in Z$, constants $\Delta > 0$, $v > 0$, a function $M(n)$, $n \geq 0$, and for any $R > 0$ there exist numbers $L \geq R$, $d_2 = d_2(R, L) > 0$ and a function $W(\cdot) \in S(V_0, R, L)$ such that*

$$A_X W(x) \leq -\Delta < 0 \text{ as } W(x) \geq 1,$$

$$\sup_{x \in X} A_X W(x) \leq v, \quad \sum_{j \geq n} \sup_{W(x) < 1} \mathbf{P}\{\eta(x) > j\} \leq M(n) \underset{n \to \infty}{\longrightarrow} 0$$

provided that $\mu_2(\xi(k), \xi^*(k)) < d_2(R, L)$ where $\eta(x) = \min\{n > 0 : W(x(n)) < 1\}$. Then a process $\{z(n), n \geq 0\}$ is (κ, μ_1, μ_2)-stable for

$$\mu_1(z(0), z^*(0)) = \kappa(z(0), z^*(0)) \tag{10.8}$$

and

$$\mu_1(z(0), z^*(0)) = \mathbf{E}\rho_Z(z(0), z^*(0)) \tag{10.9}$$

and with $\delta_1(\varepsilon)$ *and* $\delta_2(\varepsilon)$ *defined by relations:*

$$\delta_2(\varepsilon) = d_2(R(\varepsilon),\, L(R(\varepsilon))), \tag{10.10}$$

$$\delta_1(\varepsilon) = \begin{cases} \dfrac{\varepsilon^2\, R(\varepsilon)}{6\, L(R(\varepsilon))} & \text{if } \mu_1 \text{ is defined by (10.9)} \\[2ex] \dfrac{\varepsilon^2\, R(\varepsilon)}{6\,(L(R(\varepsilon)) + \varepsilon\, R(\varepsilon))} & \text{if } \mu_1 \text{ is defined by (10.8)} \end{cases} \tag{10.11}$$

where

$$R(\varepsilon) = \max\{R_1(\varepsilon),\, R_2(\varepsilon)\}, \quad R_1(\varepsilon) = \frac{3n_0}{\varepsilon^2}\left(1 + \frac{v}{2}\,(n_0+1)\right),$$

$$n_0 = \min\left\{n:\, M(n) < \frac{\varepsilon}{3}\right\}, \quad R_2(\varepsilon) = \frac{12}{\varepsilon^2}\, c(\varepsilon),$$

$$c(\varepsilon) = \inf\left\{c:\ \mathbf{P}\{V_0(z(0)) > c\} < \frac{\varepsilon}{12}\right\}.$$

In engineering problems the main interest is often concentrated on the $(h,\, \mu_1,\, \mu_2)$-stability formulated in terms of a distance defined only by marginal distributions.

Theorem 10.6. *Let* μ_{20} *be a minimal distance with respect to* μ_2*. Then under the hypotheses of Theorem 10.5 the process* $\{z(n),\, n \geq 0\}$ *is* $(\pi,\, \pi,\, \mu_{20})$*-stable with* $\delta_1(\varepsilon)$ *and* $\delta_2(\varepsilon)$ *defined by (10.10) and (10.11).*

Theorem 10.7. *Under the hypotheses of Theorem 10.5 the process* $\{z(n),\, n \geq 0\}$ *is asymptotically* $(\kappa,\, \mu_2)$*-stable in the following sense: for any* $\varepsilon > 0$ *and for any initial state* $(z(0),\, z^*(0))$ *there exist constants* $\delta(\varepsilon)$ *and* $n_1 = n_1(\varepsilon) > 0$ *such that the inequality*

$$\sup_{n>n_1} \kappa(z(n),\, z^*(n)) < \varepsilon$$

holds for all $\{(\xi(k),\, \xi^*(k))\}$ *satisfying the inequality* $\mu_2(\xi(k),\, \xi^*(k)) < \delta(\varepsilon)$ *with* $\delta(\varepsilon)$ *defined by (10.10).*

Along with these introduced notions there exist a variety of other definitions of stability characterized by different specific system features[4–9].

10.4 Method of Innovations

Method of innovations is one of the most widely used methods to prove ergodicity and stability theorems for random processes describing systems behavior[10–12].

Let $\{\tau_j,\ -\infty < j < \infty\}$ be a vector-valued stationary metrically transitive[11] sequence. Consider a vector-valued sequence w_n defined by its initial value w_1 and by the relation

$$w_{n+1} = f(w_n,\ \tau_n), \quad n \geq 1. \tag{10.12}$$

The problem is to find conditions providing the stability of a sequence $\{w_k\}$ by small perturbations of a controlling sequence $\{\tau_j\}$.

Let $\mathfrak{F}_{n,l},\ n \leq l$, be an σ-algebra generated by random variables $\tau_n, \ldots, \tau_l$ and T be a measure-conserving one-to-one shift operator defined on sets of σ-algebra $\mathfrak{F} = \mathfrak{F}_{-\infty,\infty}$ and such that

$$T\{\omega :\ \tau_j \in B_j,\ j = 1, \ldots, k\} = \{\omega :\ \tau_{j+1} \in B_j,\ j = 1, \ldots, k\}$$

for any collection of Borel sets B_j out of range of τ_j. Denote by U a corresponding operator over $\mathfrak{F}$-measurable functions: $\tau_{j+1}(\omega) = U\tau_j(\omega)$. Similarly it can be defined as operators T^{-1}, U^{-1}. An initial value w_1 is supposed to the fixed.

Events $A_n \in \mathfrak{F}_{-\infty,n+L},\ n = 1, 2, \ldots,$ are said to be **innovated** on the interval $[n,\ n+L]$, if random vectors $w_{n+k} = w_{n+k}(\omega)$ satisfy the relation

$$w_{n+k} = \varphi(\tau_n,\ \ldots,\ \tau_{n+k-1})$$

for $k > L$ and $\omega \in A_n$. The form of the function $\varphi(\cdot)$ depends only on the number of arguments and is determined by the choice of a sequence $\{A_n\}$. In other words the following conditions are satisfied:

(i) on a set A_n random vectors w_{n+k} for $k > L$ are measurable with respect to $\mathfrak{F}_{n,n+k-1}$, i.e., $\{\omega :\ w_{n+k} \in B\} \cap A_n \in \mathfrak{F}_{n,n+k-1}$ for any Borel set B,

(ii) if $\omega \in A_n \cap T^{-s} A_{n+s}$, then

$$w_{n+k}(\omega) = U^{-s}\, w_{n+k+s}(\omega)$$

for any $s > 0$ and $k > L$.

As an innovated event $\{w_n = x\}$ for any fixed point x can be taken. In particular $A_n = \{w_n = 0\} \in \mathfrak{F}_{1,n-1}$ is an innovated event, since for any $\omega \in A_n$

$$w_{n+1} = f(0,\ \tau_n), \quad w_{n+2} = f(f(0,\ \tau_n),\ \tau_{n+1}),\ \ldots .$$

More complex innovated events arise from the analysis of multiserver queueing systems. Assume that a controlling sequence $\{\tau_j^e,\ \tau_j^s,\ j \geq 1\}$ is defined on the main probability space. Here $\{\tau_j^e\}$ describes the input flow of customers arriving at moments 0, τ_1^e, $\tau_1^e + \tau_2^e, \ldots$ and $\{\tau_j^s\}$ are service times of corresponding customers. There are $m \le \infty$ servers. In waiting systems the service discipline is FCFS (First Come – First Serviced). A service to a customer can't be interrupted. In loss systems an arriving customer leaves the system unserved, if it finds no free servers. In order to define uniquely the system's behavior, it is necessary that along with a sequence $\{\tau_j^e,\ \tau_j^s\}$ a vector of initial states $w_1 = (w_{11}, \ldots, w_{1m})$ was given where w_{1i} is the time to the release of exactly i servers from the initial load.

Vector of waiting times $w_n = (w_{n,1}, \ldots, w_{n,m})$ is one of the main system characteristics. Here $w_{n,i}$ is the time from the arrival moment of the nth customer to the release of i servers from the customers with numbers less than n. Denote $x = (x_1, \ldots, x_m)$, $x^+ = (x_1^+, \ldots, x_m^+)$, $x_i^+ = \max\{0,\ x_i\}$, $i = 1, \ldots, m$; $e = (1, 0, \ldots, 0)$, $\mathbf{1} = (1, , \ldots, 1)$; $R(x)$ is a vector obtained of x by ordering its coordinates in the increasing order; $I(A)$ is an indicator of the set A.

For loss systems the recurrence relation

$$w_{n+1} = R(w_n + \tau_n^s\, e\, I(w_{n,1} = 0) - \tau_n^e\, \mathbf{1})^+, \quad n \ge 1,$$

holds. For waiting systems the corresponding relation is the following one:

$$w_{n+1} = R(w_n + \tau_n^s\, e - \tau_n^e\, \mathbf{1})^+, \quad n \ge 1.$$

Both these relations are of the form (10.12).

Examples of innovated events.

$$1.\ A_n = \{w_{n,1} = 0,\ w_{n,j} < \sum_{k=0}^{j-2} \tau_{n+k}^e,\ j = 2, \ldots, m\}. \qquad (10.13)$$

This event is an innovated one for $L = m - 2$. On the set A_n vectors w_{n+k}, $k > m - 2$, are functions only of values $(\tau_j^e,\ \tau_j^s)$, $j \ge n$.

$$2.\ A_n = \{w_{n+j,1} = 0,\ j = 0, \ldots, L,\ w_{n,m} \le \sum_{k=0}^{L} \tau_{n+k}^e\}. \qquad (10.14)$$

If an event A_n occurs, then a vector w_{n+L+1} is a function only of values $(\tau_j^e,\ \tau_j^s)$, $n \le j \le n + L$. An event (10.13) implies an event (10.14).

Often the distribution of the sequence $\{w_n\}$ is not known. Therefore in many cases it is advisable to use an event

$$\{v_{n,1} = 0,\ v_{n,j} < \sum_{k=0}^{j-2} \tau_{n+k}^e,\ j = 2, \ldots, m\} \qquad (10.15)$$

instead of (10.13) where $v_{n,j}$ is some known sequence being majorant for $w_{n,j}$: $w_{n,j} \le v_{n,j}$. For waiting and loss systems such majorants can be constructed.

Loss systems. Along with the original system consider a system having an infinite number of servers and controlled by the same sequence $\{\tau_j^e,\ \tau_j^s\}$. Let $q^{(k)}(x)$ be a stationary on k process

$$q^{(k)}(x) = \sum_{j=-\infty}^{k} I(\tau_j^s > \tau_j^e + \tau_{j+1}^e + \cdots + \tau_k^e + x). \tag{10.16}$$

Here $q^{(k)}(0)$ is a stationary number of occupied servers; $q^{(k)}(x)$ indicates a stationary number of servers being occupied at time x provided that we don't take into account customers arriving in a period of length x. Let $v_k(i)$ denote the time from the arrival of kth customer to the moment when no more than i servers are occupied by customers with numbers less than k. Then an event $\{v_k(i) < x\}$ is equivalent to $\{q^{(k-1)}(x) \le i\}$. For the original m-server system the inequalities $w_{n,i} \le v_n(m-i)$, $i = 0, 1, \ldots, m$, hold. Hence an event

$$q^{(n-1)}(0) \le m-1, \quad q^{(n-1)}\left(\sum_{k=0}^{j} \tau_{n+k}^e\right) \le m-2-j, \tag{10.17}$$

$$j = 0, \ldots, m-2,$$

implies the events (10.15) and (10.13) to be true. A relation (10.17) is equivalent to (10.15) for $v_{n,i} = v_n(m-i)$ and with functions $q^{(n-1)}(x)$ defined in explicit form (10.16).

An innovated event A_n is called **stationary**, if $A_n = T^n A_0$ with $A_0 \in \mathfrak{F}_{-\infty,L}$.

Theorem 10.8. *Let there exist a sequence of innovated events $\{A_j\}$ such that for some $l_0 > 0$ and $n_0 > 0$ the relation*

$$\mathbf{P}\left\{\bigcap_{l=l_0}^{\infty} \bigcup_{j=n_0}^{n} A_j T^{-l} A_{j+l}\right\} \underset{n\to\infty}{\longrightarrow} 1 \tag{10.18}$$

holds. Then a distribution of the sequence $\{w_{n+k},\ k \ge 0\}$ converges as $n \to \infty$ to the distribution of some stationary sequence $\{w^k,\ k \ge 0\}$ satisfying a recurrence relation (10.12). This sequence $\{w^k\}$ can be constructed on the same probability space so that

$$\mathbf{P}\left\{\bigcup_{k=0}^{\infty} \{w_{n+k} \ne U^n w^k\}\right\} \underset{n\to\infty}{\longrightarrow} 0. \tag{10.19}$$

If $\{A_j\}$ are stationary events, then $T^{-l} A_{j+l} = A_j$ and the relation (10.18) holds iff $\mathbf{P}\{A_0\} > 0$. Vice versa, if there exists w_0 such that relation (10.19) holds for $w^k = U^k w^0$, then there exist stationary innovated events $A_j = T^j A_0$, $\mathbf{P}\{A_0\} > 0$.

Consider how the method of innovations can be used to prove stability theorems. Let $\{\tau_j^{(r)}\}$, $r = 1, 2, \ldots,$ be a set of controlling sequences depending on a parameter r and $\{w^{(r)k},\ k \geq 0\}$, $\{w_n^{(r)},\ n \geq 1\}$, $r = 1, 2, \ldots,$ be corresponding stationary and nonstationary sequences (later on a superscript (r) means the association with a controlling sequence $\{\tau_j^{(r)}\}$). It is assumed that sequences $\{\tau_j^{(r)}\}$ satisfy the hypotheses of Theorem 10.8 and hence $\{w_{n+k}^{(r)},\ k \geq 0\}$ converge to $w^{(r)k}$ in the sense of (10.19). Introduce the following conditions.

A. Finite-dimensional distributions of $\{\tau_j^{(r)}\}$ converge to corresponding distributions of $\{\tau_j\}$ as $r \to \infty$.

B. Innovated events satisfy the following relations

$$\mathbf{P}\{A_0\} > 0, \quad \varliminf_{r\to\infty} \mathbf{P}\left\{\bigcup_{j=0}^{k} A_j^{(r)}\right\} \geq 1 - p(k) > 0$$

where $p(k) \to 0$ as $k \to \infty$.

This condition **B** follows from $\mathbf{B}_1$.

$\mathbf{B}_1$. $\mathbf{P}\{A_0\} > 0, \quad \varliminf_{r\to\infty} \mathbf{P}\left\{\bigcup_{j=0}^{k} A_j^{(r)}\right\} \geq \mathbf{P}\left\{\bigcup_{j=0}^{k} A_j\right\}$ for any $k \geq 0$.

In turn $\mathbf{B}_1$ follows from $\mathbf{B}_2$.

$\mathbf{B}_2$. Events A_k, $\mathbf{P}\{A_k\} > 0$ can be represented in the form of open sets $A_k = \{\gamma_{k1} > 0, \ldots, \gamma_{ks} > 0\}$ for some $s \geq 1$ where γ_{kj} are measurable with respect to $\mathfrak{F}_{-\infty,k+L}$, $U\gamma_{kj} = \gamma_{k+1j}$ and distributions of $(\gamma_{01}^{(r)}, \ldots, \gamma_{0s}^{(r)})$ weakly converge as $r \to \infty$ to the distribution of $(\gamma_{01}, \ldots, \gamma_{0s})$ (convergence is required only in the neighborhood of zero).

C. Functions $\varphi(y_1, \ldots, y_k)$ in the definition of innovated events are continuous almost surely with respect to the distribution of $(\tau_1, \ldots, \tau_k)$.

Theorem 10.9. *Under the hypotheses* **A**, **B**, **C** *the finite-dimensional distributions of $\{w_{n+k}^{(r)},\ k \geq 0\}$ weakly converge as $n \to \infty$, $r \to \infty$ to finite-dimensional distributions of $\{w^k,\ k \geq 0\}$.*

Consider stability theorems for waiting and loss systems. Suppose the following conditions are satisfied.

A*. The finite-dimensional distributions of stationary sequences $\{\tau_j^{(r)e},\ \tau_j^{(r)s}\}$ weakly converge as $r \to \infty$ to finite-dimensional distributions of $\{\tau_j^e,\ \tau_j^s\}$.

C*. $\mathbf{P}\{\tau_1^s - \tau_1^e - \cdots - \tau_k^e = 0\} = 0$ for any $k \geq 1$.

These conditions A^* and C^* provide the fulfillment of conditions A and C in the Theorem 10.9.

Waiting systems. Assume that $w_1 = 0$. Let A_n mean a stationary innovated event of the form

$$A_n = \{\tau_{n-1}^e > W_{n-1},\ \tau_n^e + \cdots + \tau_{n+m-2}^e > \tau_n^s\}$$

where W_n is a majorant defined by the relation

$$\sum_{i=1}^{m} w_{n,i} \leq W_n = \sup_{k \leq n} \left\{ (m-1)^2\, \tau_{k-1}^s + (m-1)\, v_{k-1} + \sum_{j=k}^{n-1} \xi_j \right\},$$

$$\xi_k = \tau_k^s - m\,\tau_k^e,\ \ v_k = \sup_{s \leq k} \left(\sum_{j=s}^{k} a_j \right)^+,\ \ a_j = -\tau_j^s + (m-1)\,(\tau_{j-1}^s - \tau_j^s).$$

Theorem 10.10. *Suppose conditions* A^*, $\mathbf{P}\{A_0\} > 0$, $\mathbf{E}\,\tau_1^s < m\,\mathbf{E}\,\tau_1^e$ *and* $\mathbf{E}\,\tau_1^{(r)s} \to \mathbf{E}\,\tau_1^s$ *as* $r \to \infty$ *are satisfied. Then multidimensional distributions of* $\{w^{(r)k}\}$ *weakly converge as* $r \to \infty$ *to corresponding distributions of* $\{w^k\}$.

Loss systems. As stationary innovated events we take

$$A_n = \{\tau_{n-1}^e > b_{n-2}^+ + V_{n-2},\ \sum_{j=1}^{m-2} \tau_{n+j}^e > \tau_n^s\}$$

where

$$b_n = \tau_n^s - \tau_n^e,\ \ V_n = \sup_{s \leq n} \left\{ \sum_{j=s}^{n} \eta_j,\ 0 \right\},\ \ \eta_n = b_{n-1}^+ - b_n^+ - \tau_n^e,\ \ \mathbf{E}\,\eta_n < 0.$$

Values V_n and b_n satisfy the inequality $w_{n,m} \leq b_{n+1}^+ + V_{n-1}$ for the initial state $w_1 = 0$.

Theorem 10.11. *Suppose conditions* A^*, C^*, $\mathbf{P}\{A_0\} > 0$, $\mathbf{E}\,\tau_1^s < \infty$ *and* $\mathbf{E}\,\tau_1^{(r)s} \to \mathbf{E}\,\tau_1^s$ *as* $r \to \infty$ *are satisfied. Then multidimensional distributions of* $\{w^{(r)k}\}$ *weakly converge as* $r \to \infty$ *to corresponding distributions of* $\{w^k\}$.

10.5 References

1. Loynes, R.M., The stability of a queue with non-independent inter-arrival and service times, *Proc. Cambr. Phil. Soc.*, 58, 497, 1962.
2. Loynes, R.M., The stability of a system of queues in series, *Proc. Cambr. Phil. Soc.*, 60, 569, 1964.
3. Bukan, D., Kenigsberg, E., *Scientific Control of Stores,* Nauka, Moscow, 1967 (in Russian).
4. Kalashnikov, V. V., *Qualitative Analysis of Complex Systems Behavior by the Method of Test Functions,* Nauka, Moscow, 1978 (in Russian).
5. Ljapunov, A. M., *Stability of Motion,* Acad. Pr., New York, 1966.
6. LaSalle, J. P., Lefschetz, S., *Stability by Ljapunov Direct Method: With Applications,* Acad. Pr., New York, 1963.
7. Cesari, L., *Asymptotic Behavior and Stability Problems in Ordinary Differential Equations,* Springer, Berlin, 1959.
8. Buslenko, N. P., Kalashnikov, V. V., Kovalenko, I. N., *Lectures in Theory of Complex Systems,* Sov. Radio, Moscow, 1973 (in Russian).
9. Borovkov, A. A., *Probabilistic Processes in Queueing Theory,* Nauka, Moscow, 1972 (in Russian).
10. Borovkov, A. A., On the ergodicity and stability of a sequence $w_{n+1} = f(w_n, \xi_n)$: applications to communication networks, *Teor. Veroyatn. Primen.*, 17, 160, 1972 (in Russian).
11. Borovkov, A. A., Ergodicity and stability theorems for a class of stochastic equations and their applications, *Teor. Probab. Appl.*, 23, 227, 1978.
12. Borovkov, A. A., *Asymptotic Methods in Queueing Theory,* Wiley, New York, 1984.

Chapter 11

Random Processes of the Statistical Radio Engineering

11.1 Power Spectrum of a Stationary Random Process

A power spectrum[1-7] is one of the most important characteristics of a stationary (in a wide or strict sense) random process $\xi(t)$, $\mathbf{E}\,\xi(t) = 0$. Let $\xi^{(k)}(t)$ be some trajectory of the process $\xi(t)$ and $\xi_T^{(k)}(t)$ be a truncated trajectory, i.e., $\xi_T^{(k)}(t)$ equals zero for $|t| > T/2$ and coincides with $\xi^{(k)}(t)$ for $|t| \le T/2$. A spectrum (a Fourier transformation) of a function $\xi_T^{(k)}(t)$ is defined by the relation

$$Z_T^{(k)}(\omega) = \int_{-\frac{T}{2}}^{\frac{T}{2}} \xi_T^{(k)}(t)\, e^{-i\omega t}\, dt.$$

Suppose $\xi_T^{(k)}(t)$ is interpreted as a voltage or a current corresponding to a loading with a resistance of one ohm. Then a mean capacity on a frequency ω with respect to a frequency band $\Delta f = \frac{1}{T}$ is defined by the relation

$$G_T^{(k)}(\omega) = \frac{2}{T}\,|\,Z_T^{(k)}(\omega)\,|^2 =$$

$$= \frac{2}{T} \int_{-\frac{T}{2}}^{\frac{T}{2}} \int_{-\frac{T}{2}}^{\frac{T}{2}} \xi_T^{(k)}(t_1)\, \xi_T^{(k)}(t_2)\, e^{-i\omega(t_1-t_2)}\, dt_1\, dt_2. \tag{11.1}$$

For any ω the value $G_T^{(k)}(\omega)$ converges as $T \to \infty$ to a random variable $G^{(k)}(\omega)$ being measured as a capacity per a unit band of frequencies. By averaging over a set of trajectories $\xi^{(k)}(t)$, we have from (11.1) that

$$F_T(\omega) = \mathbf{E}\, G_T(\omega) = 2 \int_{-T}^{T} \left(1 - \frac{|\tau|}{T}\right) B(\tau)\, e^{-i\omega\tau}\, d\tau,$$

$$F(\omega) = \lim_{T\to\infty} F_T(\omega) = 2 \int_{-\infty}^{\infty} B(\tau)\, e^{-i\omega\tau}\, d\tau$$

where $B(\tau)$ is a covariance function of a stationary process $\xi(t)$. It is supposed that $\int_0^{\infty} |B(\tau)|\, d\tau < \infty$.

A frequency function $F(\omega)$ is called **a power spectrum** of a stationary (in a wide or strict sense) process. This spectrum gives an averaged picture of the process's power distribution over the frequencies of elementary harmonic constituents, but it does not take into account their phase structure.

A power spectrum $F(\omega)$ and a covariance function $B(\tau)$ of a stationary process are connected by the relations

$$F(\omega) = 2 \int_{-\infty}^{\infty} B(\tau)\, e^{-i\omega\tau}\, d\tau = 4 \int_{0}^{\infty} B(\tau)\, \cos(\omega\tau)\, d\tau,$$

$$B(\tau) = \frac{1}{4\pi} \int_{-\infty}^{\infty} F(\omega)\, e^{i\omega\tau}\, d\omega = \frac{1}{2\pi} \int_{0}^{\infty} F(\omega)\, \cos(\omega\tau)\, d\omega. \tag{11.2}$$

It is supposed that

$$\int_0^{\infty} |B(\tau)|\, d\tau < \infty \quad \text{and} \quad \int_0^{\infty} |F(\omega)|\, d\omega < \infty. \tag{11.3}$$

A power spectrum is a nonnegative even function. A mean capacity of a stationary process and the area under the curve of its power spectrum are connected by the relation

$$B(0) = \frac{1}{2\pi} \int_0^\infty F(\omega)\, d\omega.$$

Moreover a spectral density of a mean capacity for $\omega = 0$ equals a quadruplicate area under the curve of a covariance function:

$$F(0) = 4 \int_0^\infty B(\tau)\, d\tau.$$

The value

$$\tau_0 = \frac{1}{B(0)} \int_0^\infty B(\tau)\, d\tau = \frac{F(0)}{4\,B(0)}$$

is often taken as **the process correlation time** (this means that in engineering problems we can consider $\xi(t)$ and $\xi(t+\tau)$ as independent random variables for any $t \in \mathbb{R}$ and $\tau > \tau_0$).

A ratio of the area under the curve of a power spectrum to a spectral density on some typical frequency ω_0 is called **a band width of a power spectrum**:

$$\Delta_b = \frac{1}{2\pi\, F(\omega_0)} \int_0^\infty F(\omega)\, d\omega = \frac{B(0)}{F(\omega_0)}.$$

This value can be interpreted as the width of the uniform in a band Δ_b power spectrum of a process having the same mean capacity $B(0)$ as the original one.

11.2 Wide-Band and Narrow-Band Processes

A random process with a continuous power spectrum is called a **narrow-band** one if its power spectrum is concentrated mainly on a relatively narrow frequency band about some fixed frequency ω_0 and a **wide-band** one otherwise. For a narrow-band process $\Delta_b \ll \omega_0$.

Consider a narrow-band process with a power spectrum $F(\omega)$ concentrated about some high frequency ω_0 where a spectral density reaches its maximum. A covariance function $B(\tau)$ of such a process is of the form

$$B(\tau) = \frac{1}{2\pi}\int_0^\infty F(\omega)\cos(\omega\tau)\,d\omega = \left(\frac{1}{2\pi}\int_{-\infty}^{\omega_0} F(\omega_0-\omega)\cos(\omega\tau)\,d\omega\right)\times$$

$$\times\cos(\omega_0\tau) + \left(\frac{1}{2\pi}\int_{-\infty}^{\omega_0} F(\omega_0-\omega)\sin(\omega\tau)\,d\omega\right)\sin(\omega_0\tau). \qquad (11.4)$$

Since we assume that a band width of a spectrum is negligibly small compared with ω_0, the upper limits of the integration on the right hand side of (11.4) can be extended to infinity without any essential loss of the accuracy. If

$$F^*(\omega) = F(\omega_0-\omega), \quad a(\tau) = \frac{1}{2\pi}\int_{-\infty}^{\infty} F^*(\omega)\cos(\omega\tau)\,d\omega,$$

$$b(\tau) = \frac{1}{2\pi}\int_{-\infty}^{\infty} F^*(\omega)\sin(\omega\tau)\,d\omega,$$

then

$$B(\tau) \approx a(\tau)\cos(\omega_0\tau) + b(\tau)\sin(\omega_0\tau).$$

If $F(\omega)$ is a symmetrical function with respect to a frequency ω_0, then

$$b(\tau) = 0, \quad B(\tau) = a(\tau)\cos(\omega_0\tau) = \left(\frac{1}{\pi}\int_0^\infty F^*(\omega)\cos(\omega\tau)\,d\omega\right)\cos(\omega_0\tau).$$

Hence a covariance function of a narrow-band process with a power spectrum symmetrically disposed about some high frequency ω_0 is approximately equal to a product of $\cos(\omega_0\tau)$ on a covariance function $a(\tau)$ corresponding to the spectrum $F^*(\omega)$ obtained by shifting ω_0 of an original spectrum in the direction of low frequencies.

Consider a power spectrum of a wide-band random process. Suppose a spectral density $F(\omega)$ of a mean capacity of a process preserves a constant value to very high frequencies. It follows from (11.2) that a covariance function does not vanish only in some small interval about the origin of coordinates, i.e., for small values of τ. A power spectrum $F(\omega) = F_0 = const$ being uniform on all frequencies is the mathematical idealized model for spectrums of such type. A random process with a uniform spectrum on all frequencies is called **a white noise**. Its covariance function is defined by the relation

$$B(\tau) = \frac{1}{4\pi} \int_{-\infty}^{\infty} F_0 \, e^{i\omega\tau} \, d\omega = \frac{F_0}{2} \, \delta(\tau)$$

where $\delta(\tau)$ is a function vanishing for all $\tau \neq 0$ and equal to infinity at $\tau = 0$ so that $\int_{-\infty}^{\infty} \delta(\tau) \, d\tau = 1$. A coefficient of correlation for a white noise is defined by

$$R(\tau) = \begin{cases} 1 & \text{if } \tau = 0 \\ 0 & \text{if } \tau \neq 0. \end{cases}$$

Thus a white noise $\xi(t)$ is characterized by the property that values of $\xi(t)$ at any two (even arbitrarily close) moments are not correlated. The notion that "a white noise" gives only a spectral picture of a random process leaves open the question of distributions, i.e., a power spectrum can't be used to define uniquely distributions. Hence as white noises we can name random processes with a uniform power spectrum and with different distributions.

In reality a white noise is an idealized model that never can be realized since values of the process at two close moments are practically always dependent, and, moreover, real processes have a finite capacity unlike an infinite total capacity of a white noise. But the boundedness of transmission bands of radio engineering devices enables use of a white noise as a model for the input processes which essentially simplifies mathematical analysis without any appreciable errors.

11.3 Random Process with a Discrete Spectrum

Parallel with stationary processes with a continuous spectrum there exist in the broad sense stationary processes with periodic covariance functions or with covariance functions converging to a constant value as $\tau \to \infty$. In this case a condition (11.3) is not satisfied and a power spectrum is not now a continuous function.

Let for example

$$\zeta(t) = \xi \cos(\omega_1 t) + \eta \sin(\omega_1 t) \tag{11.5}$$

where ξ and η are independent of t random variables and ω_1 is a constant. Then $\zeta(t)$ represents "harmonic oscillations" with a random amplitude $\alpha = -\sqrt{\xi^2 + \eta^2}$ and a phase $\varphi = \operatorname{arctg} \frac{\eta}{\xi}$ having independent of t distributions.

A random process $\zeta(t)$ is stationary in a broad sense if $\mathbf{E}\,\xi = \mathbf{E}\,\eta = \mathbf{E}\,\xi\,\eta = 0$, $\mathbf{E}\,\xi^2 = \mathbf{E}\,\eta^2 = \sigma^2/2$. Then

$$B(\tau) = \frac{\sigma^2}{2}\cos(\omega_1\tau) \tag{11.6}$$

is a covariance function of the process $\xi(t)$. A condition (11.3) is not satisfied for this function. Nevertheless the notion of a power spectrum can be extended to the considered case by means of a delta-function. A Fourier transformation of a covariance function (11.6) (i.e., a power spectrum of an oscillation with random amplitude and phase) can be written in the form

$$F(\omega) = \pi\,\sigma^2\,(\delta(\omega+\omega_1)+\delta(\omega-\omega_1)).$$

This spectrum consists of two discrete lines of the infinite intensity on the frequencies $\pm\omega_1$.

As a more general case we consider in a broad sense stationary process formed by a superposition of random processes (11.5):

$$\zeta(t) = \sum_{k=1}^{n}(\xi_k\cos(\omega_k t)+\eta_k\sin(\omega_k t))$$

where

$$\mathbf{E}\,\xi_k = \mathbf{E}\,\eta_k = \mathbf{E}\,\xi_k\,\xi_j = \mathbf{E}\,\eta_k\,\eta_j = \mathbf{E}\,\xi_k\,\eta_j = 0,$$
$$\mathbf{E}\,\xi_k^2 = \mathbf{E}\,\eta_k^2 = \frac{\sigma_k^2}{2},\quad k,\, j = 1,\,\ldots,\, n,\;\; k \neq j.$$

Then a corresponding covariance function is defined by the relation

$$B(\tau) = \sum_{k=1}^{n}\frac{\sigma_k^2}{2}\cos(\omega_k\tau).$$

A power spectrum is a sum of delta-functions on discrete frequencies:

$$F(\omega) = \sum_{k=1}^{n}\pi\,\sigma_k^2\,(\delta(\omega+\omega_k)+\delta(\omega-\omega_k)).$$

Such stationary in a broad sense processes with power spectrums consisting of sequences of spectral lines of delta-functions concentrated on discrete frequencies are called **processes with a discrete spectrum**. Values $\{\sigma_k^2\}$ indicate the distribution of a spectrum's total power along separate discrete frequencies $\{\omega_k\}$.

Every stationary in a broad sense process can be approximated by a sum of uncorrelated harmonic oscillations with random amplitudes and phases, i.e., by a stationary in a broad sense process with a discrete spectrum. Such approximation is often used in applications of the theory of random processes.

11.4 Mutual Power Spectrum

Let $\xi_T^{(k)}(t)$ and $\eta_T^{(k)}(t)$ be truncated trajectories of stationary and stationary connected (see Section 2.1) processes $\xi(t)$ and $\eta(t)$. If $Z_{T,\xi}(\omega)$ and $Z_{T,\eta}(\omega)$ are Fourier transformations of these truncated trajectories, then **mutual power spectrums** are defined by the relations:

$$F_{\xi,\eta}(\omega) = \lim_{T\to\infty} \mathbf{E}\left\{\frac{2}{T}\, Z_{T,\xi}(\omega)\,\overline{Z_{T,\eta}(\omega)}\right\}, \qquad F_{\eta,\xi}(\omega) = \overline{F_{\xi,\eta}(\omega)}$$

where a line over Z and F denotes a complex conjugation.

Mutual power spectrums and mutual covariance functions are connected by the relations

$$F_{\xi,\eta}(\omega) = 2\int_{-\infty}^{\infty} B_{\xi,\eta}(\tau)\, e^{-i\omega\tau}\, d\tau, \qquad B_{\xi,\eta}(\tau) = \int_{-\infty}^{\infty} F_{\xi,\eta}(\omega)\, e^{i\omega\tau}\, d\omega,$$

$$F_{\eta,\xi}(\omega) = 2\int_{-\infty}^{\infty} B_{\eta,\xi}(\tau)\, e^{-i\omega\tau}\, d\tau, \qquad B_{\eta,\xi}(\tau) = \int_{-\infty}^{\infty} F_{\eta,\xi}(\omega)\, e^{i\omega\tau}\, d\omega.$$

As opposed to a power spectrum of a stationary process that is a real even function, a mutual power spectrum of two processes is a complex-valued function (since a mutual covariance function is not even) with an even real part and an odd imaginary part. Let

$$\zeta(t) = \sum_{r=1}^{n} \xi_r(t)$$

where $\xi_r(t)$, $r = 1, \ldots, n$, are stationary and stationary connected random processes. Then a power spectrum of the process $\zeta(t)$ equals

$$F_{\zeta}(\omega) = \sum_{r=1}^{n} F_{\xi_r}(\omega) + \sum_{i,j=1,\, i\neq j}^{n} F_{\xi_i,\xi_j}(\omega).$$

If random processes are noncoherent (i.e., their mutual covariance functions are equal to constants) and have zero mean values, then their mutual power spectrums vanish on all frequencies and hence a power spectrum of the sum of random processes equals the sum of power spectrums.

11.5 Envelope and Phase of a Random Process

Let $S(t)$ be a real function belonging to $L_p(-\infty, \infty)$, i.e., $\int_{-\infty}^{\infty} |S(t)|^p \, dt < \infty$. Then by Hilbert integral transformation one can define (for $p \geq 1$) a function $\sigma(t)$ known as **a conjugate function** to $S(t)$:

$$\sigma(t) = -\frac{1}{\pi} \int_{-\infty}^{\infty} \frac{S(\tau)}{\tau - t} \, d\tau \tag{11.7}$$

with

$$S(t) = \frac{1}{\pi} \int_{-\infty}^{\infty} \frac{\sigma(\tau)}{\tau - t} \, d\tau \tag{11.8}$$

(for $\tau = t$ we take main (in a Cauchy meaning) values of integrals). Relations (11.7) and (11.8) can be rewritten in the form

$$\sigma(t) = -\frac{1}{\pi} \int_{0}^{\infty} \frac{1}{\tau} (S(t+\tau) - S(t-\tau)) \, d\tau,$$

$$S(t) = \frac{1}{\pi} \int_{0}^{\infty} \frac{1}{\tau} (\sigma(t+\tau) - \sigma(t-\tau)) \, d\tau.$$

For example, $1/(1+t^2)$ and $-t/(1+t^2)$, $\cos t$ and $-\sin t$ are conjugate functions.

Under some hypotheses[8] a real function $S(t)$ can be presented in the form

$$S(t) = a(t) \cos \varphi(t)$$

where

$$a(t) = \sqrt{S^2(t) + \sigma^2(t)}, \quad \varphi(t) = \operatorname{arctg} \frac{\sigma(t)}{S(t)}$$

and $\sigma(t)$ is a conjugate function to $S(t)$ that is uniquely defined by the relation (11.7).

Similarly under some general hypotheses (for example, for ergodic processes it is sufficient to assume $\mathbf{E}\,\xi(t) = 0$) for a given stationary random process $\xi(t)$ with $\mathbf{E}\,\xi(t) = 0$ one can construct by Hilbert transformation a new stationary random process $\eta(t)$ **conjugate** to $\xi(t)$:

$$\eta(t) = -\frac{1}{\pi} \int_{-\infty}^{\infty} \frac{\xi(\tau)}{\tau - t}\, d\tau \tag{11.9}$$

(for $\tau = t$ we take a main value of the integral). A convergence of the integral on the right hand side of (11.9) is understood in a mean square meaning. A random process $\xi(t)$ and a conjugate process $\eta(t)$ can be presented in the form:

$$\xi(t) = E(t) \cos \Phi(t), \qquad \eta(t) = E(t) \sin \Phi(t). \tag{11.10}$$

Processes $E(t)$ and $\Phi(t)$ are called respectively **an envelope** and **a phase** of a random process $\xi(t)$. They are defined by formulae

$$E(t) = \sqrt{\xi^2(t) + \eta^2(t)}, \qquad \Phi(t) = \operatorname{arctg} \frac{\eta(t)}{\xi(t)}. \tag{11.11}$$

It follows from (11.11) that $E(t) \ge |\xi(t)|$, i.e., random functions $\xi(t)$ and $E(t)$ are nowhere intersecting. If $\xi(t)$ and $\eta(t)$ are differentiated processes, then $E(t)\,E'(t) = \xi(t)\,\xi'(t) + \eta(t)\,\eta'(t)$. Therefore if $\xi(t) = E(t)$ at some moment t (i.e., $\eta(t) = 0$), then $E'(t) = \xi'(t)$. Hence $\xi(t)$ and $E(t)$ are nonintersecting functions with common tangents at points of touching. This property explains the name an envelope for $E(t)$.

A random process of the form (11.10) can be considered as a harmonic oscillation modulated on the amplitude and phase by random functions $E(t)$ and $\Phi(t)$.

A covariance function and a power spectrum of a conjugate random process $\eta(t)$ coincide with a covariance function $B_\xi(\tau)$ and a power spectrum $F_\xi(\omega)$ of the process $\xi(t)$. A mutual covariance function of two conjugate processes is defined by the relation

$$B_{\xi,\eta}(\tau) = \mathbf{E}\,\xi(t)\,\eta(t+\tau) = \frac{1}{\pi} \int_{-\infty}^{\infty} \frac{B_\xi(v)}{\tau - v}\, dv,$$

i.e., a mutual covariance function $B_{\xi,\eta}(\tau)$ and a covariance function $B_\xi(\tau)$ are a pair of Hilbert transformations. A mutual power spectrum of two conjugate processes is defined by the relation

$$F_{\xi,\eta}(\omega) = -i \operatorname{sign} \omega\, F_\xi(\omega).$$

Mutual covariance function $B_{\xi,\eta}(\tau)$ and a power spectrum $F_\xi(\omega)$ are connected by the relation

$$B_{\xi,\eta}(\tau) = -B_{\eta,\xi}(\tau) = \frac{1}{2\pi} \int_0^\infty F_\xi(\omega) \sin(\omega\tau)\, d\omega.$$

This relation shows that a mutual covariance function of conjugate processes is odd and for $\tau = 0$ (i.e., in coinciding moments) these random processes are noncoherent.

If $\xi(t)$ is a stationary Gaussian process, then $\eta(t)$ is also a stationary Gaussian process such that $\xi(t)$ and $\eta(t)$ have a joint normal distribution. In coinciding moments these processes are independent.

11.6 Representation of Narrow-Band Processes

The possibility of presenting a random process $\xi(t)$ in the form (11.8) does not impose any essential restrictions on a power spectrum of the process. Such representation is the most visual for narrow-band processes. An envelope $E(t)$ of the process $\xi(t)$ is a slowly changing (with respect to $\cos \Phi(t)$) in time function, i.e., values of $E(t)$ are strongly correlated, its covariance function is slowly changing on τ and its power spectrum is concentrated mainly in a low-frequency domain. Let ω_0 be a central frequency, in the neighborhood of which a narrow-band spectrum of $\xi(t)$ is disposed. Then a phase $\Phi(t)$ can be presented in a form

$$\Phi(t) = \omega_0 t - \varphi(t)$$

with a random function $\varphi(t)$ proven to be slowly changing in time[1–4]. By substituting (11.11) in (11.8) we have

$$\xi(t) = A(t) \cos(\omega_0 t) + C(t) \sin(\omega_0 t),$$

$$\eta(t) = A(t) \sin(\omega_0 t) - C(t) \cos(\omega_0 t)$$

where

$$A(t) = E(t) \cos \varphi(t), \quad C(t) = E(t) \sin \varphi(t).$$

Hence a narrow-band process has the nature of a high-frequency oscillation with a carrying frequency ω_0 and with a slowly changing envelope and phase.

Let $B_A(\tau)$, $B_C(\tau)$, $B_{A,C}(\tau)$, $B_{C,A}(\tau)$ be covariance and mutual covariance functions of processes $A(t)$ and $C(t)$. Then the following relations

$$B_A(\tau) = B_C(\tau) = B_\xi(\tau)\cos(\omega_0\tau) + B_{\xi,\eta}(\tau)\sin(\omega_0\tau),$$

$$B_{A,C}(\tau) = -B_{C,A}(\tau) = B_\xi(\tau)\sin(\omega_0\tau) - B_{\xi,\eta}(\tau)\cos(\omega_0\tau),$$

$$B_A(\tau) = B_C(\tau) = \frac{1}{2\pi}\int_0^\infty F_\xi(\omega)\cos((\omega-\omega_0)\tau)\,d\omega,$$

$$B_{A,C}(\tau) = -B_{C,A}(\tau) = \frac{1}{2\pi}\int_0^\infty F_\xi(\omega)\sin((\omega-\omega_0)\tau)\,d\omega,$$

$$B_A(0) = B_C(0) = B_\xi(0), \quad \mathbf{E}\,E^2(t) = 2\,B_\xi(0)$$

hold. In coinciding time moments (i.e., for $\tau = 0$) the random processes $A(t)$ and $C(t)$ are noncoherent. If $\xi(t)$ is a stationary Gaussian process, then both $A(t)$ and $C(t)$ are stationary Gaussian processes independent in coinciding time moments.

If a power spectrum $F(\omega)$ of a narrow-band random process $\xi(t)$ is a symmetrical function with respect to a central frequency ω_0, then one can use approximate relations

$$B_A(\tau) = B_C(\tau) \approx \frac{1}{\pi}\int_0^\infty F_\xi^*(\omega)\cos(\omega\tau)\,d\omega = a(\tau),$$

$$B_{A,C}(\tau) = -B_{C,A}(\tau) \approx 0$$

where $F_\xi^*(\omega) = F_\xi(\omega_0 - \omega)$ and $a(\tau)$ is a slowly changing function, i.e., $a(x\tau)/a(\tau) \underset{\tau\to\infty}{\longrightarrow} 1$ for any $x > 0$.

11.7 Envelope and Phase of a Gaussian Process

Consider a narrow-band linear system with a resonance frequency ω_0. Let the input process be a superposition of a determinate process $S(t)$ (a signal) with a stationary Gaussian process $\xi(t)$, $\mathbf{E}\,\xi(t) = 0$, $\mathbf{Var}\,(\xi^2(t)) = \sigma^2$ (a noise). At the output of this system a process $\xi(t)$ can be represented in the form

$$\xi(t) = A(t)\cos(\omega_0 t) + C(t)\sin(\omega_0 t)$$

where $A(t)$ and $C(t)$ are stationary and stationary connected random processes having normal joint distributions. Let a signal $S(t)$ passed through a linear system be a high-frequency oscillation of the frequency ω_0 modulated on amplitude and phase, i.e.,

$$S(t) = u(t)\cos(\omega_0 t) + v(t)\sin(\omega_0 t) = a(t)\cos(\omega_0 t - \theta(t))$$

where

$$a(t) = \sqrt{u^2(t) + v^2(t)}, \quad \theta(t) = \operatorname{arctg}\frac{v(t)}{u(t)}$$

are an envelope and a phase of a narrow-band signal. A process at the output of a linear system has the form

$$\eta(t) = (A(t)+u(t))\cos(\omega_0 t) + (C(t)+v(t))\sin(\omega_0 t) = E(t)\cos(\omega_0 t - \varphi(t))$$

where $E(t)$ and $\varphi(t)$ are an envelope and a phase of a random process $\eta(t)$ determined by the relations

$$E(t) = \sqrt{(A(t)+u(t))^2 + (C(t)+v(t))^2}, \quad \varphi(t) = \operatorname{arctg}\frac{C(t)+v(t)}{A(t)+u(t)}.$$

Since $\xi(t)$ is a Gaussian process, both $A(t)$ and $C(t)$ are Gaussian processes independent in coinciding time moments and having the same variance σ^2 as the process $\xi(t)$. Therefore a density of a joint distribution of $A(t)$ and $C(t)$ at time t has the form

$$w_2(x, y) = \frac{1}{2\pi\sigma^2}\exp\left\{-\frac{x^2+y^2}{2\sigma^2}\right\}.$$

A density of a joint distribution of an envelope $E(t)$ and a phase $\varphi(t)$ at time t has the form

$$W_2(r, \theta, t) = \frac{r}{2\pi\sigma^2}\exp\left\{-\frac{1}{2\sigma^2}\left((r\cos\theta - u(t))^2 + (r\sin\theta - v(t))^2\right)\right\}.$$

One-dimensional distributions of an envelope and a phase can be found by integration of the last relation with respect to θ and r. If a signal is absent (i.e., $u(t) = v(t) = 0$), then $E(t)$ has a Rayleigh distribution.

11.8 Impulse Random Processes

Many problems in impulse engineering require the investigation of sequences of identical impulses. The main parameters determining a geometric form and a disposition of these impulses (an amplitude, a duration time, an appearance moment of the leading edge, etc.) can vary according to a given law or be random functions. A sequence of impulses with random parameters is called **an impulse process**.

If all impulses have a given shape but random parameters, then with a sequence of impulses one can associate a sequence of multidimensional random variables. Consider a sequence of $2N+1$ impulses disposed on both sides of a zero impulse (origin of a reference). Let T be a mean time between an appearance of impulses. If $Z_N(\omega)$ is a spectral density (Fourier transformation) of the function describing this sequence of impulses, then a power spectrum of the impulse process is defined by the relation

$$F(\omega) = \lim_{N\to\infty} \frac{2}{(2N+1)T}\,\mathbf{E}\,|Z_N(\omega)|^2. \tag{11.12}$$

Let $\xi(t)$ be a realization of an impulse process with a given shape of impulses but with random amplitudes, appearance and termination moments. Let t_{2n} and t_{2n+1} be moments of appearance and termination of nth impulse ($n = 0, \pm 1, \pm 2, \ldots$) and ξ_n be its amplitude. Denote by $u(t)$ a determinate function of the unit amplitude concentrated on $(0,\ 1)$ (i.e., $u(t) = 0$ for $t \notin (0,\ 1)$) and describing a shape of impulses. Let $\zeta_n(t)$, $t \in (t_{2n},\ t_{2n+1})$, be a function describing the time evolution of the nth impulse. Then

$$\zeta_n(t) = \xi_n\, u\left(\frac{t - t_{2n}}{\tau_n}\right), \quad t \in \mathbb{R},$$

where $\tau_n = t_{2n+1} - t_{2n}$. A sequence of $2N+1$ impulses is described analytically by the expression

$$\sum_{n=-N}^{N} \xi_n\, u\left(\frac{t - t_{2n}}{\tau_n}\right), \quad t \in \mathbb{R}.$$

If $g(\omega)$ is a spectral density (Fourier transformation) of the function $u(t)$, i.e.,

$$g(\omega) = \int_0^1 u(t)\, e^{-i\omega t}\, dt,$$

then a spectral density $Z_N(\omega)$ of a sequence of $2N+1$ impulses can be written in the form

$$Z_N(\omega) = \sum_{n=-N}^{N} \xi_n\, \tau_n\, g(\omega\tau_n)\, e^{-i\omega t_{2n}}. \tag{11.13}$$

Assume that probability characteristics of impulses are independent of their numbers and probability characteristics of a collection of impulses depend only on the mutual disposal of their numbers. Then the expression

$$K(\omega) = \mathbf{E}\{(\xi_n \tau_n)^2 |g(\omega \tau_n)|^2\}$$

does not depend on the number of the nth impulse and

$$h_{n-j}(\omega) = \mathbf{E}\{\xi_n \xi_j \tau_n \tau_j g(\omega \tau_n) \overline{g(\omega \tau_j)} \exp\{-i\omega(t_{2n} - t_{2j})\}\}$$

depends only on the difference $n - j$ of numbers of these two impulses. It follows from (11.12) and (11.13) that a power spectrum of such an impulse process is defined by the relation

$$F(\omega) = \frac{2}{T}\left\{K(\omega) + 2\sum_{p=1}^{\infty} \operatorname{Re} h_p(\omega)\right\}$$

where $T = \mathbf{E}(t_{2n+2} - t_{2n})$ and $\operatorname{Re} b$ denotes a real part of the number b.

Let random parameters (amplitude, duration time, appearance moment) of the same impulse be mutually independent (but similar parameters of different impulses can be dependent). Denote: a, σ^2 are a mean value and a variance of the amplitude, R_p is a coefficient of correlation of amplitudes of nth and $(n+p)$th impulses, $w_{1,\tau}(x)$, $w_{2,\tau}(x, y; p)$ are one-dimensional and two-dimensional (of nth and $(n+p)$th impulses) densities of impulses' duration times. Then

$$K(\omega) = (a^2 + \sigma^2) K_0(\omega), \quad h_p(\omega) = (\sigma^2 R_p + a^2) K_p(\omega) H_p(\omega), \quad p \geq 1,$$

where

$$K_0(\omega) = \int_0^{\infty} x^2 |g(\omega x)|^2 w_{1,\tau}(x)\, dx, \tag{11.14}$$

$$H_p(\omega) = \mathbf{E} \exp\{-i\omega(t_{2(n+p)} - t_{2n})\},$$

$$K_p(\omega) = \int_0^{\infty}\int_0^{\infty} x\, y\, g(\omega x) \overline{g(\omega y)}\, w_{2,\tau}(x, y; p)\, dx\, dy, \quad p \geq 1.$$

11.9 Some Types of Impulse Processes

Impulse random processes can be divided into two groups. The first group contains processes with impulses arising only in determinate periodic intervals (their other parameters are random variables). A normed covariance function of such processes can take any value from zero to unit by the fixed difference between two moments. The second group contains impulse processes without such a periodic nature of intervals where impulses can arise.

Impulse processes with determinate intervals of impulse existence. A moment of the nth impulse origin can be presented as $t_{2n} = nT + \nu_n$ where T is a length of the interval and ν_n is a random shift with a zero mean and such that $\mathbf{P}\{|\nu_n| \le T/2\} = 1$.

Let impulses be mutually independent and $\varphi_\nu(\omega)$ be a characteristic function of a random variable ν_n. Then a power spectrum of such an impulse process can be presented in the form

$$F(\omega) = \frac{2}{T}\left\{(a^2+\sigma^2)\,K_0(\omega) - a^2\,|\varphi_\nu(\omega)|^2\,K_\infty(\omega) + \right.$$

$$\left. + a^2\,|\varphi_\nu(\omega)|^2\,K_\infty(\omega)\,\frac{2\pi}{T}\sum_{r=-\infty}^{\infty}\delta\left(\omega-\frac{2\pi r}{T}\right)\right\} \tag{11.15}$$

where $\delta(\cdot)$ is a delta-function, $K_0(\omega)$ is defined by (11.14) and

$$K_\infty(\omega) = \left|\int_0^\infty x\,g(\omega x)\,w_{1,\tau}(x)\,dx\right|^2 .$$

A power spectrum in (11.15) consists of continuous and discrete parts; the last one contains discrete lines on frequencies multiple to $2\pi/T$.

Special cases of impulse processes with determinate intervals of impulse existence.

1. A sequence of equidistant impulses of a given shape having an identical duration and a random amplitude.

Let τ_0 and T be a duration and a period of impulses respectively. Assume that random amplitudes of any two impulses are not correlated. Then a power spectrum of the process has a form:

$$F(\omega) = \frac{2\tau_0^2}{T}\,|g(\omega\tau_0)|^2\left\{\sigma^2+\frac{2\pi}{T}\,a^2\sum_{r=-\infty}^{\infty}\delta\left(\omega-\frac{2\pi r}{T}\right)\right\}.$$

2. A sequence of impulses of a given shape having the same amplitude a and duration τ_0, but a random appearance time in a given interval of the length T.

If the appearance moments of any two impulses are independent, then a power spectrum of a process is determined by the relation:

$$F(\omega) = \frac{2\tau_0^2 a^2}{T}\,|g(\omega\tau_0)|^2\,\left\{1-|\varphi_\nu(\omega)|^2 + \right.$$

$$+\frac{2\pi}{T}\,|\varphi_\nu(\omega)|^2 \sum_{r=-\infty}^{\infty} \delta\left(\omega-\frac{2\pi r}{T}\right)\Bigg\}.$$

3. A sequence of equidistant impulses of a given shape having the same amplitude a and a random duration.

If durations of impulses are independent random variables, then a power spectrum of the process has a form:

$$F(\omega) = \frac{2a^2}{T}\left(K_0(\omega) - K_\infty(\omega) + \frac{2\pi}{T}\,K_\infty(\omega) \sum_{r=-\infty}^{\infty} \delta\left(\omega-\frac{2\pi r}{T}\right)\right).$$

Aperiodic impulse random processes. These processes have no such determinate periodic intervals containing impulses. Let $\{\tau_n\}$ and $\{z_n\}$ be independent sequences of independent positive random variables. Suppose the characteristic functions $\varphi_\tau(\omega)$ and $\varphi_z(\omega)$ of random variables τ_n and z_n do not depend on n. Here τ_n denotes a duration of the nth impulse and z_n denotes a length of a pause between nth and $(n+1)$th impulses. Then $\{\mu_n,\ n = 0, \pm1, \pm2, \ldots\}$ with $\mu_n = t_{2n+2} - t_{2n} = \tau_n + z_n$ is a sequence of lengths of intervals between the appearance of successive impulses, $T = \mathbf{E}\,\mu_n$, and $\varphi_\mu(\omega)$ is a characteristic function of μ. We suppose the impulse process to be stationary in a wide sense. Then a power spectrum of the described aperiodic impulse random process has a form:

$$F(\omega) = \frac{2\,a^2}{T}\left\{\left(1+\frac{\sigma^2}{a^2}\right) K_0(\omega) + \right.$$

$$\left. +2\,\mathrm{Re}\left(\frac{Q(-\omega)\,Q_1(-\omega)\,\varphi_z(\omega)}{1-\varphi_\mu(\omega)}\right) + \frac{g^2(0)\,\tau_0^2}{T}\,\delta(\omega)\right\} \tag{11.16}$$

where $K_0(\omega)$ is defined by (11.14),

$$Q(\omega) = \mathbf{E}\{\tau_n\, g(\omega\tau_n)\}, \quad Q_1(\omega) = \mathbf{E}\{\tau_n\, g_1(\omega\tau_n)\},$$

$$g_1(\omega) = g(-\omega)\, e^{-i\omega}, \quad \tau_0 = \mathbf{E}\,\tau_n.$$

It follows from (11.16) that power spectrums of an aperiodic impulse process and an impulse process with a determinate periodic interval essentially differ one from another. A spectrum (11.16) does not contain a discrete part typical for a process with a determinate interval (besides one discrete component corresponding to a constant constituent of a process).

Impulse random processes of a mixed type. In some problems of impulse engineering processes of a mixed type having both random parameters and a random number of impulses on given intervals arise. Suppose the following conditions are satisfied:

(i) all impulses have the same shape defined by the function $u(t)$ of the unit amplitude vanishing outside of the interval (0, 1),

(ii) there is no correlation between impulses concentrated on different intervals,

(iii) parameters of impulses (amplitudes, durations, moments of the appearance) are independent random variables,

(iv) several impulses in the same interval of a length T can arise.

Let $\tau_1, \tau_2, \ldots, z_1, z_2, \ldots$ be independent sequences of independent positive random variables, τ_i denotes a duration of the ith impulse in the interval, z_i is a length of a pause between $(i-1)$th and ith impulses $(i = 1, 2, \ldots)$, and z_1 is the time to the beginning of the first impulse in the interval. Take such m that $\tau_1 + \cdots + \tau_{(m+1)} + z_1 + \cdots + z_{(m+1)} > T$ with the probability close to 1. Suppose one interval can contain k impulses $(k = 0, 1, \ldots, m)$ with the probability p_k, $\sum_{k=0}^{m} p_k = 1$. Assume moreover that impulses have random amplitudes, constant durations τ_0 and constant pauses of the length z_0 between two successive impulses. Set $m = [T/(\tau_0 + z_0)]$. Then a power spectrum of the process contains a continuous part $F_c(\omega)$ and a discrete part $F_d(\omega)$:

$$F(\omega) = F_c(\omega) + F_d(\omega)$$

where

$$F_c(\omega) = \frac{2\tau_0^2\sigma^2}{T} |g(\omega\tau_0)|^2 \sum_{r=0}^{m} r\, p_r,$$

$$F_d(\omega) = \frac{4\pi a^2 \tau_0^2}{T^2} \frac{|g(\omega\tau_0)|^2}{\sin^2(\omega(\tau_0 + z_0))} \times$$

$$\times \sum_{r=0}^{m} p_r \sin^2(r\omega(\tau_0 + z_0)) \sum_{k=-\infty}^{\infty} \delta\left(\omega - \frac{2\pi k}{T}\right).$$

11.10 References

1. Levin, B. R., *Theory of Random Processes and Its Applications in Radio Engineering,* Sov. Radio, Moscow, 1957 (in Russian).
2. Levin, B. R., *Theoretical Foundations of Statistical Radio Engineering,* 1, Sov. Radio, Moscow, 1966 (in Russian).
3. Rozanov, Yu., A., *Stationary Random Processes,* Holden-Day, San Francisco, 1967.

4. Levin, B. R., *Theoretical Foundations of the Statistical Radio Engineering,* 2, Sov. Radio, Moscow, 1968 (in Russian).
5. Ross, S. M., *A First Course in Probability,* Macmillan Publ., New York, 1976.
6. Koroliouk, V. S., Portenko, N. I., Skorokhod, A. V., Tourbine, A. F., *Aide Mémoire de Théorie des Probabilités et de Statistique Mathématique,* Mir, Moscou, 1983.
7. Kelly, D. G., *Introduction to Probability,* Macmillan Publ., New York, 1994.
8. Titchmarsh, E. C., *Introduction to the Theory of Fourier Integrals,* Clarendon Pr., Oxford, 1975.

Chapter 12

Renewal Theory

12.1 Renewal Equation

From the analytical viewpoint renewal theory[1–7] investigates the asymptotic behavior of solutions of **the renewal equation**

$$z(t) = f(t) + \int_0^t z(t-u)\, dP(u), \quad t \geq 0, \tag{12.1}$$

where $f(t)$ is a given Borel function, $z(t)$ is a Borel function to be obtained, and $P(t)$ is a distribution function of a nonnegative random variable, $P(+0) < 1$. If $f(t)$ is locally bounded, then this equation has a unique solution locally bounded on $[0, \infty)$ which can be expressed via **the renewal function** defined as

$$H(t) = 1 + \sum_{n \geq 1} P^{n*}(t)$$

where $P^{n*}(t)$ is an n-fold convolution of the distribution function $P(t)$ so that

$$P^{1*}(t) = P(t), \quad P^{n+1*}(t) = \int_0^t P^{n*}(t-u)\, dP(u).$$

The renewal function satisfies the renewal equation

$$H(t) = 1 + \int_0^t H(t-u)\, dP(u)$$

whereas

$$z(t) = \int_0^t f(t-u)\,dH(u)$$

satisfies the equation (12.1). If $P(t) = 1 - e^{-ct}$, then $H(t) = 1 + ct$. Hence

$$z(t) = f(t) + c\int_0^t f(u)\,du.$$

Therefore, if moreover $f(t)$ is integrable in $[0, \infty)$ and $\lim_{t\to\infty} f(t) = 0$, then

$$z(t) \to c\int_0^\infty f(u)\,du$$

as $t \to \infty$. Under mild additional assumptions concerning $f(t)$ and $P(t)$ this formula is valid in a general case also.

Theorem 12.1. (The Elementary Renewal Theorem). *If*

$$a = \int_0^\infty t\,dP(t) < \infty, \tag{12.2}$$

then

$$\lim_{t\to\infty} \frac{1}{t} H(t) = \frac{1}{a}.$$

A distribution function $P(t)$ is called **latticed** if all its increase points belong to a lattice $0,\ \delta,\ 2\delta,\ \ldots,\ n\delta,\ \ldots$ for a $\delta > 0$. This property can be expressed by the equation

$$\sum_{n\geq 0} \{P(n\delta+) - P(n\delta-)\} = 1. \tag{12.3}$$

The largest δ satisfying this condition is called **a step** of the lattice. If on the contrary

$$\sum_{n\geq 0} \{P(n\delta+) - P(n\delta-)\} < 1$$

for any $\delta > 0$, then P is called **nonlatticed.**

In the latticed case the equation (12.1) is reduced to a linear algebraic relation. Indeed, if $P(t)$ is latticed with a step δ, then

$$z(t) = \sum_{k\geq t/\delta} f(t-k\delta)\,h_k$$

where the sequence $h_k = H(k\delta+) - H(k\delta-)$ is defined by the equation

$$h_n = 1 + \sum_{k=0}^{n} h_{n-k}\, p_k$$

where $p_k = P(k\delta+) - P(k\delta-)$. For any fixed $s \in [0, \delta)$ the sequence $z_n = z(n\delta + s)$ is the solution of the (discrete) renewal equation

$$z_n = f_n + \sum_{k=0}^{n} z_{n-k}\, p_k \tag{12.4}$$

with $f_n = f(n\delta + s)$.

The Elementary Renewal Theorem is considerably improved with the Theorem by **Blackwell** and its discrete analogue known as Theorem by **Feller-Erdös-Pollard**.

Theorem 12.2. *If $P(t)$ is a nonlatticed distribution function and a relation (12.2) is satisfied, then*

$$\lim_{t\to\infty} (H(t+s) - H(t)) = \frac{s}{a}.$$

Theorem 12.3. *If $P(t)$ is a latticed distribution function, δ being its step, and a relation (12.2) is satisfied, then*

$$\lim_{t\to\infty} (H(n\delta+\delta) - H(n\delta)) = \frac{\delta}{a}.$$

If the relation (12.3) is satisfied for some $\delta > 0$, then the step of the distribution $P(t)$ equals $d\delta$, where d is the greatest common divisor of those n for which $P(n\delta+) - P(n\delta-) > 0$.

The Feller-Erdös-Pollard Theorem can be reformulated via the equation (12.4).

Theorem 12.4. *If $A = \sum_{k\geq 0} k\, p_k < \infty$ and the greatest common divisor of $\{n :\ p_n > 0\}$ equals 1, then*

$$\lim_{n\to\infty} h_n = \frac{1}{A}.$$

The Blackwell Theorem is equivalent to the following one called **the Key Renewal Theorem.**

Theorem 12.5. *Let $P(t)$ be a nonlatticed distribution function and a relation (12.2) is satisfied. Then, if a bounded monotone function $f(t)$ is* *integrable over* $[0, \infty)$, *we have*

$$\lim_{t\to\infty} z(t) = \frac{1}{a}\int_0^\infty f(t)\,dt.$$

In the latticed version of this theorem the monotonicity is not required.

Theorem 12.6. *Let* $P(t)$ *be a latticed distribution function,* δ *being its step, and a relation (12.2) is satisfied. Then*

$$\lim_{n\to\infty} z(n\delta + s) = \frac{\delta}{a}\sum_{n\geq 0} f(n\delta + s)$$

if the series in a right hand side is absolutely convergent.

An equivalent formulation of the Theorem 12.6 is as follows.

Theorem 12.7. *If the greatest common divisor of* $\{n:\ p_n > 0\}$ *equals 1 and*

$$A = \sum_k k\,p_k < \infty, \quad \sum_{n\geq 0} |f_n| < \infty$$

then

$$\lim_{n\to\infty} z_n = \frac{1}{A}\sum_{n\geq 0} f_n.$$

In the nonlatticed case the monotonicity of the function $f(t)$ is not necessary for the existence of a limit of the solution of the renewal equation (12.1). A stronger statement is based on the concept of the direct Riemann-integrability.

A function $f(t)$ is called **directly Riemann-integrable** over $[0, \infty)$, if

$$\sum_{n\geq 0} \sup_{n\leq t<n+1} |f(t)| < \infty,$$

$$\delta\sum_{n\geq 0}\{\sup_{n\delta\leq t<n\delta+\delta} f(t) - \inf_{n\delta\leq t<n\delta+\delta} f(t)\} \underset{\delta\downarrow 0}{\longrightarrow} 0.$$

In particular, a monotone bounded integrable function is directly Riemann-integrable. A Riemann-integrable function majorized by a directly integrable function is a directly Riemann-integrable function.

Theorem 12.8. *Let* $P(t)$ *be a nonlatticed distribution function and a relation (12.2) is satisfied. Then, if a function* $f(t)$ *is directly Riemann-integrable over* $[0, \infty)$, *then*

$$\lim_{t\to\infty} z(t) = \frac{1}{a}\int_0^\infty f(t)\,dt.$$

This result is a very strong one.

Theorem 12.9. *Assume that the convolution $P^{n*}(t)$ is nonsingular for some n and a bounded $[0, \infty)$-integrable function $f(t)$ tends to zero as $t \to \infty$. If moreover a relation (12.2) is satisfied, then*

$$\lim_{t\to\infty} z(t) = \frac{1}{a}\int_0^\infty f(t)\,dt.$$

If $P(t)$ is absolutely continuous, $p(t)$ being its density, then

$$H(t) = 1 + \int_0^t h(u)\,du$$

where

$$h(t) = \sum_{n\geq 1} p^{n*}(t), \quad p^{1*}(t) = p(t), \quad p^{n+1*}(t) = \int_0^t p^{n*}(t-u)\,p(u)\,du.$$

The renewal density $h(t)$ satisfies the renewal equation (12.1) with $f(t) = p(t)$, i.e.,

$$h(t) = p(t) + \int_0^t h(t-u)\,p(u)\,du.$$

Theorem 12.10. *Let $P(t)$ be absolutely continuous, $p(t)$ being its density. If moreover*

$$a = \int_0^\infty t\,p(t)\,dt < \infty, \quad \lim_{t\to\infty} p(t) = 0 \ \text{ and } \ \sup_{t>0} p^{n*}(t) < \infty \ \text{ for some } n,$$

then

$$\lim_{t\to\infty} h(t) = \frac{1}{a}.$$

The assumption of the boundedness of the $p^{n*}(t)$ can be changed to the assumption of the integrability of $p(t)$ over $(0, \infty)$ with some order greater than one. Moreover, the boundedness of $p^{n*}(t)$ implies

$$h(t) - \sum_{k=1}^{n} p^{k*}(t) \underset{t\to\infty}{\longrightarrow} \frac{1}{a}.$$

The cited theorems are valid for an infinite mean renewal time as well.

A nonnegative function $L(t)$, $t \geq 0$, is called **slowly varying** at infinity if

$$\lim_{t\to\infty} \frac{L(ts)}{L(t)} = 1$$

for any $s > 0$. A function $f(t)$, $t \geq 0$, is called **regularly varying** at infinity with an order α if $t^{-\alpha} f(t)$ is a slowly varying function at infinity. A function $g(t)$, $t > 0$, is called **regularly varying** at zero if $g(1/t)$ is regularly varying at infinity. Making use of these concepts we have the following theorems.

Theorem 12.11. *Assume that* $1 - P(t) = t^{-\alpha} L(t)$ *where* $0 \leq \alpha \leq 1$ *and* $L(t)$ *is slowly varying at infinity. Then*

$$H(t) \underset{t\to\infty}{\sim} \begin{cases} 1/L(t), & \text{if } \alpha = 0 \\ t^{\alpha} \sin(\alpha\pi)\,(\alpha\pi\, L(t))^{-1}, & \text{if } 0 < \alpha < 1 \\ t\left(\int\limits_0^t (1 - P(u))\,du\right)^{-1}, & \text{if } \alpha = 1. \end{cases}$$

This theorem is a generalization of the Elementary Renewal Theorem. If $\alpha = 1$, then the finiteness of the mean $\int\limits_0^\infty t\,dP(t)$ is not excluded, as for example in the case $1 - P(t) \sim (\log t)^{\beta}/t$ and $-\beta > 1$. The following theorem supplements the Blackwell Theorem.

Theorem 12.12. *Assume that under the hypotheses of Theorem 12.11 a function* $P(t)$ *is nonlatticed. If* $0 < \alpha \leq 0.5$, *then*

$$\liminf_{t\to\infty} t^{1-\alpha} L(t)\,(H(t+s) - H(t)) = s \sin(\alpha\pi)\,\pi^{-1};$$

if $0.5 < \alpha < 1$, *then*

$$\liminf_{t\to\infty} t^{1-\alpha} L(t)\,(H(t+s) - H(t)) = s \sin(\alpha\pi)\,\pi^{-1}; \tag{12.5}$$

if $\alpha = 1$, *then*

$$\lim_{t\to\infty} (H(t+s) - H(t)) \int_0^t (1 - P(u))\,du = s.$$

If $P(t)$ is absolutely continuous and its density is strictly monotone, then (12.5) is valid also for $0 < \alpha \leq 0.5$.

It is convenient to designate

$$M(t) = \begin{cases} 1/L(t), & \text{if } 0 < \alpha < 1 \\ \left(\int_0^t (1 - P(u))\, du\right)^{-1}, & \text{if } \alpha = 1. \end{cases}$$

The function $M(t)$ is slowly varying at infinity and has a bounded variation $\operatorname{Var} M(t)$.

Theorem 12.13. *Let the assumptions of Theorem 12.11 be satisfied with $0 < \alpha \leq 1$ and a finite limit $\lim_{t\to\infty} (M(t))^{-1} \operatorname{Var} M(t)$ exists. If moreover a nonnegative monotone function $f(t)$ is regularly varying at infinity with an order $\beta \in (-1,\ 0]$, then*

$$\int_0^t f(t-u)\, dH(u) \underset{t\to\infty}{\sim} \{(2+\beta)\, B(\alpha+\beta+1,\ 2-\alpha)\}^{-1} \times$$

$$\times \int_0^t f(u)\, du \left(\int_0^t (1 - P(u))\, du\right)^{-1}$$

where

$$B(x,\ y) = \int_0^1 t^{x-1} (1-t)^{y-1}\, dt$$

is the Euler beta-function.

The cited theorem also enables investigation of the asymptotic behavior of the solution of the equation

$$z(t) = f(t) + \lambda \int_0^t z(t-u)\, dP(u) \tag{12.6}$$

reducing the solution of (12.1) as $\lambda = 1$. If $\lambda > 1$, then there exists a unique positive μ such that

$$\lambda \int_0^\infty e^{-\mu t}\, dP(t) = 1.$$

If $\lambda = 1$, then such μ exists, is unique and equals 0. If $\lambda < 1$, then such μ (a negative one) exists iff the integral $\int_0^\infty e^{\varepsilon t}\, dP(t)$ converges for some $\varepsilon > 0$. In this case the equation (12.6) can be reduced to a usual renewal equation, making use of the multiplication by $e^{-\mu t}$. Indeed, the function $z^{\cdot}(t) = e^{-\mu t} z(t)$ satisfies the equation

$$z^{\cdot}(t) = e^{-\mu t} f(t) + \int_0^t z^{\cdot}(t-u)\, dP^{\cdot}(u)$$

where

$$P^{\cdot}(t) = \lambda \int_0^t e^{-\mu u}\, dP(u), \quad P^{\cdot}(\infty) = 1.$$

12.2 Renewal Process

A renewal process[5,7,8] is defined as a sequence $\zeta_n = \tau_1 + \cdots + \tau_n$ of sums of nonnegative independent identically distributed random variables $\tau_1, \tau_2, \ldots$. The term "renewal process" is due to reliability interpretation: ζ_n is called the nth renewal time. We have $\zeta_n \underset{n\to\infty}{\longrightarrow} \infty$ almost surely and so the number ν_t of renewals within the interval $[0, t]$ including the first renewal at time 0 is finite almost surely for any finite t.

The mean number of renewals during an interval $[0, t]$ equals the renewal function $H(t)$ associated with the distribution function $P(t) = \mathbf{P}\{\tau \le t\}$, i.e.,

$$\mathbf{E}\nu_t = 1 + \sum_{n\ge 1} \mathbf{P}\{\zeta_n \le t\} = 1 + \sum_{n\ge 1} P^{n*}(t) = H(t).$$

The process ν_t is called sometimes a renewal process.

The moments of the distribution of ν_t satisfy the renewal equation:

$$M_r(t) = L_r(t) + \int_0^t M_r(t-u)\, dP(u)$$

where

$$M_r(t) = \mathbf{E}\nu_t^r, \quad L_r(t) = \sum_{k=1}^{r} \binom{r}{k} (-1)^{k-1} M_{r-k}(t).$$

By the Elementary Renewal Theorem

$$\lim_{t\to\infty} t^{-r}\, \mathbf{E}\nu_t^r = a^{-r}$$

where

$$a = \mathbf{E}\,\tau_1 = \int_0^\infty t\,dP(t).$$

The most important variables associated with the renewal process are $\xi_t^- = t - \zeta_{\nu_t}$ and $\xi_t^+ = \zeta_{\nu_t+1} - t$. The former is interpreted as the age of a component in operation at time t; the latter is interpreted as the time to its failure. The joint distribution of both can be expressed via that of ξ_t^+ by the relation

$$\mathbf{P}\{\xi_t^- \geq u,\ \xi_t^+ \geq v\} = \mathbf{P}\{\xi_{t-u}^+ \geq u+v\} \text{ for } u \leq t.$$

The distribution function $\mathbf{P}\{\xi_t^+ \leq u\}$ satisfies the renewal equation (12.1) where $f(t) = P(t+u) - P(t)$. The limit theorems for the renewal equation imply the following results.

Theorem 12.14. *If $P(t)$ is a nonlatticed distribution function and a relation (12.2) is satisfied, then*

$$\lim_{t\to\infty} \mathbf{P}\{\xi_t^- \geq u,\ \xi_t^+ \geq v\} = \frac{1}{a}\int_{u+v}^\infty (1 - P(t))\,dt.$$

The processes ξ_t^- and ξ_t^+ are homogeneous Markov processes. The Theorem 12.14 means that their transition probabilities converge in the weak sense to the stationary distribution with the probability density function $(1 - P(t))/a$. The convergence holds true for every Borel set and is uniform iff some convolution of $P(t)$ is nonsingular.

Theorem 12.15. *If the distribution $P(t)$ is latticed with the step δ and $0 < a = \int_0^\infty t\,dP(t) < \infty$, then*

$$\lim_{n\to\infty} \mathbf{P}\{\xi_{n\delta}^- = k\delta,\ \xi_{n\delta}^+ = l\delta\} = \delta\,\frac{p_{k+l}}{a}$$

where $p_k = P(k\delta+) - P(k\delta-)$.

Theorem 12.16. *If the distribution $P(t)$ is nonlatticed and $1 - P(t) = t^{-\alpha} L(t)$ where $0 < \alpha < 1$ and $L(t)$ is slowly varying at infinity, then*

$$\lim_{t\to\infty} \mathbf{P}\left\{\frac{1}{t}\,\xi_t^- \leq x\right\} = \frac{\sin(\pi\alpha)}{\pi}\int_0^x u^{-\alpha}(1-u)^{\alpha-1}\,du, \quad 0 \leq x \leq 1,$$

$$\lim_{t\to\infty} \mathbf{P}\left\{\frac{1}{t}\,\xi_t^- \leq x\right\} = \frac{\sin(\pi\alpha)}{\pi}\int_0^x u^{-\alpha}(1+u)^{-1}\,du, \quad x \geq 0,$$

$$\lim_{t\to\infty} \mathbf{P}\left\{\frac{1}{t}\xi_t^- \geq x,\ \frac{1}{t}\xi_t^+ \geq y\right\} = \frac{\sin(\pi\alpha)}{\pi} \int_{\frac{x+y}{1-x}} u^{-\alpha}(1+u)^{-1}\,du,$$

$$0 \leq x < 1,\ y \geq 0.$$

Theorem 12.17. *If $t\,(1-P(t))$ is slowly varying at infinity and $P(t)$ is nonlatticed, then*

$$\lim_{t\to\infty} \mathbf{P}\left\{\frac{m(\xi_t^-)}{m(t)} \leq x,\ \frac{m(\xi_t^+)}{m(t)} \leq y\right\} = \min(x, y), \quad 0 \leq x, y \leq 1$$

where $m(t) = \int_0^t (1-P(s))\,ds$.

The cited theorems hold also in the latticed case provided that t tends to infinity along the points of the lattice.

12.3 Rate of the Convergence

Consider some estimates of the convergence rate[3,5,7,9–13]. Let $P(t)$ be a nonlatticed distribution function. Then

$$H(t) - \frac{t}{a} \underset{t\to\infty}{\longrightarrow} \frac{b}{2a^2}, \quad \text{if } b = \int_0^\infty t^2\,dP(t) < \infty,$$

$$H(t) - \frac{t}{a} \underset{t\to\infty}{\sim} \frac{t^2}{a^2\varepsilon(1-\varepsilon)}(1-P(t)) \text{ if } t^{1+\varepsilon}(1-P(t))$$

is slowly varying at infinity and the variation of the function $H(t)/t$ over an interval $[0, s]$ is equivalent to $cH(s)/s$ as $s \to \infty$.

Assume that $P(t)$ is a nonlatticed distribution with a nonsingular convolution $P^{n*}(t)$ for some n. If moreover $\int_0^\infty t^r\,dP(t) < \infty$, then

$$H(t+s) - H(t) = \frac{s}{a} + o(t^{1-r}) \quad \text{as } t \to \infty.$$

Theorem 12.18. *Assume that $\int_0^\infty t^r\,dP(t) < \infty$ and a convolution $P^{n*}(t)$ is nonsingular for some $n \geq 1$. Then the variation of the function*

$$H(u) - \frac{u}{a} - \frac{1}{a^2}\int_0^u dx \int_x^\infty (1-P(y))\,dy, \quad u \geq 0,$$

over the interval $[t,\, t+s]$ *is infinitely small with respect to*

$$\begin{cases} t^{2(1-r)}, & \text{if } 1 \le p < 2 \\ t^{-r}, & \text{if } p \ge 2. \end{cases}$$

If $r \ge 2$, then the following statement holds true.

Theorem 12.19. *Let a convolution* $P^{n*}(t)$ *be nonsingular for some* n. *If moreover* $\int_0^\infty t^r\, dP(t) < \infty$ *for some integer* $r \ge 2$, *then*

$$H(t) = \int_0^t g(u)\, du + F(t)$$

where the function $g(t)$ *is continuous, the variation of the function* $F(t)$ *is finite,*

$$\int_0^\infty t^r\, d\mathrm{Var}F(t) < \infty \text{ and } g(t) = a^{-1} + \int_t^\infty dx \int_x^\infty (1-P(y))\, dy + o(t^{-r}) \text{ as } t \to \infty.$$

If $1 - P(t) = o(e^{-\alpha t})$ *for some* $\alpha > 0$, *then*

$$g(t) = a^{-1} + o(e^{-\beta t}) \quad \text{and} \quad \int_0^\infty e^{\beta t}\, d\mathrm{Var}F(t) < \infty$$

for some $\beta > 0$.

Let the assumptions of the Theorem 12.19 be satisfied and $\beta \in (0,\, \alpha]$ is small enough so that $\int_0^\infty e^{\beta t}\, dP_0^{n*}(t) < 1$ where $P_0^{n*}(t)$ is the singular component of the nonsingular convolution $P^{n*}(t)$. Then the function

$$\varphi(t) = \left\{ 1 - \int_0^\infty e^{\beta t}\, dP(t) \right\}^{-1}$$

has a finite number of poles $z_1, z_2, \dots, z_m$ in the band $0 < \operatorname{Re} z < \beta$ and

$$H(t) = \frac{t}{a} + \frac{b}{2a^2} + \sum_{i=1}^m \frac{c_i}{z_i}\, e^{-x_i t} + o(e^{-\beta t}) \quad \text{as} \quad t \to \infty$$

where c_i is the residue of $\varphi(z)$ in a pole z_i.

The following approach leads to the derivation of a wide class of estimates for the rate of convergence. Let a nonnegative function $V(t)$ be nondecreasing and $V(t+u) \le V(t)\,V(s)$, $V(0+) = 1$. Let also a convolution $P^{n*}(t)$ be nonsingular. If moreover a function $f(t)$ is bounded,

$$V(t)\left\{|f(t)| + \int_t^\infty |f(s)|\,ds\right\} \underset{t\to\infty}{\longrightarrow} 0 \quad \text{and} \quad \int_0^\infty V(s)\,(1-P(s))\,ds < \infty,$$

then

$$V(\varepsilon t)\left|z(t) - \frac{1}{a}\int_0^\infty f(s)\,ds\right| \underset{t\to\infty}{\longrightarrow} 0 \tag{12.7}$$

for sufficiently small $\varepsilon > 0$.

If the distribution $P(t)$ is absolutely continuous, $p(t)$ being its probability density function, and the difference $R(t) = \alpha\,(1-P(t)) - p(t)$ is nondecreasing for some $\alpha > 0$ and $\int_0^\infty t^r\,p(t)\,dt < \infty$ for some $r > 2$, then

$$\left|H(t) - \frac{t}{a} - a^{-2}\int_0^t ds\int_s^\infty (1-P(u))\,du\right| \le C\,(1-\varepsilon\,t)^{1-r} \tag{12.8}$$

where

$$C = \left(1 + \frac{\alpha\,2^r}{\varepsilon\,(r-2)}\right)\frac{1+l}{1-l}\,\frac{\left\{a^{-1}\int_0^\infty (1+\varepsilon t)^{r-1}\,(1-P(t))\,dt\right\}^2}{1 + \int_0^\infty (1+\varepsilon t)^{r-1}\,R(t)\,dt}$$

and a positive number ε is small enough so that

$$l = \int_0^\infty (1+\varepsilon t)^{r-1}\,R(t)\,\exp\left\{-\int_0^t R(u)\,du\right\}dt < 1.$$

The difference $\alpha\,(1-P(t)) - p(t)$ is nondecreasing iff

$$p(t) = \alpha\,e^{-\alpha t}\int_0^t e^{\alpha s}\,dQ(s)$$

where $Q(s)$ is a distribution function.

Assume that the distribution $P(t)$ possesses an absolutely continuous component with the density function $q(t) \geq \beta e^{-\alpha t}$, the β and α being positive constants. If moreover $\int_0^\infty t^r q(t)\,dt < \infty$ for some $r > 2$, then

$$\left| H(t) - \frac{t}{a} - a^{-2} \int_0^t ds \int_s^\infty (1 - P(u))\,du \right| \leq C\,(1 + \varepsilon t)^{1-r} \tag{12.9}$$

where

$$C = \frac{1+l}{1-l} \left(1 + \frac{\alpha - \beta}{\varepsilon\,(r-1)}\right) \left(\frac{\varepsilon r \alpha}{\beta} + \frac{2\alpha^2}{\alpha^2 - \beta^2 + a\alpha^2\beta}\right) \times$$

$$\times \left(1 + \int_0^\infty (1 + \varepsilon t)^r\, q(t)\,dt\right) \left\{a^{-1} \int_0^\infty (1 + \varepsilon t)^{r-1}\,(1 - P(t))\,dt\right\}^2$$

and $\varepsilon > 0$ is so small that

$$\varepsilon\,(r-2) \int_0^\infty (1 + \varepsilon t)^{r-1}\,(1 - P(t))\,dt < \frac{\beta}{\alpha}$$

and

$$0 < 1 - l = \exp\{(\beta - \alpha)\,(a\alpha^2/\beta - 1)\} -$$

$$- \frac{\varepsilon \alpha r}{\beta} \left(\frac{\beta}{\alpha} - \varepsilon\,(r-1) \int_0^\infty (1 + \varepsilon t)^{r-1}\,(1 - P(t))\,dt\right)^{-1} \times$$

$$\times \left(a\alpha - \frac{\beta}{\alpha} \int_0^\infty (1 + \varepsilon t)^{r-2}\,(1 - P(t))\,dt + \right.$$

$$\left. + \beta \int_0^\infty (1 + \varepsilon t)^{r-2} \int_t^\infty (1 - P(u))\,du\,dt - \frac{a\alpha}{\beta}\right).$$

The inequalities (12.8) and (12.9) are special cases of the following theorems.

Theorem 12.20. *Assume that $P(t)$ is absolutely continuous, $p(t)$ is its probability density function, and the difference $R(t) = \alpha\,(1 - P(t)) - p(t)$ is nondecreasing for some $\alpha > 0$. If moreover*

$$\int_0^\infty t^r\, p(t)\,dt < \infty \quad \text{and} \quad \sup_{t \geq 0}\,(1 + t)^r\,|f(t)|\,dt < \infty \quad \text{for some } r > 2,$$

then

$$\left| z(t) - \frac{1}{\alpha} \int_0^\infty f(u)\,du \right| \le C\,(1+\varepsilon\, t)^{1-r}$$

where

$$C = \frac{1+l}{1-l} \left(1 + \frac{\alpha}{\varepsilon\,(r-1)}\right) \left(1 + \int_0^\infty (1+\varepsilon t)^{r-1} R(t)\,dt\right)^{-1} \times$$

$$\times \sup_{t\ge 0} \left\{ (1+\varepsilon t)^r \left| f(t) - a^{-1}(1-P(t)) \int_0^\infty f(u)\,du \right| \right\}$$

and a, ε, l *are the same as in (12.8).*

Theorem 12.21. *Let the distribution* $P(t)$ *have an absolutely continuous component with the density* $q(t) \ge \beta\, e^{-\alpha t}$ *for some* $\alpha > 0$, $\beta > 0$. *Assume also that*

$$\int_0^\infty t^r\, q(t)\,dt < \infty \quad \text{and} \quad \sup_{t\ge 0}\,(1+t)^r\,|f(t)| < \infty \;\; \text{for some } r > 2.$$

Then

$$\left| z(t) - \frac{1}{a} \int_0^\infty f(u)\,du \right| \le C\,(1+\varepsilon\, t)^{1-r}$$

holds with

$$C = \frac{1+l}{1-l} \left(1 + \frac{\alpha-\beta}{\varepsilon\,(r-1)}\right) \left(\frac{\varepsilon\, r\, \alpha}{\beta} + \frac{2\alpha^2}{\alpha^2 - \beta^2 + a\alpha^2\beta}\right) \times$$

$$\times \left(1 + \int_0^\infty (1+\varepsilon t)^r\, q(t)\,dt\right) \sup_{t\ge 0} \left\{ (1+\varepsilon t)^{r-1} \left| f(t) - \frac{1-P(t)}{\alpha} \int_0^\infty f(u)\,du \right| \right\}$$

and a, ε, l *are the same as in (12.9).*

12.4 Uniform Theorems

Consider some uniform theorems[11,13–16]. Let $\mathcal{D}$ be a class of distribution functions $P(t)$ with probability density functions $p(t)$.

Theorem 12.22. *Assume that*

$$\sup_{\mathcal{D}} p(t) \underset{t\to\infty}{\longrightarrow} 0, \quad \sup_{\mathcal{D}} \int_t^\infty (1 - P(u))\, du \underset{t\to\infty}{\longrightarrow} 0, \quad \sup_{\mathcal{D}} \sup_{t>0} p^{n*}(t) < \infty$$

for some integer $n > 0$. Then

$$\sup_{\mathcal{D}} | h(t) - a^{-1} | \underset{t\to 0}{\longrightarrow} 0.$$

If the above mentioned conditions of the validity of the relation (12.7) are satisfied uniformly in $\mathcal{D}$ and in a class $\mathcal{F}$ of functions $f(t)$, then under mild additional assumptions this relation is satisfied uniformly too.

A class $\mathcal{F}$ of functions $f(t)$ is called **uniformly directly Riemann-integrable**, if

$$\sup_{\mathcal{F}} \sup_{t\geq 0} |f(t)| < \infty, \quad \sup_{\mathcal{F}} \sum_{k\geq n} \sup_{n\leq t<n+1} |f(t)| \underset{n\to\infty}{\longrightarrow} 0,$$

$$\sup_{\mathcal{F}} \sum_{k\geq 0} \delta \{ \sup_{k\delta\leq t<k\delta+\delta} |f(t)| - \inf_{k\delta\leq t<k\delta+\delta} |f(t)| \} \underset{\delta\downarrow 0}{\longrightarrow} 0.$$

A class **D** of distribution functions $P(t)$ is called **uniformly nonlatticed**, if

$$\lim_{c\downarrow 0} \sup_{\mathcal{D}} \sum_{n\geq 0} \{P(n\delta + c) - P(n\delta - c)\} < 1$$

for any $\delta > 0$.

Theorem 12.23. *Let a class of distribution functions $\mathcal{D}$ be uniformly nonlatticed and*

$$\sup_{\mathcal{D}} \int_t^\infty (1 - P(s))\, ds \underset{t\to\infty}{\longrightarrow} 0.$$

If moreover a class $\mathcal{F}$ of functions is uniformly directly Riemann-integrable, then

$$\sup_{\mathcal{D}} \sup_{\mathcal{F}} \left| z(t) - \frac{1}{a} \int_t^\infty f(s)\, ds \right| \underset{t\to\infty}{\longrightarrow} 0.$$

The uniform nonlatticedness of a class $\mathcal{D}$ means that if a sequence of distributions which belong to $\mathcal{D}$ has a limiting distribution, then the latter one is nonlatticed.

Cumbersome conditions of the uniform direct Riemann-integrability become very simple in the case if all the functions f of the class $\mathcal{F}$ are monotone and nonnegative. Indeed, they reduce to the following ones:

$$\sup_{\mathcal{F}} f(0) < \infty, \quad \sup_{\mathcal{F}} \int_t^\infty f(s)\,ds \underset{t\to\infty}{\longrightarrow} 0.$$

12.5 Transient Phenomena

Transient phenomena[17,18] in renewal theory means a variety of properties connected to the asymptotics of the solution $z(t)$ of the equations (12.6) as $t \to \infty$ and $\lambda \to 1$. The distribution function $P(u)$ and the free term $f(t)$ can vary within the classes $\mathcal{D}$ and $\mathcal{F}$ respectively.

Theorem 12.24. *If*

$$\sup_{\mathcal{D}} \int_t^\infty (1 - P(s))\,ds \underset{t\to\infty}{\longrightarrow} 0 \quad \text{and} \quad \inf_{\mathcal{D}} \int_t^\infty (1 - P(s))\,ds > 0,$$

then

$$\sup_{\mathcal{D}} \left| \frac{1}{t}\,H(t) - \frac{1}{a} \int_0^1 e^{u\,c/a}\,du \right| \to 0$$

provided $t \to \infty, \;\; \lambda \to 1, \;\; t\,(\lambda - 1) \to c.$

Theorem 12.25. *Assume that a class $\mathcal{F}$ of functions $f(t)$ is uniformly directly Riemann-integrable, a class $\mathcal{D}$ of distribution functions $P(t)$ is uniformly nonlatticed and*

$$\lim_{t\to\infty} \sup_{\mathcal{D}} \int_t^\infty (1 - P(s))\,ds = 0.$$

Then

$$\sup_{\mathcal{F}} \sup_{\mathcal{D}} \left| z(t) - \frac{1}{a}\,e^{c/a} \int_t^\infty f(s)\,ds \right| \to 0$$

as soon as $t \to \infty, \;\; \lambda \to 1, \;\; t\,(\lambda - 1) \to c.$

If moreover $\lambda > 1$, then the following stronger conclusion holds.

Theorem 12.26. *If*

$$\sup_{\mathcal{D}} \int_t^\infty (1 - P(s))\,ds \underset{t\to\infty}{\longrightarrow} 0 \quad \text{and} \quad \inf_{\mathcal{D}} \int_t^\infty (1 - P(s))\,ds > 0,$$

then

$$\sup_{\mathcal{D}} \left| \frac{\lambda - 1}{e^{t(\lambda-1)/a} - 1} H(t) - 1 \right| \to 0$$

as soon as $t \to \infty$, $\lambda \downarrow 1$.

Theorem 12.27. *Assume that a class $\mathcal{F}$ of functions $f(t)$ is uniformly directly Riemann-integrable over $[0, \infty)$, a class $\mathcal{D}$ of distribution functions is uniformly nonlatticed and*

$$\lim_{t\to\infty} \sup_{\mathcal{D}} \int_t^\infty (1 - P(s))\, ds = 0.$$

Then

$$\sup_{\mathcal{F}} \sup_{\mathcal{D}} \left| e^{t(1-\lambda)/a} z(t) - \frac{1}{a} \int_t^\infty f(s)\, ds \right| \to 0$$

as $t \to \infty$, $\lambda \downarrow 1$.

Similar statements hold also in the case $\lambda \leq 1$ though under much stronger assumptions.

Theorem 12.28. *If*

$$\inf_{\mathcal{D}} \int_0^\infty (1 - P(s))\, ds > 0 \quad \text{and} \quad \sup_{\mathcal{D}} e^{\varepsilon t} (1 - P(t)) \underset{t\to\infty}{\longrightarrow} 0 \ \text{ for some } \varepsilon > 0,$$

then

$$\sup_{\mathcal{D}} \left| \frac{\lambda - 1}{e^{t(\lambda-1)/a} - 1} H(t) - 1 \right| \to 0$$

as $t \to \infty$, $\lambda \uparrow 1$.

Theorem 12.29. *Let a class $\mathcal{F}$ of functions $f(t)$ be uniformly directly Riemann-integrable over $[0, \infty)$, and a class $\mathcal{D}$ of distribution functions be uniformly nonlatticed. Assume also that*

$$\sup_{\mathcal{F}} e^{\varepsilon t} |f(t)| \underset{t\to\infty}{\longrightarrow} 0 \quad \text{and} \quad \sup_{\mathcal{D}} e^{\varepsilon t} (1 - P(t)) \underset{t\to\infty}{\longrightarrow} 0 \ \text{ for some } \varepsilon > 0.$$

Then

$$\sup_{\mathcal{F}} \sup_{\mathcal{D}} \left| e^{t(\lambda-1)/a} z(t) - \frac{1}{a} \int_0^\infty f(s)\, ds \right| \to 0$$

as $t \to \infty$, $\lambda \uparrow 1$.

If a class $\mathcal{D}$ of distributions is relatively compact in the sense of the convergence in variation, uniformly nonlatticed and

$$\sup_{\mathcal{D}} \int_t^\infty (1 - P(s))\, ds \underset{t\to\infty}{\longrightarrow} 0,$$

then

$$\sup_{\mathcal{D}} |t\,\{H(t+u+v) - H(t+u) - H(t+v) + H(t)\}| - c\,e^{c/a}\,\frac{uv}{a^2} \to 0 \quad (12.10)$$

as $t \to \infty,\ \lambda \to 1,\ t\,(\lambda - 1) \to c.$

Note that in a traditional situation the second difference of the renewal function is a variable of $o(t^{-r})$ provided $\int_0^\infty t^r\, dP(t)\, dt < \infty$. The relation (12.10) is a special case of the following theorem.

Theorem 12.30. *Let a class $\mathcal{F}$ of functions $f(t)$ have a uniform bounded majorant intergrable over $(0, \infty)$, a class $\mathcal{D}$ of distributions $P(t)$ be relatively compact in variation and*

$$\inf_{\mathcal{D}} \int_0^\infty \frac{dP^{n*}}{dt}\, dt > 0 \ \text{ for some } n = 1, 2, \dots .$$

Assume also that the variations of all the functions $\Phi(t)$ of a class $\mathcal{V}$ are uniformly bounded and

$$\sup_{\mathcal{V}} \int_t^\infty s\, d\mathrm{Var}\,\Phi(s) \underset{t\to\infty}{\longrightarrow} 0, \quad \sup_{\mathcal{D}} \int_t^\infty s\, dP(s) \underset{t\to\infty}{\longrightarrow} 0.$$

Then

$$\sup_{\mathcal{F}} \sup_{\mathcal{D}} \sup_{\mathcal{V}} \Bigg| t \int_0^t z(t-u)\, d\Phi(u) - $$

$$- \frac{d}{a}\, e^{c/a} \int_0^\infty f(s)\, ds + \frac{c}{d^2}\, e^{c/a} \int_0^\infty f(s)\, ds \int_0^\infty s\, d\Phi(s) \Bigg| \to 0$$

as $t \to \infty,\ \lambda \to 1,\ t\,(\lambda - 1) \to c,\ t\,(\Phi(\infty) - \Phi(0)) \to d.$

12.6 Markov Renewal Equation

Main theorems concerning the asymptotic behavior of the solution of the renewal equation admit a generalization to the so-called **Markov renewal equation**[7,17–22] defined as the system of integral equations

$$z_i(t) = f_i(t) + \sum_{j=1}^{d} \int_0^t dP_{ij}(u)\, z_j(t-u), \quad i = 1, \dots, d, \tag{12.11}$$

where $f_i(t)$, $i = 1, \dots, d$, are given functions, $z_i(t)$, $i = 1, \dots, d$, are the functions to be obtained and nondecreasing functions $P_{ij}(t)$ are such that

$$\sum_{j=1}^{d} P_{ij}(\infty) = 1, \quad i = 1, \dots, d, \qquad \min_i \sum_{j=1}^{d} P_{ij}(0+) < 1.$$

The equation (12.11) arises while investigating the so-called **renewal Markov process**, i.e., an homogeneous Markov chain $(\kappa_n,\, \zeta_n)$ additive (homogeneous) and nondecreasing in the second component, the first one possessing a finite number of states $1, \dots, d$. The additivity in the second component means that

$$\mathbf{P}\{\kappa_{n+1} = j,\, \zeta_{n+1} \le t \mid \kappa_n = i,\, \zeta_n = s\} =$$

$$= \mathbf{P}\{\kappa_1 = j,\, \zeta_1 \le t - s \mid \kappa_0 = i,\, \zeta_0 = 0\} = P_{ij}(t-s), \quad 0 \le s \le t.$$

The sequence $\{\kappa_n\}$ is a homogeneous Markov chain with one step transition probabilities $\{P_{ij}(\infty)\}$.

A Markov renewal process can be defined in a constructive way via sums of random variables conditionally independent given a path-function of the Markov chain $\{\kappa_n\}$. Assume that for each $i,\, j = 1, \dots, d$ the random variables $\tau_1(i,\, j)$, $\tau_2(i,\, j)$, $\dots$ are independent and have an identical distribution function $P_{ij}(t) = \mathbf{P}\{\tau_n(i,\, j) \le t\}$. One can set

$$\zeta_n = \zeta_0 + \sum_{k=1}^{n} \tau_k(\kappa_{k-1},\, \kappa_k).$$

The equation (12.11) with $f_i(t) = P_{ij}(\infty) - P_{ij}(t+s)$ considered for a fixed $j = 1, \dots, d$ and $s > 0$ is satisfied with the probability

$$z_i(t) = \mathbf{P}\{\kappa_{\nu_t+1},\, \zeta_{\nu_t+1} \ge t + s \mid \kappa_0 = i,\, \zeta_0 = 0\}$$

where $\nu_t = \max\{n : \zeta_n \le t\}$ is the number of renewals in $[0, t)$. The equation (12.11) can be written in a matrix form

$$\mathbf{z}(t) = \mathbf{f}(t) + \int_0^t d\mathbf{P}(u)\,\mathbf{z}(t-u)$$

where

$$\mathbf{z}(t) = (z_1(t), \ldots, z_d(t)), \;\; \mathbf{f}(t) = (f_1(t), \ldots, f_d(t)), \;\; \mathbf{P}(u) = (\,P_{ij}(u)\,)_{i,j=1}^d .$$

If the functions $f_i(t)$ are locally bounded, then, similarly to the classical case, the equation (12.11) has a unique locally bounded solution

$$z_i(t) = \sum_j \int_0^t dH_{ij}(u)\, f_j(t-u)$$

which can be written in a matrix form as

$$\mathbf{z}(t) = \int_0^t d\mathbf{H}(u)\,\mathbf{f}(t-u).$$

Here

$$(\,H_{ij}(t)\,)_{i,j=1}^d = \mathbf{H}(t) = \mathbf{E} + \sum_{n\ge 1} \mathbf{P}^{n*}(t)$$

is a matrix analog of the renewal function, $\mathbf{E} = (\,\delta_{ij}(t)\,)_{i,j=1}^d$ is the unit matrix,

$$\mathbf{P}^{1*}(t) = \mathbf{P}(t), \quad \mathbf{P}^{n+1*}(t) = \int_0^t \mathbf{P}^{n*}(t-u)\,d\mathbf{P}(u).$$

The asymptotic behavior of the solution of the renewal equation can be expanded to Markov renewal ones also though in the case of a decomposable matrix $\mathbf{P}(\infty)$. This means that the set $\{1, \ldots, d\}$ can't be divided into two nonempty sets I_1 and I_2 such that $P_{ij}(\infty) = 0$ as soon as $i \in I_1$ and $j \in I_2$.

A matrix $P(t)$ is called **latticed** if there exist $\delta > 0$ and $c_1, \ldots, c_d$ such that all the points of the increase of $P_{ij}(t)$ are concentrated in the shifted lattices $\{c_j - c_i + n\delta,\; n = 0, \pm 1, \pm 2, \ldots\}$. The largest such δ (if it exists) is called **the step** of the latticed matrix $\mathbf{P}(t)$. The vector $\mathbf{c} = (c_1, \ldots, c_d)$ is called the shift vector. Otherwise the matrix $\mathbf{P}(t)$ is called **nonlatticed**.

If a matrix $\mathbf{P}(\infty)$ is indecomposable, then a unique vector $\pi = (\pi_1, \ldots, \pi_d)$ exists such that

$$\sum_{j=1}^{d} \pi_j = 1 \quad \text{and} \quad \pi_j = \sum_{i=1}^{d} \pi_i P_{ij}(\infty).$$

Theorem 12.31. *If the matrix $\mathbf{P}(\infty)$ is irreducible and*

$$a = \sum_{i,j} \pi_i \int_0^\infty t\, dP_{ij}(t) < \infty, \tag{12.12}$$

then

$$\lim_{t\to\infty} \frac{1}{t} H_{ij}(t) = \frac{\pi_j}{a}.$$

Theorem 12.32. *If the matrix $\mathbf{P}(t)$ is nonlatticed, $\mathbf{P}(\infty)$ is indecomposable and the relation (12.12) is satisfied, then*

$$\lim_{t\to\infty} (H_{ij}(t+s) - H_{ij}(t)) = \frac{\pi_j s}{a}.$$

Theorem 12.33. *Let the matrix $\mathbf{P}(t)$ be latticed with a step δ and a shift vector $(c_1, \ldots, c_d)$. Assume also that the matrix $\mathbf{P}(\infty)$ is indecomposable and the relation (12.12) is satisfied. Then*

$$\lim_{n\to\infty} (H_{ij}(c_j - c_i + n\delta + \delta) - H_{ij}(c_j - c_i + n\delta)) = \frac{\delta}{a}.$$

Theorem 12.34. *If the matrix $\mathbf{P}(t)$ is nonlatticed, the matrix $\mathbf{P}(\infty)$ is indecomposable, the relation (12.12) is satisfied and functions $f_i(t)$, $i = 1, \ldots, d$, are directly Riemann-integrable over $[0, \infty)$, then*

$$\lim_{t\to\infty} z_i(t) = \frac{1}{a} \sum_j \pi_j \int_0^\infty f_j(t)\, dt.$$

Theorem 12.35. *Let the matrix $\mathbf{P}(\infty)$ be indecomposable, and the matrix $\mathbf{P}(t)$ be latticed with a step δ and a shift vector $(c_1, \ldots, c_d)$. Assume that the relation (12.12) is satisfied and*

$$\sum_n |f_i(n\delta - c_i)| < \infty, \quad i = 1, \ldots, d.$$

Then

$$\lim_{n\to\infty} z_i(n\delta - c_i) = \frac{1}{a} \sum_j \pi_j \sum_n f_j(n\delta - c_j).$$

Let Δ be the imprimitivity index of the matrix $\mathbf{P}(\infty)$ defined as the number of its eigenvalues on the unit circle $|z| = 1$ of a complex plane. The Δ also equals the greatest common divisor of such n for which the first diagonal element of the matrix $\mathbf{P}^n(\infty)$ is positive.

Theorem 12.36. *Let the matrix $\mathbf{P}(\infty)$ be indecomposable and the diagonal elements of the matrix $\mathbf{P}^{n\Delta *}(t)$ nonsingular for some positive integer n. If moreover the relation (12.12) is satisfied and bounded $[0, \infty)$-integrable functions $f_i(t)$, $i = 1, \ldots, d$, have a zero limit as $t \to \infty$, then*

$$\lim_{t\to\infty} z_i(t) = \frac{1}{a} \sum_j \pi_j \int_0^\infty f_j(t)\, dt.$$

In the application to Markov renewal processes Theorem 12.34 enables obtainment of the limiting distribution of the process state at the time of a level t crossing.

Theorem 12.37. *Assume that the Markov chain $\{\kappa_n\}$ is irreducible and possesses an invariant probability distribution $(\pi_0, \ldots, \pi_d)$. If the matrix $\mathbf{P}(t)$ is nonlatticed and the relation (12.12) is satisfied, then*

$$\lim_{t\to\infty} \mathbf{P}\{\kappa_{\nu_t+1} = j,\ \zeta_{\nu_t+1} - t \geq s \mid \kappa_0 = i\} = \frac{1}{a} \sum_k \pi_k \int_s^\infty (P_{kj}(\infty) - P_{kj}(t))\, dt.$$

In the branching processes theory the equation

$$y_i(t) = g_i(t) + \sum_{j=1}^d \int_0^t y_j(t-u)\, dM_{ij}(u) \qquad (12.13)$$

is often considered, the $M_{ij}(u)$ being bounded nondecreasing functions satisfying the equality $M_{ij}(0) = 0$. Analytical properties of the solution of this equation depend on the maximal eigenvalue **(the Perronian root)** λ of the matrix $\mathbf{M}(\infty) = (M_{ij}(\infty))$. If the Perronian root of an indecomposable matrix $\mathbf{M}(\infty)$ equals 1, then this matrix has left $\mathbf{v} = (v_1, \ldots, v_d)$ and right $\mathbf{u} = (u_1, \ldots, u_d)$ eigenvectors with positive components, i.e.,

$$\sum_j M_{ij}(\infty)\, u_j = u_i, \quad \sum_i v_i\, M_{ij}(\infty) = v_j.$$

In such a case, if $y_i(t)$ is a solution of the equation (12.13), then

$$z_i(t) = \frac{1}{u_i}\, g_i(t), \quad P_{ij}(t) = \frac{1}{u_i}\, M_{ij}(t)\, u_j.$$

A left invariant vector π of the matrix $\mathbf{P}(\infty)$ has components

$$\pi_i = \frac{v_i u_i}{\sum_j v_j u_j} .$$

For example, an analogue of the Theorem 12.34 for the equation (12.13) is as follows.

Theorem 12.38. *Let the matrix $\mathbf{M}(\infty)$ be indecomposable, its Perronian root equals 1, the matrix $\mathbf{M}(t)$ be nonlatticed. Assume that*

$$0 < m = \sum_{i,j} v_i \int_0^\infty t\, dM_{ij}(t)\, u_j < \infty$$

and functions $g_i(t)$, $i = 1, \ldots, d$, are directly Riemann-integrable over $[0, \infty)$. Then

$$\lim_{t\to\infty} y_i(t) = u_i \sum_j v_j \int_0^\infty g_j(t)\, dt \left(m \sum_j v_j u_j \right)^{-1} .$$

12.7 References

1. Karlin, S., On the renewal equation, *Pacif. J. Math.*, 5, 229, 1955.
2. Smith, W. L., On necessary and sufficient conditions for the convergence of the renewal density, *Trans. Amer. Math. Soc.*, 104, 79, 1962.
3. Tengels, J. L., Renewal theorems when the first or the second moment is infinite, *Ann. Math. Statist.*, 39, 1210, 1968.
4. Feller, W., *An Introduction to Probability Theory and Its Applications,* 1, Wiley, New York, 1968.
5. Feller, W., *An Introduction to Probability Theory and Its Applications,* 2, Wiley, New York, 1971.
6. Erickson, K. B., A renewal theorem for distribution on $\mathbb{R}^1$ without expectation, *Bull. Amer. Math. Soc.*, 77, 406, 1971.
7. Sevastyanov, B. A., A renewal theory, *Itogi Nauki i Tekhniki, Teoriya Veroyatn., Mat. Statist., Teoret. Kibern.*, 19, 15, 1974 (in Russian).
8. Erickson, K. B., Strong renewal theorems with infinite mean, *Trans. Amer. Math. Soc.*, 151, 263, 1970.
9. Stone, C., On absolutely continuous components and renewal theory, *Ann. Math. Statist.*, 37, 271, 1966.
10. Van der Genugten, B. B., Asymptotic expansions in renewal theory, *Compositio Math.*, 21, 331, 1969.

11. Zubkov, A. M., On the convergence rate of a renewal density, *Mat. Sbornik,* 98, 143, 1975 (in Russian).
12. Rogozin, B. A., Asymptotics of renewal functions, *Theor. Probab. Appl.,* 21, 669, 1976.
13. Kartashov, N. V., A uniform asymptotic renewal theorem, *Theor. Probab. Appl.,* 25, 589, 1980.
14. Silvestrov, D. S., One generalization of a renewal theorem, *Dopovidi Akad. Nauk Ukr.SSR,* A, 978, 1976 (in Russian).
15. Silvestrov, D. S., Renewal theorem in a scheme of series, *Teor. Veroyatn. Mat. Statist.,* 18, 144, 1978 (in Russian).
16. Kartashov, N. V., Power estimates for the convergence rate in a renewal theorem, *Teor. Probab. Appl.,* 24, 606, 1979.
17. Shurenkov, V. M., Transitional phenomena of renewal theory in asymptotic problems of the theory of random processes, 1, *Math. USSR Sbornik,* 40, 211, 1981.
18. Shurenkov, V. M., Transitional phenomena of renewal theory in asymptotic problems of the theory of random processes, 2, *Math. USSR Sbornik,* 40, 211, 1981.
19. Sevast'janov, B. A., *Verzweigungsprozesse,* Oldenburg, München, 1975.
20. Korolyuk, V. S., Borovskikh, Yu. V., *Analytical Problems of the Asymptotics of Probability Distributions,* Naukova Dumka, Kiev, 1981 (in Russian).
21. Shurenkov, V. M., Note on the multidimensional renewal equation, *Teor. Veroyatn. Primen.,* 20, 848, 1975 (in Russian).
22. Shurenkov, V. M., Transition phenomena in multidimensional renewal theorems, *Teor. Probab. Appl.,* 24, 442, 1979.

Chapter 13

Branching Processes

13.1 Galton-Watson Processes

The simplest (and hence the most studied) branching processes are **Galton-Watson processes**[1–6] describing the discrete time $t = 0, 1, 2, \ldots$ behavior of a population consisting of random number of particles. Namely, each of the existing at a current moment t particles independent of its origin, moment t and other existing at this moment particles transforms at moment $t + 1$ into some (possibly empty) population of particles of the same type which are called the descendants. The probability law of the population development is described by a set $\{p_k,\ k = 0, 1, 2, \ldots\}$ of probabilities that the number of particle's descendants equals k. The main object of investigations is the asymptotic behavior of the distribution of a random variable ζ_n, which means the number of particles at time $t = n$ provided that there was only one particle at $t = 0$.

A random sequence $\{\zeta_n,\ n = 0, 1, 2, \ldots\}$ forms a homogeneous Markov chain with one-step transition probabilities

$$p_{ij} = \sum_{k_1 + \cdots + k_i = j} p_{k_1} \cdots p_{k_i}, \quad i = 1, 2, \ldots, \quad j = 1, 2, \ldots, \quad p_{00} = 1.$$

This property of the sequence $\{\zeta_n,\ n = 0, 1, 2, \ldots\}$ is used sometimes as a formal definition of the Galton-Watson branching process.

A recursive relation

$$f_n(z) = f(f_{n-1}(z)), \quad f_0(z) - 1,$$

for the generating function

$$f_n(z) = \sum_{k\geq 0} z^k \mathbf{P}\{\zeta_n = k\} = \mathbf{E}\, z^{\zeta_n}, \quad |z| \leq 1,$$

is the basis for the analytical investigation of such processes. Here $f(z)$ is a generating function for the number of one-particle direct descendants

$$f(z) = \sum_{k\geq 0} p_k z^k.$$

Galton-Watson processes can be associated with random walks. Let $\sigma_n = 1 + \eta_1 + \cdots + \eta_n$ where $\eta_1, \eta_2, \ldots$ are independent random variables with the same distribution as $\zeta_1 - 1$, i.e., $\mathbf{P}\{\eta_n = k\} = p_{k+1}$, $k = -1, 0, 1, 2, \ldots$. Then distributions of random sequences $\xi_0 = 1$, $\xi_n = \sigma_{\xi_0+\cdots+\xi_{n-1}}$, $n \geq 1$, and ζ_n, $n \geq 0$, coincide.

If birth probabilities p_k form a geometric progression for $k \geq 1$, then a function $f_n(z)$ can be found in explicit form. Let

$$p_k = p_1 c^{k-1}, \quad k = 1, 2, \ldots, \quad p_0 = \frac{1 - p_1 - c}{1 - c}.$$

If $p_1 \neq (1 - c)^2$, then

$$f_n(z) = 1 - \frac{m^n (1 - z)}{1 - (m^n - 1)(1 - z)/(1 - p)}. \tag{13.1}$$

If $p_1 = (1 - c)^2$, then

$$f_n(z) = 1 - \frac{1}{1 + n b (1 - z)/2}. \tag{13.2}$$

Here the following notations are used:

$$p = \frac{1 - p_1 - c}{c(1 - c)}, \quad m = \frac{p_1}{(1 - c)^2}, \quad b = \frac{2 p_1 c}{(1 - c)^3}.$$

The asymptotic behavior of a Galton-Watson process significantly depends on the mean number of one-particle direct descendants $m = \sum_{k\geq 0} k p_k = f'(1)$. Thus, the mean number of one-particle descendants in the nth generation equals m^n, i.e., $\mathbf{E}\,\zeta_n = m^n$. The second factorial moment is determined by the formula

$$\mathbf{E}\,\xi_n (\xi_n - 1) = \begin{cases} \frac{b\, m^n (m^n - 1)}{m (m-1)} & \text{if } m \neq 1 \\ b n & \text{if } m = 1 \end{cases}$$

where

$$b = \mathbf{E}\,\xi_1\,(\xi_1 - 1) = f''(1) = \sum_{k\ge 1} k\,(k-1)\,p_k.$$

A branching process is called **subcritical**, if $m < 1$, **critical**, if $m = 1$, $p_1 < 1$, and **supercritical**, if $m > 1$. A branching process is called **degenerating**, if at some step the resulting population is empty for any initial population.

Theorem 13.1. *A branching Galton-Watson process is degenerating with probability 1 iff the mean number of one-particle direct descendants is less or equal than 1.*

If the probability of degeneration is less than 1, then it can be determined from the equation that is written via generating function $f(z)$ for the number of one-particle direct descendants. Let q be the probability of degeneration of a branching process starting with one particle, i.e.,

$$q = \lim_{n\to\infty} \mathbf{P}\{\zeta_n = 0\} = \mathbf{P}\{\lim_{n\to\infty} \zeta_n = 0\}.$$

Theorem 13.2. *The probability q of degeneration is equal to the nearest to 0 solution of the equation $z = f(z)$ within the unit interval $0 \le z \le 1$.*

The probability of degeneration of the branching process starting with k particles equals q^k.

In the case of subcritical and critical processes the asymptotic behavior of the probability of branching process extension after n steps

$$q_n = \mathbf{P}\{\zeta_n > 0\} = 1 - f_n(0)$$

is also of great interest.

Theorem 13.3. *If $m < 1$, then there exists a positive limit $\lim_{n\to\infty} m^{-n}\,q_n$ iff the inequality*

$$\sum_{k\ge 1} p_k\, k \log k < \infty$$

holds.

Theorem 13.4. *If $m = 1$ and $0 < b = f''(1) = \sum_{k\ge 1} k\,(k-1)\,p_k < \infty$, then $q_n \sim 2/(bn)$ as $n \to \infty$.*

Theorem 13.5 (generalization of Theorem 13.4). *If for some $0 < \alpha \le 1$*

$$f(z) = z + (1-z)^{1+\alpha}\, l(1-z)$$

and $l(\cdot)$ is a slowly changing at zero function, then

$$q_n^\alpha \, l(q_n) \sim \frac{1}{\alpha n} \quad \text{as } n \to \infty. \tag{13.3}$$

It follows from (13.3) that

$$q_n = n^{-1/\alpha} L(n) \tag{13.4}$$

where $L(\cdot)$ is a slowly changing at infinity function. Moreover this function has a finite positive limit at infinity iff a function $l(\cdot)$ has a finite positive limit at zero, i.e., $q_n \sim \text{const} \cdot n^{-1/\alpha}$ for $z \uparrow 1$.

Subcritical and critical processes also have quite different asymptotic behavior of conditional distribution of ζ_n provided that $\zeta_n > 0$.

Theorem 13.6. *If $m < 1$, then there exists a limit probability distribution*

$$p_k^* = \lim_{n \to \infty} \mathbf{P}\{\zeta_n = k \mid \zeta_n > 0\}, \quad k \geq 1,$$

with a generating function $f^*(z) = \sum_{k \geq 1} p_k^* z^k$ satisfying a functional equation

$$1 - f^*(f(z)) = m(1 - f^*(z)). \tag{13.5}$$

A limit distribution p_k^*, $k = 1, 2, \ldots$, has a finite mean value iff

$$\sum_{k \geq 1} p_k \, k \log k < \infty.$$

If the last condition is satisfied, then the equation (13.5) has a unique solution in the class of all differentiated at $z = 1$ generating functions which equal zero at $z = 0$.

Theorem 13.7. *Under the hypotheses of Theorem 13.4 the following relation*

$$\lim_{n \to \infty} \mathbf{P}\left\{\frac{1}{n}\zeta_n \geq x \mid \zeta_n > 0\right\} = e^{-2x/b}, \quad x \geq 0,$$

holds.

Theorem 13.8. *Under the hypotheses of Theorem 13.5 there exists a limit probability distribution*

$$\lim_{n \to \infty} \mathbf{P}\{q_n \zeta_n \leq x \mid \zeta_n > 0\} = \Pi(x),$$

$$\int_0^\infty e^{-sx} \, d\Pi(x) = 1 - s(1 + s^\alpha)^{-1/\alpha}, \quad s \geq 0.$$

A distribution function $\Pi(x)$ can be expressed via a density $p(y)$ of the stable distribution with a parameter α: if

$$\int_0^\infty e^{-sy}\, p(y)\, dy = e^{-s^\alpha},$$

then

$$\Pi(x) = 1 - \frac{\alpha}{\Gamma(1/\alpha)} \int_0^\infty e^{-(x/y)^\alpha}\, p(y)\, \frac{dy}{y}$$

where $\Gamma(z) = \int_0^\infty t^{z-1}\, e^{-t}\, dt$ is the Euler gamma-function.

Theorem 13.9 (inverse to Theorem 13.8). *If at every point $x > 0$ where a limit function is a continuous one, there exists a limit*

$$\lim_{n\to\infty} \mathbf{P}\{q_n\, \zeta_n \le x \mid \zeta_n > 0\} = \Pi(x)$$

and $\Pi(+0) < 1$, then $f(z) = z + (1-z)^\alpha\, l(1-z)$ for some $0 < \alpha \le 1$ and slowly changing at zero function $l(\cdot)$. Moreover

$$\int_0^\infty e^{-sx}\, d\Pi(x) = 1 - s\,(1 + s^\alpha)^{-1/\alpha}.$$

A distinctive feature of supercritical processes is the possibility of geometric growth of the number of particles provided that the nondegeneracy condition is satisfied. So, a sequence $m^{-n}\, \zeta_n$, $n = 0, 1, 2, \ldots$, forms a nonnegative martingale and hence it has a finite limit

$$\eta = \lim_{n\to\infty} m^{-n}\, \zeta_n \tag{13.6}$$

almost surely. It is evident that a limit value η vanishes with probability not less than the probability of degeneration q. The case $\eta = 0$ with probability 1 is not excluded. If $m > 1$, then a limit value of η in (13.6) is positive with a positive probability iff

$$\sum_{k\ge1} p_k\, k \log k < \infty. \tag{13.7}$$

If the condition (13.7) is satisfied, then $\mathbf{P}\{\eta = 0\} = q$, $\mathbf{E}\,\eta = 1$, and a Laplace transform $\varphi(s) = \mathbf{E}\, e^{-s\eta}$ is the unique solution of the equation

$$\varphi(s) = f\left(\varphi\left(\frac{s}{m}\right)\right), \quad s \ge 0, \quad \varphi(0) = \varphi'(0) = 1. \tag{13.8}$$

If $\sum_{k \geq 1} k^2 p_k < \infty$, then $m^{-n} \zeta_n$ converges to η in a mean square as $n \to \infty$. Moreover, regardless of the condition (13.7), it is always possible to find a sequence $c_n \uparrow \infty$ such that $c_{n+1}/c_n \underset{n \to \infty}{\longrightarrow} m$ and a limit

$$\zeta = \lim_{n \to \infty} \frac{\zeta_n}{c_n} \tag{13.9}$$

exists with probability 1, vanishes with probability of degeneration q and has a Laplace transform $\Psi(s) = \mathbf{E}\, e^{-s\zeta}$ satisfying an equation

$$\Psi(s) = f\left(\Psi\left(\frac{s}{m}\right)\right), \quad \Psi(\infty) = q.$$

Any two solutions $\Psi_1(\cdot)$ and $\Psi_2(\cdot)$ of this equation (in the class of Laplace transforms of nonnegative random variables) satisfy the relation $\Psi_1(k s) = \Psi_2(s)$, $s \geq 0$, with some positive constant k.

Limit distributions of the distance to the nearest mutual ancestor is an example of the application of the above stated results. More exactly, let the initial population consist of one particle. Then all existing at moment n particles are the descendants of one of the previously existing particles, which is a mutual ancestor. Among all mutual ancestors there exists an ancestor that appeared later than others, i.e., a nearest mutual ancestor. Let τ_n be a number of the generation where the nearest mutual ancestor of all particles existing at time n appeared. A random variable τ_n is defined under the condition $\zeta_n > 0$.

Theorem 13.10. *If $m < 1$, then*

$$\lim_{n \to \infty} \mathbf{P}\{n - \tau_n \leq k \mid \zeta_n > 0\} = \frac{p_1^*}{p_1(k)}$$

where

$$p_1(k) = \mathbf{P}\{\zeta_k = 1 \mid \zeta_k > 0\}, \quad p_1^* = \lim_{n \to \infty} p_1(n).$$

Theorem 13.11. *Under the hypotheses of Theorem 13.4 the following relation*

$$\lim_{n \to \infty} \mathbf{P}\left\{\frac{\tau_n}{n} \geq x \mid \zeta_n > 0\right\} = 1 - x, \quad 0 \leq x \leq 1,$$

holds.

Theorem 13.12. *If $m > 1$, then*

$$\lim_{n \to \infty} \mathbf{P}\{\tau_n \geq k \mid \zeta_r > 0, \ r = 1, 2, \ldots\} = (f'(q))^k.$$

By Theorem 13.7 critical processes have the most interesting asymptotic properties. But the question, whether the observed branching process is a critical one or not, often has no simple and unique answer. From this point of view it is very important to study an asymptotic behavior of branching processes as the critical parameter (in considered case the mean number of one-particle direct descendants) tends to unit as $n \to \infty$. The phenomena arising in this case are called transient ones. The strict setting of a problem is the following one.

Let $\zeta_n^{(r)}$ (for any positive integer r) be the number of one-particle descendants in nth generation in a Galton-Watson branching process with the following generating function for direct descendants

$$f^{(r)}(z) = \sum_{k \geq 0} p_k^{(r)}\, z^k \quad \text{where } m_r = \sum_{k \geq 1} k\, p_k^{(r)} \underset{r \to \infty}{\longrightarrow} 1.$$

The problem on transient phenomena is to study the asymptotic distribution of $\zeta_n^{(r)}$ as $r, n \to \infty$. Solely the condition of uniform on r integrability of the second factorial moment of the number of one-particle direct descendants and its boundedness away from zero provides the relations

$$\mathbf{P}\{\zeta_n^{(r)} > 0\} \underset{n,r \to \infty}{\longrightarrow} 0, \quad f_n^{(r)}(z) = \mathbf{E}\, z^{\zeta_n^{(r)}} \underset{n,r \to \infty}{\longrightarrow} 1$$

hold true. Moreover, a difference $1 - f_n^{(r)}(z)$ is asymptotically equivalent (uniformly on $0 \leq z < 1$) to $h_n^{(r)}(z)$ as $r, n \to \infty$ where

$$h_n^{(r)}(z) = \begin{cases} m^n\,(1-z)\left(1 + \frac{b_r}{2m_r}\,\frac{m_r^n - 1}{m_r - 1}\,(1-z)\right)^{-1} & \text{if } m_r \neq 1 \\ (1-z)\,\left(1 + n\,\frac{b_r}{2}\,(1-z)\right)^{-1} & \text{if } m_r = 1 \end{cases}$$

and

$$b_r = \sum_{k \geq 1} k\,(k-1)\, p_k^{(r)}.$$

If in the formulae (13.1) and (13.2) we replace m by m_r and b by b_r, then $1 - h_n^{(r)}(z)$ coincides with the right hand side of one of these formulae. For $z = 0$ we receive the asymptotic expression for the probability of the process extension

$$\mathbf{P}\{\zeta_n^{(r)} > 0\} \underset{n,r \to \infty}{\sim} \begin{cases} 2\,b_r^{-1}\,(m_r - 1)\,(1 - m_r^{-n})^{-1} & \text{if } m_r \neq 1, \\ 2\,(n b_r)^{-1} & \text{if } m_r = 1. \end{cases}$$

This statement is equivalent to the following one: if n and r are changing in such a way that there exists a finite or infinite limit $c = \lim n\,(m_r \quad 1)$,

then

$$\mathbf{P}\{\zeta_n^{(r)} > 0\} \sim \begin{cases} 2\,b_r^{-1}\,(1-m_r)\,m_r^n & \text{if } c = -\infty \\ 2\,(n b_r)^{-1}\,c\,(1-e^{-c})^{-1} & \text{if } -\infty < c < +\infty \\ 2\,b_r^{-1}\,(m_r - 1) & \text{if } c = +\infty \end{cases}$$

(here, as usual, we set $0/(1-e^{-0}) = 1$).

The following theorem determines the asymptotic behavior of conditional distribution of $\zeta_n^{(r)}$ provided that a branching process is nondegenerate. Without loss of generality we suppose that either $m_r = 1$ for all r or $m_r \neq 1$ also for all r.

Theorem 13.13. *If*

$$\sup_r \sum_{k \geq l} k^2\, p_k^{(r)} \underset{l \to \infty}{\longrightarrow} 0$$

and there exists a positive limit

$$b = \lim_{r \to \infty} \sum_{k \geq 1} k\,(k-1)\,p_k^{(r)},$$

then

$$\lim_{n,r \to \infty} \mathbf{P}\left\{\frac{m_r - 1}{m_r^n - 1}\,\zeta_n^{(r)} \geq x \mid \zeta_n^{(r)} > 0\right\} = e^{-2x/b}$$

for $m_r \neq 1$, *and*

$$\lim_{n,r \to \infty} \mathbf{P}\left\{\frac{1}{n}\,\zeta_n^{(r)} \geq x \mid \zeta_n^{(r)} > 0\right\} = e^{-2x/b}$$

for $m_r \equiv 1$.

Theorem 13.14 (an equivalent form of Theorem 13.13). *Let under the hypotheses of Theorem 13.13 parameters n and r change in such a way that there exists a finite or infinite limit* $c = \lim n\,(m_r - 1)$. *If* $c = -\infty$, *then*

$$\lim_{n,r \to \infty} \mathbf{P}\left\{(1-m_r)\,\zeta_n^{(r)} \geq x \mid \zeta_n^{(r)} > 0\right\} = e^{-2x/b};$$

if $-\infty < c < +\infty$, *then*

$$\lim_{n,r \to \infty} \mathbf{P}\left\{\frac{1}{n}\,\zeta_n^{(r)} \geq x \mid \zeta_n^{(r)} > 0\right\} = \exp\left\{-2\,(bc)^{-1}\,(e^c - 1)\,x\right\};$$

if $c = +\infty$, *then*

$$\lim_{n,r\to\infty} \mathbf{P}\left\{(m_r - 1)\, m_r^{-n}\, \zeta_n^{(r)} \geq x \mid \zeta_n^{(r)} > 0\right\} = e^{-2x/b}.$$

13.2 Bellman-Harris Processes

The only difference between Bellman-Harris[3,5,7–13] and Galton-Watson processes is a random time, called a lifetime, measured from the particle's birth to its death. The lifetime distribution is the same for all particles and does not depend on other particles in the population and on their number. The distribution of the number of direct descendants does not depend on the lifetime when the transformation occurred.

Thus **the Bellman-Harris branching process** is defined by a distribution function $G(t)$ (a probability that a lifetime does not exceed t) and by a generating function $f(z) = \sum_{k\geq 0} p_k\, z^k$ for the number of one-particle direct descendants. For such branching process we use an abridged notation (G, f)-process. Galton-Watson process is a special case of a Bellman-Harris process with a lifetime identically equal to unit.

A generating function $F_t(z) = \mathbf{E}\, z^{\zeta(t)}$ of the number $\zeta(t)$ of particles arising in a (G, f)-process during a time period t from one newly born particle satisfies the following nonlinear integral equation:

$$F_t(z) = z\,(1 - G(t)) + \int_0^t f(F_{t-u}(z))\, dG(u).$$

If we search through all possible distribution functions $G(\cdot)$ and all possible probability generating functions $f(\cdot)$, then among resulting branching (G, f)-processes it is possible to find processes having an infinite number of particles at some fixed time t with a positive probability, i.e., $F_t(1-) < 1$. If such phenomenon of "explosion" is absent, then a corresponding branching process is called **regular**.

Sufficient conditions of the regularity: a continuity of the distribution function $G(\cdot)$ at point $t = 0$, i.e., $G(0+) = 0$, and a finiteness of a mean number $m = \sum_{k\geq 1} k\, p_k$ of one-particle direct descendants. For a special class of distribution functions $G(\cdot)$ there were obtained necessary and sufficient conditions of regularity.

Theorem 13.15. *If the inequalities*

$$0 < \varliminf_{t\downarrow 0} t^{-\alpha}\, G(t) \leq \varlimsup_{t\downarrow 0} t^{-\alpha}\, G(t) < \infty$$

hold for some $\alpha > 0$, then a branching (G, f)-process is a regular one iff

$$\int_0^1 (1-x)^{(1-\alpha)/\alpha}\,(1-f(x))^{-1/\alpha}\,dx = \infty.$$

Any Bellman-Harris branching process is associated with a corresponding Galton-Watson process. Thus, the degeneration of a Bellman-Harris process does not depend on the particle's lifetime and therefore the probability of (G, f)-process degeneration is equal to the nearest to 0 solution of the equation $z = f(z)$ within unit interval $0 \le z \le 1$ (cf. Theorem 13.2).

Moreover a generating function $F_t(z)$ and a generating function $f_n(z)$ for the number of particles in nth generation of a Galton-Watson process satisfy the following two-sided inequalities. Namely, if $0 \le z \le q$, then

$$f_n(z) - (q-z)\sum_{k=0}^{n-1} (f'(q))^k \,(G^{k*}(t) - G^{k+1*}(t)) \le$$

$$\le F_t(z) \le f_n(z) + (q-z)\,(f'(q))^n\, G^{n*}(t);$$

if $q \le z \le 1$, then

$$f_n(z) - (z-q)\,m^n\, G^{n*}(t) \le F_t(z) \le$$

$$\le f_n(z) + (z-q)\sum_{k=0}^{n-1} m^k\,(G^{k*}(t) - G^{k+1*}(t)).$$

Here n is an arbitrary positive integer number, t is an arbitrary nonnegative number, and $G^{k*}(t)$ is a k-multiple convolution of a distribution function $G(\cdot)$ with itself. These inequalities have the most simple form for a critical case. Then $m = q = 1$ and hence

$$f_n(z) - (1-z)\,(1 - G^{n*}(t)) \le F_t(z) \le f_n(z) + (1-z)\,G^{n*}(t)$$

for all $0 \le z \le 1$, $n = 0, 1, 2, \ldots$, $t \ge 0$.

This inequality and Theorem 13.4 imply an asymptotic expression for the probability $Q_t = \mathbf{P}\{\zeta(t) > 0\} = 1 - F_t(0)$ of an extension of the Bellman-Harris critical process with a finite variance of the number of one-particle direct descendants.

Theorem 13.16. *If $m = 1$, $b = f''(1) = \sum_{k \ge 1} k\,(k-1)\,p_k < \infty$ and*

$$t^2\,(1 - G(t)) \underset{t\to\infty}{\longrightarrow} 0, \tag{13.10}$$

then $Q_t \sim 2a/(bt)$ *as* $t \to \infty$ *where* $a = \int\limits_0^\infty t\, dG(t)$.

If a condition (13.10) is replaced by

$$\lim_{t\to\infty} t^2\,(1 - G(t)) = c\,\frac{b}{2} \tag{13.11}$$

for some constant $c \geq 0$, then the probability of extension is of the same order of decrease t^{-1}, but with an another constant:

$$Q_t \sim \frac{1}{bt}\,(a + \sqrt{a^2 + 4c})$$

Considered cases can be obtained from the following theorem.

Theorem 13.17. *Let* $f(z) = z + (1-z)^\alpha\, l(1-z)$ *where* $0 < \alpha \leq 1$ *and* $l(t)$ *is a slowly changing at* $t = 0$ *function. If*

$$\lim_{t\to\infty} t^{1+1/\alpha}\,(1 - G(t))/L(t) = c$$

for some $c \geq 0$, *then* $Q_t \sim \omega\, t^{-1/\alpha}\, L(t)$ *as* $t \to \infty$; *here* $L(t)$ *is a slowly changing function from (13.4) and* $\omega > 0$, $\omega^{1+\alpha} - a\,\omega = \alpha\, c$. *If*

$$\lim_{t\to\infty} t^{1+1/\alpha}\,(1 - G(t))/L(t) = \infty \tag{13.12}$$

for some $\beta \geq 0$ *and* $t^\beta\,(1 - G(t))$ *is a slowly changing at infinity function, then* $t^{\beta/(1+\alpha)}\, Q_t$ *is a slowly changing function as* $t \to \infty$.

Assume that there exists a number μ such that

$$m \int\limits_0^\infty e^{-\mu t}\, dG(t) = 1.$$

This number μ is called **a Maltus index.** Then the probability Q_t of the extension of a subcritical process is an exponentially decreasing function. If there exists a Maltus index for $m < 1$, then $\mu < 0$.

Theorem 13.18. *If*

$$m < 1, \quad \int\limits_0^\infty t\, e^{-\mu t}\, dG(t) < \infty, \quad \sum_{k\geq 1} p_k\, k\, \log k < \infty$$

and a distribution function $G(t)$ *is nonlatticed, then there exists a finite and positive limit* $\lim\limits_{t\to\infty} e^{-\mu t}\, Q_t$.

If $m > 1$, then there exists a positive Maltus index which gives an exponent of exponential growth of the mean number of particles, i.e., if $m > 1$ and a distribution function $G(t)$ is nonlatticed, then

$$\lim_{t\to\infty} e^{-\mu t}\,\mathbf{E}\,\zeta(t) = (m-1)\left(\mu\, m^2 \int_0^\infty u\, e^{-\mu u}\, dG(u)\right)^{-1}.$$

This formula is satisfied also for subcritical processes (naturally under the assumption that a Maltus index exists and an integral in the right hand side of the formula is finite).

The mean number of particles in a critical Bellman-Harris process starting from one newly born particle is identical to unit. The asymptotical behavior of conditional distributions of the particles number $\zeta(t)$ provided that $\zeta(t) > 0$ is approximately the same for Bellman-Harris and Galton-Watson models both for subcritical and supercritical processes.

Theorem 13.19. *Under the hypotheses of Theorem 13.18 there exists a limit probability distribution*

$$p_k^* = \lim_{t\to\infty} \mathbf{P}\,\{\zeta(t) = k \mid \zeta(t) > 0\},\quad k = 1, 2, \ldots,\quad \sum_{k\geq 1} k\, p_k^* < \infty.$$

Theorem 13.20. *If $m > 1$ and a distribution function $G(t)$ is nonlatticed, then*

$$\lim_{t\to\infty} \mathbf{P}\,\{e^{-\mu t}\,\zeta(t) \leq x\} = \Phi(x),\quad \Phi(\infty) = 1.$$

A limit distribution $\Phi(x)$ is not concentrated at zero iff $\sum_{k\geq 1} p_k\, k \log k < \infty$. If the last condition is satisfied, then

$$\Phi(0+) = q,\quad \int_0^\infty x\, d\Phi(x) = (m-1)\left(\mu\, m^2 \int_0^\infty u\, e^{-\mu u}\, dG(u)\right)^{-1}$$

and a Laplace transform $\varphi(s) = \int_0^\infty e^{-sx}\, d\Phi(x)$ is the unique solution (in the class of Laplace transforms of probability distributions on $[0, \infty)$ with a given mean value) of the equation

$$\varphi(s) = \int_0^\infty f(\varphi(s\, e^{-\mu u}))\, dG(u).$$

If $\sum\limits_{k\geq 1} k^2 p_k < \infty$, *then* $e^{-\mu t}\zeta(t)$ *converges in mean square as* $t\to\infty$.

A new problem (compared with a Galton-Watson process) is the investigation of the distribution of a part of particles with a given age. Let $\zeta_v(t)$ be a number of particles that appear from one newly born particle during a time t and have at this time an age not exceeding v. Assume that the degeneration probability of a branching process equals zero. The last condition means that the number of one-particle direct descendants is not less than unit. Thus a ratio $\zeta_v(t)/\zeta(t)$ is defined with probability 1. If in addition we suppose the lifetime distribution function $G(t)$ is nonlatticed and a sum $\sum\limits_{k\geq 1} p_k\, k\log k$ is finite, then the following relation

$$\sup_{v>0}\left|\frac{\zeta_v(t)}{\zeta(t)} - H(v)\right| \underset{t\to\infty}{\longrightarrow} 0$$

holds with probability 1 where

$$H(v) = \frac{\int\limits_0^v e^{-\mu u}\,(1-G(u))\,du}{\int\limits_0^\infty e^{-\mu u}\,(1-G(u))\,du}.$$

In the case of critical processes the presence of a lifetime generates some new effects even for a finite variance of the number of one-particle direct descendants. Thus it follows from Theorems 13.7 and 13.16 that the relation

$$\lim_{t\to\infty}\mathbf{P}\left\{\frac{1}{t}\,\zeta(t)\geq x \mid \zeta(t)>0\right\} = e^{-2ax/b}$$

holds for

$$f'(1) = 1,\quad b = f''(1)<\infty,\quad t^2\,(1-G(t)) \underset{t\to\infty}{\longrightarrow} 0$$

and for a nonlatticed distribution function $G(t)$. If the condition (13.10) is replaced by (13.11) with $c>0$, then

$$\lim_{t\to\infty}\mathbf{P}\left\{\frac{1}{t}\,\zeta(t)\geq x \mid \zeta(t)>0\right\} = \frac{2a}{a+\sqrt{a^2+4c}}\,e^{-2ax/b},\quad x>0,$$

i.e., such a replacement gives a limit distribution with an atom at zero:

$$\frac{-a+\sqrt{a^2+4c}}{a+\sqrt{a^2+4c}}. \tag{13.13}$$

Moreover, this atom at zero can be interpreted as follows: there exists a discrete limit distribution

$$\lim_{t\to\infty} \mathbf{P}\left\{\frac{1}{t}\,\zeta(t) = k \mid \zeta(t) > 0\right\} =$$

$$= (-1)^{k-1}\,\frac{\sqrt{a^2+4c}}{a+\sqrt{a^2+4c}}\,\binom{1/2}{k}\left(\frac{4\,c}{a^2+4c}\right), \quad k = 1, 2, \ldots, \tag{13.14}$$

having a full mass (13.13). Here

$$\binom{x}{k} = \frac{x\,(x-1)\,\cdots\,(x-k+1)}{k!}.$$

Thus, under the condition of nondegeneracy, a population consists of a large number of "young" particles, which has a limit exponential distribution under linear norming, and of a relatively small number of "old" particles, which has a nonsingular limit distribution.

If a condition (13.11) is replaced by (13.12) and $1 - G(t)$ is a proper changing at infinity function, then the number $\zeta(t)$ of particles, provided that $\zeta(t) > 0$, has a nonsingular limit distribution $(-1)^{k-1}\binom{x}{k}$, $k = 1, 2, \ldots$. Such a distribution can be obtained by the passage to the limit in (13.14) as $c \to \infty$, $a/c \to 0$.

Theorem 13.21. *Let a distribution function $G(t)$ be nonlatticed and $f(z) = z + (1-z)^{1+\alpha}\,l(1-z)$ where $0 < \alpha \leq 1$ and $l(t)$ is a slowly changing at zero function. If the relation*

$$\lim_{t\to\infty} t^{1+1/\alpha}\,(1 - G(t))\,/\,L(t) = 0 \tag{13.15}$$

holds for some slowly changing function $L(t)$ of (13.4), then

$$\lim_{t\to\infty} \mathbf{P}\,\{Q_t\,\zeta(t) \leq x \mid \zeta(t) > 0\} = \Pi(x)$$

where

$$\int_0^\infty e^{-s\,x}\,d\Pi(x) = 1 - s\,(1+s^\alpha)^{-1/\alpha}.$$

If the relation

$$\lim_{t\to\infty} t^{1+1/\alpha}\,(1 - G(t))\,/\,L(t) = c \tag{13.16}$$

holds, then

$$\lim_{t\to\infty} \mathbf{P}\,\{Q_t\,\zeta(t) \leq x \mid \zeta(t) > 0\} = 1 - \frac{a^{1/\alpha}}{\omega} + \frac{a^{1/\alpha}}{\omega}\,\Pi\left(\frac{a^{1/\alpha}}{\omega}\,x\right)$$

where $\omega > 0$, $\omega^{1+\alpha} - a\,\omega = \alpha\, c$, *and*

$$\lim_{t\to\infty} \mathbf{P}\{\zeta(t) = k \mid \zeta(t) > 0\} = \Pi_k, \quad k = 1, 2, \ldots,$$

where

$$\sum_{k\geq 1} \Pi_k\, z^k = 1 - \frac{\omega(z)}{\omega}, \quad \omega(z) > 0 \;\; \textit{for} \;\; 0 \leq z \leq 1$$

and

$$(\omega(z))^{1+\alpha} - a\,\omega(z) = \alpha\, c\,(1 - z).$$

If

$$\lim_{t\to\infty} t^{1+1/\alpha}\,(1 - G(t))\,/\,L(t) = \infty \tag{13.17}$$

and $1 - G(t)$ *is a proper changing at infinity function, then*

$$\lim_{t\to\infty} \mathbf{P}\{\zeta(t) = k \mid \zeta(t) > 0\} = (-1)^{k-1}\binom{1/(1+\alpha)}{k}, \quad k = 1, 2, \ldots .$$

The following relation

$$\lim_{t\to\infty} \frac{t^{1/\alpha}}{L(t)}\, \mathbf{P}\{t^{-1/\alpha}\, L(t)\, \zeta(t) \geq x\} = a^{1/\alpha}\,(1 - \Pi(a^{1/\alpha}x)), \quad x > 0,$$

holds independently of the conditions (13.15)–(13.17), but assumes the finiteness of the mean value $a = \int_0^\infty t\, dG(t)$.

Stated results hold true also for a latticed distribution function $G(t)$, if a time parameter t tends to infinity along the points of corresponding lattice.

13.3 Markov Branching Processes

Processes with exponential distributed lifetime (i.e., $G(t) = 1 - e^{-t/a}$) form an important special case of Bellman-Harris branching processes. Thus a number of particles $\zeta(t)$, $t \geq 0$, form a homogeneous Markov process with transition probabilities $\{p_{ij}(t)\}$ at a small time $t \downarrow 0$:

$$p_{ij}(t) = \frac{i}{a}\, p_{j-i+1}\, t + o(t), \quad j \neq i, \quad p_{ii}(t) = 1 - \frac{i}{a}\,(1 - p_1)\, t + o(t).$$

The inverse statement is also true. If a Bellman-Harris process possesses a homogeneous Markov property, then the lifetime is an exponential distributed random variable. Such a **Markov branching process**[3–6,14,15] can be obtained from an integer-valued random process $\eta(t)$ with independent increments, $\eta_0 = 1$ and

$$\mathbf{E}\, z^{\eta(t)-\eta(0)} = \exp\{t\, f(z) - (a\, z)^{-1}\}, \quad |z| = 1,$$

by a random change of time $t \to \tau(t)$ where $\tau(t)$ is an inverse function to

$$\varphi(t) = \int_0^t \frac{du}{\eta(u)}.$$

More precisely, if $t < \varphi(\theta)$ where $\theta = \inf\{t : \eta(t) = 0\}$, then $\tau(t)$ is the unique solution of the equation

$$t = \int_0^{\tau(t)} \frac{du}{\eta(u)},$$

and the distribution of a Markov process $\zeta(t)$, $t \geq 0$, coincides with a distribution of the process

$$\xi(t) = \begin{cases} \eta(\tau(t)) & \text{if } t < \varphi(\theta) \\ 0 & \text{if } t \geq \varphi(\theta). \end{cases}$$

A generating function $F_t(z)$ satisfies an ordinary differential equation

$$\frac{dF_t}{dt} = g(F_t), \quad F_0 = z, \tag{13.18}$$

with $g(z) = a^{-1}\,(f(z) - z)$, and a partial differential equation

$$\frac{\partial F_t(z)}{\partial t} = g(z)\,\frac{\partial}{\partial z} F_t(z). \tag{13.19}$$

Moreover $F_{t+u}(z) = F_t(F_u(z))$.

By Theorem 12.15 the necessary and sufficient condition of the regularity of the Markov branching process is the divergence of the integral

$$\int_0^1 \frac{dy}{1 - f(y)}.$$

If $p_0 + p_1 + p_2 = 1$, then equations (13.18) and (13.19) can be solved in the explicit form:

$$F_t(z) = \begin{cases} 1 - e^{\mu t}\,(1 - z)(1 + p_2\,(e^{\mu t} - 1)\,(1 - z)/(m - 1))^{-1} & \text{if } m \neq 1 \\ 1 - (1 - z)\,(1 + p_2\, t\,(1 - z)/a)^{-1} & \text{if } m = 1. \end{cases}$$

Here as usual $m = f'(1) = p_1 + 2p_2$ is a mean number of one-particle direct descendants and $\mu = (m-1)/a$ is a Maltus index. If we in addition assume that $p_2 = 1$ (and consequently $p_0 = p_1 = 0$), then the formula for $F_t(z)$ can be simplified:

$$F_t(z) = \frac{z\,e^{-t/a}}{1 - z\,(1 - e^{-t/a})}.$$

Such a branching process is called **a Yule process.**

For an arbitrary Markov branching process the two first factorial moments can be written in an explicit form

$$\mathbf{E}\,\zeta(t) = F_t'(1) = e^{\mu t},$$

$$\mathbf{E}\,\zeta_t\,(\zeta_t - 1) = F''_t(1) = \begin{cases} b\,e^{\mu t}\,(e^{\mu t} - 1)\,\mu^{-1}\,a^{-2} & \text{if } m \neq 1 \\ b\,t\,a^{-2} & \text{if } m = 1 \end{cases}$$

where $\mu = (m-1)/a$ is a Maltus index.

An investigation of Markov branching processes is simpler compared with Galton-Watson processes. Therefore in several cases it is possible to derive more complete results. So, the equations (13.18) and (13.19) can be solved in quadratures:

$$\int_z^{F_t(z)} \frac{du}{g(u)} = 1.$$

Moreover, for any $\delta > 0$ a random variable $\zeta(n\delta)$ can be considered as the number of particles in the nth generation of a Galton-Watson branching process with a generating function $F_\delta(t)$ of one-particle direct descendants. Therefore some results for Markov branching processes arise from corresponding results for Galton-Watson processes and the following theorem.

Theorem 13.22. *Let $h(\cdot)$ be continuous on $(0, \infty)$ function satisfying the condition: for any $\delta > 0$ a sequence $h(n\delta)$, $n = 1, 2, \ldots$, has a limit as $n \to \infty$. Then $h(x)$ has a limit as $x \to \infty$.*

Theorem 13.23. *If $m < 1$, then*

$$\lim_{t\to\infty} e^{-\mu t}\,\mathbf{P}\,\{\zeta(t) > 0\} = \exp\left\{ - \int_0^1 \frac{f(1-u) - 1 + mu}{f(1-u) - 1 + u}\,\frac{du}{u} \right\} \tag{13.20}$$

and there exists a limit probability distribution

$$\lim_{t\to\infty} \mathbf{P}\,\{\zeta(t) = k \mid \zeta(t) > 0\} = p_k^*, \quad k \geq 1,$$

with a generating function

$$\sum_{k\geq 1} z^k\, p_k^* \;=\; 1 \;-\; \exp\left\{(m-1)\int\limits_0^z \frac{du}{f(u)-u}\right\}.$$

The necessary and sufficient condition of the convergence of the integral in (13.20) is the convergence of the series $\sum\limits_{k\geq 1} p_k\, k\log k$. This condition is necessary and sufficient for the finiteness of the mean value of a limit distribution.

Theorem 13.24. *If* $m=1$ *and* $0<b=f''(1)<\infty$, *then*

$$\lim_{t\to\infty} t\,\mathbf{P}\left\{\zeta(t)>0\right\} \;=\; \frac{2a}{b}, \qquad \lim_{t\to\infty} \mathbf{P}\left\{\frac{1}{t}\,\zeta(t)\geq x \mid \zeta(t)>0\right\} \;=\; e^{-2ax/b}.$$

This theorem can be generalized as follows.

Theorem 13.25. *If*

$$f(z) \;=\; z \;+\; (1-z)^{1+\alpha}\, l(1-z)$$

where $0<\alpha\leq 1$ *and* $l(t)$ *is a slowly changing at* $t=0$ *function, then the relations*

$$Q_t \;=\; \mathbf{P}\left\{\zeta(t)>0\right\} \sim \left(\frac{a}{t}\right)^{1/\alpha} L(t) \quad \text{as } t\to\infty,$$

$$\lim_{t\to\infty} \mathbf{P}\left\{Q_t\,\zeta(t)<x \mid \zeta(t)>0\right\} \;=\; \Pi(x)$$

hold for some slowly changing function $L(t)$ *from (13.4) and with*

$$\int\limits_0^\infty e^{-sx}\, d\Pi(x) \;=\; 1-s\,(1+s^\alpha)^{-1/\alpha}.$$

The inverse statement is also true.

Theorem 13.26. *If at every continuity point of the limit function* $\Pi(x)$ *there exists a limit*

$$\lim_{t\to\infty} \mathbf{P}\left\{Q_t\,\zeta(t)\leq x \mid \zeta(t)>0\right\} \;=\; \Pi(x)$$

and $\Pi(+0)<1$, *then*

$$f(z) \;=\; z \;+\; (1-z)^{1+\alpha} l(1-z), \qquad \int\limits_0^\infty e^{-sx}\, d\Pi(x) \;=\; 1-s\,(1+s^\alpha)^{-1/\alpha}$$

where $0 < \alpha \le 1$ *and* $l(t)$ *is slowly changing at* $t = 0$ *functions.*

Theorem 13.27. *Let* $1 < m < \infty$. *If* $\sum_{k\ge 1} p_k\, k \log k = \infty$, *then*

$$\lim_{t\to\infty} e^{-\mu t}\,\zeta(t) = 0$$

with probability 1. If $\sum_{k\ge 1} p_k\, k \log k < \infty$, *then a limit*

$$\eta = \lim_{t\to\infty} e^{-\mu t}\,\zeta(t)$$

exists almost surely with $\mathbf{E}\,\eta = 1$, $\mathbf{P}\{\eta = 0\} = q$ *(the probability of degeneration). The Laplace transform* $\varphi(s) = \mathbf{E}\,e^{-s\eta}$ *satisfies a differential equation*

$$\varphi' = \frac{1}{s\,\mu}\,g(\varphi), \quad \varphi(0) = \varphi'(0) = 1,$$

which has a solution that can be written in an implicit form:

$$s = (1-\varphi)\,\exp\left\{-\int_{\varphi}^{1}\left[\frac{m-1}{f(u)-u} + \frac{1}{1-u}\right] du\right\}. \tag{13.21}$$

Similar to the case of Galton-Watson processes, the stated above theorems make it possible to obtain limit distributions of the distance to the nearest mutual ancestor. Let τ_t be a moment of the appearance of the nearest mutual ancestor of the existing at moment t particles provided that there was only one particle at initial moment.

Theorem 13.28. *If* $m < 1$, *then*

$$\lim_{t\to\infty} \mathbf{P}\{t - \tau_t \le u \mid \zeta(t) > 0\} = \frac{p_1^*}{p_1(u)}$$

where

$$p_1^* = \lim_{t\to\infty} \mathbf{P}\{\zeta(t) = 1 \mid \zeta(t) > 0\}, \quad p_1(u) = \lim_{t\to\infty} \mathbf{P}\{\zeta(u) = 1 \mid \zeta(u) > 0\}.$$

Theorem 13.29. *Under the hypotheses of Theorem 13.24 the following relation*

$$\lim_{t\to\infty} \mathbf{P}\left\{\frac{\tau_t}{t} \ge x \mid \zeta(t) > 0\right\} = 1 - x, \quad 0 \le x \le 1,$$

holds.

Theorem 13.30. *If* $1 < m < \infty$, *then*

$$\lim_{t\to\infty} \mathbf{P}\{\tau_t \ge x \mid \zeta(u) > 0,\ u \ge 0\} = \exp\left\{x\,\frac{f'(q)-1}{a}\right\}.$$

Transition phenomena for Galton-Watson branching processes can be completely transferred on Markov branching processes. Let $\varsigma^{(r)}(t)$ (for any positive integer r) be the number of particles arising at time t from one particle in a Markov branching process with the generating function of one-particle direct descendants $f^{(r)}(z) = \sum_{k\geq 0} p_k^{(r)} z^k$ and a mean particle's lifetime a_r. Let also $m_r = \sum_{k\geq 1} k\, p_k^{(r)}$ and $\mu_r = (m_r - 1)/a_r \underset{r\to\infty}{\longrightarrow} 0$. If

$$\sup_r \frac{1}{a_r} \sum_{k\geq n} k^2\, p_k^{(r)} \underset{n\to\infty}{\longrightarrow} 0, \quad \inf_r \frac{1}{a_r} \sum_{k\geq 1} k\,(k-1)\, p_k^{(r)} > 0, \tag{13.22}$$

then

$$\mathbf{P}\{\varsigma^{(r)}(t) > 0\} \underset{r,t\to\infty}{\sim} \begin{cases} 2\, b_r^{-1}\,(m_r - 1)\,(1 - e^{-\mu_r t})^{-1} & \text{if } m_r \neq 1 \\ 2\, a_r\,(b_r\, t)^{-1} & \text{if } m_r = 1 \end{cases}$$

where $b_r = \sum_{k\geq 1} k\,(k-1)\, p_k^{(r)}$. The last relation is equivalent to the following one. Suppose both r and t tend to infinity in such a way that there exists a finite or infinite limit $c = \lim t\,\mu_r$. Then

$$\mathbf{P}\{\varsigma^{(r)}(t) > 0\} \sim \begin{cases} 2\, b_r^{-1}\,(1 - m_r)\, e^{\mu_r t} & \text{if } c = -\infty \\ 2\, a_r\,(b_r\, t)^{-1}\, c\,(1 - e^{-c})^{-1} & \text{if } -\infty < c < \infty \\ 2\, b_r^{-1}\,(m_r - 1) & \text{if } c = \infty. \end{cases}$$

In the next theorem we suppose without loss of generality that either $m_r = 1$ for all r or $m_r \neq 1$ for all r.

Theorem 13.31. *If the condition (13.22) is satisfied and there exists a positive limit*

$$d = \lim_{r\to\infty} \frac{b_r}{a_r},$$

then

$$\lim_{r,t\to\infty} \mathbf{P}\left\{\frac{\mu_r}{e^{\mu_r t} - 1}\, \varsigma^{(r)}(t) \geq x \mid \varsigma^{(r)}(t) > 0\right\} = e^{-2x/d} \quad \textit{if } m_r \neq 1,$$

$$\lim_{r,t\to\infty} \mathbf{P}\left\{\frac{1}{t}\, \varsigma^{(r)}(t) \geq x \mid \varsigma^{(r)}(t) > 0\right\} = e^{-2x/d} \quad \textit{if } m_r = 1.$$

Theorem 13.32 (an equivalent form of Theorem 13.31). *Under the hypotheses of Theorem 13.31 assume that parameters r and t tend to infinity in such a way that there exists a finite or infinite limit $c = \lim t\,\mu_r$.*

If $c = -\infty$, *then*

$$\lim \mathbf{P}\{(1-\mu_r)\,\zeta^{(r)}(t) \geq x \mid \zeta^{(r)}(t) > 0\} \;=\; e^{-2x/d}.$$

If $-\infty < c < \infty$, *then*

$$\lim \mathbf{P}\left\{\frac{1}{t}\,\zeta^{(r)}(t) \geq x \mid \zeta^{(r)}(t) > 0\right\} \;=\; \exp\left\{-\frac{2}{d}\,\frac{e^c-1}{c}\,x\right\}.$$

If $c = \infty$, *then*

$$\lim \mathbf{P}\{(\mu_r - 1)\,e^{-\mu_r t}\,\zeta^{(r)}(t) \geq x \mid \zeta^{(r)}(t) > 0\} \;=\; e^{-2x/d}.$$

13.4 Sevastyanov Model

The lifetime of a particle and the number of its direct descendants are independent random variables in the case of Bellman-Harris branching processes. **A Sevastyanov model**[5,16–22] supposes an arbitrary dependence between them. Such a branching process is defined by a distribution function $G(t)$ of the particle lifetime and by conditional probabilities $\{p_k(u)\}$ that the number of one-particle direct descendants equals k provided that this transformation occurred at the age u.

The unconditional distribution of the number of one-particle direct descendants is

$$p_k \;=\; \int_0^\infty p_k(u)\,dG(u), \quad k = 0, 1, 2, \ldots\,.$$

A generating function $F_t(z) \;=\; \mathbf{E}\,z^{\zeta(t)}$ for the number $\zeta(t)$ of particles arising at time t from one newly born particle satisfies the nonlinear integral equation

$$F_t(z) \;=\; z\,(1 - G(t)) \;+\; \int_0^t f(u,\, F_{t-u}(z))\,dG(u)$$

where $f(u,z) = \sum_{k\geq 0} z^k\,p_k(u)$.

Sufficient conditions of the regularity of the Sevastyanov process are the following ones:

(i) the distribution function $G(t)$ of a lifetime is continuous at $t = 0$, i.e., $G(+0) = 0$,

(ii) the mean number of one-particle direct descendants is finite, i.e., $m = \sum_{k\geq 1} k\,p_k < \infty$.

Just as in the case of Bellman-Harris processes, the probability q of the Sevastyanov process degeneration equals the nearest to zero solution of the equation

$$z = f(z) = \sum_{k \geq 0} z^k \, p_k$$

in a unit interval $0 \leq z \leq 1$.

The finiteness of the mean number of one-particle direct descendants ensures the finiteness of the mean number $A(t) = \mathbf{E}\,\zeta(t)$ of the descendants at time t. This mean number satisfies a renewal equation

$$A(t) = 1 - G(t) + \int_0^t A(t-u)\, m(u)\, dG(u)$$

where $m(u) = \sum_{k \geq 1} k\, p_k(u)$ is the conditional mean number of one-particle direct descendants provided that the transformation occurs at the age u.

A unique root (if it exists) of the equation

$$\int_0^\infty e^{-\mu t}\, m(u)\, dG(u) = 1$$

is called **a Maltus index** μ of the Sevastyanov process. Just as in the case of Bellman-Harris processes, a Maltus index gives an asymptotic of the mean number of particles. Let the distribution function (not necessarily probability distribution function) $M(t) = \int_0^t m(u)\, dG(u)$ be nonlatticed. If

$$\int_0^\infty e^{-\mu t}\,(1 - G(t))\, dt < \infty, \qquad \int_0^\infty t\, e^{-\mu t}\, m(t)\, dG(t) < \infty,$$

then

$$\lim_{t \to \infty} e^{-\mu t}\, A(t) = \frac{\int_0^\infty e^{-\mu t}\,(1 - G(t))\, dt}{\int_0^\infty t\, e^{-\mu t}\, m(t)\, dG(t)}.$$

Unlike the Bellman-Harris model the condition of nonlatticedness is imposed on $M(t)$ but not on $G(t)$.

An asymptotic behavior of the distribution of the number of particles and of the probability of extension is not so profoundly studied for a Sevastyanov model as for a Bellman-Harris process. The most advanced results are obtained for a critical process with a finite variance.

Theorem 13.33. *Let $m = 1$ and $M(t)$ be a nonlatticed distribution function. If*

$$a = \int_0^\infty t\, dG(t) < \infty, \qquad c = \int_0^\infty t\, m(t)\, dG(t) < \infty,$$

$$0 < b = \sum_{k \geq 1} k\,(k-1)\, p_k = f''(1) < \infty,$$

then

$$\lim_{t \to \infty} t\, \mathbf{P}\left\{\frac{1}{t}\, \zeta(t) \geq x\right\} = \frac{2c}{b} \exp\left\{-\frac{2c^2 x}{ab}\right\}, \quad x > 0.$$

Theorem 13.34. *Let the hypotheses of Theorem 13.33 be satisfied and the equality*

$$\lim_{t \to \infty} t^2 \int_t^\infty (1 + m(u))\, dG(u) = 0$$

holds. Then

$$\lim_{t \to \infty} t\, \mathbf{P}\{\zeta(t) > 0\} = \frac{2c}{b}.$$

In particular, Theorems 13.33 and 13.34 imply that under their hypotheses the distribution of $\zeta(t)/t$, provided that $\zeta(t) > 0$, weakly converges to the exponential distribution with a parameter $2c^2 a^{-2} b^{-1}$, i.e., under the hypotheses of Theorem 13.33 and 13.34 the following relation

$$\lim_{t \to \infty} \mathbf{P}\left\{\frac{1}{t}\, \zeta(t) \geq x \mid \zeta(t) > 0\right\} = \exp\left\{-\frac{2c^2 x}{ab}\right\}$$

holds. If a critical parameter m of the Sevastyanov branching process is close to the unit (but possibly not equal to unit), then the statement of Theorem 13.33 can be modified as follows. Let $\zeta^{(r)}(t)$ (for every integer positive r) be a number of descendants arising at time t from one newly born particle in a Sevastyanov branching process defined by the distribution function $G_r(t)$ of the particle lifetime and by conditional transformation probabilities $\{p_k^{(r)}(u)\}$. Let also the relation

$$m_r = \int_0^\infty m_r(u)\, dG_r(u) \underset{r \to \infty}{\longrightarrow} 1$$

be satisfied where

$$m_r(u) = \sum_{k \geq 1} k\, p_k^{(r)}(u)$$

and a sequence of functions $M_r(t) = \int_0^t m_r(u)\, dG_r(u)$, $r = 1, 2, \ldots$, converges to the limit function $M(t)$ at every point of its continuity, $M(\infty) = 1$. Let moreover there exist positive limits

$$a = \lim_{r\to\infty} \int_0^\infty t\, dG_r(t), \qquad c = \lim_{r\to\infty} \int_0^\infty t\, m_r(t)\, dG_r(t) < \infty,$$

$$b = \lim_{r\to\infty} \sum_{k\geq 1} k\,(k-1)\, p_k^{(r)}$$

where

$$p_k^{(r)} = \int_0^\infty p_k^{(r)}(u)\, dG_r(u).$$

Let finally

$$\sup_r \int_t^\infty u\, dG_r(u) \underset{t\to\infty}{\longrightarrow} 0, \qquad \sup_r \int_t^\infty u\, m_r(u)\, dG_r(u) \underset{t\to\infty}{\longrightarrow} 0,$$

$$\sup_r \sum_{k\geq n} k^2\, p_k^{(r)} \underset{n\to\infty}{\longrightarrow} 0.$$

If r and t tend to infinity in such a way that $t\,(m_r - 1) \to d$, then

$$t\,\mathbf{P}\left\{\frac{1}{t}\,\zeta^{(r)}(t) \geq x\right\} \longrightarrow \frac{2}{b}\,\frac{d}{1 - e^{-d/c}}\,\exp\left\{-\frac{2c}{ab}\,\frac{d}{e^{d/c} - 1}\,x\right\}, \quad x > 0.$$

For a supercritical process a weaker version of Theorem 13.20 holds.

Theorem 13.35. *Let $m > 1$ and $M(t)$ be a nonlatticed distribution function. If*

$$\int_0^\infty e^{-2\mu t} \sum_{k\geq 1} k^2\, p_k(t)\, dG(t) < \infty,$$

then $e^{-\mu t}\zeta(t)$ converges in a mean square to a random variable η as $t \to \infty$. This limit value η equals zero with the probability of degeneration q. The Laplace transform $\varphi(s) = \mathbf{E}\,e^{-s\eta}$ is a unique solution (in the class of Laplace transforms of nonnegative random variables) of the equation

$$\varphi(s) = \int_0^\infty f(u, \varphi(e^{-\mu u}\, s))\, dG(u)$$

with an additional condition

$$\lim_{s\downarrow 0} \frac{1-\varphi(s)}{s} = \frac{\int_0^\infty e^{-\mu t}\,(1-G(t))\,dt}{\int_0^\infty t\,e^{-\mu t}\,(1-G(t))\,dt}.$$

From the practical point of view the limit distribution of the age of an arbitrarily taken particle is of great interest. Let $\zeta_v(t)$ be the number of descendants arising at time t from one newly born particle and having the age not greater than v. Then the conditional mean value

$$\mathbf{E}\left(\frac{\zeta_v(t)}{\zeta(t)} \mid \zeta(t) > 0\right)$$

is the conditional probability that the age of an arbitrarily taken at t particle is not greater than v provided that the condition of nondegeneracy is satisfied.

Theorem 13.36. *Under the hypotheses of Theorem 13.34 a limit probability (as $t \to \infty$) that a randomly chosen at t particle has an age less than v equals*

$$\frac{\int_0^v (1-G(u))\,du}{\int_0^\infty (1-G(u))\,du}.$$

Theorem 13.37. *Under the hypotheses of Theorem 13.35 a limit probability (as $t \to \infty$) that a randomly chosen at t particle has an age less than v equals*

$$\frac{\int_0^v e^{-\mu u}\,(1-G(u))\,du}{\int_0^\infty e^{-\mu u}\,(1-G(u))\,du}.$$

If the Maltus index exists, then limit theorems for subcritical Sevastyanov processes are similar to corresponding theorems for Bellman-Harris processes.

Theorem 13.38. *If $m < 1$,*

$$\int_0^\infty t\,e^{-\mu t}\,(1+m(t))\,dG(t) < \infty, \qquad \int_0^\infty e^{-\mu t} \sum_{k\ge 1} p_k(t)\,k\,\log k\,dG(t) < \infty$$

and a distribution function $M(t)$ is not latticed, then there exists a finite and positive limit

$$\lim_{t\to\infty} e^{-\mu t}\,\mathbf{P}\{\zeta(t) > 0\}.$$

Theorem 13.39. *Under the hypotheses of Theorem 13.38 there exists a limit probability distribution*

$$\lim_{t\to\infty} \mathbf{P}\{\zeta(t) = k \mid \zeta(t) > 0\}, \quad k = 1, 2, \ldots .$$

The above stated theorems hold true also in the case of latticed distribution function, if a time parameter t tends to infinity along the points of the corresponding lattice.

13.5 Processes with Several Types of Particles

A rigorous definition of branching processes with several types of particles[3–5,12,15,17,21–25] with transformations being dependent on the age is highly cumbersome. On the other hand, a visual description is very short and simple. Every such branching process is connected to the population consisting of several types of particles and evolving in such a way: every existing at present particle independent of its origin and other existing at this moment particles is transformed at the end of its lifetime into some (possibly empty) population of newly born particles. The descendants of this particle depend only on its type and age when the transformation occurs.

A branching process with a finite number of particle types $T_1, \ldots, T_d$ is defined (for every $i = 1, \ldots, d$) by distribution function $G_i(t)$ of the lifetime of the particle of type T_i and by conditional probability $p_i(u; k_1, \ldots, k_d)$ that one-particle of type T_i direct descendants consist of k_j particles of the type T_j, $j = 1, \ldots, d$, provided that this transformation occurs at the age u. This is a Sevastyanov model.

Sufficient conditions of regularity of such a branching process are the following ones:

(i) the distribution functions $\{G_i(t)\}$ are continuous at time $t = 0$, i.e., $G_i(0+) = 0$, $i = 1, \ldots, d$,

(ii) a mean number m_{ij} of one-particle of type T_i direct descendants of type T_j is finite, i.e.,

$$m_{ij} = \sum_{k_1, \ldots, k_d} k_j\, p_i(k_1, \ldots, k_d) < \infty$$

where

$$p_i(k_1, \dots, k_d) = \int_0^\infty p_i(u;\, k_1, \dots, k_d)\, dG_i(u)$$

is the unconditional distribution of the number of one-particle of type T_i direct descendants.

A system of integral equations for multidimensional generating functions

$$F_i(t;\, \mathbf{z}) = F_i(t;\, z_1, \dots, z_d) = \mathbf{E}\, z_1^{\zeta_{i1}(t)} \dots z_d^{\zeta_{id}(t)}, \quad \mathbf{z} = (z_1, \dots, z_d),\ |z_j| \le 1,$$

forms an analytical basis for the investigation of such branching processes. Here $\zeta_{ij}(t)$ is the number of particles of type T_j arising at time t from one newly born particle of type T_i. The system of equations is of the form

$$F_i(t;\, \mathbf{z}) = z_i\,(1-G_i(t)) + \int_0^t f^i(u;\, \mathbf{F}(t-u;\, \mathbf{z}))\, dG_i(u), \quad i = 1, \dots, d, \qquad (13.23)$$

where

$$f^i(u;\, \mathbf{z}) = \sum_{k_1, \dots, k_d} z_1^{k_1} \dots z_d^{k_d}\, p_i(u;\, k_1, \dots, k_d),$$

$$\mathbf{F}(t;\, \mathbf{z}) = (F_1(t,\, \mathbf{z}),\, \dots,\, F_d(t,\, \mathbf{z})).$$

A matrix of mean values $M = (m_{ij})_{i,j=1}^d$ is one of the most important characteristics of the branching process. This matrix has a positive eigenvalue λ (a Perronian root) such that absolute values of all other eigenvalues don't exceed λ. The right and left eigenvectors associated with λ have nonnegative components.

A set of types $\mathcal{K} \subset \{T_1, \dots, T_d\}$ is called **a final class**, if every particle with the type from $\mathcal{K}$ can generate with probability 1 exactly one particle with the type from $\mathcal{K}$. In other words, a set $\mathcal{K}$ is called a final class, if $T_i \in \mathcal{K}$ implies the equality

$$\sum_{j:\, T_j \in \mathcal{K}} p_i(\underbrace{0,\, \dots,\, 0}_{j-1}, 1, \underbrace{0,\, \dots,\, 0}_{d-j}) = 1.$$

Theorem 13.40. *A branching process degenerates with probability 1 iff a set of types has no final classes and a Perronian root of the matrix M is not greater than unit.*

The probabilities of the degeneration of branching processes starting from one newly born particle of the type T_i $(i = 1, \dots, d)$ satisfy the equations containing generating functions

$$f^i(\mathbf{z}) = f^i(z_1, \dots, z_d) = \sum_{k_1, \dots, k_d} z_1^{k_1} \dots z_d^{k_d}\, p_i(k_1, \dots, k_d)$$

for the number of one-particle direct descendants.

Theorem 13.41. *A vector of degeneration probabilities* $(q_1, \dots, q_d)$ *equals the nearest to* $(0, \dots, 0)$ *solution of the system of equations*

$$z_i = f^i(z_1, \dots, z_d), \quad i = 1, \dots, d,$$

in a unit cube $0 \le z_j \le 1, \;\; j = 1, \dots, d.$

The probability of degeneration of the branching process starting from $k_1, \dots, k_d$ particles of types $T_1, \dots, T_d$ equals $q_1^{k_1} \dots q_d^{k_d}$. A branching process is called **subcritical**, if a Perronian root of the matrix M is less than unit, **critical**, if $\lambda = 1$, and **supercritical**, if $\lambda > 1$. A branching process is called **decomposable**, if a set of types can be divided into two nonempty disjointed subsets $\mathcal{K}_1$ and $\mathcal{K}_2$ such that the descendants of the particles with types from $\mathcal{K}_1$ don't contain particles with types from $\mathcal{K}_2$. If a branching process is indecomposable, then a Perronian root λ of the matrix M is a prime root of its characteristic polynomial and both right eigenvector $\mathbf{u} = (u_1, \dots, u_d)$ and left eigenvector $\mathbf{v} = (v_1, \dots, v_d)$ associated with this Perronian root have positive components.

An indecomposable branching process is called **nonlatticed**, if a matrix $M(t) = (M_{ij}(t))_{i,j=1}^d$ is nonlatticed. Here

$$M_{ij}(t) = \int_0^t m_{ij}(u)\, dG_i(u)$$

are monotone nondecreasing functions with $m_{ij}(u)$ being a conditional mean number of one-particle of type T_i direct descendants of type T_j provided that the transformation occurs at the age u, i.e.,

$$m_{ij}(u) = \sum_{k_1, \dots, k_d} k_j\, p_i(u; k_1, \dots, k_d).$$

In this case a matrix M of mean numbers of direct descendants coincides with the matrix $M(\infty)$.

The finiteness of the mean number of direct descendants implies the finiteness of the conditional mean number $A_{ij}(t)$ of particles of type T_j at time t provided that there was only one newly born particle of type T_i at time $t = 0$. These mean values satisfy the multidimensional renewal equation

$$A_{ij}(t) = \delta_{ij}\,(1 - G_i(t)) + \sum_{k=1}^d \int_0^t m_{ik}(u)\, A_{kj}(t-u)\, dG_i(u), \quad i, j = 1, \dots, d.$$

If $\lambda > 1$, then there exists exactly one positive number μ (a Maltus index) such that a Perronian root of the matrix

$$M_\mu = \left(\int_0^\infty e^{-\mu t} m_{ij}(t)\, dG_i(t) \right)_{i,j=1}^d = \left(\int_0^\infty e^{-\mu t}\, dM_{ij}(t) \right)_{i,j=1}^d$$

equals unit. Such a number exists also in a case of a critical process. This number is unique and equals zero. If μ exists for a subcritical process (what is supposed, but is not necessarily), then $\mu < 0$.

All limit theorems of Section 13.4 can be generalized on the case of indecomposable processes with a finite number of particle types.

Theorem 13.42. *Let a branching process be nonlatticed and indecomposable. If*

$$\int_0^\infty e^{-\mu t} (1 - G_i(t))\, dt < \infty, \quad i = 1, \ldots, d,$$

$$\sigma = \sum_{i,j} v_i^\mu \int_0^\infty t\, e^{-\mu t} m_{ij}(t)\, dG_i(t)\, u_j^\mu < \infty,$$

then

$$\lim_{t\to\infty} e^{-\mu t} A_{ij}(t) = \frac{1}{\sigma}\, u_i^\mu\, v_j^\mu \int_0^\infty e^{-\mu t} (1 - G_i(t))\, dt$$

where $\mathbf{u}^\mu = (u_1^\mu, \ldots, u_d^\mu)$ *and* $\mathbf{v}^\mu = (v_1^\mu, \ldots, v_d^\mu)$ *are right and left invariant vectors of the matrix* M_μ.

So under the hypotheses of this theorem the mean number of particles at time t in a nonlatticed indecomposable branching process with a nonlatticed matrix exponentially increases to infinity, if a process is supercritical, or tends to a finite limit, if a process is critical, or exponentially decreases to zero, if a process is subcritical.

Theorem 13.43. *Let a critical branching process be nonlatticed and indecomposable. If*

$$a_i = \int_0^\infty t\, dG_i(t) < \infty, \quad i = 1, \ldots, d,$$

$$\sigma = \sum_{i,j} v_i \int_0^\infty t\, dM_{ij}(t)\, u_j < \infty, \quad 0 < b = \sum_{i,j,k} v_i\, b_{jk}^i\, u_j\, u_k < \infty,$$

then

$$\lim_{t\to\infty} t\,\mathbf{P}\left\{\frac{1}{t}\,\zeta_{i1}(t) \geq x_1,\ \ldots,\ \frac{1}{t}\,\zeta_{id}(t) \geq x_d\right\} =$$

$$= u_i\,\frac{2\sigma}{b}\,\exp\left\{-\frac{2\sigma^2}{b}\,\max\left(\frac{x_1}{v_1 a_1},\ \ldots,\ \frac{x_d}{v_d a_d}\right)\right\}$$

for $x_1 \geq 0,\ \ldots,\ x_d \geq 0,\ x_1 + \cdots + x_d > 0$. *Here*

$$b^i_{jk} = \sum_{r_1,\ldots,r_d} r_j\,(r_k - \delta_{jk})\,p_i(r_1,\ldots,r_d).$$

Theorem 13.44. *If under the hypotheses of Theorem 13.43*

$$\lim_{t\to\infty} t^2 \int_t^{\infty} (1 + m_{ij}(u))\,dG_i(u) = 0,$$

then

$$\lim_{t\to\infty} t\,\mathbf{P}\left\{\sum_{j=1}^{d} \zeta_{ij}(t) > 0\right\} = u_i\,\frac{2\sigma}{b}.$$

Theorems 13.43 and 13.44 state a weak convergence of a joint distribution of random variables $\zeta_{i1}(t)/t,\ \ldots,\ \zeta_{id}(t)/t$, provided that $\sum_{j=1}^{d} \zeta_{ij}(t) > 0$, to a joint distribution of random variables $v_1 a_1 \xi,\ \ldots,\ v_d a_d \xi$ where ξ is a random variable having an exponential distribution with a parameter $2\sigma^2/b$.

In other words, under the hypotheses of Theorems 13.43 and 13.44 the following equality

$$\lim_{t\to\infty} \mathbf{P}\left\{\frac{1}{t}\,\zeta_{i1}(t) \geq x_1,\ \ldots,\ \frac{1}{t}\,\zeta_{id}(t) \geq x_d \;\middle|\; \sum_{j=1}^{d} \zeta_{ij}(t) > 0\right\} =$$

$$= \exp\left\{-\frac{2\sigma^2}{b}\,\max\left(\frac{x_1}{v_1 a_1},\ \ldots,\ \frac{x_d}{v_d a_d}\right)\right\}$$

holds.

Transient phenomena in a process with the finite number of particle types arise for $\lambda \to 1$ and $t \to \infty$. Let $\zeta^{(r)}_{ij}(t)$ be a number of particles of type T_j arising at time t from one newly born particle of type T_i in a branching process defined by distribution functions of particles' lifetimes $\{G^{(r)}_i(t)\}$ and conditional transformation probabilities $\{p^{(r)}_i(u;\,k_1,\ldots,k_d)\}$. Let also

$$m^{(r)}_{ij}(u) = \sum_{k_1,\ldots,k_d} k_j\,p^{(r)}_i(u;\,k_1,\ldots,k_d)$$

and for any $i, j = 1, \ldots, d$ a sequence of nondecreasing functions

$$M_{ij}^{(k)}(t) = \int_0^t m_{ij}^{(r)}(u)\, dG_i^{(r)}(u), \quad r = 1, 2, \ldots,$$

converges to a limit function $M_{ij}(t)$ at every point of its continuity. Let moreover $M_{ij}^{(r)}(\infty) \underset{r\to\infty}{\longrightarrow} M_{ij}(\infty)$, a matrix $M(t) = (M_{ij}(t))_{i,j=1}^d$ be non-latticed, a matrix $M(\infty)$ be indecomposable and its Perronian root be equal to unit. Under these assumptions it follows that a Perronian root λ_r of the matrix $M_r = (M_{ij}^{(r)}(\infty))_{i,j=1}^d$ tends to unit as $r \to \infty$. Moreover corresponding right and left eigenvectors of the matrix M_r can be chosen converging to right and left invariant vectors $\mathbf{u} = (u_1, \ldots, u_d)$ and $\mathbf{v} = (v_1, \ldots, v_d)$ of the matrix $M(\infty)$ as $r \to \infty$.

Assume that there exist positive limits

$$a_j = \lim_{r\to\infty} \int_0^\infty t\, dG_j^{(r)}(t), \qquad \sigma = \lim_{r\to\infty} \sum_{i,j} v_i \int_0^\infty t\, dM_{ij}^{(r)}(t)\, u_j,$$

$$b = \lim_{r\to\infty} \sum_{i,j,k} v_i \sum_{l_1,\ldots,l_d} l_j\, (l_k - \delta_{jk})\, p_i^{(r)}(l_1, \ldots, l_d)\, u_j\, u_k$$

where

$$p_i^{(r)}(l_1, \ldots, l_d) = \int_0^\infty p_i^{(r)}(u; l_1, \ldots, l_d)\, dG_i^{(r)}(u).$$

Suppose also that

$$\sup_r \int_t^\infty u\, dG_i^{(r)}(u) \underset{t\to\infty}{\longrightarrow} 0, \quad i = 1, \ldots, d,$$

$$\sup_r \int_t^\infty u\, dM_{ij}^{(r)}(u) \underset{t\to\infty}{\longrightarrow} 0, \quad i, j = 1, \ldots, d,$$

$$\sup_r \sum_{k_1+\cdots+k_d \geq n} (k_1 + \cdots + k_d)^2\, p_i^{(r)}(k_1, \ldots, k_d) \underset{n\to\infty}{\longrightarrow} 0, \quad i = 1, \ldots, d.$$

If r and t tend to infinity in such a way that $t\,(\lambda_r - 1) \to c$, then

$$t\,\mathbf{P}\left\{\frac{1}{t}\,\zeta_{i1}^{(r)}(t) \geq x_1, \ \ldots, \ \frac{1}{t}\,\zeta_{id}^{(r)}(t) \geq x_d\right\} \to$$

$$\to u_i \frac{2}{b} \frac{c}{1 - e^{-c/\sigma}} \exp\left\{ -\frac{2\sigma}{b} \frac{c}{e^{c/\sigma} - 1} \max\left(\frac{x_1}{v_1 a_1}, \ldots, \frac{x_d}{v_d a_d} \right) \right\}$$

for $x_1 \geq 0, \ldots, x_d \geq 0, \; x_1 + \cdots + x_d > 0$.

Theorem 13.45. *Let a supercritical branching process be nonlatticed and indecomposable. If*

$$\int_0^\infty e^{-2\mu t} \sum_{k_1,\ldots,k_d} k_j^2 \, p_i(t; k_1, \ldots, k_d) \, dG_i(t) < \infty, \quad i, j = 1, \ldots, d,$$

then for every $i = 1, \ldots, d$, random variables

$$\frac{\mu \, e^{-\mu t} \, \zeta_{ij}(t)}{v_j^\mu \int_0^\infty (1 - e^{-\mu u}) \, dG_j(u)}, \qquad j = 1, \ldots, d,$$

converge in a mean square to a joint limit η_i. A Laplace transform $\varphi_i(s) = \mathbf{E} \, e^{-s\eta_i}$ satisfies the system of equations

$$\varphi_i(s) = \int_0^\infty f^i(u; \varphi_1(e^{-\mu u} s), \ldots, \varphi_d(e^{-\mu u} s)) \, dG_i(u), \quad i = 1, \ldots, d, \tag{13.24}$$

with additional conditions

$$\lim_{s \downarrow 0} \frac{1 - \varphi_i(s)}{s} = \frac{u_i^\mu}{\sigma}, \qquad i = 1, \ldots, d. \tag{13.25}$$

The system of equations (13.24) with additional conditions (13.25) has a unique solution in the class of Laplace transforms of nonnegative random variables.

Let $\zeta_{ij}(t, v)$ be the number of existing at time t descendants of type T_j with an age less than v provided that at the initial moment there was only one newly born particle of type T_i. Then the conditional mean value

$$\mathbf{E} \left(\frac{\zeta_{ij}(t, v)}{\zeta_{ij}(t)} \,\middle|\, \sum_j \zeta_{ij}(t) > 0 \right)$$

is the conditional probability that a particle chosen at random at time t is of the type T_j and its age is less than v provided that the initial population consists of one newly born particle of the type T_i and a process does not degenerate up to time t.

Theorem 13.46. *Let a critical branching process be nonlatticed and indecomposable. If*

$$\lim_{t\to\infty} t^2 \int_t^\infty (1 + m_{ij}(u))\, dG_i(u) = 0, \quad i, j = 1, \ldots, d,$$

and

$$\sum_{r_1,\ldots,r_d} r_j^2\, p_i(r_1, \ldots, r_d) < \infty, \quad i, j = 1, \ldots, d,$$

then the probability that a particle chosen at random at time t is of the type T_j and its age is less than u, provided that a process does not degenerate up to time t, converges to

$$\frac{v_j \int_0^u (1 - G_j(s))\, ds}{\sum_i v_i \int_0^\infty s\, dG_i(s)}$$

as $t \to \infty$.

Theorem 13.47. *Let a supercritical branching process be nonlatticed and indecomposable. If*

$$\int_0^\infty t\, e^{-\mu t}\, dM_{ij}(t) < \infty,$$

$$\int_0^\infty e^{-2\mu t} \sum_{r_1,\ldots,r_d} r_j^2\, p_i(t; r_1, \ldots, r_d)\, dG_i(t) < \infty,$$

then the probability that a particle chosen at random at time t is of the type T_j and its age is less than u, provided that a process does not degenerate up to time t, converges to

$$\frac{v_j^\mu \int_0^u (1 - G_j(s))\, e^{-\mu s}\, ds}{\sum_i v_i \int_0^\infty (1 - G_i(s))\, e^{-\mu s}\, ds}$$

as $t \to \infty$.

Theorems 13.42–13.47 hold true also in the case of the latticed process, if a time parameter t tends to infinity along the points of a corresponding lattice.

More profound results can be formulated for the Bellman-Harris model of branching processes with a finite number of particle types. A Bellman-Harris model is a special case of the described one. A distinctive property of this model is that the transformation probabilities don't depend on the age when the transformation occurs, i.e.,

$$p_i(u; k_1, \dots, k_d) \equiv p_i(k_1, \dots, k_d).$$

Theorem 13.48. *Let a critical Bellman-Harris branching process be nonlatticed, indecomposable and*

$$0 < b = \sum_{i,j,l} v_i \sum_{k_1,\dots,k_d} k_j\,(k_l - \delta_{jl})\, p_i(k_1, \dots, k_d)\, u_j\, u_l < \infty.$$

Then: (i) if

$$\lim_{t\to\infty} t^2\,(1 - G_i(t)) = c_i\,\frac{b}{2},$$

then

$$\lim_{t\to\infty} t\,\mathbf{P}\left\{\sum_j \zeta_{ij}(t) > 0\right\} = u_i\,\frac{a + \sqrt{a^2+4c}}{b}$$

where

$$a = \sum_i u_i\, v_i\, a_i, \quad c = \sum_i v_i\, c_i, \quad a_i = \int_0^\infty t\, dG_i(t);$$

(ii) the following relations hold:

$$\lim_{t\to\infty} \mathbf{P}\left\{\zeta_{i1}(t) = k_1, \dots, \zeta_{id}(t) = k_d \,\middle|\, \sum_j \zeta_{ij} > 0\right\} =$$

$$= \frac{(-1)^{k-1}\,(a^2+4c)^{1/2-k}}{a + (a^2+4c)^{1/2}} \binom{1/2}{k} \frac{k!}{k_1! \dots k_d!}\,(4v_1 c_1)^{k_1} \dots (4 v_d c_d)^{k_d},$$

$$\lim_{t\to\infty} \mathbf{P}\left\{\frac{1}{t}\,\zeta_{i1}(t) \ge x_1, \dots, \frac{1}{t}\,\zeta_{id}(t) \ge x_d \,\middle|\, \sum_j \zeta_{ij} > 0\right\} =$$

$$= \frac{2a}{a+\sqrt{a^2+4c}} \exp\left\{-\frac{2a^2}{b} \max\left(\frac{x_1}{v_1 a_1}, \dots, \frac{x_d}{v_d a_d}\right)\right\}$$

where $k = k_1 + \dots + k_d$*;*
(iii) if

$$\lim_{t\to\infty} t^2 \sum_j v_j\,(1 - G_j(t)) = \infty,$$

$$\lim_{t\to\infty} \frac{1 - G_i(t)}{\sum\limits_j v_j\,(1 - G_j(t))} = c_i, \qquad \sum_i c_i > 0,$$

and $\sum\limits_j v_j\,(1 - G_j(t))$ *is a properly changing function as* $t \to \infty$, *then*

$$\lim_{t\to\infty} \mathbf{P}\left\{ \zeta_{ij}(t) = k_j,\ j = 1, \ldots, d \;\middle|\; \sum_j \zeta_{ij} > 0 \right\} =$$

$$= (-1)^{k-1} \binom{1/2}{k} \frac{k!}{k_1! \ldots k_d!} \left(\frac{v_1 c_1}{c}\right)^{k_1} \cdots \left(\frac{v_d c_d}{c}\right)^{k_d}$$

where $k = k_1 + \cdots + k_d$, $c = v_1 c_1 + \cdots + v_d c_d$.

Consider a generalization of Theorem 13.48 as a multidimensional analog of Theorems 13.17 and 13.5. Introduce a vector function

$$\mathbf{f}(\mathbf{z}) = (f^1(\mathbf{z}), \ldots, f^d(\mathbf{z}))$$

where

$$f^i(\mathbf{z}) = f^i(z_1, \ldots, z_d) = \sum_{k_1, \ldots, k_d} z_1^{k_1} \ldots z_d^{k_d}\, p_i(k_1, \ldots, k_d)$$

is a generating function of the number of direct descendants of a one-particle of type T_i . Let

$$\mathbf{f}_1(\mathbf{z}) = \mathbf{f}(\mathbf{z}), \quad \mathbf{f}_n(\mathbf{z}) = (f_n^1(\mathbf{z}), \ldots, f_n^d(\mathbf{z})) = \mathbf{f}(\mathbf{f}_{n-1}(\mathbf{z})) =$$

$$= (f^1(\mathbf{f}_{n-1}(\mathbf{z})), \ldots, f^d(\mathbf{f}_{n-1}(\mathbf{z}))), \quad q_n = \sum_i v_i\,(1 - f_n^i(0, \ldots, 0)).$$

Theorem 13.49. *Let a critical Bellman-Harris branching process be nonlatticed and indecomposable. Suppose moreover that for some slowly changing at zero function* $l(\cdot)$ *the relation*

$$\sum_i v_i\,(1 - f^i(1 - u_1 z, \ldots, 1 - u_d z)) = z - z^{1+\alpha}\, l(z)$$

holds where $z \geq 0$, $0 < \alpha \leq 1$. *Then* $L(n) = n^{1/\alpha} q_n$ *is a slowly changing function as* $n \to \infty$ *and* $L(n) \sim (\alpha l(q_n))^{-1/\alpha}$. *If*

$$\lim_{n\to\infty} \frac{n}{q_n} \sum_i v_i\,(1 - G_i(n)) = c,$$

then

$$Q_i(t) = \mathbf{P}\left\{\sum_j \zeta_{ij}(t) > 0\right\} \sim u_i\,\omega\,t^{-1/\alpha}\,L(t) \quad \text{as } t \to \infty$$

where

$$\omega > 0, \;\; \omega^{1+\alpha} - a\omega = \alpha c, \;\; a = \sum_i v_i\,u_i\,a_i, \;\; a_i = \int_0^\infty t\,dG_i(t).$$

If

$$\lim_{n\to\infty} \frac{n}{q_n} \sum_i v_i\,(1 - G_i(n)) = \infty$$

and $\sum_i v_i\,(1 - G_i(n))$ *is a properly changing function with an index* β *as* $t \to \infty$, *then* $t^{\beta/(1+\alpha)}\,Q_i(t)$ *is a slowly changing function as* $t \to \infty$ *and*

$$Q_i(t)/Q_j(t) \underset{t\to\infty}{\longrightarrow} u_i/u_j.$$

Theorem 13.50. *If under the hypotheses of Theorem 13.49*

$$\lim_{n\to\infty} \frac{n}{q_n} \sum_i v_i\,(1 - G_i(n)) = c,$$

then

$$\lim_{t\to\infty} \mathbf{P}\left\{t^{-1/\alpha}\,L(t)\,\zeta_{ij}(t) \le x_j, \;\; j = 1,\ldots,d \;\middle|\; \sum_j \zeta_{ij}(t) > 0\right\} =$$

$$= 1 - \frac{a^{1/\alpha}}{\omega} + \frac{a^{1/\alpha}}{\omega}\,\Pi\left(a^{1+1/\alpha}\min\left(\frac{x_1}{v_1\,a_1},\ldots,\frac{x_d}{v_d\,a_d}\right)\right)$$

where

$$\int_0^\infty e^{-sx}\,d\Pi(x) = 1 - s\,(1 + s^\alpha)^{-1/\alpha}, \quad s \ge 0.$$

Theorem 13.51. *If under the hypotheses of Theorem 13.49*

$$\lim_{n\to\infty} \frac{n\,(1 - G_i(n))}{q_n} = c_i, \quad \sum_j c_j > 0,$$

then

$$\lim_{t\to\infty} \mathbf{P}\left\{ \zeta_{ij}(t) = k_j,\ \ j = 1,\dots,d \ \middle|\ \sum_j \zeta_{ij}(t) > 0 \right\} =$$

$$= \pi_k \, \frac{k!}{k_1! \dots k_d!} \left(\frac{v_1 c_1}{c}\right)^{k_1} \cdots \left(\frac{v_d c_d}{c}\right)^{k_d}$$

where

$$k = k_1 + \dots + k_d, \quad c = \sum_i v_i c_i, \quad \sum_{k\ge 1} z^k \pi_k = 1 - \frac{\omega(z)}{\omega}, \quad \omega(z) > 0,$$

$$(\omega(z))^{1+\alpha} - a\,\omega(z) = \alpha\, c\,(1-z).$$

If

$$\lim_{n\to\infty} \frac{n}{q_n} \sum_i v_i\,(1 - G_i(n)) = \infty,$$

$$\lim_{t\to\infty} \frac{1 - G_i(t)}{\sum_j v_j\,(1 - G_j(n))} = c_i, \quad \sum_i c_i > 0,$$

and $\sum_j v_j\,(1 - G_j(n))$ *is a properly changing function as* $t \to \infty$, *then*

$$\lim_{t\to\infty} \mathbf{P}\left\{ \zeta_{ij}(t) = k_j,\ \ j = 1,\dots,d \ \middle|\ \sum_j \zeta_{ij}(t) > 0 \right\} =$$

$$= (-1)^{k-1} \binom{1/(1+\alpha)}{k} \frac{k!}{k_1! \dots k_d!} \left(\frac{v_1 c_1}{c}\right)^{k_1} \cdots \left(\frac{v_d c_d}{c}\right)^{k_d}$$

where $k = k_1 + \dots + k_d$, $c = v_1 c_1 + \dots + v_d c_d$.

Theorems 13.19 and 13.20 can be generalized on processes with several types of particles. For the subcritical process the existence of the Maltus index μ is additionally supposed.

Theorem 13.52. *Let a subcritical Bellman-Harris process be nonlatticed and indecomposable. If*

$$\int_0^\infty t\, e^{-\mu t}\, dG_i(t) < \infty, \quad i = 1,\dots,d,$$

$$\sum_{k_1,\dots,k_d} p_i(k_1,\dots,k_d)\, k_j \log k_j < \infty, \quad i,j = 1,\dots,d,$$

then for all $i = 1, \dots, d$ *there exists a finite positive limit*

$$\lim_{t\to\infty} e^{-\mu t} \mathbf{P}\left\{\sum_j \zeta_{ij}(t) > 0\right\}$$

and a limit probability distribution

$$\pi(k_1, \dots, k_d) = \lim_{t\to\infty} \mathbf{P}\left\{\zeta_{ij}(t) = k_j,\ j = 1, \dots, d \,\middle|\, \sum_j \zeta_{ij}(t) > 0\right\},$$

which does not depend on i.

Theorem 13.53. *Let a supercritical Bellman-Harris process be non-latticed and indecomposable. If*

$$\sum_{k_1,\dots,k_d} k_j^2 p_i(k_1, \dots, k_d) < \infty, \quad i, j = 1, \dots, d,$$

$$G_i(t) = \int_0^t g_i(u)\, du, \quad \int_0^\infty g_i^2(u)\, du < \infty, \quad i = 1, \dots, d,$$

then for every $i = 1, \dots, d$ *the random variables*

$$\frac{\mu\, e^{-\mu t} \zeta_{ij}(t)}{v_j^\mu \int_0^\infty (1 - e^{-\mu u})\, dG_j(u)}, \quad j = 1, \dots, d,$$

converge almost surely to a joint limit η_i *as* $t \to \infty$. *The Laplace transforms* $\varphi_i(s) = \mathbf{E} e^{-s\eta_i}$, $i = 1, \dots, d$, *are the unique solution (in the class of Laplace transforms of nonnegative random variables) of the system of equations*

$$\varphi_i(s) = \int_0^\infty f^i(\varphi_1(e^{-\mu u} s), \dots, \varphi_d(e^{-\mu u} s))\, dG_i(u), \quad i = 1, \dots, d,$$

with additional conditions $\varphi_i(\infty) = q_i$,

$$\lim_{s\downarrow 0} \frac{1 - \varphi(s)}{s} = u_i^\mu \left(\sum_j v_j^\mu \frac{\int_0^\infty t\, e^{-\mu t} G_j(t)}{\int_0^\infty e^{-\mu t} G_j(t)} u_j^\mu\right)^{-1}, \quad i = 1, \dots, d.$$

These given theorems hold true also for the latticed process, when a time parameter t tends to infinity along the points of the corresponding lattice.

For Markov branching processes with a finite number of particle types Theorems 13.52 and 13.53 can be formulated more precisely. We obtain a Markov process when lifetimes of particles have either exponential or singular distribution.

Let $G_i = 1 - e^{-t/a_i}$. A Belmann-Harris branching process with a finite number of particle types and exponential distributions of particle lifetimes is called a **Markov branching process with a continuous time** (and with a finite number of particle types). A corresponding system of equations (13.23) transforms into a system of ordinary differential equations

$$\frac{\partial F_i(t, \mathbf{z})}{\partial t} = g_i(\mathbf{F}(t, \mathbf{z})), \quad F_i(0, \mathbf{z}) = z_i, \quad i = 1, \dots, d,$$

where

$$g_i(\mathbf{z}) = g_i(z_1, \dots, z_d) = \frac{1}{a_i}\,(f^i(z_1, \dots, z_d) - z_i).$$

This system is equivalent to the system of partial differential equations

$$\frac{\partial}{\partial t} F_i(t, z) = \sum_j g_j(\mathbf{z}) \frac{\partial}{\partial z_j} F_i(t, \mathbf{z}).$$

Moreover a vector $\mathbf{F}(t, \mathbf{z})$ satisfies a semigroup relation $\mathbf{F}(t + s, \mathbf{z}) = \mathbf{F}(t, \mathbf{F}(s, \mathbf{z}))$. Hence it follows that the matrix of mean values can be represented in the form

$$(\mathbf{E}\,\zeta_{ij}(t))_{i,j=1}^d = e^{tA}$$

where $A = (a_{ij})_{i,j=1}^d$, $a_{ij} = (m_{ij} - \delta_{ij})/a_i$. The matrix A has a real eigenvalue μ, which is greater than real parts of all other eigenvalues. This eigenvalue is a Maltus index of the Markov branching process with a continuous time.

A right eigenvector of the matrix A corresponding to this eigenvalue μ is proportional to the right invariant vector $\mathbf{u}^\mu$ of the matrix M_μ. A left eigenvector of the matrix A corresponding to eigenvalue μ has the components $v_i^\mu a_i/(1 + \mu a_i)$, $i = 1, \dots, d$, where $(v_1^\mu, \dots, v_d^\mu)$ is the left invariant vector of the matrix M_μ.

Theorem 13.54. *Let a supercritical Markov branching process with a continuous time be indecomposable. Then for any $i = 1, \dots, d$ the random variables*

$$\frac{1 + \mu\, a_j}{v_j^\mu\, a_j}\, e^{-\mu t}\, \zeta_{ij}(t), \quad j = 1, \dots, d,$$

converge almost surely to a joint limit η_i. *An inequality* $\mathbf{P}\{\eta_i > 0\} > 0$ *holds for some* $i = 1, \ldots, d$ *iff*

$$\sum_{k_1,\ldots,k_d} p_i(k_1,\ldots,k_d)\, k_j \log k_j < \infty$$

for any $i, j = 1, \ldots, d$. *If the last condition is fulfilled, then a vector* $(\mathbf{P}\{\eta_1 = 0\}, \ldots, \mathbf{P}\{\eta_d = 0\})$ *coincides with the vector of degeneration probabilities, and Laplace transforms* $\varphi_i(s) = \mathbf{E}\, e^{-s\eta_i}$ *are the unique solution of the system of ordinary differential equations*

$$\frac{d\varphi_i}{ds} = \frac{1}{\mu\, s}\, g_i(\varphi_1,\ldots,\varphi_d), \quad i = 1,\ldots,d,$$

with additional conditions

$$\lim_{s\downarrow 0} \frac{1-\varphi_i(s)}{s} = \frac{u_i^\mu}{\sigma}, \quad i = 1,\ldots,d,$$

where

$$\sigma = \sum_j \frac{v_j^\mu\, a_j\, u_j^\mu}{1+\mu\, a_j}.$$

Theorem 13.55. *Let a subcritical Markov branching process with a continuous time be indecomposable. Then there exists the limit*

$$\lim_{t\to\infty} \frac{1}{u_i^\mu}\, e^{-\mu t}\, \mathbf{P}\left\{\sum_j \zeta_{ij}(t) > 0\right\}$$

which does not depend on i. *This limit is a positive one iff*

$$\sum_{k_1,\ldots,k_d} p_i(k_1,\ldots,k_d)\, k_j \log k_j < \infty \tag{13.26}$$

for any $i, j = 1, \ldots, d$.

Theorem 13.56. *Let a subcritical Markov branching process with a continuous time be indecomposable. Then there exists a limit probability distribution*

$$P^*(k_1,\ldots,k_d) = \lim_{t\to\infty} \mathbf{P}\left\{\zeta_{ij}(t) = k_j,\ j = 1,\ldots,d \;\middle|\; \sum_j \zeta_{ij}(t) > 0\right\}$$

and a generating function

$$F^*(\mathbf{z}) = F^*(z_1,\ldots,z_d) = \sum_{k_1,\ldots,k_d} z_1^{k_1} \ldots z_d^{k_d}\, P^*(k_1,\ldots,k_d)$$

is a solution of the problem

$$\sum_j g_i(\mathbf{z}) \frac{\partial}{\partial z_i} F^*(\mathbf{z}) = -\mu (1 - F^*(\mathbf{z})), \quad F^*(\mathbf{0}) = 0. \tag{13.27}$$

A limit distribution has a finite mean value iff the condition (13.26) is fulfilled. Moreover this condition ensures the uniqueness of the solution of the problem (13.27) in the class of probability generating functions with a finite first moment.

Suppose the distributions of lifetimes are concentrated in one point being the same for all types. Without loss of generality we can assume that lifetimes of all particles are identical to unit. Such a branching process is a multidimensional analog of a Galton-Watson process and is called **a Markov branching process with a discrete time** (and with a finite number of particle types). A system of equations (13.23) is equivalent to the following recursive relation

$$F_i(1, \mathbf{z}) = f^i(\mathbf{z}), \quad F_i(n+1, \mathbf{z}) = f^i(\mathbf{F}(n, \mathbf{z})).$$

Hence the vector $\mathbf{F}(n, \mathbf{z})$ is the nth iteration of the vector $\mathbf{f}(\mathbf{z})$, i.e., $\mathbf{F}(n, \mathbf{z}) = = \mathbf{f}(\mathbf{z})$. A matrix of mean numbers of descendants in nth generation is equal to the nth power of the matrix M, i.e., $(\mathbf{E}\, \zeta_{ij}(n))_{i,j=1}^d = M^n$.

A Markov branching process with a discrete time is called **nonperiodic,** if the modules of all eigenvalues of the matrix M are less than its Perronian root.

Theorem 13.57. *Let a subcritical Markov branching process be indecomposable and nonperiodic. Then there exists a limit*

$$\lim_{n\to\infty} \lambda^{-n} u_i^{-1} \mathbf{P} \left\{ \sum_j \zeta_{ij}(n) > 0 \right\},$$

which does not depend on i. This limit is positive iff

$$\sum_{k_1,\ldots,k_d} p_i(k_1, \ldots, k_d)\, k_j \log k_j < \infty \tag{13.28}$$

for any $i, j = 1, \ldots, d$.

Theorem 13.58. *Under the hypotheses of Theorem 13.57 there exists an independent of i limit probability distribution*

$$P^*(k_1, \ldots, k_d) = \lim_{n\to\infty} \mathbf{P} \left\{ \zeta_{ij}(n) = k_j,\ j = 1, \ldots, d \,\middle|\, \sum_j \zeta_{ij}(n) > 0 \right\}$$

with a generating function

$$F^*(\mathbf{z}) = F^*(z_1,\ldots,z_d) = \sum_{k_1,\ldots,k_d} z_1^{k_1}\ldots z_d^{k_d}\, P^*(k_1,\ldots,k_d)$$

satisfying an equation

$$1 - F^*(f(\mathbf{z})) = \lambda\,(1 - F^*(\mathbf{z})). \tag{13.29}$$

Moreover a limit distribution has a finite mean value iff the condition (13.28) is fulfilled. This condition ensures the uniqueness of the solution of the equation (13.29) in the class of probability generating functions vanishing at zero and with a finite first moment.

Theorem 13.59. *Let a supercritical Markov branching process with a discrete time be indecomposable and nonperiodic. Then for any* $i = 1,\ldots,d$ *the random variables*

$$\lambda^{-n}\, v_j^{-1}\, \zeta_{ij}(n), \quad j = 1,\ldots,n,$$

converge with the probability 1 to a joint limit η_i. *An inequality* $\mathbf{P}\{\eta_i > 0\} > 0$ *holds for some* $i = 1,\ldots,d$ *iff*

$$\sum_{k_1,\ldots,k_d} p_i(k_1,\ldots,k_d)\, k_j \log k_j < \infty$$

for any $i, j = 1,\ldots,d$. *If the last condition is fulfilled, then a vector* $(\mathbf{P}\{\eta_1 = 0\},\ldots,\mathbf{P}\{\eta_d = 0\})$ *coincides with the vector of degeneration probabilities, and Laplace transforms* $\varphi_i(s) = \mathbf{E}\, e^{-s\eta_i}$ *are the unique solution (in the class of Laplace transforms of nonnegative random variables) of the system of equations*

$$\varphi_i(\lambda s) = f^i(\varphi_1(s),\ldots,\varphi_d(s)), \quad i = 1,\ldots,d,$$

with additional conditions

$$\lim_{s\downarrow 0} \frac{1-\varphi_i(s)}{s} = \frac{u_i}{\sum_j v_j\, u_j}, \quad i = 1,\ldots,d.$$

In the case of a critical process a Markov property makes it possible to formulate a statement being inverse to Theorem 13.50.

Theorem 13.60. *Let a critical Markov branching process with a discrete time be indecomposable and nonperiodic. Let moreover for some* $i = 1,\ldots,d$ *there exists a limit*

$$\lim_{n\to\infty} \mathbf{P}\left\{ q_n\, \zeta_{ij}(n) \le x_j,\ j = 1,\ldots,d \;\middle|\; \sum_j \zeta_{ij}(n) > 0 \right\} = \Pi(x_1,\ldots,x_d)$$

at every continuity point of the limit function where

$$q_n = \sum_i v_i \mathbf{P}\left\{\sum_j \zeta_{ij}(n) > 0\right\}, \quad \Pi(0+,\dots,0+) < 1.$$

Then

$$\sum_i v_i\,(1 - f^i(1 - u_1 z,\,\dots,\,1 - u_d z)) = z - z^{1+\alpha}\,l(z),$$

$$\Pi(x_1,\dots,x_d) = \Pi(\min(x_1,\dots,x_d)), \quad \int_0^\infty e^{-sy}\,d\Pi(y) = 1 - s\,(1 + s^\alpha)^{-1/\alpha}$$

where $0 < \alpha \le 1$ and $l(\cdot)$ is a slowly changing at zero function.

Theorem 13.61. *Let a critical Markov branching process with a continuous time be indecomposable. Let moreover for some $i = 1,\dots,d$ there exists a limit*

$$\lim_{t\to\infty} \mathbf{P}\left\{Q_t\,\zeta_{ij}(t) \le x_j,\ j = 1,\dots,d \;\middle|\; \sum_j \zeta_{ij}(t) > 0\right\} = \Pi(x_1,\dots,x_d)$$

at every continuity point where

$$Q_t = \sum_i v_i \mathbf{P}\left\{\sum_j \zeta_{ij}(t) > 0\right\}, \quad \Pi(0+,\dots,0+) < 1.$$

Then

$$\sum_i v_i\,(1 - f^i(1 - u_1 z,\,\dots,\,1 - u_d z)) = z - z^{1+\alpha}\,l(z),$$

$$\Pi(x_1,\dots,x_d) = \Pi\left(a\,\min\left(\frac{x_1}{v_1 a_1},\,\dots,\,\frac{x_d}{v_d a_d}\right)\right),$$

$$\int_0^\infty e^{-sy}\,d\Pi(y) = 1 - s\,(1 + s^\alpha)^{-1/\alpha}$$

where $a = \sum_i v_i a_i u_i$, $0 < \alpha \le 1$ and $l(\cdot)$ are slowly changing at zero functions.

13.6 Jiřina Processes

Jiřina processes[3,26–32] are homogeneous Markov processes $\xi(t)$ taking real nonnegative values and with transition probabilities $P_t(x, A)$ satisfying the following equation:

$$\int_0^\infty P_t(x+y, dz)\, f(z) = \int_0^\infty \int_0^\infty P_t(x, dz_1)\, P_t(y, dz_2)\, f(z_1+z_2) \tag{13.30}$$

for any $x, y \geq 0$ and for any measurable bounded functions $f(\cdot)$. This relation reflects the property of branching. It means that a conditional distribution of $\xi(t)$, provided $\xi(0) = x + y$, coincides with the distribution of the sum of two independent random variables with the same conditional distributions of $\xi(t)$, provided $\xi_0 = x$ and $\xi_0 = y$ respectively. A time parameter t can take either integer positive values (a discrete time) or real nonnegative values (a continuous time).

Relation (13.30) is equivalent to the following one

$$\mathbf{E}\,(e^{-s\xi(t)} \mid \xi(0) = x) = \int_0^\infty e^{-sy}\, P_t(x, dy) = e^{-x\,K_t(s)}$$

where $s \geq 0$, $x \geq 0$ and for any $t \geq 0$ the function $K_t(s)$ is the logarithm, taken with a negative sign, of the Laplace transform of the infinitely divisible distribution on $[0, \infty)$, i.e.,

$$K_t(s) = a_t\, s + \int_0^\infty (1 - e^{-sy})\, \Pi_t(dy) \tag{13.31}$$

where

$$a_t \geq 0, \quad \Pi_t\{0\} = 0, \quad \int_0^\infty (y \wedge 1)\, \Pi_t(dy) < \infty.$$

Functions $K_t(s)$ satisfy a semigroup equation $K_{t+u}(s) = K_t(K_u(s))$. The following iterations

$$K_n(s) = K(K_{n-1}(s)), \quad K(s) = a\,s + \int_0^\infty (1 - e^{-sy})\, \Pi(dy) = K_1(s)$$

are the basis for studying of Jiřina processes with discrete time. An important characteristic of stochastic continuous Jiřina processes with continuous time is the so-called cumulant

$$H(s) = \lim_{t\downarrow 0} \frac{K_t(s) - s}{t} = \alpha s - \beta s^2 + \int_0^\infty (1 - s(y \wedge 1) - e^{-sy}) \Lambda(dy) \quad (13.32)$$

where

$$s \geq 0, \quad \beta \geq 0, \quad \Lambda\{0\} = 0, \quad \int_0^\infty (y^2 \wedge 1) \Lambda(dy) < \infty.$$

The function $K_t(s)$ can be defined in terms of $H(s)$ by the following differential equations

$$\frac{\partial K_t(s)}{\partial t} = H(K_t(s)), \ K_0(s) = s, \ \frac{\partial K_t(s)}{\partial t} = H(s) \frac{\partial K_t(s)}{\partial s}, \ K_0(s) = s.$$

A stochastic continuous Ǐirina process with continuous time can be obtained from the process with independent increments by a random change of time. Let $\eta(t)$ be a random process with independent increments such that $\eta(0) \geq 0$ and

$$\mathbf{E} e^{-s(\eta(t) - \eta(0))} = e^{-tH(s)}, \quad s \geq 0.$$

Let also θ be the first moment, when $\eta(t)$ is vanishing, i.e., $\theta = \inf\{t : \eta(t) = 0\}$. Then a random process

$$\xi(t) = \begin{cases} \eta(\tau(t)) & \text{if } t < \varphi(\theta) \\ 0 & \text{if } t \geq \varphi(\theta) \end{cases}$$

is a Ǐirina process with a cumulant $H(s)$. Here

$$\varphi(t) = \int_0^t \frac{du}{\eta(u)}, \quad \varphi(\tau(t)) = \tau(\varphi(t)) = t.$$

The necessary and sufficient condition of the regularity of a Ǐirina process with a cumulant $H(s)$ is the divergence of the integral

$$\int_0^\varepsilon \frac{ds}{H(s)}$$

for any $c > 0$.

Jirina branching process is the approximation of a Galton-Watson branching process, the origin of a great number of particles. More precisely, let $\zeta_n^{(r)}$ (for any integer positive r) be a number of particles in the nth generation of a Galton-Watson process with a generating function $f^{(r)}(z)$ of one-particle direct descendants starting as B_r particles where $B_r \uparrow \infty$ as $r \to \infty$. Let also m_r be a mean number of one-particle direct descendants in the rth process.

Theorem 13.62. *Let $l_r \to \infty$ and $l_r\,(m_r - 1) \to c$ as $r \to \infty$. If*

$$\lim_{r\to\infty} l_r\, B_r\, (f^{(r)}(e^{-s/B_r}) - e^{-s/B_r}) = -H(s), \quad s \geq 0, \quad H(0+) = 0, \quad (13.33)$$

then $H(s)$ is a cumulant of a Jirina process and marginal distributions of processes $\xi^{(r)}(t) = \zeta^{(r)}_{[tl_r]}/B_r$ weakly converge as $r \to \infty$ to marginal distributions of a Jirina process with a cumulant $H(s)$.

Theorem 13.63 (inverse to Theorem 13.62). *Assume that $l_r \to \infty$ and $l_r\,(m_r - 1) \to c$ as $r \to \infty$. If for any $t > 0$ the distribution of $\xi^{(r)}(t) = \zeta^{(r)}_{[tl_r]}/B_r$ weakly converges to the distribution of $\xi(t)$ with $\mathbf{P}\,\{\xi(t) > 0\} > 0$, then $\xi(t)$ is a stochastically continuous Jirina process and the relation (13.33) is fulfilled.*

Moreover, under the hypotheses of Theorem 13.62 the following relation

$$\lim_{r\to\infty} B_r\,\mathbf{P}\left\{\frac{1}{B_r}\,\zeta^{(r)}_{[tl_r]} \geq x\right\} = \Pi_t[x,\,\infty] \qquad (13.34)$$

holds at every continuity point $x > 0$ of a limit function. A measure $\Pi_t(\cdot)$ is defined by the expansion (13.31).

Any function of the form (13.32) can be used in Theorems 13.62 and 13.63 as a limit cumulant $H(s)$. Theorems on the convergence to continuous Jirina processes with $H(s) = \alpha\, s - \beta\, s^2$ are of a particular interest. This is the case, when the function $K_t(s)$ and coefficients in its expansion (13.31) can be found in explicit form. If $H(s) = \alpha\, s - \beta\, s^2$, then

$$K_t(s) = \begin{cases} s\, e^{\alpha t}\,(1 + s\,\beta\,(e^{\alpha t} - 1)/\alpha)^{-1}, & \text{if } \alpha \neq 0 \\ s\,(1 + s\,\beta t)^{-1}, & \text{if } \alpha = 0, \end{cases} \qquad a_t = 0,$$

$$\Pi_t(dy) = \begin{cases} \alpha^2 e^{\alpha t}/(\beta\,(e^{\alpha t} - 1))^2 \exp\{-y\,\alpha/(\beta\,(e^{\alpha t} - 1))\}\, dy & \text{if } \alpha \neq 0 \\ \beta^{-2}\, t^{-2} \exp\{-y\,\beta^{-1}\, t^{-1}\}\, dy & \text{if } \alpha = 0. \end{cases}$$

For example, if

$$f^{(r)}(z) = 1 + m_r\,(z - 1) + \frac{b_r}{2}\,(z - 1)^2 - \varepsilon_r(z)\,(z - 1)^2$$

with $b_r \to b > 0$ and $\sup_r \varepsilon_r(z) \underset{z\uparrow 1}{\longrightarrow} 0$, then the conditions of Theorem 13.62 are fulfilled with $B_r = l_r$ and $H(z) = c\,z + b\,z^2/2$. In this case the relation (13.34) transforms into relation

$$\lim_{r\to\infty} l_r \mathbf{P}\left\{\frac{1}{l_r}\,\zeta^{(r)}_{[tl_r]} \geq x\right\} = \frac{2}{b}\,\frac{c}{1-e^{-ct}}\,\exp\left\{-\frac{2}{b}\,\frac{c}{e^{ct}-1}\,x\right\}.$$

The first two moments of the Ǐirina process $\xi(t)$ can be obtained in the explicit form. In the case of a continuous time we have:

$$\mathbf{E}\,(\xi(t)\mid\xi(0)=x) = x\,e^{\mu t},$$

$$\mathbf{E}\,(\xi^2(t)\mid\xi(0)=x) = \begin{cases} x^2\,e^{2\mu t} + x\,\sigma\,(e^{2\mu t} - e^{\mu t})/\mu, & \text{if } \mu\neq 0 \\ x^2 + x\,\sigma\,t & \text{if } \mu = 0 \end{cases}$$

where

$$\mu = H'(0) = \alpha + \int_1^\infty (y-1)\,\Lambda(dy),$$

$$\sigma = -H''(0) = 2\,\beta + \int_0^\infty y^2\,\Lambda(dy).$$

If the time is discrete, then

$$\mathbf{E}\,(\xi_n\mid\xi_0=x) = x\,m^n,$$

$$\mathbf{E}\,(\xi_n^2\mid\xi_0=x) = \begin{cases} x^2\,m^{2n} + x\,b\,m^{-1}\,(m^{2n}-m^n)\,(m-1)^{-1}, & \text{if } m\neq 1 \\ x^2 + x\,b\,n & \text{if } m = 1 \end{cases}$$

where

$$m = K'(0) = a + \int_0^\infty y\,\Pi(dy),$$

$$b = -K''(0) = \int_0^\infty y^2\,\Pi(dy).$$

An asymptotic behavior of trajectories of the Ǐirina process is similar to Galton-Watson processes, but at the same time it has some peculiarities. So, we use not only the mean value m, but also a parameter a giving a lower bound for ξ_n : $\xi_n \geq \xi_0\,a^n$.

Theorem 13.64. *Jirina branching process ξ_n with a discrete time converges to zero with probability 1 iff $m \le 1$. If $m > 1 > a$ and $\xi_0 = x$, then $\xi_n \to 0$ with probability e^{-qx} and $\xi_n \to \infty$ with probability $1-e^{-qx}$ as $n \to \infty$. Here q is the unique positive solution of the equation $s = K(s)$. If $a \ge 1$ and $\xi_0 > 0$, then $\xi_n \to \infty$ with probability 1 as $n \to \infty$.*

A similar statement holds in the case of a continuous time. In the next theorem we use the following (not necessary finite) parameter

$$\gamma = \lim_{s\to\infty} H'(s) = \begin{cases} \alpha - \int\limits_0^\infty \min(y, 1)\, \Lambda(dy) & \text{if } \beta = 0 \\ -\infty & \text{if } \beta > 0. \end{cases}$$

Theorem 13.65. *A stochastically continuous separable Jirina branching process $\xi(t)$ with a continuous time converges to zero with probability 1 iff $\mu \le 0$. If $\mu > 0 > \gamma$ and $\xi(0) = x$, then $\xi(t) \to 0$ with probability e^{-qx} and $\xi(t) \to \infty$ with probability $1 - e^{-qx}$ as $t \to \infty$. Here q is the unique positive solution of the equation $H(s) = 0$. If $\gamma \ge 0$ and $\xi(0) > 0$, then $\xi(t) \to \infty$ with probability 1 as $t \to \infty$.*

Limit theorems for the Jirina process with discrete time are very similar to corresponding limit theorems for Galton-Watson processes.

Theorem 13.66. *If $m > 1$, then a limit*

$$\eta = \lim_{n\to\infty} m^{-n} \xi_n$$

exists and is finite almost surely. Moreover $\mathbf{P}\{\eta > 0 \mid \xi_0 = 1\} > 0$ iff the relation

$$\int_1^\infty x \log x\, \Pi(dx) < \infty \tag{13.35}$$

holds. If this condition is satisfied, then

$$\mathbf{P}\{\eta = 0 \mid \xi_0 = x\} = e^{-qx}, \qquad \mathbf{E}(e^{-s\eta} \mid \xi_0 = x) = e^{-x\Psi(s)}$$

where $\Psi(s)$ is the unique (in the class of logarithms, taken with a negative sign, of Laplace transforms of nonnegative random variables) solution of the equation

$$\Psi(ms) = K(\Psi(s))$$

with the additional condition $\lim_{s\downarrow 0} \Psi(s)/s = 1$.

For a supercritical Jirina process (as in a Galton-Watson model) it can be found independently of the condition (13.35) a normalizing sequence

$c_n \uparrow \infty$ such that $c_{n+1}/c_n \to m$ and almost surely there exists a finite limit $\lim_{n\to\infty} \xi_n/c_n$. This limit is positive with the probability $1 - e^{-qx}$ provided that $\xi_0 = x$. A Laplace transform of a limit value, provided that $\xi_0 = x$, equals $e^{-x\,\Psi(s)}$, where $\Psi(s)$ is the unique (up to the linear transformation of the argument) solution (in the class of logarithms, taken with a negative sign, of Laplace transforms of nonnegative random variables) of the equation

$$\Psi(ms) = K(\Psi(s)).$$

Theorem 13.67. *If* $m = 1$ *and* $0 < b = \int_0^\infty y^2\,\Pi(dy) < \infty$, *then*

$$\lim_{n\to\infty} n\,\mathbf{P}\{\xi_n \geq x \mid \xi_0 = 1\} = \frac{2}{b} \quad \text{for } x > 0,$$

$$\lim_{n\to\infty} \mathbf{P}\left\{\frac{\xi_n}{n} \geq x \mid \xi_0 = 1,\ \xi_n \geq 1\right\} = e^{-2x/b} \quad \text{for } x \geq 0.$$

Theorem 13.68. *If* $m < 1$, *then*

$$\lim_{n\to\infty} \frac{K_n(s)}{K_n(1)} = \int_0^\infty (1 - e^{-sx})\,M(dx),$$

$$\lim_{n\to\infty} \frac{1}{K_n(1)}\,\mathbf{P}\{\xi_n \geq x \mid \xi_0 = 1\} = M[x, \infty)$$

at every continuity point $x > 0$ *of a limit function. Here*

$$M\{0\} = 0, \qquad \int_0^\infty \min(x, 1)\,M(dx) < \infty$$

and $M[x, \infty) > 0$ *for* $x > 0$. *In this case* $\int_0^\infty x\,M(dx) < \infty$ *iff*

$$\int_1^\infty x \log x\,\Pi(dx) < \infty,$$

which is equivalent to the finiteness and positiveness of the limit

$$\lim_{n\to\infty} m^{-n} K_n(1).$$

A simple corollary of Theorem 13.68 is the existence of the conditional limit probability distribution of the subcritical Jirina process ξ_n, provided that $\xi_n \geq 1$, i.e.,

$$\lim_{n\to\infty} \mathbf{P}\{\xi_n \geq x \mid \xi_0 = 1,\ \xi_n \geq 1\} = \frac{M[x,\infty)}{M[1,\infty)}$$

at every continuity point $x \geq 1$ of a limit function. This theorem implies that a mean value of the limit distribution is finite iff the integral

$$\int_1^\infty x \log x\, \Pi(dx)$$

is finite. The measure $M(dx)$ is defined by the equation $\varphi(K(s)) = m\,\varphi(s)$, $\varphi(1) = 1$, which is satisfied by the function

$$\varphi(s) = \int_0^\infty (1 - e^{-sx})\, M(dx).$$

This equation has a unique solution in the class of functions $\varphi(s)$ such that $\varphi(s)/s$ is a monotone function.

The measure $M(dx)$ is finite iff $K(\infty) < \infty$, which is equivalent to the inequality $\mathbf{P}\{\xi_1 = 0 \mid \xi_0 = 1\} > 0$. In this case a statement analogous to Theorem 13.68 can be formulated. If $m < 1$ and $K(\infty) < \infty$, then

$$\lim_{n\to\infty} \mathbf{P}\{\xi_n > x \mid \xi_0 = 1,\ \xi_n > 0\} = \frac{M[x,\infty)}{M[0,\infty)}$$

at every continuity point $x \geq 0$ of the limit function.

The most essential distinction of Jirina processes from Galton-Watson processes is illustrated by the following theorem, which can't have an analog in the Galton-Watson model.

Theorem 13.69. *If*

$$0 < a < m < 1 \quad \text{and} \quad \int_0^1 x \log x\, \Pi(dx) > -\infty,$$

then there exists a limit probability distribution

$$\lim_{n\to\infty} \mathbf{P}\{a^{-n}\xi_n \leq x \mid \xi_0 = 1\} = F(x), \quad F(0+) = 0, \quad F(\infty) = 1,$$

$$\int_0^\infty e^{-sx}\, dF(x) = e^{-\chi(s)}$$

where $\chi(s)$ is the unique (up to the linear transformation of the argument) solution (in the class of logarithms, taken with a negative sign, of Laplace transforms of nonsingular nonnegative random variables) of the equation $\chi(as) = K(\chi(s)),\ s \geq 0$.

This theorem can be formulated more precisely. If the integral

$$\int_0^1 x \log x\, \Pi(dx)$$

diverges, then a normalizing sequence $c_n \downarrow 0$ can be found such that $c_{n+1}/c_n \to a$ and a distribution of the random variable ξ_n/c_n weakly converges to a nonsingular probability distribution with a Laplace transform $e^{-x\,\chi(s)}$ provided that $\xi_0 = x$. Here $\chi(s)$ is the unique (up to the linear transformation of the argument) solution (in the class of logarithms, taken with a negative sign, of Laplace transforms of nonsingular nonnegative random variables) of the equation $\chi(as) = K(\chi(s)),\ s \geq 0$. Moreover the condition $a > 0$ is proved to be necessary for the existence of some normalizing sequence $c_n \downarrow 0$ ensuring the convergence of the distribution of the random variable ξ_n/c_n to a nonsingular probability distribution.

13.7 Some Other Models of Branching

Consider some models of branching[3,5,17–18,33–48].

Immigration. The simplest example of a decomposable branching process is a branching with immigration – a Markov branching process with a discrete time with two types T_0 and T_1 of particles. At time of transformation the particle of type T_0 reproduces itself and some particles of type T_1. The particles of type T_1 can reproduce only the particles of type T_1. In other words, the descendants of the particle of type T_0 immigrate into the population and adjoin to be a branching process parallel with already existing particles. If at the initial moment the population does not contain particles of the type T_0, then we have an ordinary Galton-Watson branching process.

Let $f(z) = \sum_{k\geq 0} p_k z_k$ and $g(z) = \sum_{k\geq 0} g_k z_k$ be generating functions for the number of direct descendants of type T_1 reproduced by one particle of the type T_1 and one particle of the type T_0 respectively. According to general equations (13.23) a generating function $F_n(z)$ for the number η_n of particles of the type T_1 originated in n steps from one particle of type T_0 satisfies the following recursive relation

$$F_n(z) = g(f_{n-1}(z))\, F_{n-1}(z),$$

where $f_n(z) = f(f_{n-1}(z))$ is a generating function for the number of particles in the nth generation of the corresponding Galton-Watson process starting as one particle. Hence

$$F_n(z) = \prod_{k=0}^{n-1} g(f_k(z)).$$

The asymptotic behavior of the distribution of η_n essentially differs from the asymptotic behavior of the distribution of the number of particles in indecomposable processes.

Theorem 13.70. *If* $m = \sum_k k\, p_k \le 1$, *then there exist limits*

$$\pi_k = \lim_{n\to\infty} \mathbf{P}\{\eta_n = k\}, \quad k = 0, 1, 2, \dots,$$

with $\sum_{k\ge 0} \pi_k = 1$ *iff*

$$\int_0^1 \frac{1 - g(z)}{f(z) - z}\, dz < \infty.$$

If $m > 1$, then the finiteness of this integral is equivalent to the existence of the logarithmic moment for the number of immigrating particles, i.e., $\sum_{k\ge 0} q_k \log k = \infty$. A generating function for the sequence $\{\pi_k\}$ equals $\prod_{k\ge 0} g(f_k(z))$.

Theorem 13.71. *Let* $m > 1$ *and* $\sum_{k\ge 0} p_k\, k \log k < \infty$. *If*

$$\sum_{k\ge 0} q_k \log k = \infty,$$

then

$$\lim_{n\to\infty} m^{-n} \eta_n = \infty$$

with probability 1. If $\sum_{k\ge 0} q_k \log k < \infty$, *then there exists almost surely a finite limit*

$$\xi = \lim_{n\to\infty} m^{-n} \eta_n \quad \text{and} \quad \mathbf{E}\, e^{-s\xi} = \prod_{k\ge 1} g(\varphi(s m^{-k}))$$

where $\varphi(s)$ is the solution of the problem (13.8).

If $m > 1$ and $\sum\limits_{k \geq 1} p_k\, k \,\log k = \infty$, then analogously to Galton-Watson processes a normalizing sequence $c_n \uparrow \infty$ can be found such that $c_{n+1}/c_n \to$ $\to m$ and

$$\lim_{n\to\infty} \frac{\eta_n}{c_n} < \infty \quad \text{almost surely, if} \quad \sum_k q_k \log k < \infty,$$

$$\lim_{n\to\infty} \frac{\eta_n}{c_n} = \infty \quad \text{almost surely, if} \quad \sum_k q_k \log k = \infty.$$

If we choose the sequence $\{c_n\}$ coinciding with normalizing sequence in (13.9), then the Laplace transform of a limit value equals $\prod\limits_{k\geq 1} g(\Psi(sm^{-k}))$ where $\Psi(s)$ is a Laplace transform of a limit value in (13.9).

Theorem 13.72. *If*

$$m = 1, \quad b = \sum_{k\geq 1} k\,(k-1)\, p_k < \infty, \quad 0 < a = \sum_{k\geq 1} k\, q_k < \infty,$$

then

$$\lim_{n\to\infty} \mathbf{P}\left\{\frac{\eta_n}{n} \leq x\right\} = \left(\frac{2}{b}\right)^{2a/b} \frac{1}{\Gamma(2a/b)} \int\limits_0^x y^{2a/b-1}\, e^{-2y/b}\, dy.$$

Nonhomogeneous in time processes generalize a Galton-Watson model by supposing that the distribution of the number of one-particle direct descendants depends on the number of generation. Such a process is determined by the sequence of generating functions

$$f^{(r)}(z) = \sum_{k\geq 0} z^k\, p_k^{(r)}, \quad r = 1, 2, \ldots,$$

defining the distribution of the number of one-particle direct descendants in rth generation. A generating function of the number ζ_n of one-particle descendants within n generations equals $f^{(1)}(f^{(2)}(\ldots(f^{(n)}(z))\ldots))$.

Theorem 13.73. *For any $k = 0, 1, 2, \ldots$ there exists a limit*

$$\pi_k = \lim_{n\to\infty} \mathbf{P}\{\zeta_n = k\}$$

and $\sum\limits_{k\geq 1} \pi_k > 0$ *iff* $\sum\limits_r (1 - p_1^{(r)}) < \infty$.

Bounded branching processes. In such processes the number ζ_n of one-particle descendants in nth generation is bounded by a nonnegative sequence c_n. In all other respects the evolution of the process is the same as the evolution of a Galton-Watson process with a generating function $f(z)$ for the number of one-particle direct descendants. The sequence $\{\zeta_n\}$ is a Markov chain with $\zeta_0 = 1$ and transition probabilities

$$\mathbf{P}\{\zeta_n = r \mid \zeta_{n-1} = k\} = \mathbf{P}\{\min(c_n, \xi_1 + \cdots + \xi_k) = r\}$$

where $\xi_1, \xi_2, \ldots$ are independent random variables with the same distribution and a generating function $f(z)$.

Let q be a minimal root of the equation $z = f(z)$ in the interval $0 \le z \le 1$, i.e., q is the degeneration probability of the corresponding Galton-Watson process.

Theorem 13.74. *A bounded branching process is almost surely degenerating, iff* $\sum\limits_{n\ge1} q^{c_n} = \infty$.

Controlled branching processes. The number of particles in such a process is controlled by an integer-valued positive sequence $\{c_r\}$. For example, if the number of particles equals r and $c_r < r$, then $r - c_r$ particles are removed from the population and don't take part in the further evolution of the process. Remaining c_r particles generate descendants according to the rules of the Galton-Watson process. If $c_r > r$, then the population is supplemented by $c_r - r$ new particles and all c_r particles take part in the further evolution of the process.

Strictly speaking the number ζ_n of particles in nth generation is a homogeneous Markov chain with transition probabilities

$$\mathbf{P}\{\zeta_n = k \mid \zeta_{n-1} = r\} = \mathbf{P}\{\xi_1 + \cdots + \xi_{c_r} = k\}$$

where $\xi_1, \xi_2, \ldots$ are independent integer-valued positive random variables with the same distribution p_k, $k = 0, 1, \ldots$. Let m be a mean number of one-particle direct descendants.

Theorem 13.75. *If* $p_0 > 0$, $c_0 = 0$ *and* $\overline{\lim\limits_{r\to\infty}}\, c_r/r < 1/m$, *then for any* $k \ge 0$

$$\mathbf{P}\{\lim_{n\to\infty} \zeta_n = 0 \mid \zeta_0 = k\} = 1.$$

Theorem 13.76. *Let* $\inf\limits_{r\ge l} c_r/r > 1/m$. *If*

$$\sup_n \mathbf{P}\{\zeta_n \ge l \mid \zeta_0 = k\} > 0,$$

then

$$\mathbf{P}\{\lim_{n\to\infty} \zeta_n = \infty \mid \zeta_0 = k\} > 0,$$

$$\mathbf{P}\{\overline{\lim_{n\to\infty}}\,\zeta_n \le l-1 \mid \zeta_0 = k\} + \mathbf{P}\{\lim_{n\to\infty}\zeta_n = \infty \mid \zeta_0 = k\} = 1.$$

In the case of $l = 1$ the statement of Theorem 13.76 means that the degeneration probability is less than unit and almost all nondegenerating trajectories tend to infinity.

The next theorems give a partial answer to the question concerning the asymptotic behavior of ζ_n as $n \to \infty$ in the case when $1/m$ can be a limit point of the sequence c_r/r, $r = 1, 2, \ldots$.

Theorem 13.77. *Let*

$$\overline{\lim_{n\to\infty}}\,(c_r - \frac{r}{m}) < \frac{b}{2m^2}$$

where

$$b = \sum_{k\ge 1} k^2 p_k - m^2 \le \infty.$$

Then

$$\mathbf{P}\{\lim_{n\to\infty}\zeta_n = \infty \mid \zeta_0 = k\} = 0, \quad k \ge 0.$$

If $c_0 = 0$ *and* $\sup_n \mathbf{P}\{\zeta_n = 0 \mid \zeta_0 = k\} > 0$ *for any* $k \ge 0$, *then*

$$\mathbf{P}\{\lim_{n\to\infty}\zeta_n = 0 \mid \zeta_0 = k\} = 1, \quad k \ge 0.$$

Theorem 13.78. *Let*

$$b = \sum_{k\ge 1} k^2 p_k - m^2 < \infty \quad \text{and} \quad c_r - \frac{r}{m} > \frac{b}{2m^2} \quad \text{for } r \ge 1.$$

If $\sup_n \mathbf{P}\{\zeta_n \ge l \mid \zeta_0 = k\} > 0$, *then*

$$\mathbf{P}\{\lim_{n\to\infty}\zeta_n = \infty \mid \zeta_0 = k\} > 0.$$

Theorem 13.79 (similar to the limit theorem for a subcritical Galton-Watson branching process with immigration). *If* $\overline{\lim_{r\to\infty}}\,(c_r - r/m) < 0$, *then for some integer positive d and for any integer nonnegative r, l there exists a limit probability distribution*

$$\lim_{n\to\infty}\mathbf{P}\{\zeta_{nd+r} = k \mid \zeta_0 = l\} = \pi_k(r, l).$$

A theorem similar to a limit theorem for a supercritical Galton-Watson process holds true.

Theorem 13.80. *Let $w(x)$, $x > 0$, be a nonnegative upwards convex function and*

$$\sum_{n\geq 1} \frac{1}{n w(n)} < \infty, \qquad \sum_{k\geq 1} p_k\, k\, w(k) < \infty.$$

If $\inf\limits_{r\geq l} c_r/r > 1/m$ *and*

$$\sup_n \mathbf{P}\{\zeta_n \geq l \mid \zeta_0 = k\} > 0, \qquad \frac{c_r}{r} = \frac{a}{m} + O\left(\frac{1}{w(r)}\right) \quad \text{as } r \to \infty,$$

then almost surely there exists a limit $\eta = \lim\limits_{n\to\infty} a^{-n}\,\zeta_n$ *for* $\zeta_0 = k$ *and*

$$\mathbf{P}\{\eta > 0 \mid \zeta_0 = k\} > 0, \quad \mathbf{E}(\log \eta \mid \eta > 0,\ \zeta_0 = k) < \infty.$$

Branching in a random environment generalizes a Galton-Watson model by supposing that the distribution of the number of one-particle direct descendants depends on the state of the external environment. Successive states $\kappa_0,\ \kappa_1,\ \kappa_2,\ \dots$ of this environment form a stationary ergodic sequence taking values in a measurable space $(X,\ \mathcal{A})$.

Let $p_k(x)$, $x \in X$, be a conditional probability that the number of one-particle direct descendants equals k, provided that the state of the environment is x. Probabilities $p_k(x)$ are supposed to be measurable on x (with respect to σ-algebra $\mathcal{A}$) functions.

By a given trajectory of the environment states $\kappa_0,\ \kappa_1,\ \dots,\ \kappa_n,\ \dots$ the number ζ_n of particles in nth generation forms a Markov chain with transition probabilities

$$\mathbf{P}\{\zeta_{n+1} = r \mid \zeta_n = k,\ \kappa_0,\ \kappa_1,\ \dots,\ \kappa_n,\ \dots\} =$$

$$= \mathbf{P}\{\zeta_{n+1} = r \mid \zeta_n = k,\ \kappa_n\} = \sum_{l_1+\cdots+l_k=r} p_{l_1}(\kappa_n)\dots p_{l_k}(\kappa_n).$$

An equivalent definition is the following one:

$$\mathbf{E}\left(z^{\zeta_{n+1}} \mid \kappa_0,\ \kappa_1,\ \dots,\ \kappa_n,\ \zeta_0,\ \zeta_1,\ \dots,\ \zeta_n\right) = f_{\kappa_n}(z^{\zeta_n})$$

with the probability 1 where $f_x(z) = \sum\limits_{k\geq 0} z^k\, p_k(x)$ is a generating function for the number of one-particle direct descendants under the environment state x.

A generating function for the number of one-particle descendants in nth generation equals $\mathbf{E}\, f_{\kappa_0}(f_{\kappa_1}(\dots(f_{\kappa_{n-1}}(z))\dots))$. A degeneration probability of the process starting with k particles equals $\mathbf{E}\,\rho^k$ where

$$\rho = \lim_{n\to\infty} f_{\kappa_0}(\dots(f_{\kappa_n}(z))\dots).$$

The stationarity and ergodicity of the sequence $\{\kappa_n\}$ imply that the event $\{\rho = 1\}$ can be of probability either 0 or 1.

Let $m(x)$ be a mean number of one-particle direct descendants under the environment state x, i.e.,

$$m(x) \;=\; \sum_{k\geq 1} k\, p_k(x) \;=\; f_x'(1).$$

Theorem 13.81. *If*

$$\mathbf{E}\,(\log m(\kappa_0),\;\; m(\kappa_0) > 1) < \infty \quad \textit{and} \quad -\infty \leq \mathbf{E}\,\log m(\kappa_0) \;\leq 0,$$

then $\mathbf{P}\,\{\rho = 1\} \;=\; 1.$

Theorem 13.82. *If* $\mathbf{E}\,\log m(\kappa_0) > 0,$

$$\mathbf{E}\,\log(1 - p_0(\kappa_0)) \;>\; -\infty, \quad \mathbf{E}\,(\log m(\kappa_0), \quad m(\kappa_0) < 1) \;>\; -\infty,$$

then $\mathbf{P}\,\{\rho = 1\} \;=\; 0.$

Theorem 13.83. *If* $\mathbf{E}\,\log(1 - p_0(\kappa_0)) \;>\; -\infty$ *and*

$$\mathbf{E}\,(\log m(\kappa_0),\;\; m(\kappa_0) > 1) \;=\; \mathbf{E}\,(-\log m(\kappa_0),\;\; m(\kappa_0) < 1) \;=\; \infty,$$

then $\mathbf{P}\,\{\rho = 1\} \;=\; 0.$

Branching processes with a varying regime. This model is proposed only in the case of a continuous time. It is assumed that transformation probabilities depend on the size of population, provided that the number of particles in the population does not exceed a given level N. If the number of particles is greater than N, then a process develops according to the rules of a Markov branching process. The number $\zeta(t)$ of particles at time t in a branching process with a varying regime is a homogeneous Markov process with transition probabilities at the small time t:

$$p_{ij}(t) \;=\; \frac{i}{a_i}\, p^i_{j-i+1}\, t \;+\; o(t), \quad j \neq i \leq N,$$

$$p_{ii}(t) \;=\; 1 - \frac{i}{a_i}\,(1 - p^i_1)\, t \;+\; o(t), \quad i \leq N,$$

$$p_{ij}(t) \;=\; \frac{i}{a}\, p^i_{j-i+1}\, t \;+\; o(t), \quad j \neq i > N,$$

$$p_{ii}(t) \;=\; 1 - \frac{i}{a}\,(1 - p_1)\, t \;+\; o(t), \quad i > N.$$

Here a_i is the mean lifetime of the particle, provided that the total number of particles in the population equals i, and p^i_k, $k \;=\; 0, 1, 2, \ldots,$ is the conditional distribution of the number of one-particle direct descendants, provided that the total number of particles in the population equals i.

Theorem 13.84. *Let $\min_i p_0^i > 0$. If $m = \sum_{k\geq 1} k\, p_k \leq 1$, then a branching process with a varying regime is degenerating with probability 1. If $m > 1$, then degeneration probability equals*

$$\frac{A_1^2}{A_1^1} - \frac{A_1^{N+1}}{A_1^1} \frac{\sum_{i=1}^{n} i\, q^{i-N-1} \sum_{j>N-i+1} q^j\, p_j^i\, B_i^1/a_i}{A_1^1 + \sum_{i=1}^{n} i\, q^{i-N-1} \sum_{j>N-i+1} q^j\, p_j^i\, B_i^N/a_i}$$

where $q < 1$, $\sum_{k\geq 0} q^k\, p_k = q$, A_j^i is a cofactor of the (i, j)th component of the $(N+1) \times (N+1)$-matrix

$$\begin{pmatrix} 0 & p_0^1/a_1 & 0 & 0 & \cdots & 0 \\ 0 & p_1^1/a_1 & 2p_0^2/a_2 & 0 & \cdots & 0 \\ 0 & p_2^1/a_1 & 2p_1^2/a_2 & 3p_0^3/a_3 & \cdots & 0 \\ \vdots & \vdots & \vdots & \vdots & \vdots & \vdots \\ 0 & p_N^1/a_1 & 2p_{N-1}^2/a_2 & 3p_{N-2}^3/a_3 & \cdots & Np_1^N/a_n \end{pmatrix},$$

and B_j^i is a cofactor of the (i, j)th component of the $N \times N$-matrix, which can be obtained from the given above matrix by deletion of the first line and the first column.

If

$$\sum_k k^2\, p_k^i < \infty, \quad i = 1, \ldots, N, \quad \sum_k k^2\, p_k < \infty,$$

then an asymptotic behavior of distributions of branching processes with the varying regime is approximately the same as for Markov branching processes. If $m > 1$, then $e^{-\mu t}\,\zeta(t)$ converges in the mean square as $t \to \infty$ to a random variable with a Laplace transform (for $\zeta(0) = 1$):

$$\varphi(s) - \int_{\varphi(s)}^{1} \frac{a}{f(y) - y} \sum_{j=1}^{N} j\, y^{j-1} \left(\frac{f^{(j)}(y) - y}{a_j} - \right.$$

$$\left. - \frac{f(y) - y}{a} \right) p_j \left(-a \int_{\varphi(s)}^{y} \frac{dx}{f(x) - x} \right)$$

where $\mu = (m-1)/a$, $p_j(t) = \mathbf{P}\{\zeta(t) = j \mid \zeta(0) = 1\}$ and $\varphi(s)$ is determined by the relation (13.21).

If $m = 1$, then the probability of the process extension is asymptotic inversely proportional to the time parameter, and the distribution of $\zeta(t)/t$, provided $\zeta(t) > 0$, converges to the exponential distribution. If $m < 1$, then under the existence of sufficiently large number of moments, the degeneration probability exponentially decreases and there exists a limit probability distribution of $\zeta(t)$, provided that $\zeta(t) > 0$. In this case the parameter corresponding to the exponentially decreasing nondegeneration probability does not necessarily coincide with $\mu = (m-1)/a$.

A Crump-Mode-Jagers model (unlike the Sevastyanov model) permits the reproduction of the particle during its whole life (and not only at the end of the lifetime). In other words, with any particle one can associate a nondecreasing integer-valued random process $\nu(t)$ defined on the interval $[0,\ \tau]$ of the particle lifetime and determine the number of descendants at time t reproduced by the particle in $[0,\ t]$. It is supposed that processes associated with different particles are independent and have the same distributions.

If the mean number $\mathbf{E}\,\nu(\tau)$ of all one-particle descendants is finite and $\mathbf{E}\,\nu(0) < 1$, then a Crump-Mode-Jagers process is regular. The same hypotheses are sufficient for the finiteness of the mean number $A(t) = \mathbf{E}\,\zeta(t)$ of particles $\zeta(t)$ present at time t, provided that the population consists only of one newly born particle at the initial time. A mean value $A(t)$ satisfies the following renewal equation

$$A(t) = 1 - G(t) + \int_0^t A(t-u)\, dM(u)$$

where $G(t) = \mathbf{P}\{\tau \le t\}$, $M(t) = \mathbf{E}\,\nu(\min(t,\tau))$. Consequently the relation

$$\lim_{t\to\infty} e^{-\mu t} A(t) = \frac{\int_0^\infty e^{-\mu t}(1 - G(t))\, dt}{\int_0^\infty t\, e^{-\mu t}\, dM(t)}$$

holds, provided that there exists a Maltus index μ as the root of the equation

$$\int_0^\infty e^{-\mu t}\, dM(t) = 1.$$

The integrals in the numerator and denominator are finite values and a distribution function $M(t)$ is nonlatticed.

The degeneration probability of the Crump-Mode-Jagers process (as in a Galton-Watson model) equals the minimal root of the equation $z = f(z)$ in a unit interval $0 \le z \le 1$ where $f(z) = \mathbf{E}\, z^{\nu(\tau)}$ is a generating function of the total number of one-particle descendants.

Theorem 13.85. *If*

$$\mathbf{E}\,\nu(0) < \mathbf{E}\,\nu(\tau) = 1, \quad 0 < b = \mathbf{E}\,\nu(\tau)\,(\nu(\tau) - 1) < \infty,$$

$$\lim_{t\to\infty} t^2\,(1 - G(t)) = \lim_{t\to\infty} t^2\,(1 - M(t)) = 0,$$

then

$$\lim_{t\to\infty} t\,\mathbf{P}\,\{\zeta(t) > 0\} = \frac{2\,c}{b}$$

where $c = \int\limits_0^\infty t\,dM(t)$.

Theorem 13.86. *If* $\mathbf{E}\,\nu(\tau) < 1$, $\mathbf{E}\,\tau\,e^{-\mu\tau} < \infty$,

$$\mathbf{E}\int\limits_0^\tau t\,e^{-\mu\tau}\,d\nu(t) < \infty, \quad \mathbf{E}\left(\int\limits_0^\tau e^{-\mu t}\,d\nu(t)\,\log\nu(\tau)\right) < \infty$$

and a distribution function $M(t)$ *is nonlatticed, then a limit*

$$\lim_{t\to\infty} e^{-\mu t}\,\mathbf{P}\,\{\zeta(t) > 0\}$$

exists. Moreover it is a positive and finite limit.

An asymptotic behavior of distributions of the number of particles in a Crump-Mode-Jagers process, provided that the condition of the nondegeneracy holds, is analogous to the Sevastyanov model.

Theorem 13.87. *Under the hypotheses of Theorem 13.86 there exists a limit probability distribution*

$$\lim_{t\to\infty} \mathbf{P}\,\{\zeta(t) = k \mid \zeta(t) > 0\}, \quad k = 1, 2, \dots .$$

Theorem 13.88. *If under the hypotheses of Theorem 13.85 a distribution function* $M(t)$ *is nonlatticed, then*

$$\lim_{t\to\infty} \mathbf{P}\left\{\frac{1}{t}\,\zeta(t) \ge x \mid \zeta(t) > 0\right\} = \exp\left\{-\frac{2c^2}{ab}\,x\right\}, \quad x \ge 0,$$

where $a = \int\limits_0^\infty t\,dG(t)$.

Theorem 13.89. *If* $\mathbf{E}\nu(\tau) > 1 > \mathbf{E}\nu(0)$, $\mathbf{E}\nu^2(\tau) < \infty$ *and a distribution function* $M(t)$ *is nonlatticed, then* $e^{-\mu t}\zeta(t)$ *converges in a mean square as* $t \to \infty$ *to limit random variable with the unit mean and the variance*

$$\frac{\mathbf{E}\left(\int\limits_0^\tau e^{-\mu t}\,d\nu(t)\right)^2 - 1}{1 - \int\limits_0^\infty e^{-2\mu t}\,dM(t)}.$$

13.8 References

1. Slack, R. S., A branching process with mean one and possibly infinite variance, *Z. Wahrscheinlichkeitstheor. und verw. Geb.*, 9, 139, 1968.
2. Slack, R. S., Further notes on branching process with mean one, *Z. Wahrscheinlichkeitstheor. und verw. Geb.*, 25, 31, 1972.
3. Athreya, K. B., Ney, P. E., *Branching Processes,* Springer, Berlin, 1972.
4. Chistyakov, V. P., Transition phenomena in branching processes with several types of particles, *Teor. Veroyatn. Primen.*, 17, 669, 1972 (in Russian).
5. Sevast'janov, B. A., *Verzweigungsprozesse,* Oldenburg, München, 1975.
6. Zubkov, A. M., Limit distributions of a distance to a nearest mutual ancestor, *Teor. Veroyatn. Primen.*, 20, 614, 1975 (in Russian).
7. Goldstein, M. I., Critical age-dependent branching processes: single and multitype, *Z. Wahrscheinlichkeitstheor. und verw. Geb.*, 17, 74, 1971.
8. Athreya, K. B., Kaplan, N., Convergence of the age distribution in the one-dimensional supercritical age-dependent branching process, *Ann. Probab.*, 4, 38, 1976.
9. Vatutin, V. A., Condition of Bellman-Harris branching process regularity, *Doklady Akad. Nauk SSSR,* 230, 15, 1976 (in Russian).
10. Vatutin, V. A., Limit theorems for a critical Bellman-Harris branching process with infinite variance, *Teor. Probab. Appl.*, 21, 839, 1976.
11. Vatutin, V. A., Discrete limit distributions of number of particles in critical Bellman-Harris branching processes, *Teor. Probab. Appl.,* 22, 146, 1977.
12. Vyugin, O. V., Transition phenomena in Bellman-Harris branching processes, *Teor. Probab. Appl.,* 22, 836, 1977.
13. Vatutin, V. A., A new limit theorem for the critical Bellman-Harris branching processes, *Math. USSR Sbornik,* 37, 411, 1979.

14. Kingman, J. F. C., Continuous time Markov processes, *Proc. London Math. Soc.,* 13, 539, 1963.
15. Vatutin, V. A., Limit theorems for critical Markov branching processes with several types of particles and infinite second moments, *Math. USSR Sbornik,* 32, 215, 1977.
16. Jagers, P., *Branching Processes with Biological Applications,* Wiley, New York, 1976.
17. Shurenkov, V. M., Two limit theorems for critical branching processes, *Teor. Probab. Appl.,* 21, 534, 1976.
18. Green, P. J., Conditional limit theorems for general branching processes, *J. Appl. Probab.,* 14, 451, 1977.
19. Vatutin, V. A., Asymptotic behavior of non-extension probability of a critical branching process, *Teor. Probab. Appl.,* 22, 140, 1977.
20. Shurenkov, V. M., On additive functionals of branching processes, *Teor. Probab. Appl.,* 24, 396, 1979.
21. Shurenkov, V. M., Transitional phenomena of renewal theory in asymptotic problems of the theory of random processes, 1, *Math. USSR Sbornik,* 40, 107, 1981.
22. Shurenkov, V. M., Transitional phenomena of renewal theory in asymptotic problems of the theory of random processes, 2, *Math. USSR Sbornik,* 40, 211, 1981.
23. Vatutin, V. A., Limit theorem for a critical Bellman-Harris branching process with several types of particles and infinite second moments, *Teor. Probab. Appl.,* 23, 776, 1978.
24. Vatutin, V. A., Discrete limit distributions of the number of particles in a Bellman-Harris branching process with several types of particles, *Teor. Probab. Appl.,* 24, 509, 1979.
25. Vatutin, V. A., On a class of critical Bellman-Harris branching processes with several types of particles, *Teor. Probab. Appl.,* 25, 760, 1980.
26. Seneta, E., Vere-Jones, D., On the asymptotic behavior of subcritical branching processes with continuous state space, *Z. Wahrscheinlichkeitstheor. und verw. Geb.,* 10, 212, 1968.
27. Seneta, E., Vere-Jones, D., On a problem of M. Jǐrina concerning continuous state branching processes, *Czech. Math. J.,* 19, 277, 1969.
28. Ryzhov, Yu. M., Skorohod, A. V., Homogeneous branching processes with a finite number particle types and a continuously changing mass, *Teor. Veroyatn. Primen.,* 15, 722, 1970 (in Russian).
29. Kawazu, K., Watanabe, S., Branching processes with immigration and related limit theorems, *Teor. Veroyatn. Primen.,* 16, 34, 1971.
30. Shurenkov, V. M., An asymptotic behavior of vector branching processes with a continuous state space, *Preprint No. 72-5,* Institute Math., Kiev, 1972 (in Russian).

31. Shurenkov, V. M., An asymptotic behavior of branching processes with a continuous state space and a finite number of types, *Teor. Veroyatn. Mat. Statist.*, 8, 169, 1973 (in Russian).
32. Seneta, E., Some supplementary notes on one-type continuous state branching processes, *Z. Wahrscheinlichkeitstheor. und verw. Geb.*, 34, 87, 1976.
33. Crump, K. S., Mode, C. J., A general age-dependent branching processes, I, *J. Math. Anal. Appl.*, 24, 494, 1968.
34. Crump, K. S., Mode, C. J., A general age-dependent branching processes, II, *J. Math. Anal. Appl.*, 25, 8, 1969.
35. Zubkov, A. M., Conditions of degeneracy of a bounded branching process, *Mat. Zametki*, 8, 9, 1970 (in Russian).
36. Zubkov, A. M., Conditions of degeneracy of a bounded branching process with a continuous time, *Teor. Veroyatn. Primen.*, 17, 296, 1972 (in Russian).
37. Boyko, R. V., A controlled branching process, *Ukr. Mat. Shurn.*, 26, 237, 1974 (in Russian).
38. Boyko, R. V., A steady-state distribution of a controlled branching process and an optimization problem, *Ukr. Mat. Shurn.*, 26, 522, 1974 (in Russian).
39. Hotlte, J. M., Extinction probability for a critical general branching process, *Stochast. Processes and Appl.*, 2, 303, 1974.
40. Zubkov, A. M., Analogs between Galton-Watson processes and φ-branching processes, *Teor. Veroyatn. Primen.*, 19, 319, 1974 (in Russian).
41. Sevastyanov, B. A., Zubkov, A. M., Controlled branching processes, *Teor. Veroyatn. Primen.*, 19, 15, 1974 (in Russian).
42. Boyko, R. V., Limit theorems for branching processes with a varying regime (a subcritical case), *Ukr. Mat. Shurn.*, 27, 750, 1975 (in Russian).
43. Vatutin, V. A., Critical Bellman-Harris branching process with immigration and several types of particles, *Teor. Probab. Appl.*, 21, 435, 1976.
44. Boyko, R. V., A limit theorem for branching processes with a varying regime (a critical case), *Ukr. Mat. Shurn.*, 29, 89, 1977 (in Russian).
45. Topchii, V. A., Integral limit theorem for critical Crump-Mode-Jagers branching processes with a discrete time, *Preprint,* Institute Math., Novosibirsk, 1977 (in Russian).
46. Badalbaev, I. S., Rakhimov, I., Critical branching processes with immigration of decreasing intensity, *Teor. Probab. Appl.*, 23, 259, 1978.
47. Boyko, R. V., A supercritical branching process with a varying regime, *Ukr. Mat. Shurn.*, 32, 201, 1980 (in Russian).

48. Topchii, V. A., On the asymptotics of probability of the extension of critical general branching processes, *Doklady Akad. Nauk SSSR,* 252, 55, 1980 (in Russian).

Chapter 14

Ergodic Theory and Stationary Processes

14.1 Stationary Processes

In the framework of random processes theory, ergodic theory deals with conditions of the existence of time-averages of random processes and of their coincidence with phase-averages. More precisely speaking, the following problems are of interest[1−6]. Consider a measurable random process $\{\zeta(t),\ t \geq 0\}$ over a measurable phase space $(E,\ \mathcal{A})$. Ergodic theorems give conditions under which time-averages

$$\lim_{t\to\infty} \frac{1}{t} \int_0^t f(\zeta(u+s_1),\ \ldots,\ \zeta(u+s_n))\, du \tag{14.1}$$

exist almost surely and moreover they are degenerated, i.e., they reduce to nonrandom constants for any $s_1, \ldots, s_n \geq 0$ and any $\mathcal{A}^n$-measurable functions $f(x_1, \ldots, x_n)$. If all such limits exist and are constants almost surely, then the process $\{\zeta(t),\ t \geq 0\}$ is called **ergodic**.

Similarly, an $(E,\ \mathcal{A})$-valued sequence $\{\zeta_n\}$ is called **ergodic** if limits

$$\lim_{n\to\infty} \frac{1}{n} \sum_{k=0}^{n-1} f(\zeta_{k+s_1},\ \ldots,\ \zeta_{k+s_n}) \tag{14.2}$$

exist and are degenerated almost surely for any natural numbers $(s_1, \ldots, s_n)$ and any $\mathcal{A}^n$-measurable function $f(x_1, \ldots, x_n)$.

Theorems of the existence of limit distributions

$$\lim_{t\to\infty} \mathbf{P}\left\{\zeta(t+s_1)\in A_1,\ \ldots,\ \zeta(t+s_n)\in A_n\right\}$$

are also called **ergodic theorems**.

Any statement concerning the existence of limit (14.1) or (14.2) leads to the concept of a strictly stationary process or sequence. Indeed, if the limit (14.1) exists and is degenerated almost surely for any $\mathcal{A}^n$-measurable function $f(x_1,\ldots,x_n)$ and any $s_1,\ldots,s_n\geq 0$, then its value can be expressed in the form

$$\int\limits_E \cdots \int\limits_E f(x_1,\ldots,x_n)\, Q_{s_1,\ldots,s_n}(dx_1,\ldots,dx_n).$$

The family of probability measures $Q_{s_1,\ldots,s_n}(A_1,\ldots,A_n)$ over $(E^n,\ \mathcal{A}^n)$ is a conforming family of finite-dimensional distributions in Kolmogorov's sense, thus

$$Q_{t+s_1,\ldots,t+s_n}(A_1,\ldots,A_n)\ =\ Q_{s_1,\ldots,s_n}(A_1,\ldots,A_n)$$

for any $t\geq 0,\ s_1,\ldots,s_n\geq 0$.

Therefore, if a phase-space $(E,\mathcal{A})$ satisfies the conditions of the Kolmogorov Theorem concerning the construction of a random process given its finite-dimensional distributions $Q_{s_1,\ldots,s_n}(A_1,\ldots,A_n)$, then the process will be stationary. Moreover, if the initial process is stationary and limits of time-averages exist and are degenerated, then the family $Q_{s_1,\ldots,s_n}(A_1,\ldots,A_n)$ will coincide with the family of finite-dimensional distributions of the process $\zeta(t)$ so that

$$Q_{s_1,\ldots,s_n}(A_1,\ldots,A_n)\ =\ \mathbf{P}\left\{\zeta(s_1)\in A_1,\ \ldots,\ \zeta(s_n)\in A_n\right\}.$$

Hence in the ergodic theory a stationary process is usually considered from the very beginning. The following statement is of principal importance; it is called the **"Maximal Ergodic Theorem"**.

Theorem 14.1. *If a sequence $\{\xi_n,\ n\geq 0\}$ of real-valued random variables is stationary and $\mathbf{E}\,|\xi_0|<\infty$, then*

$$\mathbf{E}\left(\xi_0;\ \sup_{n\geq 0}\sum_{k=0}^{n}\xi_k\geq 0\right)\geq 0.$$

The Theorem 14.1 implies the following **Birkhoff-Chinčin Theorem**.

Theorem 14.2. *If a sequence $\{\xi_n,\ n \geq 0\}$ of real-valued random variables is stationary and* $\mathbf{E}\,|\xi_0| < \infty$, *then the limit*

$$\eta = \lim_{n\to\infty} \frac{1}{n} \sum_{k=1}^{n-1} \xi_k$$

exists almost surely and $\mathbf{E}\,\eta = \mathbf{E}\,\xi_0$.

Theorem 14.3. *If a stationary real-valued random process* $\{\xi(t), t \geq 0\}$ *is measurable and* $\mathbf{E}\,|\xi(0)| < \infty$, *then the limit*

$$\eta = \lim_{t\to\infty} \frac{1}{t} \int_0^t \xi(u)\,du$$

almost surely exists and $\mathbf{E}\,\eta = \mathbf{E}\,\xi(0)$.

If moreover the random variable $\xi(0)$ has a finite moment of an order $p \geq 1$, then the convergence of averages to their limits holds true in the mean of the order p.

If an $(E,\ \mathcal{A})$-valued process $\zeta(t)$ is stationary and a real-valued function $f(x_1, \dots, x_n)$ is bounded over E^n and $\mathcal{A}^n$-measurable, then the real-valued random process $\xi(t) = f(\zeta(t+s_1), \dots, \zeta(t+s_n))$ is bounded and stationary. Hence the limits (14.1) exist for a stationary process. A degeneration criterion for such processes is based on the concept of **a shift operator**. Such operators can be defined for any random process $\zeta(t)$ possessing the property that the set of its path functions is closed under time shifts. This means that for any $s \geq 0$ and any elementary event $\omega \in \Omega$ such $\omega' \in \Omega$ exists that $\zeta(t+s,\ \omega) = \zeta(t,\ \omega')$ for all $t \geq 0$.

Let the space of path functions of the process $\zeta(t)$ be closed under time shifts. Then by a choice axiom such a mapping $T_s : \ \Omega \to \Omega$ exists that $\zeta(t,\ T_s\omega) = \zeta(t+s,\ \omega)$. Any such mapping T_s is measurable with respect to the σ-algebra $\mathcal{M}$ generated by the random process $\zeta(t)$. Moreover, if an event Γ belongs to $\mathcal{M}$, then the event $T_s^{-1}\Gamma$ does not depend on a concrete choice of shift operators T_s. Hence for arbitrary $\mathcal{M}$-measurable random variable ξ the equality $\xi\,T_s(\omega) = \xi(T_s\omega)$ well defines a **shifted** $\mathcal{M}$-measurable variable ξT_s. If ξ is the indicator of a $\mathcal{M}$-measurable event Γ, then ξT_s is the indicator of the $\mathcal{M}$-measurable event $T_s^{-1}\Gamma$.

The shift operators T_s defined as above are permutable with arbitrary functional operations over random variables, and also with the limit operation. More strictly speaking, if a sequence of $\mathcal{M}$-measurable random variables $\{\xi_n\}$ converges at each point of Ω, then

$$(\lim \xi_n)\,T_s = \lim\,(\xi_n T_s) \quad \text{and} \quad f(\xi_1, \dots, \xi_n)\,T_s = f(\xi_1 T_s, \dots, \xi_n T_s)$$

for whatever a Borel function f of n real variables is.

Shift operators are connected to the σ-algebra $\mathcal{J}$ of **invariant events** defined as the set of events $\Gamma \in \mathcal{M}$ satisfying the condition $T_s^{-1}\Gamma = \Gamma$ for every $t \geq 0$.

When solving ergodic theory problems, it is worthwhile to use a stricter concept of the measurability of a random process than the usual one. A random process $\{\zeta(t),\, t \geq 0\}$ over a probability space $(\Omega,\, \mathcal{F},\, \mathbf{P})$ taking values in a phase-space $(E,\, \mathcal{A})$ is called **measurable** if the mapping $(\omega,\, t) \to \zeta(t,\, \omega)$ is measurable as the mapping of the product $(\Omega \times \mathcal{R}_+,\, \mathcal{M} \times \mathcal{B}_+)$ of the measurable space $(\Omega,\, \mathcal{M})$ with the σ-algebra $\mathcal{M}$ generated by the process $\{\zeta(t),\, t \geq 0\}$ and the real half-line $R_+ = [0,\, \infty)$ with the Borel σ-algebra $\mathcal{B}_+$ into the measurable space $(E,\, \mathcal{A})$. The usual concept of the measurability of a random process $\{\zeta(t),\, t \geq 0\}$ would be obtained if $\mathcal{M}$ is changed by $\mathcal{F}$, and $\mathcal{B}_+$ is completed in the sense of Lebesgue length.

The measurability of a random process $\zeta(t)$ with a shift-closed space of path-functions implies the $\mathcal{M} \times \mathcal{B}_+$-measurability of the mapping $(\omega,\, t) \to \to \xi T_t(\omega)$ for any $\mathcal{M}$-measurable random variable ξ.

Theorem 14.4. *Let a stationary random process $\zeta(t)$ be measurable and let the set of its path functions be shift-closed. Then, provided that the mean value of a $\mathcal{M}$-measurable random variable ξ is finite, the equality*

$$\lim_{t \to \infty} \frac{1}{t} \int_0^t \xi\, T_u\, du = \mathbf{E}\,(\xi \,|\, \mathcal{J})$$

holds true almost surely.

In the terms of shift operators T_t the stationarity of a process $\zeta(t)$ means that the shift operators conserve the probability measure over $\mathcal{M}$, i.e., $\mathbf{P}\{T_t^{-1}\Gamma\} = \mathbf{P}\{\Gamma\}$ for any $t \geq 0$, $\Gamma \in \mathcal{M}$.

In some cases the probability $\mathbf{P}$ is invariant under a random shift $T_\tau \omega = T_{\tau(\omega)}\omega$ where τ is a nonnegative $\mathcal{M}$-measurable random variable. The measurability of a random process $\zeta(t)$ implies the $\mathcal{M}$-measurability of the event $T_\tau^{-1}\Gamma$ for $\Gamma \in \mathcal{M}$. Hence, if ξ is $\mathcal{M}$-measurable, then ξT_τ is also measurable.

The invariance of the probability $\mathbf{P}$ with respect to the operator T_τ means that $\mathbf{P}\{T_\tau^{-1}\Gamma\} = \mathbf{P}\{\Gamma\}$, $\Gamma \in \mathcal{M}$, or, in a detailed form,

$$\mathbf{P}\{\zeta(\tau + t_1) \in A_1,\, \ldots,\, \zeta(\tau + t_n) \in A_n\} = \mathbf{P}\{\zeta(t_1) \in A_1,\, \ldots,\, \zeta(t_n) \in A_n\}$$

for $n = 1, 2, \ldots,\ t_1, \ldots, t_n \geq 0,\ A_1, \ldots, A_n \in \mathcal{A}$.

This definition can be also used in case of the shift τ is finite almost surely, but not everywhere. The shift operator T_τ will be defined only for those ω for which $\tau(\omega) < \infty$. The random variable ξT_τ is also defined for the case $\tau < \infty$. The invariance of the reliability $\mathbf{P}$ with respect to T_τ means that

$$\mathbf{P}\{T_\tau^{-1}\Gamma,\ \tau < \infty\} = \mathbf{P}\{\Gamma\}, \quad \Gamma \in \mathcal{M},$$

or, in other words,

$$\mathbf{P}\{\zeta(\tau + t_1) \in A_1, \ldots, \zeta(\tau + t_n) \in A_n,\ \tau < \infty\} =$$

$$= \mathbf{P}\{\zeta(t_1) \in A_1, \ldots, \zeta(t_n) \in A_n\}$$

for $n = 1, 2, \ldots,\ t_1, \ldots, t_n \geq 0,\ A_1, \ldots, A_n \in \mathcal{A}$.

Theorem 14.5. *Let a random process $\zeta(t)$ be measurable and the set of its path functions be shift-closed. Then, if a shift operator T_τ conserves the probability $\mathbf{P}$ over $\mathcal{M}$ and, moreover, $\mathbf{P}\{\tau > 0\} = 1$ and $\mathbf{E}\tau < \infty$, then*

$$\lim_{t\to\infty} \frac{1}{t} \int_0^t \xi T_u \, du = \frac{1}{\mathbf{E}(\tau \,|\, \mathcal{J})} \mathbf{E}\left(\int_0^\tau \xi T_u \, du \,\Big|\, \mathcal{J}\right)$$

almost surely, the random variable ξ being $\mathcal{M}$-measurable and

$$\mathbf{E} \int_0^\tau |\xi T_u| \, du < \infty.$$

Theorem 14.6. *Let a random process $\zeta(t)$ be measurable and the set of its path functions be shift-closed. If, moreover, a shift operator T_τ conserves the probability $\mathbf{P}$ over $\mathcal{M}$ and $\mathbf{P}\{0 < \tau < \infty\} = 1$, then*

$$\lim_{t\to\infty} \frac{\int_0^t \xi T_u \, du}{\int_0^t \eta T_u \, du} = \frac{\mathbf{E}\left(\int_0^\tau \xi T_u \, du \,\big|\, \mathcal{J}\right)}{\mathbf{E}\left(\int_0^\tau \eta T_u \, du \,\big|\, \mathcal{J}\right)}$$

almost surely where

$$\mathbf{E} \int_0^\tau (|\xi T_u| + |\eta T_u|) \, du < \infty \quad \text{and} \quad \mathbf{E}\left(\int_0^\tau \eta T_u \, du \,\Big|\, \mathcal{J}\right) \neq 0$$

almost surely.

In particular, if $\xi = f(\zeta(0))$ and $\eta = g(\zeta(0))$, then the stated theorem gives conditions for the equality

$$\lim_{t\to\infty} \frac{\int_0^t f(\zeta(u))\,du}{\int_0^t g(\zeta(u))\,du} = \frac{\mathbf{E}\left(\int_0^T f(\zeta(u))\,du \,|\, \mathcal{J}\right)}{\mathbf{E}\left(\int_0^t g(\zeta(u))\,du \,|\, \mathcal{J}\right)}$$

(almost surely).

Theorems 14.4 and 14.5 show that under their hypotheses the ergodicity of a process $\zeta(t)$ is equivalent to the triviality of the σ-algebra $\mathcal{J}$ of invariant events (a σ-algebra is called **trivial** if all its events have probabilities equaling either 0 or 1). If the σ-algebra $\mathcal{J}$ is nontrivial and is generated by a countable set of its elements, then, provided the conditional probability $\mathbf{P}\{\cdot\,|\,\mathcal{J}\}$ is regular, there exists at most continuum set of disjoint sets $\{\Lambda\}$ of $\mathcal{J}$ such that $\cup\Lambda = \Omega$ and a probability $\mathbf{P}_\Lambda$ over $\mathcal{M}$ exists satisfying the equality

$$\mathbf{P}\{\Gamma\,|\,\mathcal{J}\} = \sum_{\{\Lambda\}} \mathbf{P}_\Lambda(\Gamma)\, I_\Lambda, \quad \Gamma \in \mathcal{M},$$

with probability 1. The stated equality is called **ergodic decomposition.** It is due to the fact that under the hypotheses of Theorem 14.6

$$\lim_{t\to\infty} \frac{\int_0^t \xi T_u\,du}{\int_0^t \eta T_u\,du} = \frac{\mathbf{E}\int_0^T \xi T_u\,du}{\mathbf{E}\int_0^T \eta T_u\,du}$$

$\mathbf{P}_\Lambda$-almost surely for each Λ of the defined set.

For the σ-algebra $\mathcal{J}$ to be countably generated it suffices that σ-algebra $\mathcal{M}$ should be countably generated. The latter property in its turn is implied by the structure of a complete metric separable space over a phase-space of the process $\zeta(t)$ and right/left continuity of its path functions. The stated condition implies also the regularity of the conditional probability $\mathbf{P}\{\cdot\,|\,\mathcal{J}\}$.

If an $(E,\,\mathcal{A})$-valued stationary sequence $\{\zeta_n\}$ is ergodic, then, in particular,

$$\frac{1}{m}\sum_{k=0}^{m-1} I_A(\zeta_k) \underset{n\to\infty}{\longrightarrow} \mathbf{P}\{\zeta_0 \in A\}, \quad A \in \mathcal{A}.$$

The left-hand side equals the frequency of the event $\{\zeta_k \in A\}$ within first m steps. This means that the frequency tends to the probability as soon as the sequence is ergodic.

The simplest example of an ergodic random sequence is a sequence of independent identically distributed random variables. A homogeneous finite-state irreducible Markov chain with a stationary initial distribution is ergodic as well. A stationary Gaussian sequence $\{\xi_n\}$ is ergodic iff its spectral measure is deprived of atoms, i.e., its covariance function has the form $\int_{-\pi}^{\pi} e^{inx}\, dF(x)$, the function $F(x)$ being continuous. Similarly, a stationary Gaussian process $\xi(t)$ is ergodic iff its covariance function has the form $\int_{-\infty}^{\infty} e^{itx}\, dF(x)$, the spectral function $F(x)$ being continuous.

A regenerative process is an example of a stationary ergodic process admitting a probability $\mathbf{P}$ conserving transform of a random shift. A characteristic property of an $(F, \mathcal{A})$-valued regenerative process $\{\zeta(t),\ t \geq 0\}$ consists of the independence of process $\{\zeta(t),\ t < \tau\}$ and $\{\zeta(t+\tau),\ t \geq 0\}$ if a positive random variable τ is appropriately chosen, and also of the coincidence of finite-dimensional distributions of the process $\{\zeta(t+\tau),\ t \geq 0\}$ with those of the process $\{\zeta(t),\ t \geq 0\}$. The σ-algebra of invariant events of a regenerative process is trivial. Without loss of generality the random variable τ can be assumed to be $\mathcal{M}$-measurable. Hence the shift operator T_τ is probability $\mathbf{P}$ conserving over $\mathcal{M}$.

Therefore, if $\mathbf{E}\tau < \infty$, then the regenerative process is ergodic. If, moreover, the distribution of the random variable τ is nonlatticed and an $\mathcal{A}$-measurable bounded function $f(x)$ is such that the mean $\mathbf{E} f(\zeta(t))$ is continuous in t almost everywhere, then

$$\lim_{t\to\infty} \mathbf{E} f(\zeta(t)) = \frac{1}{\mathbf{E}\tau}\, \mathbf{E} \int_0^\tau f(\zeta(t))\, dt.$$

Given a shift operator T_τ conserving the probability $\mathbf{P}$ over σ-algebra $\mathcal{M}$, some other shift operators conserving some probability measures can be defined.

For example, let a $\mathcal{M}$-measurable integer-valued positive random variable ν be such that $\mathbf{P}\{\nu < \infty\} = 1$ and $T_\tau \nu = \nu - 1$ for $\nu > 1$. Then a shift operator T_{τ_ν}, where $\tau_1 = \tau,\ \tau_{k+1} = \tau + \tau_k T_\tau,\ k = 1, 2, \ldots,$ conserves the probability defined by

$$\mathbf{Q}\{\Gamma\} = \frac{1}{\mathbf{P}\{\nu = 1\}}\, \mathbf{P}\{T_\tau^{-1}\Gamma,\ \nu = 1\}, \quad \Gamma \in \mathcal{M}.$$

The probability $\mathbf{P}$ can be expressed via $\mathbf{Q}$ by the formula

$$\mathbf{P}\{\Gamma\} = \frac{1}{c} \sum_{k=0}^{\infty} \mathbf{Q}\{T_\tau^{-k}\Gamma,\ k < \nu\}, \quad \Gamma \in \mathcal{M},$$

where $c = \sum_{k=0}^{\infty} \mathbf{Q}\{k < \nu\}$. An example of ν: the time of the occurrence of the first of the events

$$T_\tau^{-1}\Delta,\ T_\tau^{-2}\Delta, \ldots, T_\tau^{-n}\Delta, \ldots, \quad \text{where} \quad \Delta \in \mathcal{M},\ \mathbf{P}\{\Delta\} > 0,$$

or, more formally speaking,

$$\nu(\omega) = \begin{cases} \min\{n \geq 1 :\ \omega \in T_\tau^{-n}\} & \text{if this set of } n \text{ is nonempty} \\ \infty & \text{otherwise.} \end{cases}$$

Under such a choice of ν the probability $\mathbf{Q}$ is the conditional probability $\mathbf{P}$ under the condition Δ. If moreover the process $\zeta(t)$ is ergodic, then $\mathbf{P}\{\nu < \infty\} = 1$ and hence its finite-dimensional distributions can be reconstructed uniquely given a conditional probability $\mathbf{P}\{\cdot\,|\,\Delta\}$, $\Delta \in \mathcal{M}$, $\mathbf{P}\{\Delta\} > 0$.

14.2 Recurrent Markov Chains

Ergodic properties of a Markov chain should not depend on an initial distribution. The ergodicity of a Markov chain $\{\zeta_n,\ n \geq 0\}$ taking values in a phase-space $(E,\ \mathcal{A})$ under an arbitrary initial distribution means that for any bounded $\mathcal{A}$-measurable function $f(x)$ there exists a limit

$$\lim_{n\to\infty} \frac{1}{n} \sum_{k=0}^{n-1} f(\zeta_k) \tag{14.3}$$

almost surely in the sense of the probability $\mathbf{P}_x$ associated with the condition $\zeta_0 = x$, and moreover this limit is not random and does not depend on x.

If a Markov chain $\{\zeta_n\}$ is ergodic independently of an initial distribution, then the limit (14.3) equals $\mathbf{P}_x$-almost surely $\int_E \pi(dx)\, f(x)$ where the probability distribution $\pi(dx)$ over the phase-space $(E,\ \mathcal{A})$ is **invariant** with respect to the transition probability $P(x,\ A) = \mathbf{P}_x\{\zeta_1 \in A\}$, i.e.,

$$\int_E \pi(dx)\, P(x,\ A) = \pi(A), \quad A \in \mathcal{A}. \tag{14.4}$$

Moreover, if a Markov chain is ergodic under any initial distribution, then any state of positive π-probability is visited infinitely many times $\mathbf{P}_x$-almost surely, whatever $x \in E$ be. In other words, the condition $\pi(A) > 0$ implies

$\mathbf{P}_x\{\zeta_n \in A \text{ for infinitely many } n\} = 1$. The latter property is called **π-recurrence (Harris recurrence)** of a Markov chain $\{\zeta_n\}$.

The inverse statement is also true. If the equation (14.4) has a probability solution π and a Markov chain $\{\zeta_n\}$ with the transition probability $P(t, A)$ is π-recurrent, then it is ergodic independently of an initial distribution. If moreover the set of its path functions is closed with respect to shifts (and hence a shift operator T of one-step shift of a path function is well defined), then

$$\lim_{n\to\infty} \frac{1}{n}\sum_{k=0}^{n-1} \xi T^k = \mathbf{E}_\pi \xi \quad (\mathbf{P}_x\text{-almost surely})$$

where ξ is measurable with respect to $\zeta_0, \zeta_1, \ldots$ and integrable over the probability $\mathbf{P}_\pi = \int \pi(dx)\,\mathbf{P}_x$.

The essence of the ergodicity of a Markov chain is in its **recurrence** property[6,7]. Under mild additional conditions the recurrence of a Markov chain implies the existence and uniqueness of the solution of the equation (14.4). Such a solution is not necessarily a probabilistic one, but it is σ-finite in any case.

Theorem 14.7. *Let a homogeneous Markov chain ζ_n be recurrent, and let a σ-algebra over its phase-space be generated by a countable set of its elements. Then the Markov chain ζ_n has a unique (up to a multiplicative constant) invariant σ-finite measure π and is π-recurrent.*

Under the assumptions of this theorem the invariant measure has the widest variety of positive sets among all the possible probability inducing measures, i.e., each such measure is absolutely continuous with respect to the invariant one.

An example of a Markov chain with an infinite total invariant measure: a random walk over a real line provided that the distribution of a one-step transition is nonsingular with the Lebesgue measure. The latter is the unique (up to a multiplicative constant) invariant measure.

Theorem 14.8. *Let a σ-algebra $\mathcal{A}$ over the phase-space of a homogeneous Markov chain $\{\zeta_n\}$ be countably generated. If the Markov chain $\{\zeta_n\}$ is recurrent, π being its invariant measure, then*

$$\lim_{n\to\infty} \frac{\sum_{k=0}^{n} f(\zeta_k)}{\sum_{k=0}^{n} g(\zeta_k)} = \frac{\int \pi(dx)\, f(x)}{\int \pi(dx)\, g(x)} \quad (\mathbf{P}_x\text{-almost surely})$$

where $\mathcal{A}$-measurable functions f and g are π-integrable and

$$\int g(x)\,\pi(dx) \neq 0.$$

If a Markov chain $\{\zeta_n\}$ has an invariant probability measure π and the σ-algebra of invariant events is trivial, then under the initial distribution π the $\{\zeta_n\}$ is **a stationary ergodic sequence**, i.e.,

$$\lim_{n\to\infty} \frac{1}{n} \sum_{k=0}^{n-1} f(\zeta_k) \;=\; \int \pi(dx)\, f(x) \tag{14.5}$$

$\mathbf{P}_\pi$-almost surely under obvious measurability and summability assumptions on the function f. This means that the equality (14.5) holds $\mathbf{P}_x$-almost surely for π-almost every $x \in E$. At the same time for some $x \in E$ the conditional probability of the equality (14.5) given $\zeta_0 = x$ may be less than a unit and even vanish. Consider, for example, a random walk over a circle of a unit length. In such a case successive displacements of a particle are independent and have a common distribution function $F(x)$. If the support of the latter distribution does not belong to a set $\{k/m,\ k = 0, 1, \ldots, m\}$ for an integer m, then the (0, 1)-uniform distribution is the unique invariant distribution of such a Markov chain. Moreover, the σ-algebra of its invariant events is trivial under any given initial distribution. Thus, under the (0, 1)-uniform initial distribution the random walk is ergodic (with the exception of the latticed case). But the random walk is ergodic under arbitrary initial distribution iff some convolution of the distribution $F(x)$ is nonsingular with respect to the Lebesgue measure.

The Harris recurrency should not be confounded with the topological recurrency. An homogeneous Markov chain $\{\zeta_n\}$ is said to be **topologically recurrent** if the σ-algebra over its phase-space is generated by a separable topology and any open set is achievable from any initial state with the probability 1. The distinction of these two concepts can be seen from the random walk over a circle. In the nonlatticed case it is topologically recurrent in a natural topology, whereas Harris recurrency (with respect to Lebesgue length) of the random walk over a circle consists of the nonsingularity of some convolution of the distribution of one-step displacement. In the case of a finite or countable phase-space both concepts are equivalent.

The problem of the existence of limit distributions $\lim\limits_{n\to\infty} \mathbf{P}_x\{\zeta_n \in A\}$ for recurrent Markov chains is closely connected to the finiteness of the invariant measure and the periodicity of the chain. A recurrent Markov chain is called **positive** if its invariant measure is finite; it is called **null** otherwise. The period of a Markov chain $\{\zeta_n\}$ is defined as the largest integer d for which the phase-space equals the union of disjoint sets $D, C_0, \ldots, C_{d-1}$ such that $P(x, C_j) = 1$ for $x \in C_{j-1}, j = 1, \ldots, d-1$, $P(x, C_0) = 1$ for $x \in C_{d-1}$ and $\mathbf{P}_x\{\zeta_n \in D$ infinitely often$\} = 0$, $x \in E$. In other words, under any initial distribution the Markov chain $\{\zeta_n\}$ almost surely enters one of the sets $\{C_i\}$ and than changes sets periodically with the probability 1.

If the defined decomposition is impossible for an integer $d > 1$ or the period equals 1, then the Markov chain is called **aperiodic**. A priori the case when a period does not exist, is not excluded, even if the Markov chain is recurrent. The following theorems show that such a situation can exist in rare exceptional cases.

Theorem 14.9. *A recurrent Markov chain with a countably generated σ-algebra over the phase-space has a period.*

Theorem 14.10. *Let σ-algebra $\mathcal{A}$ over the phase-space $(E, \mathcal{A})$ of a recurrent Markov chain be countably generated. If the Markov chain is positive recurrent, has an invariant distribution π and is aperiodic, then*

$$\lim_{n\to\infty} \sup_{A\in\mathcal{A}} |\mathbf{P}_x\{\zeta_n \in A\} - \pi(A)| = 0, \quad x \in E.$$

If a Markov chain is null and an $\mathcal{A}$-measurable function is integrable with respect to the invariant measure, then

$$\lim_{n\to\infty} \mathbf{E}_x f(\zeta_n) = 0, \quad x \in E.$$

Theorem 14.11. *Let a positive recurrent Markov chain $\{\zeta_n\}$ have a period d, and a σ-algebra $\mathcal{A}$ over its phase-space $(E, \mathcal{A})$ is countably generated. If $C_0, \ldots, C_{d-1}$ form the cyclic decomposition of $\{\zeta_n\}$ and π is its invariant distribution, then*

$$\lim_{n\to\infty} \sup_{A\in\mathcal{A}} |\mathbf{P}_x\{\zeta_{nd+r} \in A\} - d\,\pi(AC_k)| = 0, \quad x \in C_j,$$

where $0 \le j \le d-1$, $0 \le k \le d-1$, $j + r \equiv k \pmod d$.

The following theorems concern some extremely useful recurrency criteria. Here the recurrency should be considered in a slightly wider sense. A Markov chain will be said to be **recurrent**, if, an appropriate nontrivial σ-finite measure ν over the phase-space is defined, such a ν-null set can be extracted from the phase-space so that the Markov chain considered over the remaining part of the phase-space would be ν-recurrent in the above defined sense. The main result is formulated via the kernel

$$G_\alpha(x, A) = \sum_{n\ge1} (1-\alpha)\,\alpha^{n-1} P^n(x, A)$$

where $0 \le \alpha < 1$ and $P^n(x, A) = \mathbf{P}_x\{\zeta_n \in A\}$ is the transition probability of the Markov chain ζ_n. The kernel $G_\alpha(x, A)$ can be interpreted stochastically. If an integer-valued positive random variable θ is geometrically distributed with a parameter α, i.e., $\mathbf{P}\{\theta = k\} = (1-\alpha)\alpha^{k-1}$, $k = 1, 2, \ldots$, and is independent of the chain $\{\zeta_n\}$, then $G_\alpha(x, A) = \mathbf{P}_x\{\zeta_0 \in A\}$.

Theorem 14.12. *Let a σ-algebra $\mathcal{A}$ over the phase-space $(E, \mathcal{A})$ of a Markov chain $\{\zeta_n\}$ be countably generated. Then for the chain to be recurrent it is necessary that for any $\alpha \in [0, 1)$, and sufficient that for at least one such α, a positive $\mathcal{A}$-measurable function $q(x)$ and a finite measure $m(A)$ over $\mathcal{A}$ exist such that*

$$G_\alpha(x, A) \ge q(x)\, m(A), \quad \sum_{n\ge 1} \int_E m(dx) \int_E P^n(x, dy)\, q(y) = \infty.$$

The theorem implies a recurrency criterion of the form like the similar criterion for a finite or countable phase-space.

Theorem 14.13. *If a σ-algebra $\mathcal{A}$ over the phase-space $(E, \mathcal{A})$ is generated by a countable set of its elements, then a Markov chain $\{\zeta_n\}$ taking values in $(E, \mathcal{A})$ is recurrent iff a nontrivial σ-finite measure m over $\mathcal{A}$ exists such that $m(A) > 0$ implies*

$$\sum_{n\ge 1} P^n(x, A) = \infty$$

for any $x \in E$.

This criterion implies, in particular, a simple sufficient condition of the ergodicity of a Markov chain $\{\zeta_n\}$ provided that a σ-algebra over its phase-space is countably generated: *a Markov chain $\{\zeta_n\}$ is ergodic for every initial distribution iff for any $x \in E$ and $A \in \mathcal{A}$ the limit*

$$\lim_{n\to\infty} \frac{1}{n} \sum_{k=1}^{n} P^k(x, A)$$

exists and is independent of x. The limit distribution is invariant.

Theorem 14.14. *A Markov chain over a phase-space $(E, \mathcal{A})$ is recurrent iff there exist a set $C \in \mathcal{A}$ achievable from any initial state, a nontrivial finite measure m over $\mathcal{A}$ and an integer n such that*

$$P^n(x, A) \ge m(A) \ \text{ for } x \in C, \ A \subset C, \ A \in \mathcal{A}.$$

A Markov chain $\{\zeta_n\}$ is called **irreducible** if there exists such a nontrivial σ-finite measure m over its phase-space $(E,\, \mathcal{A})$ that

$$\sum_{n\geq 1} P^n(x,\, A) > 0$$

as soon as $m(A) > 0$. The irreducibility of a Markov chain implies the inequality $G_\alpha(x,\, A) \geq q(x)\, m(A)$ under an appropriate measure m and a function q. Hence an irreducible chain over a phase-space $(E,\, \mathcal{A})$ with a countably generated σ-algebra is recurrent iff

$$\sum_{n\geq 1} \int_E m(dx) \int_E P^n(x,\, dy)\, q(y) \;=\; \infty$$

for any positive $\mathcal{A}$-measurable function q and an irreducibility ensuring probability measure m.

A sufficient condition for the recurrency of an irreducible (with respect to a probability measure m) Markov chain $\{\zeta_n\}$, provided that the σ-algebra over its phase-space is countably generated, consists of the existence of such $\varepsilon > 0$ and an integer n that $P^n(x,\, A) \leq 1 - \varepsilon$ as soon as $m(A) \leq \varepsilon$ or $P^n(x,\, A) > \varepsilon$ for $m(A) > 1 - \varepsilon$.

Historically this condition **(the Döblin condition)** has played the role of being one of the first sufficiently wide conditions enabling the effective investigation of a Markov chain over a general phase-space. Under this condition the convergence in Theorems 14.10 and 14.11 is uniform in $x \in E$. Moreover, independent of irreducibility conditions for the Markov chain $\{\zeta_n\}$, the Döblin condition (with a probability measure m) implies the existence of a finite number of disjoint absorbing sets $D_1, \ldots, D_r \in \mathcal{A}$ such that their union is achievable with the probability 1 from any initial state. Moreover the Markov chain $\{\zeta_n\}$ reduced to any of sets D_j, $j = 1, \ldots, r$, is ergodic for any initial distribution.

14.3 Semi-Markov Processes

A necessary condition for the ergodicity of a semi-Markov process $\zeta(t)$, $t \geq 0$, for every initial distribution consists of the recurrency of the Markov chain $\{\zeta(\tau_k),\ k = 0, 1, 2, \ldots\}$ embedded to the process at times $0 = \tau_0 < < \tau = \tau_1 < \tau_2 < \cdots$ when it changes its states.

If a σ-algebra $\mathcal{A}$ of the phase-space $(E,\, \mathcal{A})$ of the process is countably generated, then the recurrency of the embedded Markov chain is sufficient at least for the existence and degeneracy of the limits

$$\lim_{t\to\infty} \frac{\int_0^t f(\zeta(u))\,du}{\int_0^t g(\zeta(u))\,du}$$

with probability 1 with respect to each conditional probabilities $\mathbf{P}_x$, $x \in E$, ($\zeta(0) = x$ being the condition). An important and nontrivial consequence of the recurrency of the embedded Markov chain $\zeta(\tau_k)$ consists of the unboundedness almost surely of the sequence τ_k.

In all the theorems[6,8–11] cited below (with the exception of Theorem 14.5) it is assumed that a σ-algebra $\mathcal{A}$ is countably generated.

Theorem 14.15. *Let a Markov chain $\zeta(\tau_k)$ embedded in a semi-Markov process $\zeta(t)$ be ergodic, π being its invariant distribution, and $\mathbf{E}_\pi\,\tau < \infty$. Then*

$$\lim_{t\to\infty} \frac{1}{t}\int_0^t f(\zeta(u))\,du = \frac{\mathbf{E}_\pi\,\tau\, f(\zeta(0))}{\mathbf{E}_\pi\,\tau} \tag{14.6}$$

P_x-almost surely where an $\mathcal{A}$-measurable function $f(x)$ is such that

$$\mathbf{E}_\pi\,\tau\,|f(\zeta(0))| < \infty. \tag{14.7}$$

Theorem 14.16. *Let an embedded Markov chain $\zeta(\tau_k)$ be recurrent and have an invariant measure π such that $\mathbf{E}_\pi\tau < \infty$. Then the relation (14.6) holds where an $\mathcal{A}$-measurable function $f(x)$ satisfies (14.7).*

The right hand side of (14.6) can be written as

$$\frac{\int f(x)\,m(x)\,\pi(dx)}{\int m(x)\,\pi(dx)}$$

where $m(x) = \mathbf{E}_x\tau$. In particular, if the function $f(x)$ equals the indicator of a set $A \in \mathcal{A}$, then the limit of the corresponding time averages equals the relative time spent by the process in this set. By the Theorems 14.12 and 14.13 this relative time equals

$$\frac{\int_A m(x)\,\pi(dx)}{\int_E m(x)\,\pi(dx)}\,.$$

Since the invariant measure mentioned in the last theorem is not necessarily finite, the measure $\mathbf{P}_\pi = \int \pi(dx)\,\mathbf{P}_x$ is only σ-finite in a general case.

If the set of path functions of the process $\zeta(t)$ is closed with respect to shifts, and hence shift operators T_s are defined, in the assumptions of each of these theorems the equality

$$\lim_{t\to\infty} \frac{1}{t} \int_0^t \xi\, T_u\, du = \frac{\mathbf{E}_\pi \int_0^\tau \xi\, T_u\, du}{\mathbf{E}_\pi \tau}$$

holds P_x-almost surely where ξ is measurable with respect to a σ-algebra $\mathcal{M}$ generated by the process $\{\zeta(t),\ t \geq 0\}$ and

$$\mathbf{E}_\pi \int_0^\tau \xi\, T_u\, du < \infty.$$

In particular, in the case $\xi = I_{\{\tau \leq s\}}$ the limit in a left-hand side equals the relative time of such an event that the residual time to the next transition does not exceed s. The right-hand side fraction equals $\mathbf{E}_\pi(\tau \wedge s)/\mathbf{E}_\pi \tau$, i.e., so-called steady state distribution function of the waiting time to the next transition.

Theorem 14.17. *If the set of path functions of a semi-Markov process $\zeta(t)$ is closed with respect to shifts and an embedded Markov chain $\zeta(\tau_k)$ is recurrent, π being its invariant measure, then*

$$\lim_{t\to\infty} \frac{\int_0^t \xi\, T_u\, du}{\int_0^t \eta\, T_u\, du} = \frac{\mathbf{E}_\pi \int_0^\tau \xi\, T_u\, du}{\mathbf{E}_\pi \int_0^\tau \eta\, T_u\, du} \tag{14.8}$$

P_x-almost surely where

$$\mathbf{E}_\pi \int_0^\tau (|\xi\, T_u| + |\eta\, T_u|)\, du < \infty \quad \text{and} \quad \mathbf{E}_\pi \int_0^\tau \eta\, T_u\, du \neq 0.$$

If $\xi = f(\zeta(0))$, $\eta = g(\zeta(0))$, then the right-hand side of the equality (14.8) equals $\mathbf{E}_\pi \tau\, f(\zeta(0)) \,/\, \mathbf{E}_\pi \tau\, g(\zeta(0))$. This expression, in turn, coincides with

$$\frac{\int f(x)\, m(x)\, \pi(dx)}{\int g(x)\, m(x)\, \pi(dx)}.$$

The asymptotic behavior of the probabilities $\mathbf{P}_x\{\zeta(t) \in A\}$ as $t \to \infty$ depends in an essential way on the latticedness/nonlatticedness of the semi-Markov process $\zeta(t)$. A precise definition of this property is based on the following expression of the joint distribution of $(\tau,\ \zeta(\tau))$:

$$\mathbf{P}_x\{\tau \le t,\ \zeta(\tau) \in A\} = \int_A P(x,\, dy)\, F(x,\, y;\, t)$$

where $P(x,\, A) = \mathbf{P}_x\{\zeta(\tau) \in A\}$ is the transition probability function of the embedded Markov chain and $F(x,\, y;\, t) = \mathbf{P}\{\tau \le t \,|\, \zeta(0) = x,\ \zeta(\tau) = y\}$ is the conditional distribution function of the sojourn time of the process $\zeta(t)$ in a state x, if the next transition results in the state y. The mentioned joint distribution function is an $\mathcal{A} \times \mathcal{A}$-measurable function of $x,\, y$. The possibility of such an expression follows from the countable origination of the σ-algebra $\mathcal{A}$.

A semi-Markov process with a recurrent embedded Markov chain possessing the invariant measure π is called **latticed** if a positive δ and a $\mathcal{A}$-measurable function $c(x)$ exist such that

$$\int_E \pi(dx) = \int_E P(x,\, dy) \left| \int_0^\infty \exp\{i \frac{2\pi}{\delta} t\}\, dF(x,\, y;\, t) - \right.$$

$$\left. - \exp\{i \frac{2\pi}{\delta} (c(y) - c(x))\} \right| = 0.$$

In other words, the latticedness of a semi-Markov process $\zeta(t)$ means that for almost every $x,\, y$ in reference to the measure $\pi(dx)\, P(x,\, dy)$ all the increase points of the distribution function $F(x,\, y;\, t)$ belong to a shifted lattice

$$\{c(y) - c(x) + n\,\delta, \quad n = 0, \pm 1, \pm 2, \ldots\}.$$

The maximal δ satisfying these conditions is called **the step** of the latticed semi-Markov process. The function $c(x)$ is called **the shift function**. If such an expression exists for no $\delta > 0$, then a recurrent semi-Markov process is called **nonlatticed**.

Theorem 14.18. *Let an embedded Markov chain $\zeta(\tau_k)$ of a semi-Markov process $\zeta(t)$ be recurrent and let π be its invariant measure, $\mathbf{E}_\pi \tau < \infty$. If moreover the process $\zeta(t)$ is nonlatticed, then*

$$\lim_{t\to\infty} \mathbf{P}_x\{\zeta(t) \in A\} = \frac{1}{\mathbf{E}_\pi \tau} \int_A \pi(dy)\, \mathbf{E}_y \tau, \quad A \in \mathcal{A}.$$

If a process $\zeta(t)$ is latticed, then

$$\lim_{t\to\infty} \mathbf{P}_x\{\zeta(n\delta + s) \in A\} =$$

$$= \frac{\delta}{\mathbf{E}_\pi \tau} \int_A \pi(dy) \sum \mathbf{P}_y\{\tau > k\delta - c(y) + c(x) - s\}, \quad A \in \mathcal{A},$$

where δ and $c(x)$ were defined above, and the sum is expanded over such integer k that $k\delta \geq c(y) - c(x) + s$.

Theorem 14.19. *Let $\zeta(t)$ be a semi-Markov process with a finite set $E = \{1, 2, \ldots, m\}$ of states and let an embedded Markov chain be recurrent, π_i being its invariant probabilities $(1 \leq i \leq m)$. If the limits*

$$a_i = \lim_{t\to\infty} t^\alpha L(t)\, \mathbf{P}_i\{\tau > t\}, \quad i = 1, \ldots, m,$$

exist for some $\alpha \in [0,\ 1)$ and are slowly varying at infinity function $L(t)$, and moreover these limits are finite and at least one of them is nonzero, then

$$\lim_{t\to\infty} \mathbf{P}_i\{\zeta(t) = j\} = \frac{a_j\,\pi_j}{\sum_k a_k\,\pi_k}.$$

Under the hypotheses of Theorem 14.19 (the value $\alpha = 1$ is not excluded) there exists also the limit distribution of the random variable

$$\gamma(t) = \frac{1}{t} \int_0^t f(\zeta(u))\, du.$$

If the function $f(t)$ is nonnegative, then

$$\lim_{t\to\infty} \mathbf{P}_i\{\gamma(t) \leq u\} = G(u)$$

where the limiting distribution function is defined via its Stieltjes transform

$$\int_0^\infty \frac{1}{s+u}\, dG(u) = \frac{\sum_{i=1}^m (f(i)+s)^{\alpha-1}\, a_i\,\pi_i}{\sum_{i=1}^m (f(i)+s)^{\alpha}\, a_i\,\pi_i}.$$

The limit distribution is degenerated as $\alpha = 1$ or as only one of the values $a_1, \ldots, a_n$ is nonzero. If $\alpha = 0$, then the limit distribution gives the weight $a_i\,\pi_i \,/\, \sum_j a_j\,\pi_j$ to a point $f(i)$. If the function $f(i)$ takes only values 0 and 1, $\alpha = 0.5$ and

$$\sum_{f(i)=1} a_i\,\pi_i = \sum_{f(i)=0} a_i\,\pi_i,$$

then

$$G(u) = \frac{2}{\pi} \arcsin \sqrt{u}, \quad 0 \leq u \leq 1.$$

A semi-Markov process possesses a characteristic memoryless property at times $\tau_1, \tau_2, \dots$, i.e., its evolution starts after every change of a state. Thus, a trajectory of a semi-Markov process remains constant while the interference of a chance occurs that results in changing a state and leads to a total loss of a memory. The last property reflects the Markov nature of the interference.

Ergodic Theorems 14.12–14.14 can be carried over random processes with Markov interference of a chance. The only difference of such processes from semi-Markov processes is an arbitrary structure of trajectories between interference moments.

Let $\zeta_i(x, t)$, $t \in [0, \gamma(x)]$, $x \in E$, $i = 1, 2, \dots$, be a sequence of independent probability copies of a family of random processes $\zeta(x, t)$, $t \in [0, \gamma(x)]$, $x \in E$, with a phase-space $(E, \mathcal{A})$ defined on a random closed interval $[0, \gamma(x)]$. Here $\gamma(x) > 0$, $\zeta(x, 0) = x$ almost surely and $\zeta(x, t)$ is measurable with respect to $x \in E$, $t \geq 0$, $\omega \in \Omega$. An initial state $\zeta(0)$ of the process $\zeta(t)$ with a Markov interference of a chance is independent of the sequence $\zeta_i(x, t)$, $i = 1, 2, \dots$, and its value at time t is obtained by recursion relation

$$\zeta(t) = \zeta_k(\zeta(\tau_{k-1}), t - \tau_{k-1}),$$

$$\tau_{k-1} \leq t < \tau_k = \tau_{k-1} + \gamma_k(\zeta(\tau_{k-1})), \quad k = 1, 2, \dots, \quad \tau_0 = 0,$$

defining a process $\zeta(t)$ to the moment $t - \sup \tau_k$.

A narrower descriptive definition postulates a loss of the memory (Markov property) at time $\tau = \tau_1$ of the first interference of a chance, i.e.,

$$\mathbf{P}_x\{\zeta(\tau + t_1) \in A_1, \dots, \zeta(\tau + t_n) \in A_n \mid \mathcal{M}_\tau\} =$$
$$= \mathbf{P}_{\zeta(\tau)}\{\zeta(t_1) \in A_1, \dots, \zeta(t_n) \in A_n\} \tag{14.9}$$

$\mathbf{P}_x$-almost surely for any $t_1, \dots, t_n \geq 0$, $n = 1, 2, \dots$, $A_1, \dots, A_n \in \mathcal{A}$; $\mathbf{P}_x$ is a regular conditional probability over a σ-algebra $\mathcal{M}$ generated by a process $\{\zeta(t), t \geq 0\}$ provided that $\zeta(0) = x$; a moment τ is a stopping time with respect to a process $\zeta(t)$, i.e., for any $t \geq 0$ an event $\{\tau \leq t\}$ belongs to a σ-algebra $\mathcal{M}_t$ generated by $\{\zeta(s), s \leq t\}$; a σ-algebra $\mathcal{M}_\tau$ contains those and only those $\mathcal{M}$-measurable events Γ satisfying the condition $\{\tau \leq t\} \cap \Gamma \in \mathcal{M}_t$ for all $t \geq 0$.

If the space of trajectories of a process $\zeta(t)$ is closed with respect to shifts and T_t is a shift operator defined on it, then a relation (14.9) can be rewritten in the following form:

$$\mathbf{E}_x(\xi T_\tau \mid \mathcal{M}_\tau) = \mathbf{E}_{\zeta(\tau)}\xi \tag{14.10}$$

$\mathbf{P}_x$-almost surely for any bounded and $\mathcal{M}$-measurable random variable ξ.

A general theory of processes with Markov interference of a chance can be essentially simplified when the space of trajectories of a process $\zeta(t)$ is closed with respect to stopping times, i.e., for any $\omega \in \Omega$ and any $s \geq 0$ there exists $\omega' \in \Omega$ such that $\zeta(t \wedge s, \omega) = \zeta(t, \omega')$ for any $t \geq 0$. Since the set of trajectories of a process $\zeta(t)$ is closed with respect to shifts and stopping times, the σ-algebra $\mathcal{M}$ is generated by a stopped process $\zeta(t \wedge \tau)$, $t \geq 0$, and the equality (14.10) implies

$$\mathbf{E}_x (\xi \, T_{\tau_k} \mid \mathcal{M}_{\tau_k}) = \mathbf{E}_{\zeta(\tau_k)} \xi$$

$\mathbf{P}_x$-almost surely where $\tau_{k+1} = \tau + \tau_k T_\tau$, $k = 0, 1, 2, \ldots$, $\tau_0 = 0$, and ξ is bounded and $\mathcal{M}$-measurable. Thus every moment τ_k is a moment of Markov interference of a chance. A Markov property of a sequence $\{\zeta(\tau_k), k = 0, 1, 2, \ldots\}$ is an important consequence of this statement. Similar to a semi-Markov case this sequence is called **an embedded Markov chain**.

An advantage of a constructive approach compared to a descriptive one is that moments τ_k in a constructive definition are not supposed to be measurable with respect to process $\zeta(t)$, $t \geq 0$, but definitions in a descriptive approach are simple and visual.

In all the theorems cited below it is assumed that a σ-algebra $\mathcal{A}$ is countably generated and a set of trajectories of a process $\zeta(t)$ is closed with respect to shifts and stopping times.

Theorem 14.20. *Let a Markov chain $\{\zeta(\tau_k), k = 0, 1, 2, \ldots\}$ embedded in a process $\{\zeta(t), t \geq 0\}$ at moments $\{\tau_k, k = 0, 1, 2, \ldots\}$ of a Markov interference of a chance be recurrent, π being its invariant measure, and $\mathbf{P}_\pi \{\tau = 0\} = 0$, $\mathbf{E}_\pi \tau < \infty$. Then*

$$\lim_{t \to \infty} \frac{1}{t} \int_0^t \xi \, T_u \, du = \frac{1}{\mathbf{E}_\pi \tau} \, \mathbf{E}_\pi \int_0^\tau \xi \, T_u \, du$$

P_x-almost surely provided that ξ is $\mathcal{M}$-measurable and

$$\mathbf{E}_\pi \int_0^\tau |\xi \, T_u| \, du < \infty.$$

Theorem 14.21. *Let a Markov chain $\{\zeta(\tau_k), k = 0, 1, 2, \ldots\}$ embedded in a process $\{\zeta(t), t \geq 0\}$ at moments $\{\tau_k, k = 0, 1, 2, \ldots\}$ of a Markov interference of a chance be recurrent, π being its invariant measure,* *and* $\mathbf{P}_\pi\{\tau = 0\} = 0$. *Then*

$$\lim_{t\to\infty} \frac{\int_0^t \xi\, T_u\, du}{\int_0^t \eta\, T_u\, du} = \frac{\mathbf{E}_\pi \int_0^\tau \xi\, T_u\, du}{\mathbf{E}_\pi \int_0^\tau \eta\, T_u\, du}$$

P_x*-almost surely for* $\mathcal{M}$*-measurable random variables* ξ *and* η *such that*

$$\mathbf{E}_\pi \int_0^\tau (|\xi\, T_u| + |\eta\, T_u|)\, du < \infty \quad \text{and} \quad \mathbf{E}_\pi \int_0^\tau \eta\, T_u\, du \neq 0.$$

An unboundedness of the sequence $\{\tau_k\}$ P_x-almost surely is an important and nontrivial consequence of the recurrence of the embedded chain. Moreover, under the hypotheses of Theorem 14.21 by extending a main probability space one can construct a sequence of random moments $v_k \uparrow \infty$ such that every moment v_k, $k = 1, 2, \ldots$, is a moment of Markov interference of a chance, $\zeta(v_1)$, $\zeta(v_2)$ are independent and identically distributed, random processes $\{\zeta(t + v_k),\ t \in [0,\ v_{k+1} - v_k)\}$, have the same finite-dimensional distributions for all $k = 1, 2, \ldots$, and random processes $\{\zeta(t \wedge v_k),\ t \geq 0\}$, $\{\zeta(t + v_{k+1}),\ t \geq 0\}$ are independent for all $k = 1, 2, \ldots$.

Theorem 14.22. *Let a Markov chain* $\{\zeta(\tau_k),\ k = 0, 1, 2, \ldots\}$ *embedded in a process* $\{\zeta(t),\ t \geq 0\}$ *at moments* $\{\tau_k,\ k = 0, 1, 2, \ldots\}$ *of a Markov interference of a chance possess an invariant* σ*-finite measure* π *and a* σ*-algebra* $\mathcal{J}$ *be trivial with respect to measure* $\mathbf{P}_\pi$*. If* $\mathbf{P}_\pi\{\tau = 0\} = 0$ *and* $\mathbf{E}_\pi \tau < \infty$, *then*

$$\lim_{t\to\infty} \frac{1}{t} \int_0^t \xi\, T_u\, du = \frac{1}{\mathbf{E}_\pi \tau}\, \mathbf{E}_\pi \int_0^\tau \xi\, T_u\, du$$

P_x*-almost surely for* π*-almost all* $x \in E$ *where* ξ *is* $\mathcal{M}$*-measurable and*

$$\mathbf{E}_\pi \int_0^\tau |\xi\, T_u|\, du < \infty.$$

14.4 Level Crossings

Level crossing problems[3] are of interest, in particular, for reliability theory because a failure of the system or its component can usually be treated as the crossing of a level by a random process describing the system behavior.

The most widely used level crossing parameters are: the number of up-crossings or down-crossings, the total number of crossings during a given time interval, the time interval between two successive up-crossings or down-crossings, the total sojourn time above or below the level during a given time interval, the number and values of local extremums and so on.

Let us introduce some rigorous definitions. A continuous real-valued random process $\{\xi(t),\ t \in \mathbb{R}\}$ crosses a level c from below to above (up-crossing) at a point s, if there exists an $r > 0$ such that $\xi(t) < c$ for $s-r < t < s$ and $\xi(t) > c$ for $s < t < s+r$. A crossing from above to below (down-crossing) is defined similarly; the inequalities $\xi(t) < c$ and $\xi(t) > c$ should be interchanged. Up-crossings are also called **exits**; down-crossings are called **inputs**. A point s is called a point of **a level c crossing**, if every neighborhood contains such u and v that $(\xi(u) - c)\,(\xi(v) - c) < 0$. There are some crossings that are neither exits nor inputs. Whereas if the number of crossings during a time interval is finite, then each of them is either an exit or an input.

If a stationary random process $\{\xi(t),\ t \in \mathbb{R}\}$ has continuously differentiable path functions and $\xi(0)$ has a bounded density function, then the number of level crossings is finite with probability 1 for every given finite time interval. For a Gaussian process the mean-square differentiability implies the finiteness (almost surely) of the number of level crossings in a finite time interval.

Consider a stationary $\xi(t)$. If the number ν_t of level crossings during a time interval $(0, t)$ is finite with probability 1 for every $t > 0$, then the sequence of level crossings, as well as those of exits or inputs, form a stationary point process. Thanks to this property, we can apply ergodic theorems to investigate the asymptotic behavior of the total number $\nu(s, t]$ of level c crossings as well as the number $\nu_+(s, t]$ of up-crossings and the number $\nu_-(s, t]$ of down-crossings given a level c. So, if the process is ergodic and $\mathbf{E}\,\nu(0, 1) < \infty$, then

$$\lim_{t-s\to\infty} \frac{\nu(s, t]}{t-s} = \mathbf{E}\,\nu(0, 1],$$

$$\lim_{t-s\to\infty} \frac{\nu_+(s, t]}{t-s} = \lim_{t-s\to\infty} \frac{\nu_-(s, t]}{t-s} = \frac{1}{2}\,\mathbf{E}\,\nu(0, 1].$$

The mean number of level crossings can be expressed explicitly for a Gaussian process. If a stationary Gaussian process $\xi(t)$ has a zero mean and is mean-square differentiable, then

$$\mathbf{E}\,\nu(s, t] = \frac{t-s}{\pi} \left(\frac{b_2}{b_0}\right)^{1/2} \exp\left\{-\frac{c^2}{2b_0}\right\}$$

where

$$b_0 = B(0), \quad b_2 = -B''(0) = 2\lim_{t\downarrow 0}\frac{B(0)-B(t)}{t^2}, \quad B(t) = \mathbf{E}\,\xi(0)\,\xi(t).$$

A time interval between the up-crossing of a level c and successive down-crossing is called **an ejection**. The length of such an interval is called the ejection length. If a continuous stationary process $\xi(t)$ is ergodic and $\mathbf{E}\,\nu(0,\,1] < \infty$, then the number $\nu_+(s,\,t;\,u)$ of ejections contained in an interval $(s,\,t)$ of lengths less than u is asymptotically (almost surely) equivalent to $(t-s)\,\mathbf{E}\,\nu_+(0,\,1;\,u)$ and hence

$$\frac{\nu_+(s,\,t;\,u)}{\nu_+(s,\,t]} \underset{t-s\to\infty}{\longrightarrow} \frac{\mathbf{E}\,\nu_+(0,\,1;\,u)}{\mathbf{E}\,\nu_+(0,\,1]}$$

almost surely if $\mathbf{E}\,\nu(0,\,1] > 0$. The ratio in a right-hand side, considered as a function of u, is a distribution function of a nonnegative random variable which can be considered as a length of an ejection chosen at random. For this function

$$F^+(u) = \frac{\mathbf{E}\,\nu_+(0,\,1;\,u)}{\mathbf{E}\,\nu_+(0,\,1]}$$

more explicit expressions (in some sense) can be obtained. For example, if we suppose in addition, $\mathbf{P}\,\{\xi(0) = c\} = 0$, then

$$F^+(t) = 1 + \frac{2}{a}\,D^+\,q^+(t) \tag{14.11}$$

where $a = \mathbf{E}\,\nu(0,\,1]$, $q^+(t) = \mathbf{P}\,\{\xi(0) > c,\ \nu(0,\,t] = 0\}$ and D^+ is the right operator of differentiation, i.e.,

$$D^+\,g(t) = \lim_{s\downarrow 0}\frac{1}{s}\,(g(t+s) - g(t)).$$

Moreover, under the stated hypotheses

$$F^+(u) = \lim_{s\uparrow 0}\mathbf{P}\,\{\nu(0,\,u] \ge 1 \mid \nu_+(s,\,0] \ge 1\}. \tag{14.12}$$

Hence $F^+(u)$ is the distribution function of the length of an ejection starting at $t = t_0$. The equalities (14.11) and (14.12) hold independently of the ergodicity condition for the process $\xi(t)$. The mean length of an ejection chosen at random is defined by the expression

$$\int_0^t t\,dF^+(t) = \frac{2}{a}\,(\mathbf{P}\,\{\xi(0) > c\} - q^+(\infty)).$$

If the process $\xi(t)$ is ergodic, then $q^+(\infty) = 0$ and hence

$$\int_0^t t\,dF^+(t) = \frac{2}{a}\,\mathbf{P}\{\xi(0) > c\}.$$

Consider a more general case, when we study a length of the interval between the up-crossing of a level c and the next kth crossing of this level. The distribution function $F_k^+(t)$ of this length is defined as the ratio of the mean number of intervals of the mentioned kind contained in the interval $(0, 1)$ with lengths less than t to the mean number of exits within the interval $(0, 1)$. If $\mathbf{P}\{\xi(0) = c\} = 0$ and $0 < a = \mathbf{E}\,\nu_1 < \infty$, then

$$F_k^+(t) = \lim_{u\uparrow 0} \mathbf{P}\{\nu(0, t] \geq k \mid \nu_+(u, 0] \geq 1\},$$

$$F_{2k-1}^+(t) = 1 + \frac{2}{a}\,D^+ q_k^+(t), \quad F_{2k}^+(t) = 1 + \frac{2}{a}\,D^+ p_k^+(t), \quad k = 1, 2, \ldots,$$

where

$$q_k^+(t) = \sum_{j=0}^{k-1} \mathbf{P}\{\xi(0) > c,\ \nu(0, t]\} + \sum_{j=1}^{k} (k-j)\,\mathbf{P}\{j - 1 \leq \nu(0, t] \leq j\},$$

$$p_k^+(t) = \sum_{j=0}^{k-1} (k-j)\,\mathbf{P}\{\nu_+(0, t] = j\}.$$

The analytical distinction of $F_{2k}^+(t)$ and $F_{2k-1}^+(t)$ can be explained as follows: $F_{2k}^+(t)$ is the distribution function of the time between an arbitrary chosen exit and the successive kth exit, whereas $F_{2k-1}^+(t)$ is the distribution function of the time between an arbitrary chosen exit and the successive kth input.

Moments of the distributions $F_{2k-1}^+(t)$ and $F_{2k}^+(t)$ can be easily expressed via the functions $p_k^+(t)$ and $q_k^+(t)$. For example,

$$\int_0^\infty t\,dF_{2k-1}^+(t) = \frac{2}{a}\,((k-1) + \mathbf{P}\{\xi(0) > c\} -$$

$$-(k-1)\,q_2^+(\infty) + (k-2)\,q_1^+(\infty)),$$

$$\int_0^\infty t\,dF_{2k}^+(t) = \frac{2k}{a}\,(1 - p_1^+(\infty)).$$

If moreover the process $\xi(t)$ is ergodic, then

$$q_k^+(\infty) = p_k^+(\infty) = 0$$

and $F_k^+(u)$ equals with probability 1 the limit (as $t-s\to\infty$) of the ratio of the number of intervals between an exit and the kth consequent crossing of a length less than u, contained in the interval $(s,\,t)$, to the total number of exits in $(s,\,t)$. Expression for the first order moments of the functions $F_k^+(t)$ is of a simple form in the ergodic case. Indeed,

$$\int_0^\infty t\,dF_{2k-1}^+(t) = \frac{2}{a}\left((k-1)+\mathbf{P}\{\xi(0)>c\}\right), \qquad \int_0^\infty t\,dF_{2k}^+(t) = \frac{2k}{a}.$$

The changing of the sign of the process $\xi(t)$ converts the input below a level c to the exit over the level $-c$ and vice versa. This fact enables the reduction of formulae for $F_k^-(t)$ to appropriate formulae for $F_k^+(t)$. Thus, if a stationary process $\{\xi_t,\,t\in\mathbb{R}\}$ has continuous path functions and $\mathbf{P}\{\xi(0)=c\}=0,\ 0<a=\mathbf{E}\,\nu(0,1]<\infty$, then

$$F_k^-(t) = \lim_{s\uparrow 0}\mathbf{P}\{\nu(0,\,t]\ge k \mid \nu(s,\,0]\ge 1\},$$

$$F_{2k-1}^-(t) = 1+\frac{2}{a}\,D^+\,q_k^-(t),$$

$$F_{2k}^-(t) = 1+\frac{2}{a}\,D^+\,p_k^-(t), \quad k=1,2,\ldots,$$

where

$$q_k^-(t) = \sum_{j=0}^{k-1}\mathbf{P}\{\xi(0)<c,\ \nu(0,\,t]=2j\}+\sum_{j=1}^{k}(k-j)\,\mathbf{P}\{j-1\le\nu(0,\,t]\le j\},$$

$$p_k^-(t) = \sum_{j=0}^{k-1}(k-j)\,\mathbf{P}\{\nu_-(0,\,t]=j\}.$$

Let $G_1^-(t)$ be the distribution function of the length of a fall under a level c covering the point 0. Let also $G_2^-(t)$ be the distribution function of the interval between two successive inputs covering the point 0. Then

$$G_1^-(t) = \mathbf{P}\{\xi(0)<c\} - q_1^-(\infty) - \frac{2}{a}\int_t^\infty s\,dF_1^-(s),$$

$$G_2^-(t) = 1 - p_1^-(\infty) - \frac{2}{a}\int_t^\infty s\,dF_2^-(s).$$

We have also

$$G_1^-(\infty) = \mathbf{P}\{\xi(0) < c,\ \nu(0,\infty) \geq 1\}, \quad G_2^-(\infty) = \mathbf{P}\{\nu_-(0,\infty) \geq 1\}.$$

If the process $\xi(t)$ is ergodic, then

$$G_1^-(u) = \mathbf{P}\{\xi(0) > c\} - \frac{2}{a}\int_u^\infty s\,dF_1^-(s),$$

$$G_2^-(u) = 1 - \frac{2}{a}\int_u^\infty s\,dF_2^-(s).$$

Similarly, the function

$$G_1^+(t) = \mathbf{P}\{\xi_0 > c\} - q_1^+(\infty) - \frac{2}{a}\int_t^\infty s\,dF_1^+(s)$$

is the distribution function of the length of the ejection covering the point 0, whereas the function

$$G_2^+(t) = 1 - p_2^+(\infty) - \frac{2}{a}\int_t^\infty s\,dF_2^+(s)$$

is the distribution function of the interval between two successive inputs covering the point 0.

Let $\xi(t)$ be a stationary Gaussian process, $\mathbf{E}\,\xi(t) = 0$, and

$$\mathbf{E}\,\xi(0)\,\xi(t) = b_0 - b_2\frac{t^2}{2} + O\left(\frac{t^2}{|\log t|^p}\right)$$

for $t \downarrow 0$ and some $p > 1$. If

$$\lim_{t\to\infty} \log t\,\mathbf{E}\,\xi(0)\,\xi(t) = 0, \quad \lim_{t\to\infty} \sqrt{\log t}\,\mathbf{E}\,\xi(0)\,\xi'(t) = 0,$$

then

$$\mathbf{P}\left\{\nu\left(\frac{s_1}{a}, \frac{t_1}{a}\right] = k_1, \ldots, \nu\left(\frac{s_n}{a}, \frac{t_n}{a}\right] = k_n\right\} \underset{c\to\infty}{\longrightarrow}$$

$$\xrightarrow[c\to\infty]{} \frac{(t_1 - s_1)^{k_1}}{k_1!}\, e^{s_1 - t_1} \ldots \frac{(t_n - s_n)^{k_n}}{k_n!}\, e^{s_n - t_n}$$

where $n = 1, 2, \ldots,\ \ s_1 < t_1 < \cdots < s_n < t_n,\ \ k_1, \ldots, k_n = 0, 1, 2, \ldots,$

$$a = \frac{1}{\pi}\left(\frac{b_2}{b_0}\right)^{1/2} \exp\left\{-\frac{c^2}{2b_0}\right\}$$

and $\nu(s,\, t]$ equals the number of level crossings during the time interval $(s,\, t]$. In other words, finite-dimensional distributions are asymptotically Poisson in an appropriate time scale.

If space functions of a stationary process $\xi(t)$ are continuously differentiable and $\xi'(t)$ crosses the level 0 finitely many times in a bounded time interval, then crossing points are local extremums of the process $\xi(t)$. Exits of $\xi'(t)$ from the zero level to above are minimums of $\xi(t)$, whereas inputs are its maximums. Assume that the process $\xi(t)$ is ergodic. Then the number $\nu'_t(s,\, t;\, u)$ of exits of $\xi'(t)$ above the level 0 in a time interval $(s,\, t]$, the corresponding values of the process $\xi(t)$ being less than u, is asymptotically equivalent to $(t - s)\,\mathbf{E}\,\nu'_+(0,\, 1;\, u)$ as $(t - s) \to \infty$. If moreover $0 < \mathbf{E}\,\nu'_+(0,\, 1;\, \infty) < \infty$, then

$$\lim_{t-s\to\infty} \frac{\nu'_+(s,\, t;\, u)}{\nu'_+(s,\, t;\, \infty)} = \frac{\mathbf{E}\,\nu'_+(0,\, 1;\, u)}{\mathbf{E}\,\nu'_+(0,\, 1;\, \infty)}$$

with probability 1. The ratio in a right-hand side is a distribution function (in u) of a random variable which can be treated as the height of an arbitrarily chosen local minimum of the process $\xi(t)$. This definition of the distribution function

$$G_+(u) = \frac{\mathbf{E}\,\nu'_+(0,\, 1;\, u)}{\mathbf{E}\,\nu'_+(0,\, 1;\, \infty)}, \quad u \in \mathbb{R},$$

is valid independently of the ergodicity of the process $\xi(t)$ even if the mean in the denominator is finite and positive. If $\mathbf{P}\{\xi'(0) = 0\} = \mathbf{P}\{\xi(0) = = u\} = 0$ for any u, then

$$G_+(u) = \lim_{s\uparrow 0} \mathbf{P}\{\xi(0) \le u \mid \nu'_+(s,\, 0;\, \infty) \ge 1\},$$

so that under the stated conditions $G_+(u)$ can be considered as the distribution function of the ordinate of a local minimum of the process $\xi(t)$ with a fixed abscissa in a point t_0. The stated conditions (may be with the exception of the ergodicity) hold for a Gaussian stationary process $\xi(t)$, if

$$\mathbf{E}\,\xi(0)\,\xi(t) = b_0 - b_2\frac{t^2}{2} + b_4\frac{t^4}{24} + o(t^4) \quad \text{as } t \to 0.$$

If moreover $\mathbf{E}\,\xi(0) = 0$, then

$$G_+(u) = \Phi\left(\sqrt{\frac{b_4}{\Delta}}\,u\right) - \sqrt{\frac{2\pi b_2}{b_4 b_0}}\,\varphi(u)\,\Phi\left(\frac{b_2 u}{\sqrt{\Delta}}\right),$$

where

$$\Phi(u) = \int_{-\infty}^{u} \varphi(t)\,dt = \frac{1}{\sqrt{2\pi}} \int_{-\infty}^{u} e^{-t^2/2}\,dt.$$

This formula defines the distribution function of the height of an arbitrarily chosen local maximum or the ordinate of the local maximum with a fixed abscissa. Necessary formulae for a local minimum can be obtained from the ones for a local maximum; $\xi(t)$ should be changed to $-\xi(t)$.

14.5 Covariance Theory

If a complex-valued stationary process $\xi(t)$, $t \in \mathbb{R}$, possesses finite second-order moments, then its mean is finite and its covariance function depends only on the difference of variables, i.e., $\mathbf{E}\,\xi(t) = \mathbf{E}\,\xi(0) = a$, $\mathbf{E}\,\xi(t)\,\overline{\xi}(s) = \mathbf{E}\,\xi(t-s)\,\overline{\xi}(0) = |a|^2 + B(t-s)$; the bar hereinafter denotes the complex conjugate value. The stated properties, independent of the strict stationarity, are characteristic for **stationary in the broad sense processes**.

The most important property of the stationary in a broad sense random process $\xi(t)$ consists of an integral representation of its covariance function $B(t)$ via the Fourier integral[4,12,13].

Theorem 14.23. *A continuous function $B(t)$ is the covariance function of a mean-square continuous stationary in a broad sense random process iff*

$$B(t) = \int_{-\infty}^{\infty} e^{itx}\,dF(x), \quad i^2 = -1, \tag{14.13}$$

where the $F(x)$ is a right-continuous nondecreasing function and $0 = F(-\infty) < F(\infty) < \infty$.

The function $F(x)$ in (14.13) is defined uniquely and is called **the spectral function** of the stationary in a broad sense process. Its total variation equals the variance of the process $\xi(t)$, i.e.,

$$\mathbf{E}\,|\xi(0)|^2 - |\mathbf{E}\,\xi(0)|^2 = B(0) = F(\infty) - F(-\infty).$$

Similarly, a sequence of complex-valued random variables ξ_n, $n = 0, \pm1, \pm2, \ldots$, is called **stationary in the broad sense**, if

$$\mathbf{E}\,|\xi_n|^2 < \infty, \quad \mathbf{E}\,\xi_n = \mathbf{E}\,\xi_0, \quad \mathbf{E}\,\xi_k\,\overline{\xi}_n = \mathbf{E}\,\xi_{k-n}\,\overline{\xi}_0, \quad n, k = 0, \pm1, \pm2, \ldots.$$

If a sequence ξ_n, $n, k = 0, \pm 1, \pm 2, \dots$, is stationary in a broad sense, then

$$\mathbf{E}\,\xi_n\,\overline{\xi}_0 = \mathbf{E}\,|\xi_0|^2 + \int_{-\pi}^{\pi} e^{inx}\,dF(x)$$

where the spectral function $F(x)$ is left-continuous, nondecreasing and $F(-\pi) = 0 < F(\pi) < \infty$.

The Theorem 14.23 states the isomorphism of two Hilbert spaces associated with the process $\xi(t)$: the space $L_2(F)$ of the squared-summable (over the function $F(x)$) functions and the mean-square closure Ξ of linear combinations of the values $\xi(t)$. More definitely speaking, the space $L_2(F)$ consists of such complex-valued functions $f(x)$ that $\int_{-\infty}^{\infty} |f(x)|^2\,dF(x) < \infty$. The scalar product (f, g) of functions $f(x)$ and $g(x)$ of $L_2(F)$ is defined as $\int_{-\infty}^{\infty} f(x)\,\overline{g}(x)\,dF(x)$. If $f(x) = g(x)$ almost everywhere with respect to the measure dF, then f and g are identified.

The Hilbert space Ξ consists of finite linear combinations

$$\sum_{i=1}^{n} c_i\,(\xi(t_i) - a)$$

where $a = \mathbf{E}\,\xi(t)$ and of their mean-square limits. The scalar product of random variables ξ and η of Ξ is defined as $(\xi, \eta) = \mathbf{E}\,\xi\,\overline{\eta}$. Each two random variables ξ and η coinciding almost surely are, as usually, identified, so that they are the same point of the space Ξ. Every random variable of the space has the zero mean value.

The isomorphism of both spaces is defined in the following way. A random variable $\xi(t) - a$ maps into the function e^{itx}; a linear combination $\sum_i c_i\,(\xi(t_i) - a)$ maps into the linear combination $\sum_k c_k\,e^{it_k x}$. Finally, mean-square limits for random variables of Ξ map into the corresponding mean-square limits for functions of $L_2(F)$. In terms of these Hilbert spaces, the Theorem 14.23 states that the above defined mapping of two spaces conserves the scalar product.

The main result of the covariance theory consists of the spectral expansion for the stationary in a broad sense and mean-square continuous random process $\xi(t)$. This expansion is a consequence of the isomorphism of the spaces $L_2(F)$ and Ξ.

Let a random variable $\zeta(t)$ of Ξ map into a function of $L_2(F)$ defined as 1 for $x \le t$ and as 0 for $x > t$. The random process $\zeta(t)$, $t \in \mathbb{R}$, is a process with orthogonal increments, i.e., $\mathbf{E}\,\zeta(t) = 0$ and $\mathbf{E}\,(\zeta(t)-$

$-\zeta(s))\overline{(\zeta(v)-\zeta(u))} = 0$ for $s \le t \le u \le v$. Moreover $\mathbf{E}|\zeta(t)-\zeta(s)|^2 =$ $= F(t) - F(s)$ and for every continuous function f of $L_2(F)$ there exists a mean-square limit of integral sums

$$\sum_{k=0}^{n-1} f(t_k)\,(\zeta(t_{k+1}) - \zeta(t_k))$$

as $n \to \infty$, $t_0 < t_1 < \cdots < t_n$, $\max\limits_{1\le k\le n}(t_k - t_{k-1}) \to 0$. This limit is denoted by $\int\limits_{-\infty}^{\infty} f(t)\,d\zeta(t)$. We have

$$\mathbf{E}\int\limits_{-\infty}^{\infty} f(t)\,d\zeta(t) = 0, \quad \mathbf{E}\left|\int\limits_{-\infty}^{\infty} f(t)\,d\zeta(t)\right|^2 = \int\limits_{-\infty}^{\infty} |f(t)|^2\,dF(t).$$

Moreover, the isomorphism of the spaces $L_2(F)$ and Ξ maps a function f into the random variable $\int\limits_{-\infty}^{\infty} f(s)\,d\zeta(s)$. In particular, for $f(s) = e^{its}$ the following spectral expansion holds true:

$$\xi(t) = a + \int\limits_{-\infty}^{\infty} e^{its}\,d\zeta(s). \tag{14.14}$$

This formula is the most visual in the special case of a step function $F(x)$. Assume that the only increase points of $F(x)$ are s_k and $F(s_k+0)-$ $-F(s_k-0) = b_k$. Then

$$\xi(t) = a + \sum_k |\,\eta_k\,|\; e^{i\,(t\,s_k + \theta_k)}$$

where

$$\eta_k = \zeta(s_k^+) - \zeta(s_k^-) = |\eta_k|\,e^{i\theta_k}, \quad \mathbf{E}\,|\eta_k|^2 = b_k.$$

Therefore the process $\xi(t) - a$ is presented as a sum of harmonic oscillations of random phases and random uncorrelated amplitudes. In a general case the increment $F(t) - F(s)$ can be considered as the mean value of the sum of squares of amplitudes corresponding to frequencies between $c/(2\pi)$ and $t/(2\pi)$.

An important consequence of the spectral expansion (14.14) is **the Shannon-Kotelnikov formula** for a process with a bounded spectrum.

If all the increase points of the spectral function $F(x)$ are situated inside an interval $(-c, c)$, then

$$\xi(t) = \sum_{n=-\infty}^{\infty} \frac{\sin(ct - \pi n)}{ct - \pi n} \, \xi(\pi n/c)$$

holds true in the sense of the mean-square convergence. Therefore all the values $\xi(t)$ are well defined by their values at times $\pi n/c$, $n = 0, \pm 1, \pm 2, \ldots$.

The formula (14.14) makes it possible also to answer the question of **the ergodicity in a broad sense** of the process $\xi(t)$. This property means the existence and degeneracy of the mean-square limit of the time averages $\frac{1}{t}\int_0^t \xi(u)\,du$, the integral being considered as the mean-square limit of Riemann integral sums. The (14.14) immediately implies the mean-square equality

$$\frac{1}{t}\int_0^t \xi(u)\,du \underset{t\to\infty}{\longrightarrow} a + (\zeta(0+) - \zeta(0-)).$$

Therefore a stochastically continuous stationary in a broad sense process is ergodic in a broad sense iff its spectral function is continuous at point 0.

Similarly, for a stationary in a broad sense random sequence ξ_n, $n = 0, \pm 1, \pm 2, \ldots$, the spectral function $F(x)$ defines a random process $\zeta(t)$, $-\pi \le t \le \pi$, with orthogonal increments, i.e.,

$$\mathbf{E}\,\zeta(t) = 0, \ \ \mathbf{E}\,|\,\zeta(t) - \zeta(s)\,|^2 = F(t) - F(s) \ \ \text{for} \ -\pi \le s \le t \le \pi$$

and

$$\xi_n = \mathbf{E}\,\xi_0 + \int_{-\pi}^{\pi} e^{int}\,d\zeta(t), \quad n = 0, \pm 1, \pm 2, \ldots .$$

The mean-square limit of time averages $\frac{1}{n}\sum_{k=1}^{n} \xi_k$ equals $\mathbf{E}\,\xi_0 + (\zeta(0+) - \zeta(0-))$ almost surely. This value equals $\mathbf{E}\,\xi_0$ with probability 1 iff $F(0+) = F(0-)$.

14.6 References

1. Loève, M., *Probability Theory,* Van Nostrand, New York, 1963.
2. Neveu, J., *Mathematical Foundations of the Calculus of Probability,* Holden-Day, San Francisco, 1963.

3. Cramér, H., Leadbetter, M. R., *Stationary and Related Stochastic Processes,* Wiley, New York, 1967.
4. Rozanov, Yu. A., *Stationary Random Processes,* Holden-Day, San Francisco, 1967.
5. Franken, P., Streller, A., Stationary generalized regenerative processes, *Teor. Probab. Appl.,* 24, 79, 1979.
6. Shurenkov, V. M., *Ergodic Theorems and Related Questions in the Theory of Random Processes,* Naukova Dumka, Kiev, 1981 (in Russian).
7. Revuz, D., *Markov Chains,* Elsevier, Amsterdam, 1975.
8. Ezhov, I. I., Shurenkov, V. M., Ergodic theorems connected with Markov property of random processes, *Teor. Probab. Appl.,* 21, 620, 1976.
9. Shurenkov, V. M., Eleyko, Ya. I., Limit distributions of time means in a semi-Markov process with a finite number of states, *Ukr. Mat. Zhurn.,* 31, 598, 1979 (in Russian).
10. Atreya, K. B., McDonald, D., Ney, P. E., Limit theorems for semi-Markov processes and renewal theory for Markov chains, *Ann. Probab.,* 6, 788, 1978.
11. Atreya, K. B., Ney, P. E., Limit theorems for semi-Markov processes, *Bull. Austral. Math. Soc.,* 19, 283, 1978.
12. Kovalenko, I. N., Sarmanov, O. V., *A Short Course of the Theory of Random Processes,* Vyshcha Shkola, Kiev, 1978 (in Russian).
13. Koroliouk, V. S., Portenko, N. I., Skorokhod, A. V., Tourbine, A. F., *Aide Mémoire de Théorie des Probabilités et de Statistique Mathématique,* Mir, Moscow, 1983.

Chapter 15

Markov Processes

15.1 Transition Functions

The markovian property consists of the independence of the future of the past given the current state. Let a process $\xi(t)$ be defined in an interval T. Then the "future" with respect to a moment s is the σ-algebra $\mathcal{M}^s$ generated by the process values at $t > s$ whereas "the past" is the σ-algebra $\mathcal{M}_s$ generated by values of the process $\xi(t)$ for $t < s$. Hence the random process $\xi(t)$, $t \in T$, taking values in a measurable phase-space $(E, \mathcal{A})$, is called a Markov process if

$$\mathbf{P}\{\Gamma_1 \Gamma_2 \,|\, \xi(s)\} = \mathbf{P}\{\Gamma_1 \,|\, \xi(s)\}\, \mathbf{P}\{\Gamma_2 \,|\, \xi(s)\}$$

almost surely for every $s \in T$, $\Gamma_1 \in \mathcal{M}_s$, $\Gamma_2 \in \mathcal{M}^s$. This definition is invariant to the time-reversion and equivalent to each of the two following ones:

$$\mathbf{P}\{\xi(t) \in A \,|\, \mathcal{M}_s\} = \mathbf{P}\{\xi(t) \in A \,|\, \xi(s)\}$$

almost surely, and

$$\mathbf{P}\{\xi(s) \in A \,|\, \mathcal{M}^t\} = \mathbf{P}\{\xi(s) \in A \,|\, \xi(t)\}$$

almost surely where $s, t \in T$, $s < t$, $A \in \mathcal{A}$.

The most important characteristic of a Markov process is its **transition function**[1–9] $P(s, x;\, t, A)$ defining the conditional distribution of $\xi(t)$ with respect to $\xi(s)$, $s < t$. More precisely speaking, an x-depending probability measure $P(s, x;\, t, A)$ is called the transition function of a Markov process $\xi(t)$, $t \in T$, if it is $\mathcal{A}$-measurable in x and

$$\mathbf{P}\{\xi(t) \in A \,|\, \xi(s)\} = P(s, \xi(s);\, t, A)$$

almost surely for every $s < t$, $s, t \in T$, $A \in \mathcal{A}$. In the case of the time reversion this definition of the transition function induced the definition of **the co-transition function** $P^-(t, x; s, A)$, i.e., an $\mathcal{A}$-measurable in x probability measure in $A \in \mathcal{A}$ satisfying the equation

$$\mathbf{P}\{\xi(s) \in A \mid \xi(t)\} = P^-(t, \xi(t); s, A)$$

almost surely for every $s < t$, $s, t \in T$, $A \in \mathcal{A}$.

Not every Markov process possesses the transition function, but if such a function exists, then the following **Kolmogorov-Chapman equation** holds true:

$$\int_E P(s, x; t, dy)\, P(t, y; u, A) = P(s, x; u, A)$$

for every $s < t < u$, $s, t, u \in T$, $A \in \mathcal{A}$ and for almost every $x \in E$ with respect to the distribution of $\xi(s)$.

A substantial theory of Markov processes is possible only under the condition that the Kolmogorov-Chapman equation is satisfied identically, i.e., for every $x \in E$.

A measurable space is called **a Borel space** if it is isomorphic to a Borel subset of a complete metric separable space with the σ-algebra of Borel subsets. Every measurable Borel space is isomorphic either to, at most, a countable set with the σ-algebra of all its subsets or to the segment $[0, 1]$ of a real line with the σ-algebra of Borel sets. In particular, all non-countable Borel spaces are isomorphic.

Theorem 15.1. *A Markov process in a Borel phase-space possesses the transition function satisfying the Kolmogorov-Chapman equation identically.*

Given the transition function $P(s, x; t, A)$ of a Markov process $\xi(t)$ all the multidimensional conditional distributions can be obtained as follows:

$$\mathbf{P}\{\xi(t_1) \in A_1, \ldots, \xi(t_n) \in A_n \mid \xi(s)\} = \int_{A_1} P(s, \xi(s); t_1, dx_1) \times$$

$$\times \int_{A_2} P(t_1, x_1; t_2, dx_2) \times \cdots \times \int_{A_{n-1}} P(t_{n-2}, x_{n-2}; t_{n-1}, dx_{n-1}) \times$$

$$\times P(t_{n-1}, x_{n-1}; t_n, A_n)$$

for $s, t_1, \ldots, t_n \in T,\ s < t_1 < \cdots < t_n,\ A_1, \ldots, A_n \in \mathcal{A},\ n = 2, 3, \ldots$.

If the domain T has a minimal element, then the initial distribution and the transition function define all its multidimensional distributions. If the time interval T has no minimal point, then one-dimensional distributions $Q_t(A) = \mathbf{P}\{\xi(t) \in A\}$ should be given for the equation

$$\int_E Q_s(dx)\, P(s, x;\ t, A) = Q_t(A), \quad s < t, \ s, t \in T, \ A \in \mathcal{A},$$

to be satisfied.

A set of probability measures $Q_t(A)$ over the σ-algebra $\mathcal{A}$ satisfying this equation is called **the input law**. If the phase-space $(E, \mathcal{A})$ is Borel, then by Kolmogorov Theorem any input law defines a Markov process possessing a given transition function $P(s, x;\ t, A)$.

The concept of the transition function in a phase-space $(E, \mathcal{A})$ is of meaning even without a reference to Markov processes. A function $P(s, x;\ t, A)$ defined for every $s < t$ from a real interval T is called **a transition function** if

(i) it is a measure in $A \in \mathcal{A}$, s, x and t being fixed,

(ii) it is an $\mathcal{A}$-measurable in $x \in E$ function,

(iii) $P(s, x;\ s, E) \leq 1$,

(iv) $\displaystyle\int_E P(s, x;\ t, dy)\, P(t, y;\ u, A) = P(s, x;\ u, A)$ for every $s < t$,

$s, t \in T,\ x \in E,\ A \in \mathcal{A}$.

If moreover $P(s, x;\ t, E) = 1$, then it is called **a probability (stochastic) transition function**.

In a case where a time interval T does not contain its minimal element, the existence of a Markov process with a given probability transition function is nontrivial. However for Borel phase-spaces an affirmative answer on the question is given by the following theorem.

Theorem 15.2. *Every transition function in the uncountable Borel phase-space can be expanded to a transition function in the same phase-space defined over the entire real line* $\mathbb{R}$.

Since $\mathbb{R}$ is homomorphic to an open interval, say (0, 1), the Theorem 15.2 implies the existence of a Markov process defined for every $t \in \mathbb{R}$ and possessing a given probability transition function.

Let a Markov process $\xi(t)$, $t \in T$, with a transition function $P(s, x; t, A)$ in a Borel phase-space $(E, \mathcal{A})$ be realized (by Kolmogorov Theorem) as a probability measure $\mathbf{P}$ over the set of arbitrary functions $\omega : T \longrightarrow E$ with the σ-algebra $\mathcal{M}$ generated by sets $\{\omega : \omega(t) \in A\}$ for every $t \in T$, $A \in \mathcal{A}$, $\xi(t; \omega) = \omega(t)$. Then a Kolmogorov Theorem makes it possible to define a family of probability measures $\mathbf{P}_{s,x}$, $x \in E$, for any σ-algebra $\mathcal{M}^s$ in such a way that

$$\mathbf{P}_{s,x}\{\xi(u) \in A \,|\, \mathcal{M}_t^s\} = \mathbf{P}_{t,\xi(t)}\{\xi(u) \in A\} = P(t, \xi(t); u, A) \ \text{(a.s.)} \quad (15.1)$$

for $s < t < u$, $x \in E$, $A \in \mathcal{A}$, $\mathcal{M}_t^s = \mathcal{M}^s \cap \mathcal{M}_t$.

Often while defining a Markov process often the existence of regular conditional probabilities $\mathbf{P}_{s,x}$ over the σ-algebra $\mathcal{M}^s$ given $\xi(s) = x$ is postulated, and so the equality (15.1) is considered a Markov property.

Many problems of Markov process theory require the consideration of Markov processes in a random time interval. Such processes are reduced to Markov processes over $\mathbb{R}$ via expanding the phase-space. Indeed, one can join two outside elements $\{a\}$ and $\{b\}$ to the phase-space and define a new process

$$\xi'(t) = \begin{cases} a & \text{for } t \le \alpha \\ \xi(t) & \text{for } \alpha < t < \beta \\ b & \text{for } t \ge \beta. \end{cases}$$

A process $\xi(t)$ in a random interval (α, β) is called Markov, if the process $\xi'(t)$ is Markov.

Its transition function $P'(s, x; t, A)$ being considered for $x \in E$, $A \in \mathcal{A}$ is still a transition function but not necessarily a probability one. It is called a transition function of the process $\xi(t)$, $t \in (\alpha, \beta)$. The Markov property of the process $\xi'(t)$ in a Borel space is equivalent to each of the following equalities

$$\left.\begin{aligned} \mathbf{P}\{\xi(t) \in A,\ t < \beta \,|\, \mathcal{M}_s\} &= P(s, \xi(s); t, A) \quad \text{(a.s. as } \alpha < s) \\ \mathbf{P}\{\xi(s) \in A,\ s > \alpha \,|\, \mathcal{M}^t\} &= \hat{P}(t, \xi(t); s, A) \quad \text{(a.s. as } t < \beta) \end{aligned}\right\} \quad (15.2)$$

where $s < t$, $A \in \mathcal{A}$. Both the transition function $P(s, x; t, A)$ and the co-transition function $P^-(t, x; s, A)$ are the reductions of those functions of the process $\xi'(t)$. The σ-algebras of the future $\mathcal{M}^t$ and of the past $\mathcal{M}_s$ are generated by events $\{\xi(u) \in A,\ \alpha < u < \beta\}$ for $u > t$, $A \in \mathcal{A}$ and $u < s$, $A \in \mathcal{A}$ respectively. They coincide with those of the process $\xi'(t)$.

One-dimensional distributions

$$Q_s(A) = \mathbf{P}\{\alpha < s < \beta,\ \xi(s) \in A\} \tag{15.3}$$

of a Markov process $\xi(s),\ s \in (\alpha, \beta)$, satisfy the relation

$$\int_E Q_s(dx)\, P(s, x; t, A) \uparrow Q_t(A) \quad \text{for } s \uparrow t,\ A \in \mathcal{A}. \tag{15.4}$$

If a σ-finite measure $Q_t(A)$ over $\mathcal{A}$ depends on a real parameter t in such a way that (15.4) is satisfied, then it is called **excessive** with respect to the given transition function.

Therefore, for a Borel phase-space, according to the relations (15.2) and (15.3), a Markov process in a random time interval has the transition function and the excessive measure with an entire mass which does not exceed a unit.

The converse statement is not true, so that not every excessive measure $Q_t(A)$ with the mass ≤ 1 can be associated with a Markov process in a random time interval (α, β) over the Borel space $(E, \mathcal{A})$, which would have the given transition function $P(s, x; t, A)$. Such a Markov process exists iff

$$\sup_{n \geq 1} \sup_{t_0 < t_1 < \cdots < t_n} \left\{ q(t_0) + \sum_{k=1}^{n} (q(t_k) - q(t_{k-1}, t_k)) \right\} = 1$$

where

$$q(t) = Q_t(E), \quad q(s, t) = \int_E Q_s(dx)\, P(s, x; t, E), \quad s < t.$$

If this condition is satisfied and hence a Markov process $\xi(t),\ t \in (\alpha, \beta)$, exists such that it generated both the excessive measure $Q_t(A)$ and the transition function $P(s, x; t, A)$, then

$$q(s, t) = \mathbf{P}\{\alpha < s < t < \beta\}, \quad q(t) = \mathbf{P}\{\alpha < t < \beta\},$$

$$\mathbf{P}\{\alpha \geq s\} = \sup_{n \geq 1} \sup_{s = t_0 < t_1 < \cdots < t_n} \sum_{k=1}^{n} (q(t_k) - q(t_{k-1}, t_k)),$$

$$\mathbf{P}\{\beta \leq s\} = \sup_{n \geq 1} \sup_{t_0 < t_1 < \cdots < t_n = s} \sum_{k=1}^{n} (q(t_{k-1}) - q(t_{k-1}, t_k)).$$

In particular, the measure $Q_t(A)$ is **invariant**, i.e.,

$$\int_E Q_s(dx)\, P(s,\, x;\; t,\, A) \;=\; Q_t(A)$$

iff $\mathbf{P}\{\alpha = -\infty\} = 1;\;\; P(s,\, x;\; t,\, E) = 1$ iff $\mathbf{P}\{\beta = +\infty\} = 1.$

A dual concept to excessive measure is one of **the excessive function**. A nonnegative $\mathcal{A}$-measurable, a real parameter t-dependent function $h_t(x)$ for which the value $+\infty$ is not excluded, is said to be **excessive** with respect to a transition function $P(s,\, x;\; t,\, A)$ if

$$\int_E P(s,\, x;\; t,\, dy)\, h_t(y) \uparrow h_s(x) \quad \text{for } t \downarrow s,\;\; x \in E.$$

In particular, this means that

$$\int_E P(s,\, x;\; t,\, dy)\, h_t(y) \le h_s(x) \quad \text{for } \; s < t.$$

Each $\mathcal{A}$-measurable nonnegative function $h_t(x)$ satisfying the cited inequality is associated with the following transformation of the transition function $P(s,\, x;\; t,\, A)$ to another transition function

$$P_h(s,x;t,A) = \begin{cases} \int_A P(s,x;t,dy)\, \frac{h_t(y)}{h_s(x)} & \text{for } 0 < h_s(x) < \infty \\ 0 & \text{otherwise.} \end{cases}$$

If the measure $Q_t(A)$ is excessive with respect to the transition function $P(s,\, x;\; t,\, A)$, then the measure

$$Q_t^h(A) \;=\; \int_A Q_t(dx)\, h_t(x)$$

(in the case that it is σ-finite) is excessive with respect to the transformed transition function $P_h(s,\, x;\; t,\, A)$. Therefore in the case of a Borel phase-space $(E,\, \mathcal{A})$ the equality

$$< Q,\, h >= \sup_{n\ge 1}\; \sup_{t_0<t_1<\cdots<t_n} \left\{ \int_E Q_{t_0}(dx)\, h_{t_0}(x) + \sum_{k=1}^{n} \left(\int_E Q_{t_k}(dx)\, h_{t_k}(x) - \right.\right.$$

$$\left.\left. - \int_E Q_{t_{k-1}}(dx) \int_E P(t_{k-1},\, x;\; t_k,\, dy)\, h_{t_k}(y) \right) \right\} \;=\; 1$$

is necessary and sufficient for the existence of a Markov process such that

$$\mathbf{P}\{\xi(s) \in A,\ \xi(t) \in B,\ \alpha < s < t < \beta\} = \int_A Q_s(dx) \int_B P(s,\ x;\ t,\ dy)\, h_t(y).$$

Independent of the condition $< Q,\, h >= 1$, assuming only the finiteness of $h_t(x)$ almost everywhere according to the measure $Q_t(dx)$ over a set Ω of all possible path functions ω mapping an ω-dependent interval $(\alpha(\omega),\, \beta(\omega))$ into E, it is possible to define a σ-finite measure $\mathbf{P}$ over the σ-algebra $\mathcal{M}$ generated by sets $\{\omega :\ \omega(t) \in A,\ \alpha(\omega) < t < \beta(\omega)\}$, $t \in \mathbb{R}$, $A \in \mathcal{A}$, such that

$$\mathbf{P}\{\Omega\} = < Q,\, h >,$$

$$\mathbf{P}\{\omega :\ \omega(t_1) \in A_1, \ldots, \omega(t_n) \in A_n,\ \alpha(\omega) < t_1 < \cdots < t_n < \beta(\omega)\} =$$

$$= \int_{A_1} Q_{t_1}(dx_1) \int_{A_2} P(t_1, x_1; t_2, dx_2) \ldots \int_{A_n} P(t_{n-1}, x_{n-1}; t_n, dx_n)\, h_{t_n}(x_n).$$

15.2 Infinitesimal Characteristics

One of the actual problems in the theory of Markov processes is the description of different classes of transition functions[1,4,9,10]. The possible approach to its solution is based on the representation of the function

$$g_s(x) = \int_E P(s,\ x;\ t,\ dy)\, f(y) \tag{15.5}$$

as the solution of **the Kolmogorov equation**

$$g_s(x) = f(x) + \int_s^t Q(du)\, g_u(x), \quad s < t. \tag{15.6}$$

The integral in the right-hand side of this equation is defined as the limit of integral sums

$$\sum_{k=1}^{n} Q(u_{k-1},\ u_k)\, g_{u_k}(x)$$

as the condition $D(s,\, t) = \{s = u_0 < \cdots < u_n = t,\ \max_{1 \le k \le n} (u_k - u_{k-1}) \to 0\}$ holds. A linear operator $Q(u,\, v)$ additively depends on u and v $(u < v)$, i.e., $Q(u,\, v) + Q(v,\, w) = Q(u,\, w)$ for $u < v < w$. It acts on the function f in such a way

$$Q(s,\, t)\, f(x) \;=\; \lim_{D(s,t)} \sum_{k=1}^{n} \left(\int_E P(u_{k-1},\, x;\; u_k,\, dy)\, f(y) - f(x) \right). \tag{15.7}$$

An equation (15.6) is obtained by a formal passage to the limit in the identity

$$g_s(x) \;=\; f(x) + \sum_{k=1}^{n} \left(\int_E P(u_{k-1},\, x;\; u_k,\, dy)\, g_{u_k}(y) - g_{h_k}(x) \right) \tag{15.8}$$

where $s = u_0 < \cdots < u_n = t$. A function $g_u(x)$ for $u < t$ is connected to $f(x)$ by the equation (15.5), $g_t(x) = f(x)$. A linear operator $Q(s,t)$ is defined by the behavior of the transition function $P(u, x; v, A)$ for $u - v \to 0$, and in this sense it can be considered as its **infinitesimal** characteristic.

The main difficulty of this approach is the proof that a set of bounded $\mathcal{A}$-measurable functions f, for which a limit (15.7) exists for any $x \in E$, $s < t$ and a right-hand side of the identity (15.8) converges to a right-hand side of the equation (15.6), is a dense subset in the set of all bounded $\mathcal{A}$-measurable functions in the topology of a bounded pointwise convergence. In particular, if

$$Q(s,\, t) \;=\; \int_s^t Q_u\, du,$$

then the integral equation (15.6) transforms into a differential equation

$$-\frac{\partial}{\partial s}\, g_s(x) \;=\; Q_s\, g_s(x), \quad x \in E, \tag{15.9}$$

which is to be solved for $s < t$ with an additional condition $y_t(x) = f(x)$. In this case a linear operator Q_s is determined on its domain by the differentiation:

$$Q_s\, f(x) = \lim_{t \downarrow s} \frac{1}{t-s} \int_E P(s,\, x;\; t,\, dy)\; f(y). \tag{15.10}$$

The outlined program of the construction of local characteristics (15.7) can be realized to a sufficient extent for uniformly continuous on s, t probability transition functions $P(s,\, x;\; t,\, A)$ in an arbitrary phase space $(E,\, \mathcal{A})$. A uniformly continuous dependency of the transition function

$P(s, x; t, A)$ on its time parameters s, t is equivalent to a formally weaker condition:

$$P(s, x; t, \{x\}) \underset{t-s \to 0}{\longrightarrow} 1 \tag{15.11}$$

for all $x \in E$. Under this hypothesis it is possible to construct (for any $x \in E$) a σ-finite measure $Q(x; du \times dx)$ on the product $\mathbb{R} \times E$ and a countable $\mathcal{A}$-measurable decomposition $D_1, D_2, \ldots$ of the space E such that

$$Q_x([s, t] \times E) \le q(x; s, t) =$$

$$= \sup_{s=u_0<\cdots<u_n=t} \sum_{k=1}^{n} (1 - P(u_{k-1}, x; u_k, \{x\})),$$

$$Q(x, \{t\} \times E) = Q(x, \mathbb{R} \times \{x\}) = 0, \tag{15.12}$$

$$Q(x, [s, t] \times A) = \lim_{n \to \infty} \sum_{k=s \cdot 2^n}^{t \cdot 2^n} P\left(k \cdot 2^{-n}, x; (k+1) \cdot 2^{-n}, A\right) < \infty \tag{15.13}$$

for $x \notin A$, $A \subset D_i$, $i = 1, 2, \ldots$, $s = l \cdot 2^{-r}$, $t = m \cdot 2^{-r}$, $r = 1, 2, \ldots$, $l, m = 0, \pm 1, \pm 2, \ldots$. If moreover

$$q(x; a, b) < \infty, \tag{15.14}$$

$$Q(x, [a, b] \times E) = q(x; a, b), \tag{15.15}$$

then

$$\lim_{s=u_0<\cdots<u_n=t} \sum_{k=1}^{n} (P(u_{k-1}, x; u_k, A) - I_A(x)) =$$

$$= Q(x, [s, t] \times A) - q(x; s, t)\, I_A(x),$$

$$P(s, x; t, A) = I_A(x) + \int_s^t \int_E Q(x, du \times dy) \times$$

$$\times (P(u, y; t, A) - P(u, x; t, A)) \tag{15.16}$$

for all $s < t$ out of the interval $[a, b]$.

In the case of a Borel phase space $(E, \mathcal{A})$, conditions (15.14) and (15.15) admit a probability interpretation. Under the condition (15.14) it is possible to construct a Markov process $\xi(t)$, $t \in [a, b]$, having a transition function $P(s, x; t, A)$ and with a positive sojourn time in the state x: a conditional probability that a process that does not leave the state x to the moment t, provided that $\xi(s) = x$, is equal to $\exp\{-q(x; s, t)\}$. Condition (15.15) makes it possible to evaluate the process's state at the exit moment of the state x. It follows that from (15.15) the equality

$$Q(x;\ [s,\ t] \times A) \;=\; \int_s^t q_x(du)\, p_u(x,\ A)$$

where a density $p_u(x,\ A)$ can be chosen as **a probability measure** on A for any $x \in E$, $u \in [a,\ b]$, which can be considered as a conditional distribution of $\xi(u)$ provided that the process has left the state x at the moment u.

For any finite measure $Q(x,\ du \times dy)$ satisfying the condition (15.12) and $\mathcal{A}$-measurable depending on $x \in E$, the Kolmogorov equation (15.16) has **a minimal solution** $P(s,\ x;\ t,\ A)$ being uniformly continuous on $s,\ t$, but it is not necessarily a probability transition function. Any other non-negative solution is a majorant for a minimal solution. If a minimal solution is a probability transition function, then it is a unique solution. The sufficient condition is the uniform boundedness of a full mass of the measure $Q(x,\ du \times dy)$. For example, this is a case when the condition (15.11) holds uniformly on $x \in E$. Then the conditions (15.14) and (15.15) are also satisfied.

As an other example of the situation, when it is possible to construct infinitesimal characteristics, we can name **homogeneous transition functions**. A transition function $P(s,\ x;\ t,\ A)$ in a phase space $(E,\ \mathcal{A})$ is called homogeneous, if $P(s,\ x;\ t,\ A) = P(0,\ x;\ t-s,\ A) = P(x,\ t-s,\ A)$ for any $x \in E$, $A \in \mathcal{A}$. If the transition function is homogeneous, then the necessary condition is

$$Q(s,\ t)\, f(x) \;=\; (t-s)\, Qf(x)$$

where

$$Qf(x) \;=\; Q(0,\ 1)\, f(x) \;=\; \lim_{t \downarrow 0} \frac{1}{t} \left(\int_E P(x,\ t,\ dy)\, f(y) - f(x) \right).$$

In this case as the domain of the operator Q it is convenient to take a set $\mathfrak{D}$ of bounded $\mathcal{A}$-measurable functions f such that a limit in the last relation exists for all $x \in E$ and

$$\sup_{x \in E} \frac{1}{t} \left| \int_E P(x,\ t,\ dy)\, f(y) - f(x) \right| < \infty.$$

With a domain $\mathfrak{D}$ chosen ion such a way an operator Q is called **an infinitesimal operator** of the homogeneous transition function. A closure of the set $\mathfrak{D}$ in the topology of the bounded pointwise convergence coincides with the set $\mathfrak{G}$ of bounded $\mathcal{A}$-measurable functions satisfying the equality

$$\lim_{t\downarrow 0} \int_E P(x, t, dy)\, f(y) = f(x), \quad x \in E. \tag{15.17}$$

Along with every element f a set $\mathfrak{D}$ contains a function

$$q_t(x) = \int_E P(x, t, dy)\, f(y) \tag{15.18}$$

being the unique solution in $\mathfrak{D}$ of the differential equation

$$\frac{\partial}{\partial t}\, g_t(x) = Q\, q_t(x), \quad t > 0,$$

and satisfying additional conditions:

$$\lim_{t\downarrow 0} g_t(x) = f(x) \quad \text{for all } x \in E,$$

$$\sup_{t>0} \sup_{x\in E} |g_t(x)| < \infty,$$

$$\sup_{0\le t\le n} \sup_{x\in E} \left| \frac{\partial}{\partial t}\, g_t(x) \right| < \infty \quad \text{for all } n = 1, 2, \dots .$$

Thus in the homogeneous case the question of the reconstruction of the transition function by its local characteristics amounts to the question of the density of the set $\mathfrak{G}$ in the set $\mathfrak{B}$ of all bounded $\mathcal{A}$-measurable functions in the topology of the bounded pointwise convergence. The answer is positive, if, for example,

$$\lim_{t\downarrow 0} P(x, t, \{x\}) = 1 \tag{15.19}$$

for any $x \in E$. A formula (15.17) is a homogeneous variant of the condition (15.11). Under this condition the sets $\mathfrak{G}$ and $\mathfrak{B}$ coincide. In the homogeneous case statements (15.13), (15.16) ensure the existence of the $\mathcal{A}$-measurable decomposition $D_1, D_2, \dots$ of the space E and the existence of the $\mathcal{A}$-measurable with respect to $x \in E$ measure $Q(x, A)$ on $\mathcal{A}$ such that

$$Q(x, E) \le q(x) = \sup_{t>0} \frac{1}{t}\,(1 - P(x, t, \{x\})), \quad Q(x, \{x\}) = 0,$$

$$Q(x,\, A) \;=\; \lim_{t\downarrow 0}\; P(x,\, t,\, A) \quad \text{for all } \; x\overline{\in} A,\;\; A\subset \mathfrak{D}_i,\;\; i=1,2,\dots\,.$$

If moreover

$$Q(x,\, E) \;=\; q(x) \;<\; \infty, \tag{15.20}$$

then

$$\frac{\partial}{\partial t}\, P(x,\, t,\, A) \;=\; -q(x)\, P(x,\, t,\, A) + \int\limits_E Q(x,\, dy)\, P(y,\, t,\, A)$$

for any $A\in\mathcal{A}$.

Under the condition (15.20) the infinitesimal operator Q of the transition function $P(x,\, t,\, A)$ is a linear integral operator for any $x\in E$. Its value on the function $f\in\mathfrak{D}$ is the function taking value

$$\int\limits_E Q(x,\, dy)\, (f(y)-f(x))$$

at the point x. A domain $\mathfrak{D}$ of the operator Q contains bounded $\mathcal{A}$-measurable functions $f(\cdot)$ satisfying the inequality

$$\sup_{t>0}\, \sup_{x\in E} \left| \frac{1}{t} \int\limits_E P(x,\, t,\, dy)\, f(y) - f(x) \right| \;<\; \infty.$$

However there exist homogeneous transition functions satisfying the condition (15.19) (and hence uniquely defined by their infinitesimal operators), but not satisfying the condition (15.20). For example:

(i) a phase space E consists of rational numbers,

(ii) σ-algebra $\mathcal{A}$ contains all one-point (and hence all other) subsets,

(iii) with every rational number x one can associate a positive exponentially distributed random variable $\eta(x)$ with a mean value $a(x)$; random variables $\eta(x)$, $x\in E$, are supposed to be independent in the totality and $\sum\limits_{x\le n} a(x)<\infty$ for any positive integer n, but $\sum\limits_{x\in E} a(x)=\infty$,

(iv) a probability transition function $P(x,\, t,\, A)$ is defined by its values on one-elements sets:

$$P(x,\, t,\, \{y\}) = \begin{cases} 0 & \text{if } x>y \\ \mathbf{P}\left\{\eta(x)>t\right\} & \text{if } x=y \\ \mathbf{P}\left\{\sum\limits_{x\le r<y} \xi(r) \le t < \sum\limits_{x\le r\le y} \xi(r)\right\} & \text{if } x<y. \end{cases} \tag{15.21}$$

Here

$$P(x,\,t,\,\{x\}) = e^{-t/a(x)} \underset{t\downarrow 0}{\longrightarrow} 1, \quad q(x) = \sup_{t>0} \frac{1}{t}\,(1 - P(x,\,t,\,\{x\})) = \frac{1}{a(x)},$$

but

$$Q(x,\,\{y\}) = \lim_{t\downarrow 0} \frac{1}{t}\, P(x,\,t,\,\{y\}) \le \lim_{t\downarrow 0} \frac{1}{t}\, \mathbf{P}\left\{ \sum_{x\le r<y} \xi(r) \le t \right\} \le$$

$$\le \lim_{t\downarrow 0} \frac{1}{t}\, \mathbf{P}\left\{ \xi(x) \le t,\; \xi\left(\frac{n+y}{2}\right) \le t \right\} = \lim_{t\downarrow 0} \frac{1}{t}\,(1 - e^{-t/a(x)}) \times$$

$$\times \left(1 - e^{-t/a((x+y)/2)}\right) = 0 \quad \text{if } x \neq y,$$

and consequently $Q(x,\,E) = 0$ for all $x \overline{\in} E$. In a rough approximation a Markov process with the transition function (15.21) monotonically moves along all rational numbers staying a positive time in each of them.

Feller stochastically continuous transition functions in compact metric spaces E with a Borel σ-algebra $\mathcal{A}$ form a very important class of homogeneous transition functions allowing for an efficient description in terms of infinitesimal characteristics. A Feller property of the transition function $P(t,\,x,\,A)$ means that the right-hand side of (15.18) continuously depends on $x \in E$ for any continuous function f. A stochastic continuity means that the continuity of the function f implies the relation (15.17) to be true. Thus a stochastic continuity postulates the inclusion of continuous on E functions to the set $\mathfrak{G}$ and thereby postulates the density of the last set in the set $\mathfrak{B}$ in the topology of a bounded pointwise convergence.

A part of the domain of the infinitesimal operator of the stochastically continuous Feller transition function in a metric compact set containing continuous functions is dense in the set of all continuous functions (in the topology of a bounded pointwise convergence). Therefore by studying the infinitesimal operators of stochastically continuous Feller transition functions, we can restrict ourselves to continuous functions. So we can consider $\mathfrak{G}$ as the set of all continuous functions and the domain $\mathfrak{D}$ of the infinitesimal operator as the dense subset of $\mathfrak{G}$. This point of view is the most convenient for the description of infinitesimal operators. An infinitesimal operator Q of any transition function possesses the following properties:

(i) its domain $\mathfrak{D}$ is dense (in the topology of a bounded pointwise convergence) in $\mathfrak{G}$,

(ii) for some $c > 0$ an operator $cI - Q$ maps $\mathfrak{D}$ onto $\mathfrak{G}$,

(iii) if the function f out of $\mathfrak{D}$ reaches a nonpositive maximum at some point x, then $Qf(x) \le 0$.

The last property is called **the maximum principle**.

Given properties are sufficient for the operator Q acting in the space $\mathfrak{G}$ of continuous functions on the metric compact E to be an infinitesimal operator of the Feller stochastically continuous transition function.

15.3 Diffusion Processes

The most spread method making it possible to study transition functions is to postulate the existence of limits (15.10) for some (as simple as possible) class of functions f and then to investigate the properties of the differential equation (15.9). Transition functions of diffusion processes can be used as an example of a such approach[1,11,12].

A numerical Markov process $\xi(t)$ with a transition function $P(s, x;\ t, A)$ is called **a diffusion process**, if there exist such functions $a(t, x)$ and $b(t, x)$ that the conditions

$$\lim_{t \downarrow s} \frac{1}{t-s} \int_{|y-x| \geq c} P(s, x;\ t, dy) = 0, \tag{15.22}$$

$$\lim_{t \downarrow s} \frac{1}{t-s} \int_{|y-x| < c} (y-x)\, P(s, x;\ t, dy) = a(s, x), \tag{15.23}$$

$$\lim_{t \downarrow s} \frac{1}{t-s} \int_{|y-x| < c} (y-x)^2\, P(s, x;\ t, dy) = b(s, x) \tag{15.24}$$

hold for all $c > 0,\ s,\ x$. Functions $a(s, x)$ and $b(s, x)$ are called **shift** and **diffusion coefficients** of the process $\xi(t)$ respectively.

If shift and diffusion coefficients are continuous with respect to their variables and the function

$$g(s, x) = \int_{-\infty}^{\infty} P(s, x;\ t, dy)\, f(y)$$

is doubly continuously differentiable with respect to x for all $s < t$, then $g(s, x)$ is continuously differentiable with respect to s and

$$-\frac{\partial}{\partial s} g(s, x) = a(s, x) \frac{\partial}{\partial x} g(s, x) + \frac{1}{2} b(s, x) \frac{\partial^2}{\partial x^2} g(s, x), \quad s < t, \tag{15.25}$$

$$\lim_{s \uparrow t} g(s, x) = f(x). \tag{15.26}$$

This equation is called **a backward** (or a second) **Kolmogorov equation.**

A forward (or a first) **Kolmogorov equation** (which is sometimes called **a Fokker-Planck equation**) is the equation for the transition density of a diffusion process. Let

$$P(s, x; t, A) = \int_A p(s, x; t, y)\, dy.$$

Suppose the derivatives

$$\frac{\partial}{\partial t} p(s, x; t, y), \quad \frac{\partial}{\partial y} (a(t, y)\, p(s, x; t, y)), \quad \frac{\partial^2}{\partial y^2} (b(t, y)\, p(s, x; t, y))$$

exist and are continuous with respect to variables $t \in (s, \infty)$, $x \in \mathbb{R}$, and conditions (15.22)–(15.24) hold uniformly with respect to x. Then

$$\frac{\partial}{\partial t} p(s, x; t, y) = -\frac{\partial}{\partial y} (a(t, y)\, p(s, x; t, y)) +$$

$$+\frac{1}{2} \frac{\partial^2}{\partial y^2} (b(t, y)\, p(s, x; t, y)), \quad t > s.$$

Under some hypotheses of the smoothness type imposed on shift and diffusion coefficients, there exists exactly one absolutely continuous transition function of the diffusion process (with given shift and diffusion coefficients) having the transition density satisfying both the forward and backward Kolmogorov equations. The following theorems give an exact formulation of this statement.

Theorem 15.3. *If functions $a(t, x)$ and $b(t, x)$ are bounded and continuous with respect to their variables and*

$$|a(t, y) - a(t, x)| + |b(t, y) - b(t, x)| \le c\, |y - x|^{\alpha},$$

$$|b(t, x) - b(s, x)| \le c\, |t - s|^{\alpha}, \quad b(t, x) \ge c^{-1}$$

for some $c > 0$, $\alpha > 0$ and for all s, t, x, y, then there exists a unique absolutely continuous transition function of the diffusion process with a shift coefficient $a(t, x)$ and a diffusion coefficient $b(t, x)$. The transition density $p(s, x; t, y)$ can be continuous with respect to s, t, x, y $(s < t)$, doubly continuously differentiable with respect to x and satisfy the relation

$$-\frac{\partial}{\partial s} p(s, x; t, y) = a(s, x) \frac{\partial}{\partial x} p(s, x; t, y) +$$

$$+\frac{1}{2} b(s, x) \frac{\partial^2}{\partial x^2} p(s, x; t, y), \quad s < t.$$

Mentioned in Theorem 15.3, transition density $p(s,\, x;\; t,\, y)$ of the diffusion process satisfies the following estimates (with some positive constants k and γ):

$$p(s,\, x;\; t,\, y) \le k\,(t-s)^{-1/2}\,\exp\left\{-\gamma\,\frac{|y-x|^2}{t-s}\right\},$$

$$\left|\frac{\partial}{\partial x}\,p(s,\, x;\; t,\, y)\right| \le \frac{k}{t-s}\,\exp\left\{-\gamma\,\frac{|y-x|^2}{t-s}\right\},$$

$$\left|\frac{\partial^2}{\partial x^2}\,p(s,\, x;\; t,\, y)\right| \le k\,(t-s)^{-3/2}\,\exp\left\{-\gamma\,\frac{|y-x|^2}{t-s}\right\},$$

$$\left|\frac{\partial}{\partial t}\,p(s,\, x;\; t,\, y)\right| \le k\,(t-s)^{-3/2}\,\exp\left\{-\gamma\,\frac{|y-x|^2}{t-s}\right\}.$$

These estimates permit the statement that under the hypotheses of Theorem 15.3 the function

$$g(s,\, x) = \int p(s,\, x;\; t,\, y)\, f(y)\, dy$$

satisfies an equation (15.25) and a boundary condition (15.26) for any bounded continuous function $f(x)$.

Theorem 15.4. *If under the hypotheses of Theorem 15.3 there exist continuous and bounded with respect to their variables derivatives*

$$a'(t,\, x) = \frac{\partial}{\partial x}\,a(t,\, x), \quad b'(t,\, x) = \frac{\partial}{\partial x}\,b(t,\, x), \quad b''(t,\, x) = \frac{\partial^2}{\partial x^2}\,b(t,\, x)$$

and

$$|a'(t,\, y) - a'(t,\, x)| + |b'(t,\, y) - b'(t,\, x)| + |b''(t,\, y) - b''(t,\, x)| \le c\,|y-x|^{\alpha}$$

for some $c > 0$, $\alpha > 0$ and for all $t,\, x,\, y,$ then

$$\frac{\partial}{\partial t}\,p(s,\, x;\; t,\, y) = \frac{\partial}{\partial y}\,(a(t,\, y)\, p(s,\, x;\; t,\, y)) -$$

$$-\frac{1}{2}\,\frac{\partial^2}{\partial y^2}\,(b(t,\, y)\, p(s,\, x;\; t,\, y)), \quad t > s.$$

A class of diffusion processes is closed with respect to monotone smooth transformations of the phase space. Let the function $\varphi(t,\, x)$ be monotone doubly continuously differentiable with respect to x and continuously differentiable with respect to t and let $a(t,\, x)$ and $b(t,\, x)$ be shift and diffusion coefficients of the diffusion process $\xi(t)$. Then a process $\eta(t) = \varphi(t, \xi(t))$ is a diffusion process with shift and diffusion coefficients

$$\frac{\partial}{\partial t}\varphi(t, \psi(t, x)) + a(t, \psi(t, x))\,\frac{\partial}{\partial x}\varphi(t, \psi(t, x)) + $$
$$+\frac{1}{2}\,b(t, \psi(t, x))\,\frac{\partial^2}{\partial x^2}\varphi(t, \psi(t, x)),$$

and

$$b(t, \psi(t, x))\left(\frac{\partial}{\partial x}\varphi(t, \psi(t, x))\right)^2,$$

respectively where $\psi(t, x)$ as a function of x is an inverse function to $\varphi(t, x)$, i.e., $\varphi(t, \psi(t, x)) = \psi(t, \varphi(t, x)) = x$.

Relations (15.22)–(15.24) hold, if the more convenient conditions are satisfied: there exists $\delta > 0$ such that

$$\lim_{t \downarrow s} \frac{1}{t - s} \int_{-\infty}^{\infty} |y - x|^{2+\delta}\, P(s, x;\, t, dy) = 0,$$

$$\lim_{t \downarrow s} \frac{1}{t - s} \int_{-\infty}^{\infty} (y - x)\, P(s, x;\, t, dy) = a(t, x),$$

$$\lim_{t \downarrow s} \frac{1}{t - s} \int_{-\infty}^{\infty} (y - x)^2\, P(s, x;\, t, dy) = b(t, x).$$

A Wiener process $w(t)$ is the most important example of the diffusion process. This is a process with a zero shift coefficient and a unit diffusion coefficient. A transition density $p(s, x;\, t, y)$ of a Wiener process is a Gaussian density with a mean value x and a variance $t - s$, i.e.,

$$p(s, x;\, t, y) = \frac{1}{\sqrt{2\pi\,(t - s)}}\,\exp\left\{-\frac{(y - x)^2}{2\,(t - s)}\right\}.$$

A wide class of diffusion processes with continuous coefficients is formed by the solutions of stochastic differential equations

$$d\xi(t) = a(t, \xi(t))\,dt + \sigma(t, \xi(t))\,dw(t)$$

or in the integral form

$$\xi(t) = \xi(0) + \int_0^t a(s, \xi(s))\,ds + \int_0^t \sigma(s, \xi(s))\,dw(s). \tag{15.27}$$

The second integral in the right-hand side of this equation can be simply defined provided that an integrand continuously depends on s . Let a continuous with the probability 1 random process $\zeta(t)$, $t \geq 0$, be conforming with a Wiener process $w(t)$, $t \geq 0$, in the following sense: an increment $w(t)-w(s)$, $t > s$, does not depend on random variables $\zeta(u)$, $w(u)$, $u \leq s$. Then there exists a limit in the sense of convergence in the probability of integral sums

$$\sum_{k=1}^{n} \zeta(s_{k-1})\,(w(s_k) - w(s_{k-1})), \quad 0 = s_0 < \cdots < s_n = t,$$

for $\max\limits_{1\leq k\leq n}(s_k - s_{k-1}) \to 0$, which is denoted by integral symbol $\int\limits_0^t \zeta(s)\,dw(s)$.

Let $\xi(t)$, $t \geq 0$, be a process conforming with a Wiener process and having continuous trajectories such that the equation (15.27) holds with the probability 1 for all $t > 0$. Then this process is called the solution of an equation (15.27) with coefficients $a(t, x)$ and $\sigma(t, x)$ being continuous with respect to t and x.

Theorem 15.5. *Let $a(t, x)$ and $\sigma(t, x)$ be continuous functions with respect to t and x,*

$$\sup_{t,x} \frac{|a(t, x)| + |\sigma(t, x)|}{1 + |x|} < \infty$$

and for any $r > 0$ there exist such $c > 0$ that

$$|a(t, y) - a(t, x)| + |\sigma(t, y) - \sigma(t, x)| \leq c\,|y - x|$$

for $|x| \leq r$, $|y| \leq r$. If $\xi(0)$ is independent of a Wiener process $w(t)$, $t \geq 0$, then the equation (15.27) has a unique continuous solution $\xi(t)$, $t \geq 0$, which is a diffusion process with a shift coefficient $a(t, x)$ and a diffusion coefficient $\sigma^2(t, x)$.

A transition function $P(s, x;\, t, A)$ of the diffusion process $\xi(t)$ can be defined by the formula (see Theorem 15.5)

$$P(s, x;\, t, A) = \mathbf{P}\{\xi(t;\, s, x) \in A\}$$

where for given s, x the random process $\xi(t;\, s, x)$ is a solution (unique under the hypotheses of Theorem 15.5) of the equation

$$\xi(t;\, s, x) = x + \int\limits_0^t a(u, \xi(u;\, s, x))\,du + \int\limits_s^t \sigma(u, \xi(u;\, s, x))\,dw(u), \quad t \geq s.$$

The next theorem states conditions when a given diffusion process can be obtained as the solution of a stochastic differential equation.

Theorem 15.6. *Let a shift coefficient $a(t, x)$ and a diffusion coefficient $b(t, x)$ be continuous functions with respect to t, x and let a transition function $P(s, x; t, A)$ of a diffusion process $\xi(t)$, $t \geq 0$, have continuous trajectories. Assume that*

$$\sup_{t,x} \frac{|a(t, x)|}{1+|x|} < \infty, \qquad \inf_{t,x} b(t, x) > 0$$

and there exist continuous derivatives $\frac{\partial}{\partial t} b(t,x)$, $\frac{\partial}{\partial x} b(t, x)$ and a function $\psi(x)$ such that

$$\psi(x) \geq 1 + |x|, \qquad \sup_t \mathbf{E}\, \psi(\xi(t)) < \infty,$$

$$\left| \int (y-x)\, P(s, x; t, dy) \right| + \left| \int (y-x)^2\, P(s, x; t, dy) \right| \leq (t-s)\, \psi(s),$$

$$\int (|y| + y^2)\, P(s, x; t, dy) \leq \psi(x).$$

Then there exists a Wiener process $w(t)$ such that

$$\xi(t) = \xi(0) + \int_0^t a(s, \xi(s))\, ds + \int_0^t \sqrt{b(s, \xi(s))}\, dw(s), \quad t \geq 0,$$

with the probability 1.

15.4 References

1. Dynkin, E. V., *Markov Processes,* Springer, Berlin, 1965.
2. Dynkin, E. V., Initial and final behavior of trajectories of Markov processes, *Uspekhi Mat. Nauk,* 26, 153, 1971 (in Russian).
3. Dynkin, E. V., Integral representations of excessive measures and excessive functions, *Uspekhi Mat. Nauk,* 27, 43, 1972 (in Russian).
4. Kuznetsov, C. E., Construction of Markov processes with random birth and death times, *Teor. Veroyatn. Primen.,* 18, 596, 1973 (in Russian).
5. Skorohod, A. V., Gihman, I. I., *Theory of Stochastic Processes,* 1, Springer, Berlin, 1974.
6. Skorohod, A. V., Gihman, I. I., *Theory of Stochastic Processes,* 2, Springer, Berlin, 1975.

7. Skorohod, A. V., Gihman, I. I., *Theory of Stochastic Processes,* 3, Springer, Berlin, 1979.
8. Koroliouk, V. S., Portenko, N. I., Skorokhod, A. V., Tourbine, A. F., *Aide Mémoire de Théorie des Probabilités et de Statistique Mathématique,* Mir, Moscow, 1983.
9. Kuznetsov, C. E., Any Markov process in a Borel space has a transition function, *Teor. Probab. Appl.,* 25, 384, 1980.
10. Shurenkov, V. M., Kolmogorov equations for a class of transition functions, *Teor. Probab. Appl.,* 21, 854, 1976.
11. Skorohod, A. V., Gihman, I. I., *Introduction to the Theory of Random Processes,* Saunders, Philadelphia, 1969.
12. Skorohod, A. V., Gihman, I. I., *Stochastic Differential Equations,* Springer, Berlin, 1972.

Chapter 16

Statistics for Some Classes of Random Processes

16.1 Parameter Estimation for Flows of Homogeneous Events

Let us observe a flow of homogeneous events on $[0, \infty]$. On the basis of such observations some statistical conclusions are required. Suppose that theoretical arguments make it possible to state that a considered flow is of a specific type (for example, a nonhomogeneous Poisson) depending on several unknown parameters. A problem arises in estimating these parameters by flow observations. Another important problem consists of testing different hypotheses concerning a flow of events. In this section we briefly describe statistical estimates for characteristics of Poisson, self-exciting, doubly stochastic Poisson and some other processes[1,2].

In many applications the most natural first hypothesis is the supposition of the complete randomness of events along the time. A mathematical model for such a flow is a stationary Poisson process. It is characterized by a unique parameter λ, which can be estimated by flow observations. In this case the following properties of Poisson process are used:

(i) the probability that n events of the flow in the interval $(0, T)$ occur in $(t_1, t_1 + dt_1), \ldots, (t_n, t_n + dt_n)$, $0 < t_1 < t_2 < \cdots < t_n < T$, equals $\lambda^n e^{-\lambda T} dt_1 \ldots dt_n$,

(ii) if N_i, $i = 1, \dots, m$, are numbers of occurred events in disjoint intervals with lengths t_i, $i = 1, \dots, m$, then $N = \sum_{i=1}^{m} N_i$ has a Poisson distribution with a parameter $\lambda\,(t_1 + \cdots + t_m)$.

These properties imply that the unbiased estimate $\hat{\lambda}$ of the parameter λ can be constructed by using solely the length of observation period T and the number n of occurred events: $\hat{\lambda} = n/T$.

Let ξ denote a random variable having a Poisson distribution with a parameter $\mu = \lambda T$. A confidence interval for μ (and consequently for $\lambda = \mu/T$) is constructed in the following way. For large values of μ a random variable $(\xi - \mu)/\sqrt{\mu}$ has approximately a normal distribution with a zero mean and a unit variance. For a given confidence coefficient $1 - \alpha$ using the tables[3] for normal distribution we can choose $c(\alpha)$ such that

$$\mathbf{P}\left\{-c(\alpha) < \frac{\xi - \mu}{\sqrt{\mu}} < c(\alpha)\right\} \approx 1 - \alpha.$$

Assume that we have observed n events of the flow in the interval $(0, T)$. Then using the relation $(n - \mu)/\sqrt{\mu} = \pm c(\alpha)$ we find a confidence interval for μ:

$$(\underline{\mu},\, \overline{\mu}) = \left(n + \frac{1}{2}\,c^2(\alpha) - \sqrt{n\,c^2(\alpha) + \frac{1}{4}\,c^4(\alpha)},\right.$$

$$\left. n + \frac{1}{2}\,c^2(\alpha) + \sqrt{n\,c^2(\alpha) + \frac{1}{4}\,c^4(\alpha)}\right). \qquad (16.1)$$

Consequently if $\mu_0 = \lambda_0 T \overline{\in} (\underline{\mu},\, \overline{\mu})$, then we reject a hypothesis $H_0 = \{\mu = \mu_0\}$. If $\mu_0 \in (\underline{\mu},\, \overline{\mu})$, then available statistical data don't contradict the hypothesis H_0.

Example 16.1. Suppose that some device has failed 12 times in a period of 25 months. At the same time on the basis of previous experience we can assume that failures occur according to a Poisson process with the intensity $\lambda_0 = \frac{1}{3}$ failures per a month. So,

$$T = 25, \quad n = 12, \quad H_0 = \{\mu = \frac{1}{3} \cdot 25 = 8\frac{1}{3}\}.$$

If $\alpha = 0.05$, then $c(\alpha) = 1.96$ and according to (16.1) the confidence intervals for μ and λ are (7.13, 20.71) and (0.25, 0.84) respectively.

Assume that instead of the number of events in $[0, T)$ we observe a time to the nth event occurrence. If this time equals T, then we take T/n as the statistical estimate for λ. For large values of n a random value $2\,\lambda T$ has approximately a χ^2-distribution with $2n$ degrees of freedom. A confidence interval for λ can be obtained by using the tables of χ^2-distribution. Let χ^2_{2n} be a random variable having a χ^2-distribution with $2n$ degrees of freedom. Assume that for a given confidence coefficient $1-\alpha$ we have found $c_1(\alpha) < < c_2(\alpha)$ such that

$$\mathbf{P}\{\chi^2_{2n} > c_2(\alpha)\} \approx \frac{\alpha}{2}, \quad \mathbf{P}\{\chi^2_{2n} < c_1(\alpha)\} \approx \frac{\alpha}{2}.$$

Then we take

$$\left(\frac{1}{2T}\,c_1(\alpha),\ \frac{1}{2T}\,c_2(\alpha)\right)$$

as the confidence interval for λ.

If in example 16.1 the time to the 12th failure equals 25 months, then we use tables of χ^2-distribution with $2 \times 12 = 24$ degrees of freedom. For $\alpha = 0.05$ we obtain a confidence interval with $c_1(\alpha) = 12.40$, $c_2(\alpha) = 39.36$, i.e., $0.248 < \lambda < 0.787$.

Enumerate some results obtained by the method of maximal likelihood for such important classes of point processes as nonhomogeneous Poisson, generalized Poisson, self-exiting, doubly stochastic Poisson and marked point processes.

1. A nonhomogeneous Poisson process. The problem of a statistical estimation can be formulated in the following way. Let $\{N(t),\ t \geq 0\}$ be a nonhomogeneous Poisson process with the intensity $\lambda(t,\, X)$ where $\lambda(\cdot, \cdot)$ is some known function and X is a vector of unknown parameters. Let $D(0,\, T)$ denote statistical data obtained by the observation of a process $N(t)$ in $(0,\, T)$ and $\mathbf{P}\{D(0,\, T)\,|\,X\}$ denotes a distribution of the statistics $D(0,\, T)$ by fixed vector X of unknown parameters. The problem to be solved: having obtained concrete observations and by a known distribution $\mathbf{P}\{D(0,\, T)\,|\,X\}$ it is necessary to construct an estimate $\hat{X} = \hat{X}(D(0,\, T))$.

If the form of a function $\lambda(\cdot, \cdot)$ is known then we can use a maximal likelihood method to estimate the unknown parameters. Assume that in the half-intervals $[t_{i-1},\, t_i)$, $i = 1, \ldots, k$, $t_0 = 0$, we have observed n_i, $i = 1, \ldots, k$, events of the flow. Then as the estimate $\hat{X}$ we take a value X maximizing the probability

$$\mathbf{P}\{N(t_{i-1},\, t_i) = n_i,\ i = 1, \ldots, k \mid X\}.$$

For a nonhomogeneous Poisson process the estimate $\hat{X}$ provides the maximum of a function

$$l(X) = -\int_0^T \lambda(t,\, X)\, dt + \sum_{i=1}^{k} n_i \ln\left(\int_{t_{i-1}}^{t_i} \lambda(t,\, X)\, dt\right).$$

For example, if $\lambda(t, X) = X\,\mu(t)$, $t \geq 0$, where $\mu(t)$ is a known function, then

$$\hat{X} = (n_1 + n_2 + \cdots + n_k)\left(\int_0^T \mu(t)\,dt\right)^{-1}.$$

In most cases $\hat{X}$ can't be found in explicit form and therefore numerical methods are usually used.

2. A generalized Poisson process. This is a process of the form

$$\eta(t) = \sum_{i=1}^{N(t)} \xi_i, \quad t \geq 0,$$

where ξ_i, $i \geq 1$, are independent identically distributed random variables and $N(t)$ is a nonhomogeneous Poisson process. Assume that ξ_i, $i \geq 1$, are discrete random variables taking values $\{a_1, a_2, \dots\}$ with probabilities $\{p_1(X), p_2(X), \dots\}$ and a process $N(t)$ has an intensity $\lambda(t, X)$. Here X is a set of unknown parameters. A generalized Poisson process can be considered as a marked nonhomogeneous Poisson process with marks $\{\xi_i\}$. We have observed n events of the flow in $(0, T)$ occurred at moments $0 < t_1 < t_2 < \cdots < t_n < T$ and it was n_i events with the mark a_i, $n_1 +$ $+\, hspace0.5mmn_2 + \cdots = n$. Then an estimate $\hat{X}$ for X, obtained by the maximal likelihood method, provides the maximum of the function

$$l(X) = -\int_0^T \lambda(t, X)\,dt + \sum_{i=1}^{n} \ln \lambda(t_i, X) + \sum_{i=1}^{\infty} n_i \ln p_i(X).$$

With the exception of special cases the value $\hat{X}$ is obtained by numerical methods.

3. Self-exciting process (see Section 6.4). If we observe a self-exciting process, then the problem is to estimate the unknown parameters in the intensity function of the process. However the likelihood function $l(X)$ is of the same form as for a nonhomogeneous Poisson process. Therefore a corresponding estimate $\hat{X}$ can be constructed by analogy to the considered above case.

4. Doubly stochastic Poisson process (see Section 6.5). In view of Theorem 6.7 a doubly stochastic Poisson process $\{N(t),\ t \geq 0\}$ with a random intensity $\{\lambda(t, \xi(t)),\ t \geq 0\}$ is a self-exciting process with an intensity $\{\lambda^*(t),\ t \geq 0\}$. Here $\xi(t)$ is a so-called information process. Therefore the estimates of unknown parameters of a doubly stochastic Poisson process can be constructed in a similar way as the above considered processes.

Assume that we observe a process $\{N(t),\ t \geq 0\}$ in the interval $(0,\ t)$ and it is required to construct an estimate for some function $f(\cdot)$ of the process $\xi(t)$ (for example, $f(\xi(t)) = \xi(t)$ or $f(\xi(t)) = \lambda(t,\ \xi(t))$). This is a so-called filtration problem. For problems both linear and nonlinear, filtration estimates with minimal mean square errors can be constructed.

5. Marked point processes. Consider a point process such that with every point there is associated some mark taking values in a discrete space $\{k_1,\ k_2,\ \dots,\ k_n\}$. Let $N([a,\ b) \times \{k_i\})$ be a number of points in $[a,\ b)$ with a mark k_i and $\lambda(t,\ k_i,\ X)$ be an intensity of the process depending on the set of unknown parameters X. If we have observed the trajectory of the process in the interval $(0,\ t)$, then we can take

$$l(X) = \sum_{i=1}^{\infty} \left(- \int_0^t \lambda(v,\ k_i;\ X)\, dv + \int_0^t \ln \lambda(v,\ k_i;\ X)\, N(dv \times \{k_i\}) \right)$$

as the likelihood function. The desired estimate $\hat{X}$ provides a maximum of this function $l(X)$.

6. Renewal process. In practice the necessity often arises to verify whether statistical data satisfy the goodness-of-fit test for a renewal process. If this hypothesis is satisfied, then in order to construct an empirical distribution function, we can use usual methods for processing simple random samples. In order to verify whether statistical data do not contradict the hypothesis of a renewal process, the following goodness-of-fit tests based on second order characteristics of the sequence of intervals between events of the flow[2] can be used:

(i) tests based on coefficients of correlation of the sequence,

(ii) tests based on a spectral function of intervals.

If statistical data don't support the hypothesis of a renewal process, then it is necessary to carry out additional analysis and to choose a new hypothetical model for the observed flow of events. Then we repeat the goodness-of-fit test for this new model. One can use the following tests based on:

(i) spectral density of intervals,

(ii) spectral density of the number of events.

Investigation of queueing systems has focused much attention on the inverse problems of the theory of homogeneous events flows, i.e., problems connected with the reconstruction of statistical properties of flows by observing their superposition[4]:

(i) we observe a superposition of s independent recurrent flows defined by the same distribution function $F(x)$; the problem is to construct the estimates for s and $F(x)$;

(ii) we observe a superposition of a recurrent and a Poisson flow; the problem is to construct the estimates for $F(x)$ and for the intensity of a Poisson flow.

A method of moments was used to solve the second of these problems.

16.2 Regression Analysis

The theory of statistical regression[5,6] predicts one or several random variables $(y_1, \dots, y_q) = \mathbf{y}$ on the basis of the information obtained for some measured random variables $(x_1, \dots, x_p) = \mathbf{x}$. These last random variables are supposed to be independent and are known as **predicting** or **regression variables**. Variables $\{y_i\}$ are known as **dependent** or **criterion variables**.

The prediction is necessary in many practical situations. In meteorology it is necessary to make a weather forecast based on atmospheric measurements at different moments. A firm using some set of special tests wants to select workers who can carry out a given work the most successfully. A scientist is interested in selecting individuals who can produce descendants bringing the maximal profit. In all these examples as criteria we have used observed in the future values, which are predicted on the basis of available measurements in order to make the proper decision.

Let y be a scalar criterion variable, $\mathbf{x} = (x_1, \dots, x_p)$ be a vector of p observed regression variables and $f(\mathbf{x}) = f(x_1, \dots, x_p)$ be a function predicting a criterion variable y (here $y, x_1, \dots, x_p$ are dependent random variables).

Assume that we have observed a series of experimental values of variables $y, x_1, \dots, x_p$. The main aim of the regression analysis is to choose such function $f(\mathbf{x})$ which characterizes with maximal precision the dependence of y on $x_1, \dots, x_p$ and in the future makes it possible to evaluate y on the basis of observations over $x_1, \dots, x_p$.

Let $A(\mathbf{x}) = \mathbf{E}\{y \mid \mathbf{x}\}$. Then the minimum of $\mathbf{E}(y - f(\mathbf{x}))^2$ is reached for $f(\mathbf{x}) = A(\mathbf{x})$. This function $A(\mathbf{x})$ (called **a regression** of y on $\mathbf{x}$) has the following properties.

1. $A(\mathbf{x})$ maximizes a correlation with a criterion variable. Let $\rho(y, f)$ be a coefficient of correlation between y and f, i.e.,

$$\rho(y, f) = \frac{\mathbf{Cov}(y, f)}{\sigma_y \sigma_f}, \quad \mathbf{Cov}(y, f) = \mathbf{E}\{(y - \mathbf{E}\, y)(f(x) - \mathbf{E}\, f(x))\},$$

$$\sigma_y^2 = \mathbf{E}(y - \mathbf{E}\, y)^2, \quad \sigma_f^2 = \mathbf{E}(f(x) - \mathbf{E}\, f(x))^2, \quad \sigma_A^2 = \mathbf{E}(A(\mathbf{x}) - \mathbf{E}\, A(\mathbf{x}))^2.$$

Then $\rho(y, A(\mathbf{x})) \geq 0$ and $\rho(y, A(\mathbf{x})) \geq |\rho(y, f)|$ for any $f(\mathbf{x})$. Moreover, $\rho(y, A(\mathbf{x})) = \sigma_A/\sigma_y$. Squared maximal value of a coefficient of correlation is called **a correlation ratio** η^2_{yx}, $0 \leq \eta^2_{yx} \leq 1$. If the prediction error $\sigma^2_{yx} = \mathbf{E}(y - A(\mathbf{x}))^2 \to 0$, then $\eta^2_{yx} \to 0$, i.e., η^2_{yx} is the measure for the accuracy of prediction.

2. Suppose we want to reduce a domain of observed vector $\mathbf{x}$ to some of its part α. This reduced domain must ensure the maximal expectation of y. Let $F(\cdot)$ be a distribution function of the vector $\mathbf{x}$. Then a problem can be formulated as follows: such domain w must be chosen that $\int\limits_w dF(\mathbf{x}) = \alpha$ and an integral $\int\limits_w A(\mathbf{x})\,dF(\mathbf{x})$ takes its maximal value. This desired domain is $w = \{\mathbf{x} : A(\mathbf{x}) \geq c\}$ where c is defined by the condition $\int\limits_w dF(\mathbf{x}) = \alpha$.

If x is a scalar value, then the most often used functions in the regression analysis are the following ones: $y = \alpha + \beta x$ – linear regression, $y = \alpha + \beta_1 x + \cdots + \beta_k x^k$ – polynomial regression, $y = \alpha + \beta \ln x$, $y = \alpha + \beta_1 x + \beta_2 x^{-1}$, $y = (\alpha + \beta x)^{-1}$, $y = (\alpha + \beta_1 x + \cdots + \beta_k x^k)^{-1}$, $\ln y = \alpha + \beta x$ – exponential regression, $\ln y = \alpha + \beta \ln x$.

It is necessary to evaluate coefficients α, β, β_1, β_2, ... for a chosen function. Inadequate choice of the regression function can result in a very inaccurate prediction. Therefore in order to find an appropriate regression function, it is necessary sometimes to search through several alternative variants.

The linear regression is the most studied case. Let $f(x_1, \ldots, x_p) = \alpha + \beta_1 x + \cdots + \beta_p x_p$ where α, β_1, ..., β_p are defined by the condition

$$\mathbf{E}(y - \alpha - \beta_1 x_1 - \cdots - \beta_p x_p)^2 \Longrightarrow \min. \qquad (16.2)$$

Let $y^{(i)}$, $x_1^{(i)}$, ..., $x_p^{(i)}$, $i = 1, \ldots, n$, be series of observations over $y, x_1, \ldots, x_p$. Then as estimates for α, β_1, ..., β_p we take

$$\hat{\beta} = \mathbf{C}^{-1}\sigma, \quad \hat{\alpha} = \overline{\mathbf{y}} - (\hat{\beta})^T\,\overline{\mathbf{x}}$$

where

$$\hat{\beta} = \begin{pmatrix} \hat{\beta}_1 \\ \vdots \\ \hat{\beta}_p \end{pmatrix}, \quad \sigma = \begin{pmatrix} \sigma_1 \\ \vdots \\ \sigma_p \end{pmatrix}, \quad \overline{\mathbf{x}} = \begin{pmatrix} \overline{x}_1 \\ \vdots \\ \overline{x}_p \end{pmatrix}, \quad \mathbf{C} = \begin{pmatrix} c_{11} & \cdots & c_{1p} \\ \vdots & & \vdots \\ c_{p1} & \cdots & c_{pp} \end{pmatrix},$$

$$\overline{y} = \frac{1}{n}\sum_{i=1}^{n} y^{(i)}, \quad \overline{x}_j = \frac{1}{n}\sum_{i=1}^{n} x_j^{(i)}, \quad \sigma_j = \frac{1}{n}\sum_{i=1}^{n} (y^{(i)} - \overline{y})\,(x_j^{(i)} - \overline{x}_j),$$

$$c_{jk} = \frac{1}{n}\sum_{i=1}^{n}(x_j^{(i)} - \overline{x}_j)\,(x_k^{(i)} - \overline{x}_k), \quad j,\,k = 1,\ldots,p.$$

C^{-1} is an inverse matrix to C, "T" is an operation of transposition.

The function $A(\mathbf{x}) = \hat{\alpha} + \mathbf{x}^T\,\hat{\beta}$ is a linear regression with a minimal mean square error. Moreover this function has a maximal correlation with y among all linear functions. A minimal value of the function in (16.2) equals

$$\frac{1}{n}\sum_{i=1}^{n}(y^{(i)} - \overline{y})^2 - \sigma^T\,\mathbf{C}^{-1}\,\sigma.$$

A correlation ratio for the linear regression equals

$$\eta_{yx}^2 = \sigma^T\,\mathbf{C}^{-1}\,\sigma\left(\frac{1}{n}\sum_{i=1}^{n}(y^{(i)} - \overline{y})^2\right)^{-1}.$$

16.3 Analysis of the Trend in a Poisson Flow of Events

In many cases it is desirable to verify whether the intensity of the flow changes along the time (trend) or not. Such a test can be realized by comparison with a zero hypothesis that there is no trend. In some problems it is necessary to apply flexible methods capable of recognizing a type of trend and not only its presence. In other problems the most attention is paid to testing the significance of a trend. The main difficulties arise from the choice of the criterion. Some of them have a great capacity to discover trends of a specific type; others are not so powerful, but can be applied to a wide range of competitive hypotheses. Often the problem of choosing a proper criterion and determining the type of a trend can be solved by studying statistical data. The experience of a researcher is of decisive importance.

Two types of methods being used to investigate a nonstationary Poisson flow of events[1] are described. Methods of the first type are based on standard methods of regression analysis. The second type of methods are based on efficient theoretical analysis of special mathematical models.

1. A regression analysis of intervals. Suppose we have observed a large number of events. It is necessary to investigate the gradual variation of the intensity of their occurrence. Choose some positive integer number l (it is desirable that $l \geq 4$) such that there is no appreciable change in the flow intensity at the time between the occurrence of l successive events. Let y_i, $i = 1, 2, \ldots, r$, be a time from the occurrence of $((i-1)l+1)$th event to the occurrence of $(il+1)$th event. Here $r = [n/l]$ where n is the total number of observed events. Assume that the intensity $\lambda(t)$ is a slowly changing function along the time and in intervals with lengths $y_1, \ldots, y_r$; this function can be satisfactorily approximated by constants $\lambda_1, \ldots, \lambda_r$. Moreover we assume that with every y_i one can associate an independent random variable z_i such that in the case of the simplest model

$$\ln \lambda_i = \alpha + \beta z_i, \quad i = 1, \ldots, r. \tag{16.3}$$

For example, as z_i we can choose:

(i) a function of the center of an interval with the length y_i provided that $\lambda(\cdot)$ is considered as a function of a time t,

(ii) a function of the event number in the center of an interval with the length y_i provided that $\lambda(\cdot)$ is considered as a function of the sequence of event's indexes,

(iii) a value of some independent random variable (that controls the behavior of the flow intensity) averaged over an interval of the length y_i.

It follows[1] from the relation (16.3) that

$$\mathbf{E} \ln Y_i = -(\alpha' + \beta z_i), \quad \mathbf{Var}(\ln Y_i) = b_l$$

where

$$\alpha' = \alpha + a_l, \quad b_l \approx \left(l - \frac{1}{2}\right)^{-1}, \quad a_l \approx \ln l - \left(2l - \frac{1}{3}\right)^{-1}.$$

Thus we have obtained a lineal model with unknown parameters α and β. Now it is possible:

(i) to construct estimates for α and β by a standard method of least squares,

(ii) to test a zero hypothesis $\beta = 0$ and to construct an approximate confidence interval for β by regression methods,

(iii) to compare a mean square deviation from the regression line with a theoretical value b_l; thereby we can verify the adequacy of the supposed model.

Application of a logarithmic transformation to intervals ensures the constant known variance and makes it possible to use an exponential relation (16.3) being the most natural and simple dependence for the flow intensity. The relation (16.3) can be generalized because of a choice of more complex functions as in the right-hand side of (16.3) (for example, $\alpha + \beta z_i + \gamma z_i^2$). This method was illustrated by the example of accidents that occurred in mines in Great Britain[1].

It should be noted that methods of regression analysis can be also used in the investigations of the connection between n_i and z_i where $n_1, n_2, \ldots$ are numbers of events that occurred in successive equal time periods and $z_1, z_2, \ldots$ are some independent variables connected with corresponding intervals.

2. Special methods. These methods ensure an efficient theoretical analysis for some specific models of event's sequences which are closely connected to a Poisson flow.

Assume that a considered phenomenon of a trend is a smooth change of the flow intensity such that the relation

$$\lambda(t) = e^{\alpha + \beta t} \tag{16.4}$$

holds. Suppose we have observed n events in $[0, T]$ that occurred at moments $0 < t_1 < t_2 < \cdots < t_n < T$. An estimate for the parameter β, obtained by **the method of maximum likelihood**, is the solution of an equation

$$L(\beta) = 0$$

where

$$L(\beta) = \begin{cases} n\beta^{-1} - nT(1 - e^{-\beta T})^{-1} + \sum_{i=1}^{n} t_i & \text{if } \beta \neq 0 \\ -\frac{1}{2} nT + \sum_{i=1}^{n} t_i & \text{if } \beta = 0. \end{cases}$$

For fixed $\hat{\beta}$ a sufficient statistic for the evaluation of α is n:

$$\hat{\alpha} = \ln(\hat{\beta} n)\,(e^{\hat{\beta} T} - 1)^{-1}.$$

The presence or absence of a trend (16.4) can be detected by means of criteria for verification of whether a zero hypothesis $H_0 = \{\beta = 0\}$ holds. If statistical data don't confirm the hypothesis H_0, then we have the presence of the trend.

16.4 Statistics of Queueing Systems

By the statistical investigation of real systems we deal with the following problems:

(i) choice of the adequate model for the system's description,

(ii) evaluation of the model's parameters by empirical data,

(iii) statistical conclusions with respect to the system's future behavior.

First of all it is necessary to verify the main assumptions being used in a model[1,7,8] (for example, the independence of repair times).

Consider a queueing system $M|M|1$ with waiting (a simple input flow, exponentially distributed service times, only one service device). In order to estimate parameters of this system, five strategies[9] can be suggested:

(i) to observe a service process during a fixed time,

(ii) to observe a process to the moment when the summary busy time of a device reaches a given value,

(iii) to observe a process to the moment when the number of entered customers reaches a given value,

(iv) to observe a process to the moment when the number of serviced customers reaches a given value,

(v) to observe a process to the moment when the total number of entered and serviced customers reaches a given value.

For every one of these cases maximum likelihood estimates for the flow and service intensities were obtained. The first strategy gives the most precise estimates (in the asymptotic sense) for these parameters.

Assume that we observe a queue length at the moments of customers' arrival in a system $GI|M|1$ (a recurrent flow). Let m_N be an estimate for the mean queue length obtained averaging queue lengths at arrival moments of the first N customers. A dependence of m_N on the system's load ρ was investigated[10]. It was proved that in order to construct estimates with the same variances as $\rho \to 1$ we require N asymptotically equivalent to $c(1-\rho)^{-2}$ for some constant $c > 0$.

If a queueing system with waiting has the simplest input flow and service times are exponentially distributed random variables, then an output flow is also the simplest one with the same parameter as the input flow (provided that $\rho < 1$). This is the case when we are not able to reconstruct the service intensity by observations of the output flow. In all other cases observations of the output flow of the system $M|G|1$ (an arbitrary distribution function of service times) during an infinitely long time permit unique reconstruction of both the intensity of the input flow and the distribution of a service time[11] (provided that a system's load is less than a critical one, i.e., $\rho < 1$). The distribution of the interval between service terminations of two customers reconstructs the distribution of a service time up to the unknown parameter. This parameter can be evaluated by means of a joint distribution of two such successive intervals. A similar problem was solved for a system $M|G|1$ with a bounded sojourn time[12]. The unbiased estimates with a minimal variance were obtained[13] for main characteristics of efficiency of some simple queueing systems (such as $M|G|\infty$, $M|M|1$, $M|G|1$, $M|M|s$). It was supposed that only one system's parameter is unknown. Estimates were constructed on the basis of observations over some functions of systems' states such as the number of customers in a queue or in a system, waiting time in a system, etc. A comparison was made of the effectiveness of some estimates.

Let $\eta(t)$ be a semi-Markov birth and death process differing from the usual semi-Markov process with a countable set of states by an additional condition on the transition probabilities $\{p_{ij}\}$: $p_{ij} = 0$ for $|i-j| > 1$. An asymptotical distribution of estimates for some "profit" functionals of the system's behavior described by semi-Markov birth and death processes was investigated[14]. The estimates were constructed by observing one trajectory of a system. These results can be used to estimate the precision of the statistical simulation of queueing systems. For further interesting results concerning statistics of queueing systems (see references 15-17).

16.5 Statistical Analysis of Markov and Semi-Markov Processes

1. Markov process with a discrete time (Markov chain). Let $\{\xi_n,\, n = 1, 2, \ldots\}$ be a homogeneous Markov chain taking values in a measurable space $(X,\, \mathfrak{U})$. A transition probability

$$p_\theta(x,\, A) \;=\; \mathbf{P}\,\{\xi_{n+1} \in A\,|\,\xi_n = x\}$$

depends on some unknown parameters $\theta = (\theta_1, \ldots, \theta_r) \in \Theta \subset \mathbb{R}^r$. An initial distribution is $q_\theta(A)$, $A \in \mathfrak{U}$. It is supposed that there exists a unique stationary distribution $p_\theta(\cdot)$ satisfying the equation

$$p_\theta(A) \;=\; \int\limits_X p_\theta(x,\, A)\, p_\theta(dx), \quad A \in \mathfrak{U}.$$

From observing the states $x_1, x_2, \ldots, x_{n+1}$ of a Markov chain at moments $1, 2, \ldots, n+1$ it is necessary to construct estimates for these unknown parameters $\theta_1, \ldots, \theta_r$. This problem can be solved by a maximum likelihood method.

Let there exist a not necessarily finite measure $\lambda(\cdot)$ defined on $\mathfrak{U}$ and such that measures $q_\theta(A)$ and $p_\theta(x,\, A)$ are absolutely continuous with respect to $\lambda(\cdot)$, i.e., there exist functions $f(y;\,\theta)$ and $f(x,\,y;\,\theta)$ such that relations

$$q_\theta(A) = \int_X f(y;\,\theta)\,\lambda(dy), \quad p_\theta(x,\,A) = \int_X f(x,\,y;\,\theta)\,\lambda(dy)$$

hold for any $A \in \mathfrak{U}$. Then **a maximum likelihood function** is the following one:

$$\ln f(x_1;\,\theta) + \sum_{k=1}^{n} \ln f(x_k,\,x_{k+1};\,\theta).$$

For large values of n a term $\ln f(x_1;\,\theta)$ has an insignificant influence on the whole sum. Therefore it can be neglected. Thus a function

$$L_n(\theta) = \sum_{k=1}^{n} \ln f(x_k,\,x_{k+1};\,\theta)$$

is usually taken as a maximum likelihood function.

As an estimate $\hat{\theta}$ for a parameter θ^0 we take one of the solutions of the system of equations (here we suppose the existence of partial derivatives):

$$\frac{\partial}{\partial\theta_i} L_n(\theta) = 0, \quad i = 1,\ldots,r. \tag{16.5}$$

Under some hypotheses[7] there exists a sequence $\{\hat{\theta}^{(n)}\}$ of random vectors out of Θ, $\hat{\theta}^{(n)} = \hat{\theta}(x_1,\ldots,x_{n+1})$ such that $\{\hat{\theta}^{(n)}\}$ converges in probability to θ^0 as $n \to \infty$. Consequently there exists a consistent maximum likelihood estimate. Moreover $\hat{\theta}^{(n)}$ is a local maximum of $L_n(\theta)$ with the probability tending to 1. If $\overline{\theta}^{(n)}$ is a second consistent solution of (16.5), then $\overline{\theta}^{(n)} = \hat{\theta}^{(n)}$ with the probability tending to 1 as $n \to \infty$.

Consider a Markov chain with a finite set of states $X = \{1,\,2,\,\ldots,\,m\}$. Let $\{p_{ij}(\theta)\}$ be transition probabilities depending on unknown parameters $\theta_1,\,\ldots,\,\theta_r$. Under some hypotheses we can take one of the solutions of the system of equations

$$\sum_{i,j=1}^{m} \frac{n_{ij}}{p_{ij}(\theta)}\,\frac{\partial p_{ij}(\theta)}{\partial\theta_l} = 0, \quad l = 1,\ldots,r,$$

as the consistent estimate for θ^0. Here n_{ij} is the number of indexes k such that $1 \le k \le n$ and $x_k = i,\ x_{k+1} = j$.

Suppose the transition probabilities are unknown values and it is necessary to estimate them by observing the states of a Markov chain at moments $1, 2, \ldots, n+1$. As the maximum likelihood estimate we take

$\hat{p}_{ij} = n_{ij}/n_i$, $i, j = 1, \ldots, m$, where $n_i = \sum_{j=1}^{m} n_{ij}$. This is a consistent estimate, but generally speaking a biased one. However the bias tends to zero as $n \to \infty$. Moreover this estimate has asymptotically normal distribution.

2. Markov process with a continuous time. Let $\{\xi(t),\, t \geq 0\}$ be a Markov process taking values in a measurable space $(X, \mathfrak{U})$. A distribution function of a sojourn time of $\xi(t)$ in a state x is $1 - e^{-q(x,\,\theta)\,t}$ where $q(x, \theta) \geq 0$.

Let $q(x, \cdot\,; \theta)$ be a measure on $\mathfrak{U}$ for any $x \in X$ and $\theta \in \Theta$, $q(x, X; \theta) = q(x; \theta)$. Then the probability that $\xi(t+0) \in A$, provided that $\xi(t-0) = x$ and t is a transition moment, equals

$$\Pi_\theta(x; A) = \frac{q(x, A; \theta)}{q(x; \theta)}.$$

Here $\theta = (\theta_1, \ldots, \theta_r) \in \Theta$ is a vector of unknown parameters.

Suppose we have observed a trajectory of the process $\xi(t)$ in an interval $(0, t)$, $z_1, \ldots, z_{\nu(t)}$ is a sequence of process states and $\rho_1, \ldots, \rho_{\nu(t)}$ are the sojourn times in these states where $\nu(t) = \max\{k : \rho_1 + \cdots + \rho_k < t\}$. Moreover assume that a distribution $\Pi_\theta(x; A)$ is absolutely continuous with respect to some not necessarily finite measure $\lambda(\cdot)$ on $\mathfrak{U}$:

$$\Pi_\theta(x; A) = \int_A f(x, y; \theta)\, \lambda(dy), \quad x \in X.$$

Then we can take a function

$$L_t(\theta) = \sum_{k=1}^{\nu(t)-1} \left(\ln f(z_k, z_{k+1}; \theta) + \ln q(z_{k+1}; \theta) - \rho_{k+1}\, q(z_{k+1}; \theta)\right)$$

as the maximum likelihood function. By maximizing the function $L_t(\theta)$ with respect to θ, we can find a consistent estimate $\hat{\theta}$.

3. A semi-Markov process with a finite number of states. Consider a semi-Markov process defined by transition probabilities $\{p_{ij}\}_{i,j=1}^m$ and by distribution functions $\{F_{ij}(x)\}_{i,j=1}^m$. Without any loss of generality we can assume that $F_{ij}(x)$, $i = 1, \ldots, m$, does not depend on j, i.e., $F_{ij}(x) = F_i(x)$, $j = 1, \ldots, m$. It is required to construct estimates for $\{p_{ij}\}$ and $\{F_i(x)\}$ by process observations in the interval $(0, t)$.

Suppose we have observed that a semi-Markov process was $N_i(t)$ times in a state i in intervals with lengths $\xi_{i1}, \ldots, \xi_{iN_i(t)}$. Let $N_{ij}(t)$,

$1 \le i, j \le m$, be the number of process transitions from i to j in the interval $(0, t)$. As the estimates for $\{p_{ij}\}$ and $\{F_i(x)\}$ we take ratios

$$\hat{p}_{ij}(t) = \frac{N_{ij}(t)}{N_i(t)}, \quad \hat{F}_i(x, t) = \frac{1}{N_i(t)} \sum_{k=1}^{N_i(t)} \delta(x - \xi_{ik}), \quad t, x \ge 0, \tag{16.6}$$

where

$$\delta(x) = \begin{cases} 1 & \text{if } x \ge 0 \\ 0 & \text{if } x < 0. \end{cases}$$

Here $\hat{F}_i(x, t)$ is an empirical distribution function defined by a sample of a random size. If $N_i(t) = 0$, then we set $\hat{p}_{ij}(t)\,\hat{F}_i(x, t) = 0$.

Theorem 16.1. *Estimates (16.6) are uniformly strongly consistent in the sense that with the probability 1*

$$\max_{i,j=1,\ldots,m} \sup_{x \ge 0} |\hat{p}_{ij}(t)\,\hat{F}_i(x,t) - p_{ij}\,F_i(x)| \underset{t\to\infty}{\longrightarrow} 0.$$

Theorem 16.2. *For a fixed i, j, x a two-dimensional random variable*

$$(\, t^{1/2}\,(\hat{p}_{ij}(t) - p_{ij}), \quad t^{1/2}\,(\hat{F}_i(x, t) - F_i(x))\,)$$

converges in a distribution as $t \to \infty$ to a two-dimensional normal random variable with a zero means and a covariance matrix (σ_{ij}):

$$\sigma_{11} = \mu_i\,p_{ij}\,(1 - p_{ij}), \quad \sigma_{22} = \mu_i\,F_i(x)\,(1 - F_i(x)), \quad \sigma_{12} = \sigma_{21} = 0$$

where μ_i is a mean time between successive moments when a semi-Markov process enters the state i.

Corollary. *For a fixed i, j, x a random variable*

$$t^{1/2}\,(\hat{p}_{ij}(t)\,\hat{F}_i(x, t) - p_{ij}\,F_i(x))$$

converges in a distribution as $t \to \infty$ to a normal random variable with a zero mean and a variance $\mu_i\,F_i(x)\,p_{ij}\,(F_i(x) - 2\,F_i(x)\,p_{ij} + p_{ij})$.

16.6 References

1. Cox, D. R., Lewis, P. A., *The Statistical Analysis of Series of Events,* Methuen, London, 1966.
2. Snyder, D. L., *Random Point Processes,* Wiley, London, 1975.
3. Lindley, D. V., Scott, W. F., *New Cambridge Statistical Tables,* 2nd Edition, Cambridge University Press, Cambridge, 1995.

4. Ambartzumian, R. V., Two inverse problems concerning the superposition of recurrent point processes, *J. Appl. Probab.*, 2, 449, 1965.
5. Rao, C. R., *Linear Statistical Inference and Its Applications,* Wiley, New York, 1973.
6. Koroliouk, V. S., Portenko, N. I., Skorokhod, A. V., Tourbine, A. F., *Aide Mémoire de Théorie des Probabilités et de Statistique Mathématique,* Mir, Moscow, 1983.
7. Billingsley, P., *Statistical Inference for Markov Processes,* Univ. Chicago Press, Chicago, 1961.
8. Lee, T. C., Judge, G. G., Zellner, A., *Estimating the Parameters of the Markov Probability Model from Aggregate Time Series Data,* North-Holland Publ. Comp., Amsterdam, 1977.
9. Greenberg, I., The behavior of a simple queue at various times and epochs, *SIAM Review,* 9, 234, 1967.
10. Daley, D. J., Monte-Carlo estimation of the mean queue size in a stationary GI/M/1 queue, *Oper. Res.,* 16, 1002, 1968.
11. Kovalenko, I. N., On the reconstruction of system's characteristics by observations over the output flow, *Doklady Akad. Nauk SSSR,* 164, 979, 1965 (in Russian).
12. Ivnitsky, V. A., Reconstruction of characteristics of single-server queueing system with a bounded sojourn time by observations over the output flow, *Engin. Cybernetics,* 7, No. 3, 52, 1969.
13. Andronov, A. M., Kordonsky, K. B., Rozenblit, P. Y., Application of theory of unbiased estimates in queueing problems, *Engin. Cybernetics,* 10, 227, 1972.
14. Andronov, A. M., Rozenblit, P. Y., Statistics of semi-Markov birth and death processes with application to queueing systems, *Engin. Cybernetics,* 10, 465, 1972.
15. Bhat, U. N., Sixty years of queueing theory, *Manag. Sci.,* 15, B-280, 1969.
16. Kovalenko, I. N., Queueing theory, *Itogi Nauki i Tekhniki, Teoriya Veroyatn., Mat. Statist., Teoret. Kibern.,* 8, 1, 1971 (in Russian).
17. Barlow, R. E., Proschan, F., *Statistical Theory of Reliability and Life Testing. Probability Models,* Holt, Reinhart and Winston, New York, 1975.
18. Bartlett, M. S., The frequency goodness of fit test for probability chains, *Proc. Cambr. Phil. Soc.,* 47, 86, 1951.
19. Billingsley, P., Statistical methods in Markov chains, *Ann. Math. Statist.,* 32, 12, 1961.
20. Albert, A., Estimating the infinitesimal generator of a continuous time finite state Markov process, *Ann. Math. Statist.,* 33, 727, 1962.
21. Moore, E. H., Pyke, R., Estimation of the transition distribution of a Markov renewal process, *Ann. Inst. Statist. Math.,* 20, 411, 1968.

Chapter 17

Simulation of Random Processes

17.1 Monte Carlo Method

The Monte Carlo method (or method of simulation) is a numerical method based on a probabilistic interpretation of values to be computed and making use of the generation of random variables for their estimation[1-6].

Assume a value a is to be computed. Having investigated its nature (e.g., a system of equations satisfied by it) one tries to choose such a random variable ξ that $a = \mathbf{E}\xi$. Suppose that a sequence of independent realizations $\{\xi_i\}$ of the ξ is available. Then the law of large numbers implies

$$\hat{a}_n \stackrel{def}{=} \frac{1}{n}(\xi_1 + \cdots + \xi_n) \xrightarrow[n\to\infty]{p} a.$$

Therefore one may use the approximation of the a by $\hat{a}_n$ if n is sufficiently large.

Example 17.1 ("Buffon's needle"). It is required to compute the number π approximately. We start with the identity

$$\frac{2}{\pi} = \frac{1}{\pi}\int_{-\frac{\pi}{2}}^{\frac{\pi}{2}} \cos\varphi \, d\varphi$$

where $\frac{1}{\pi}\,d\varphi$ can be considered as the probability of the falling of a random angle uniformly distributed in the interval $(-\frac{\pi}{2},\ \frac{\pi}{2})$ into an elementary sector $(\varphi,\ \varphi + d\varphi)$. Let ξ be a random variable uniformly distributed over the interval (0, 1), and let φ and ξ be independent. Then $\cos\varphi = \mathbf{P}\{\xi < \cos\varphi\}$. Hence $\frac{2}{\pi} = \mathbf{P}\{\xi < \cos\varphi\}$. Having made n independent trials of the random variable (φ, ξ) one can compute the frequency $\hat{p}_n$ of the event $\{\xi < \cos\varphi\}$, and hence the expression $\pi \approx 2/\hat{p}_n$ is the desired approximation. It is evident that the estimate is consistent. The probability of the event $\{\xi < \cos\varphi\}$ is traditionally interpreted as the probability of the intersection of a needle made to fall at random onto a lined sheet of paper. It is assumed that the needle length equals the distance between successive lines.

Example 17.2. The integral

$$I = \int_a^b f(x)\,dx$$

is to be computed. Assume a sequence $\{\omega_k\}$ of independent trials of a random variable ω uniformly distributed over the interval (a, b) is available. Then $I = (b-a)\,\mathbf{E}\,f(\omega)$ and hence

$$(b-a)\,(f(\omega_1) + \cdots + f(\omega_n))/n \xrightarrow[n\to\infty]{P} I.$$

An alternative approach can be applied in the case of $c \le f(x) \le d$. Assume that γ is a random variable uniformly distributed in the interval (c, d) and moreover ξ and γ are independent. It is easy to show that

$$I = (b-a)\,c + (b-a)\,(d-c)\,\mathbf{P}\{\gamma < f(\omega)\}.$$

Hence

$$I \approx (b-a)\,c + (b-a)\,(d-c)\,p_n$$

where p_n is the frequency of the event $\{\gamma < f(\omega)\}$ in n independent trials.

Example 17.3. The well known Laplace equation is usually solved by the greed method. After the reduction to a finite difference scheme the system of equations

$$4\,u_{ij} = u_{i+1j} + u_{ij+1} + u_{i-1j} + u_{ij-1}$$

can be derived for (i, j) belonging to a domain G. The values $u_{ij} = f_{ij}$ are defined for (i, j) belonging to Γ, the boundary of G. Consider a symmetrical random walk $\{\xi_n, n \ge 0\}$ over the domain G absorbing at the boundary Γ. A particle can move from the position $\xi_n = (i, j)$ to each of the four positions $\xi_{n+1} = (i \pm 1, j)$ or $\xi_{n+1} = (i, j \pm 1)$ with probability

1/4. If (i^*, j^*) is the terminating point of the random walk then define the functional $f^* = f_{i^* j^*}$. We have the identity

$$u_{ij} = \mathbf{E}\{f^* \mid \xi_0 = (i, j)\}.$$

Hence the Monte Carlo method can be applied for the solution of the Laplace equation as well.

In each of the three examples the value to be computed has no intrinsic probabilistic sense. In each case we introduce a probabilistic interpretation for purely computational purposes. Contrary to these examples, in many cases a probabilistic problem (being the main purpose of the investigation) is formulated from the very beginning. In such cases a stochastic model is usually available expressed in terms of a random process. Monte Carlo method is used as the method of the modelling of the system behavior. Provided the model is absolutely adequate to a real process, statistical trials are similar to experiments with the real process. The main advantage of the simulation method consists of the relative simplicity of stochastic equations describing the system behavior, in the comparison with the equations for averaged parameters of the system.

Example 17.4. The vector of the deviations of the missile from the point target at the burst time can be expressed as (ξ_1, ξ_2, ξ_3) where ξ_i are independent Gaussian random variables with zero means and variances $\sigma_1^2, \sigma_2^2, \sigma_3^2$. It is desired to estimate the probability of the event $\xi_1^2 + \xi_2^2 + \xi_3^2 < r^2$. The Monte Carlo solution is trivial: to realize ξ_1, ξ_2, ξ_3 n times and compute the frequency of the mentioned events. At the same time an analytical expression for the probability being considered is available only in some exceptional cases (for example if $\sigma_1 = \sigma_2 = \sigma_3$).

17.2 Random and Pseudorandom Numbers

It is an unachievable ideal to have a source of independent random variables with a given distribution. Nevertheless, in the problems connected to model building it is usually assumed that such an ideal source is available. Such a source is called **a generator of random numbers**. It generates random numbers $\{\omega_n\}$ assumed to be independent and uniformly distributed in the interval $(0, 1)$. The word "numbers" instead of "variables" is due to the discrete character of numbers in a digital computer.

Some computers have physical random numbers generators in their hardware. All of them are based on vastly varying noise processes. But the problem of stability of such generators is very difficult. Instead of random numbers **pseudorandom numbers** are commonly used. Pseudorandom numbers are members of a deterministic sequence $\{x_n\}$ resembling some frequency properties of a purely random sequence[7–10].

Example 17.5. Let θ be an irrational number; $< a >$ denotes the fractional part of a number a (so that $< \pi >= 0,1415296536\ldots$ and so on). It is well known that the sequence $x_n =< n\theta >$ resembles the simplest frequency property of the sequence of uniform random numbers. Indeed, denote by $m_n(a, b)$ the frequency of the event $\{a \leq x_k < b\}$ for the trials $k = 1, 2, \ldots, n$. Then

$$\frac{1}{n}\, m_n(a, b) \underset{n\to\infty}{\longrightarrow} b - a$$

for each $a, b : 0 \leq a \leq b \leq 1$.

Each pseudorandom sequence can resemble only some, but not all, of the properties of a random sequence. Thus, in the Example 17.5 for the diagrams (x_n, x_{n+1}) we have $x_{n+1} - x_n = < \theta >$ or $< \theta > -1$.

If an applied problem is being solved by the simulation method then it is necessary to use a pseudorandom sequence resembling all those frequencies upon which the solution depends. While realizing random numbers it is required to apply special tests for checking their quality. There exist tests for uniformity, tests for the absence of a correlation of x_n and x_{n+1} and others. Some tests are based on the lengths of the series of $x_n \leq 0.5$ and series within which $x_n > 0.5$.

Linear methods of the generating of pseudorandom numbers are best known. They are based on the recurrent relation

$$X_{n+1} = \sum_{i=0}^{k-1} a_i X_{n-i} + b \pmod{p}, \quad n \geq k, \quad x_n = \frac{1}{p} X_n, \quad n \geq 1,$$

where $a_i, b, p, X_1, \ldots, X_k$ are integers[7]. **The simple multiplicative method**

$$X_{n+1} = a X_n \pmod{p}, \quad x_n = \frac{1}{p} X_n, \quad n \geq 1,$$

and **the mixed multiplicative method**

$$X_{n+1} = a X_n + b \pmod{p}, \quad n \geq 1, \quad x_n = \frac{1}{p} X_n, \quad n \geq 1,$$

are most widely used.

The sequence $\{x_n\}$ is periodic; its period does not exceed p. For the criterion for the period to equal p precisely (see Reference 8).

17.3 Transformations of Random Variables and Sequences

When modelling a random process it is often required to generate a random variable ξ with a prescribed distribution or to generate a random sequence with prescribed multidimensional properties.

Let ω be a random variable uniformly distributed in the interval (0, 1) and $F(x)$ be a distribution function. Apply the transformation $\xi = F^{-1}(x)$ where F^{-1} is the inverse to F. Then ξ is a random variable with the distribution function $F(x)$.

An alternative approach requiring no inversion of the distribution function consists of the following. Assume that the distribution F has a density $p(x)$ vanishing outside an interval $[a, b]$ and satisfying the inequality $p(x) \leq c$ within the interval. Let ν be the smallest n satisfying the condition $c\omega_{2i-1} < p(a + \omega_{2i}(b-a))$. Then we set $\xi = a + \omega_{2\nu}(b-a)$.

Let $\{\xi_k,\ 1 \leq k \leq n\}$ be an n-dimensional random variable with a density function $p(x_1, \ldots, x_n)$. A universal approach to the realization of such a random vector consists of the sequential realization of ξ_1, ξ_2 and so on. Suppose that $\xi_1, \ldots, \xi_{k-1}$ have been already realized and their values $\xi_1 = x_1, \ldots, \xi_{k-1} == x_{k-1}$ are known. Compute the conditional density function of ξ_n given $x_1, \ldots, x_{n-1}$:

$$p(x) = p(x \mid x_1, \ldots, x_{k-1})$$

and generate ξ_k similarly to ξ for the one-dimensional case.

17.4 Modelling of Markov Chains

A discrete time Markov chain can be considered as a family of random variables ξ_0, $\eta^{(i)}$ where ξ_0 is the initial state and $\eta^{(i)}$ is the state at step n provided state i at step $n-1$. Suppose that an algorithm is available for generating both ξ_0 and $\eta^{(i)}$ via a random variable ω so that $\xi_0 = f_0(\omega)$, $\eta^{(i)} = f_i(\omega)$. Then a realization $\{\xi_n,\ n \geq 0\}$ can be built via a sequence $\{\omega_n,\ n \geq 0\}$ of independent realizations of the random variable ω by the following recurrent procedure:

$$\xi_0 = f(\omega_0), \qquad \xi_n = f_i(\omega_n) \text{ if } \xi_{n-1} = i, \quad n \geq 1.$$

Example 17.6. Consider a Markov chain with states 1, 2, 3 and the transition matrix

$$\begin{pmatrix} 0.5 & 0.25 & 0.25 \\ 0.2 & 0.6 & 0.2 \\ 0.1 & 0.1 & 0.8 \end{pmatrix}.$$

Assume that the initial distribution equals (0.2; 0.3; 0.5). Then the algorithm for the generation of ξ_0 and ξ_n via ξ_{n-1} can be defined as follows:

$$f(\omega) = \begin{cases} 1 & \text{if } 0 \le \omega < 0.2, \\ 2 & \text{if } 0.2 \le \omega < 0.5, \\ 3 & \text{if } 0.5 \le \omega \le 1; \end{cases} \qquad f_2(\omega) = \begin{cases} 1 & \text{if } 0 \le \omega < 0.5, \\ 2 & \text{if } 0.5 \le \omega < 0.75, \\ 3 & \text{if } 0.75 \le \omega \le 1; \end{cases}$$

$$f_1(\omega) = \begin{cases} 1 & \text{if } 0 \le \omega < 0.2, \\ 2 & \text{if } 0.2 \le \omega < 0.8, \\ 3 & \text{if } 0.8 \le \omega \le 1; \end{cases} \qquad f_3(\omega) = \begin{cases} 1 & \text{if } 0 \le \omega < 0.1, \\ 2 & \text{if } 0.1 \le \omega < 0.2, \\ 3 & \text{if } 0.2 \le \omega \le 1. \end{cases}$$

17.5 Modelling of a Markov Process with a Finite Phase Space

Suppose that the modelling of a Markov process $\xi(t)$ with transition rates $\lambda_{ij}(t)$ and exit rates $\lambda_i(t)$ is required. The modelling can be realized step by step through a consecutive prolongation of the path function from jump point t_{n-1} to t_n. A universal algorithm can be defined as follows. Set $\xi(t_{n-1}+0) = i$. Generate an exponential random variable γ with parameter 1 and compute the integral $I(t) = \int\limits_{t_0}^{t} \lambda_i(\tau)\, d\tau$ with a varying $t \ge t_0$. As soon as it achieves the level γ we make the process pass to another state, i.e., t_n is defined by the equation $I(t_n) = \gamma$. The next state is defined at random, according to the distribution

$$\mathbf{P}\{\xi(t_n+0) = j\} = \frac{\lambda_{ij}(t_n)}{\lambda_i(t_n)}, \quad j \ne i.$$

For a homogeneous random process there is no necessity to compute the $I(t)$ since a simple formula can be applied: $t_n = t_{n-1} + \gamma/\lambda_i$.

17.6 Modelling of Semi-Markov Processes

Let $Q(t) = (Q_{ij}(t))$ be a Q-matrix of a SMP $\xi(t)$, $Q_i(t) = \sum\limits_j Q_{ij}(t)$ be the distribution of the sojourn time in a state i. If all these functions are well-computed then the following algorithm can be applied. Suppose that t_0 is a transition point of the process and $\xi(t_0+0) = i$. Generate the next state j according to the distribution $\{Q_{ij}(\infty)/Q_i(\infty)\}$. If j is already chosen then generate the sojourn time in the state i according to the distribution function $Q_{ij}(t)/Q_{ij}(\infty)$. In the new state j the situation is renewed; i should be changed to j. Thus a recursive stochastic algorithm is defined.

An alternative algorithm prescribes generation of the sojourn time first, according to $Q_i(t)$ the distribution function. If the sojourn time equals t then the next state is chosen at random, in accordance with the distribution

$$\left\{ \frac{Q'_{ij}(t)}{Q'_i(t)} \right\}$$

provided $Q_{ij}(t)$ are absolutely continuous. Though if t is a discontinuity point of $Q_i(t)$ then the mentioned distribution should be changed to

$$\left\{ \frac{Q_{ij}(t+0) - Q_{ij}(t-0)}{Q_i(t+0) - Q_i(t-0)} \right\}.$$

In case it is not convenient to deal with the Q-matrix, an infinitesimal approach can be used. Let $\lambda_{ij}(t)$ and $\lambda_i(t)$ be transition and exit rates of an SMP $\xi(t)$ at a state i after a time t spent in it. Assume that $\lambda_i(t) \leq$ $\leq a_i$ for each i. As soon as the process enters a state i, say, at time t_0, an exponentially distributed random variable γ_1 with the parameter a_i is generated, and also a (0, 1)-uniform random variable ω_1 independent of γ_1. If $a_i \omega_1 \leq \lambda_i(t_0 + \gamma_1)$, then the next transition point is defined as $t_0+\gamma_1$ and the next state is defined as random, according to the distribution $\{\lambda_{ij}(\gamma_1)/\lambda_i(\gamma_1)\}$. If $a_i \omega_1 > \lambda_i(t_0 + \gamma_1)$, then the next independent trial should be done. If $a_i \omega_{\nu-1} > \lambda_i(\gamma_1+\cdots+\gamma_{\nu-1})$ and $a_i \omega_\nu \leq \lambda_i(\gamma_1+\cdots+\gamma_\nu)$, then define the next transition point as $t_0 + \gamma_1 + \cdots + \gamma_\nu$ and choose the next state following the distribution

$$\left\{ \frac{\lambda_{ij}(\gamma_1 + \cdots + \gamma_\nu)}{\lambda_i(\gamma_1 + \cdots + \gamma_\nu)} \right\}.$$

Random variables $\gamma_1, \gamma_2, \ldots, \omega_1, \omega_2, \ldots$ should be independent.

A SMP over an arbitrary phase state can be modelled in a similar way.

17.7 Modelling of Regenerative Processes

For the realization of a regenerative process it suffices to have an algorithm of the realization of breaking processes of which it is sewed[11]. In many cases the parameter to be computed can be expressed via a parameter concerning a regeneration cycle. In such a way the process of initiation is reduced to the modelling of n independent regeneration cycles. The most typical case can be described as follows.

Let $\xi(t)$ be a stationary regenerative process, and it is required to compute the average value

$$a = \mathbf{E}\, f(\xi(t))$$

where f is a real function, provided the mean is finite. Let $z(t)$ be a breaking process, $0 \le t < \tau$ used for the construction of the regenerative process. Then

$$a = \frac{\mathbf{E}\,\zeta}{\mathbf{E}\,\tau}$$

where

$$\zeta = \int_0^\tau f(z(t))\, dt.$$

A consistent estimate of a has the form

$$\hat{a}_n = \frac{\zeta_1 + \cdots + \zeta_n}{\tau_1 + \cdots + \tau_n}$$

where ζ_i and τ_i are the values of ζ and τ computed from the ith trial. Assuming the trials to be independent it can be shown that $\sqrt{n}\,(\hat{a}_n - a)$ is an asymptotically Gaussian random variable with zero mean and variance

$$\mathbf{Var}\,\hat{a}_n \sim \frac{1}{(\mathbf{E}\,\tau)^2}\,(\mathbf{Var}\,\zeta - 2\,a\,\mathbf{cov}(\zeta,\,\tau) + a^2\,\mathbf{Var}\,\tau).$$

17.8 Modelling of Brownian Motion and Some Related Processes

Let $w(t)$ be a standard Wiener process. The realization of a stepwise function resembling some of its properties is usually reduced to the realization of a sequence of independent random variables $\{v_i\}$ such that $\mathbf{E}\,v_i = 0$, $\mathbf{E}\,v_i^2 = \Delta$. Set $s_0 = 0$, $s_n = v_1 + \cdots + v_n$ for $n \ge 1$, and consider the random process $w_\Delta(t)$ defined for each $t = n\Delta$, $n \in \mathbb{N}^+$, by the formula $w_\Delta(n\,\Delta) = s_n$. Then a joint distribution of random variables $w_\Delta([t_k/\Delta]\Delta)$, $k = 1, \ldots, m$, is close to the joint distribution of $w(t_k)$, $1 \le k \le m$, if Δ is small. The variables v_i can be chosen in different ways, for example, as $\sqrt{3\,\Delta}\,(2\,\omega_i - 1)$ where ω_i are $(0,\,1)$-uniform variables or as $\sqrt{\Delta}\,(2\,\gamma_i - 1)$ where $\{\gamma_i\}$ are binomial variables so that $\mathbf{P}\,\{\gamma_i = 0\} = \mathbf{P}\,\{\gamma_i = 1\} = 1/2$. The process $w_\Delta(t)$ can be interpolated within intervals $((n-1)\,\Delta,\; n\,\Delta)$ in different ways, according to the problem being solved. The most simple is the linear interpolation.

Solutions of stochastic differential equations are usually realized via difference schemes. Let, for example, an m -dimensional random process $\xi(t) = = (\xi_1(t), \ldots, \xi_m(t))^T$ be defined by the equation

$$d\xi(t) = a(t, \xi)\, dt + b(t, \xi)\, dw(t)$$

where $w(t) = (w_1(t), \ldots, w_r(t))^T$ is the vector composed of independent standard Wiener processes, $a(t, \xi)$ is an m-dimensional function and $b(t, \xi)$ is a matrix $(m \times r)$ function. Then the difference scheme can be defined by the equation

$$\xi(t_k) - \xi(t_{k-1}) = a(t_{k-1}, \xi(t_{k-1}))\,(t_k - t_{k-1}) + \\ + b(t_{k-1}, \xi(t_{k-1}))\,(w(t_k) - w(t_{k-1})).$$

17.9 Modelling of Processes with Independent Increments

Assume that it is required to realize an m-dimensional process $\xi(t) = = (\xi_1(t), \ldots, \xi_m(t))^T$ with independent increments characterized by the characteristic function

$$\varphi_t(\lambda) = \exp\left\{ i\,(a(t), \lambda) - \frac{1}{2}(B(t)\,\lambda, \lambda) + \int_{R^m} (e^{i(\lambda, x)} - 1)\,\Pi(t, dx) \right\}.$$

The realization can be expressed in the form

$$\xi(t) = \xi_0(t) + \xi_1(t)$$

where the right-hand side terms are realized independently, according to the following procedures.

The process $\xi_0(t)$ is defined as the solution of the stochastic differential equation

$$d\zeta(t) = a'(t)\, dt + B'(t)\, dw(t), \quad \zeta(0) = \xi(0).$$

The process $\xi_1(t)$ can be realized via a nonhomogeneous Poisson process $Y(t)$ with a leading function $\Lambda(t) = \Pi(t, \mathbb{R}^m)$ if this function is finite. In such a case jump points of $\xi_1(t)$ coincide with jump points of the Poisson process. The values of jumps should be chosen at random. If a jump occurs at time t then its value falls into a Borel set $A \subset \mathbb{R}^m$ with the probability $\Pi'(t, A)/\Pi'(t)$ provided the derivatives exist. Numerical difficulties arise in case $\Pi(t, \mathbb{R}^m)$ is either infinite or too large. In such a case it may be reasonable to realize only jumps with values outside of some neighborhood V of the origin. We thus consider $\Pi(t, A \setminus V)$ instead of $\Pi(t, A)$ and $\Pi(t, \mathbb{R}^m \setminus V)$ instead of $\Pi(t, \mathbb{R}^m)$. As to small jumps, they can be included in the diffusion component by equating moments of the first and second orders.

17.10 Modelling of Gaussian Processes

Let $\xi(t)$ be a Gaussian process. Without a loss of the generality we suppose that $\mathbf{E}\xi(t) = 0$. Consider a sequence of points $t_n, n = 1, 2, \ldots$. It is possible to express all the $\xi_n = \xi(t_n)$ via independent standard normal variables $\{\eta_i\}$ using a triangle transformation

$$\xi_n = \sum_{j=1}^{n} a_{nj}\,\eta_j.$$

Coefficients a_{nj} satisfy the system of equations

$$\sum_{j=1}^{m} a_{nj}\,a_{mj} = b_{mn}, \quad 1 \le m \le n,$$

where $b_{mn} = \mathbf{E}\,\xi_m\,\xi_n$. This system can be solved recursively. The beginning of this process is as follows.

Equations	*Results*
(1.1) $a_{11}^2 = b_{11}$	$a_{11} = \pm\sqrt{b_{11}}$
(2.1) $a_{21}a_{11} = b_{12}$	$a_{21} = b_{12}/a_{11}$
(2.2) $a_{21}^2 + a_{22}^2 = b_{22}$	$a_{22} = \pm\sqrt{b_{22} - a_{21}^2}$
(3.1) $a_{31}a_{11} = b_{13}$	$a_{31} = b_{13}/a_{11}$
(3.2) $a_{31}a_{21} + a_{32}a_{22} = b_{23}$	$a_{32} = (b_{23} - a_{31}a_{21})/a_{22}$
(3.3) $a_{31}^2 + a_{32}^2 + a_{33}^2 = b_{33}$	$a_{33} = \pm\sqrt{b_{33} - a_{31}^2 - a_{32}^2}$

For the modelling of a stationary Gaussian process its spectral development is usually applied. Let $\xi(t)$ be a complex-valued Gaussian random process with zero mean and a spectral function $F(\lambda)$. Assume that $dF(\lambda) = 0$ outside an interval Δ. Divide the interval Δ into smaller intervals Δ_k, and let λ_k be a point including to Δ_k. The process $\xi(t)$ can be approximated by the sum

$$\xi(t) \approx \sum_{k} \zeta_k \exp\{it\lambda_k\}$$

where $\{\zeta_k\}$ are independent complex Gaussian random variables characterized by the conditions $\mathbf{E}\,\zeta_k = 0$, $\mathbf{E}\,|\zeta_k|^2 = F_k$ where $F_k = \int\limits_{\Delta_k} dF(\lambda)$. Two equivalent methods for the modelling of ζ_k exist:

(i) generate independent real Gaussian random variables ξ_k and η_k with $\mathbf{E}\,\xi_k = \mathbf{E}\,\eta_k = 0$, $\mathbf{Var}\,\xi_k = \mathbf{Var}\,\eta_k = F_k/2$ and then set $\zeta_k = \xi_k + i\,\eta_k$,

(ii) generate independent real random variables ρ_k and φ_k, the former being characterized by the distribution function $1 - \exp\{-x^2/F_k\}$ and the latter being uniformly distributed in the interval $(0, 2\pi)$. Then set

$$\zeta_k = \rho_k\, e^{i\varphi_k}.$$

17.11 References

1. Hammersley, J. M., Handscomb, D. C., *Monte Carlo Methods,* Methuen, London, 1964.
2. Buslenko, N. P., *Simulation von Produktionsprozessen,* Teubner, Leipzig, 1971.
3. Pollyak, Yu. A., *Probability Simulation on Digital Computers,* Sov. Radio, Moscow, 1971 (in Russian).
4. Bakhvalov, N. S., *Numerical Methods,* Nauka, Moscow, 1975 (in Russian).
5. Ermakov, S. M., *Die Monte-Carlo-Methode und Verwandte Fragen,* Oldenbourg, München, 1975.
6. Ermakov, S. M., Mikhaylov, G. A., *A Course of Statistical Simulation,* Nauka, Moscow, 1976 (in Russian).
7. Lehmer, D. H., Mathematical methods in large-scale computing units, *Ann. Comp. Lab. Harvard Univ.,* 26, 141, 1951.
8. Gill, A., *Linear Sequential Circuits: Analysis, Synthesis and Applications,* McGraw-Hill, New York, 1966.
9. Golenko, D. I., *Simulation and Statistical Analysis of Pseudorandom Numbers on Digital Computers,* Nauka, Moscow, 1965 (in Russian).
10. Nance, R. E., Claude, O. J., A bibliography on random number generation, *Comput. Revs,* 13, 495, 1972.
11. Crane, M. A., Iglehart, D. L., Simulating stable stochastic systems. III. Regenerative processes and discrete-event simulations, *Oper. Res,* 23, 33, 1975.

Index